Ship Iconography in Mosaics

An aid to understanding ancient ships and their construction

Zaraza Friedman

BAR International Series 2202
2011

Published in 2016 by
BAR Publishing, Oxford

BAR International Series 2202

Ship Iconography in Mosaics

ISBN 978 1 4073 0758 9

COVER IMAGE *Lod - Maritime Frame with Both Ships (Avissar 1996, fig.3, p.157)*

BAR Publishing is the trading name of British Archaeological Reports (Oxford) Ltd. British Archaeological Reports was first incorporated in 1974 to publish the BAR Series, International and British. In 1992 Hadrian Books Ltd became part of the BAR group. This volume was originally published by Archaeopress in conjunction with British Archaeological Reports (Oxford) Ltd / Hadrian Books Ltd, the Series principal publisher, in 2011. This present volume is published by BAR Publishing, 2016.

Printed in England

BAR
PUBLISHING

BAR titles are available from:

BAR Publishing
122 Banbury Rd, Oxford, OX2 7BP, UK
EMAIL info@barpublishing.com
PHONE +44 (0)1865 310431
FAX +44 (0)1865 316916
www.barpublishing.com

"The sailor should scan his course from the shore,
if so be that he has the power and means;
for once he is on the high seas he must run with whatever comes."
(Athenaeus: *Deipnosophistae*, XV.695.8)

"Friend, observe here with pleasure the charming things
which art has placed in the mosaic cubes,
petrifying and repelling jealousy and the eye of envy.
You are the one who is proud of the enjoyable art."
(Greek inscription at Sheikh Zouède, mid-5th century CE)

Table of Contents

List of Illustrations

ABBREVIATION

ADRAL	Atti della R. Accademia dei Lincei
AJA	American Journal of Archaeology
AJP	American Journal of Philology
Antiquity	A Quarterly Review of Archaeology
Archaeology	
ASP	American Studies in Papyrology
BA	The Biblical Archaeologist
Boll. D'Arte	Bolletino d'Arte
CQ	The Classical Quarterly
DOP	Dumbarton Oaks Papers
FU	Forma Urbis
HESPERIA	Journal of the American School of Classical Studies at Athens
IJNA	The International Journal of Nautical Archaeology
JHS	Journal of Hellenic Studies
JNES	Journal of Near Eastern Studies
JRS	The Journal of Roman Studies
MAAR	Memoirs of the American Academy in Rome
NG	The National Geographic Magazine
OJA	Oxford Journal of Archaeology
PAPS	Proceedings of the American Philosophical Society
Rend. P.A.R.A.	Atti della Pontificia Accademia Romana di Archeologia, Rendiconti
The American Neptune	
TPAPA	Transactions and Proceedings of the American Philological Association
TROPIS	International Symposium on Ship Construction in Antiquity

Aknowledgements

This monograph is the extension and revised version of the author's PhD dissertation and a continuous research that started with my MA dissertation. The research was possible with the aid of two scholarships, several people who gave me moral supported and encouragement throughout the entire process, as well as the permits that were granted to me from the museums and the sites where the mosaics in discussion are found.

In the first place I would like to express my thanks to my parents Itzhak and Miriam for their support and encouragement together with my brother Michael, his wife Ofra, and both daughters Yasmin and Lian. Many people to whom I am very thankful have assisted in the evolution of my research. My deepest thanks go to both my advisors Prof. Michal Artzy of the Department of Maritime Civilizations, University of Haifa, and Prof. Michele Piccirillo of Studium Biblicum Franciscanum, Jerusalem (who sadly passed away before his time, 26.10.2009). Travel to the sites at Ostia in Italy, Piazza Armerina in Sicily, Kenchreai in Greece and the writing of the thesis were partially supported by the Maurice Hatter Fellowship for Maritime Studies (2001). My thanks go to the Reuven and Edith Hecht Museum, University of Haifa for granting me the Hecht Scholarship to conduct the research of ship iconography in mosaics (2000). The research carried out at the sites and the museums where mosaic with ship depictions are found was possible to the permits granted to me by the directors of these sites. My deepest thanks go to Dr. Anna Gallina Zevi, Soprintendenza of Ostia Antica, who gave me permission to work at the sites of Piazzale delle Corporazioni at Ostia. I am very grateful to Dr. Maria Luisa Veloccia Rinaldi, Soprintendente of Museo Archeologico Nazionale, Palazzo Barberini, who gave me permission to carry out a detailed study of the Palestrina mosaic. My thanks go to the curator Dr. Maria Rosario Borniello from Museo Archeologica Nazionale in Naples, Italy, who gave me the permission for the study of the mosaics with the maritime scenes and the frescoes from Pompeii that are found among the collections of the Museum. The study of the unique glass *opus sectile* maritime panels from Kenchreai, Greece, was made possible by the permission that I received from the American School of Classical Studies at Athens. My thanks go to Prof. James Redfield, then the director of the School and also to Mrs. Maria Pilali, the Administrator of ASCSA (2002). My thanks go to Mrs. Jean Perras, the Administrator of the Isthmia Museum where the Kenchreai panels are found (2002). I am deeply thankful to Prof. Levent Zoroglu of the Selçuk University at Konya, Turkey, who gave me the original picture of the Harbor mosaic from Kelenderis/Aydinçik along with the preliminary report of the site. The research of the Lod maritime mosaic panel could be carried out in *situ*, due to the permit granted to me by Mrs. Miriam Avissar of the Israel Antiquities Authority, during the excavations of the Roman Villa in 1996. I am deeply thankful to Dr. Andrea Camilli, Soprintendente per i Beni Archeologici della Toscana, then the director of the Pisa Ships excavations. Dr. Camilli was kind enough to give me the original picture of the elliptical oarport and the bow of Wreck C with the preserved projecting pointed cutwater. My thanks go to Ms. Julia Boetto from the Roman Ships Museum, within the area of Leonardo da Vinci Airport, Italy, who gave me the pictures of two Fiumicino Ships 1 and 5.

The list of people who gave me their support during the study and writing the PhD dissertation is extremely long. Their contribution has been invaluable and my thanks go to each and every one of them, even though their names have not been mentioned here.

Preface

Mosaic surfaces (floor and/or wall) are definitely one of the most accomplished arts that developed in the Mediterranean in antiquity. It is a durable art form that survived from antiquity to the preset days. Each country surrounding the Mediterranean Basin brought to the development of techniques and subjects matter that reflect cultural development and diffusion. The varied and complex designs depicted in mosaics may be classified according to subjects' representation and their architectural settings: baths, villas, shippers' squares and ecclesiastic structures. The question as to whether ship representations in mosaics, augment the data of ancient ships, is addressed, bearing in mind the limitations of the milieu. Ships depicted in mosaics were meant to be everlasting images or an allusion to the activity of the owner for those who entered spaces where such mosaics were positioned.[1] It is possible that some of the merchants who grew rich from the profits of shipping of African goods may have chosen to record the source of their wealth on their pavements, as we carefully may suggest of both ships depicted in the Lod mosaic.[2] The ships illustrated in the Great Hunt mosaic at Piazza Armerina may hint that the owner of the villa was engaged in shipping business and trading with African animals for the game held in the Colosseum and Circus Maximus at Rome, to satisfy the pleasures of the Roman Emperor.

Ship motifs do not appear as single items but rather as part of a design. They were driven from different sources such as vase paintings, reliefs, frescoes, graffiti, models, ancient literature, sketches, or from observations of vessels sailing around the Mediterranean as anchored in harbors in antiquity. Different types of vessels depicted with many details suggest a close approach to realism. The vessels are shown with their hull shape, propulsion and steering gear, or cargoes on board. They also suggest the sailing environment (river, lake, delta, harbor or open sea). Ship motifs applied as a theme for mosaic decorations provide us with a wider understanding of the techniques used for laying the stones and the planning of such designs on the mosaic surface. Although, mosaics are not an easy and flexible medium for the depiction of ships, they represent significant details that may be considered relevant information about shipbuilding traditions and types of vessels typical of the period when the mosaics with such themes were prepared.

The mosaics with ship depictions examined in this monograph are not well known and they comprise much of the knowledge that we can gather on ancient ships and their preserved shipbuilding traditions in the Mediterranean from the Hellenistic to the Byzantine periods. The mosaics were studied according to their geographical location in the Mediterranean Basin and their dates span from the late 3rd century BCE to the 6th century CE.[3] The purpose of this study is to gather data regarding types of ancient ships, their rigging and sailing environment as depicted in the mosaics, and compare them with the archaeological remains of ancient ships that have been found in the Mediterranean.

The study begins with the earliest representation of a part ship, depicted as the headdress of Berenike II. Two mosaics, dated to 200 and 175 BCE, with the same motif were discovered at Thmuis (Tell Timai) in Egypt (fig. 1.1), which probably belonged to the royal palace of Ptolemy II.

The Lod mosaic, Israel, discovered in 1996, decorated the floor of a Roman Villa (late-3rd or early 4th century CE). Within the marine scene are depicts two merchant ships, probably of western origin, from North Africa.

In the Thetis mosaic from Yakto (Antioch) a complex maritime scene inhabited by putti riding dolphins and fishing with a rod or fishing-net from a boat surround the goddess (460 – 465 CE). The boat may be associated with a unique representation of a dugout engaged in fishing.

The figurative maritime mosaic from Kelenderis (on the southern coast of Turkey) depicts a large merchantman entering or being in the process of anchoring in a harbor as towing astern one small sailing boat with fully open square sail and a rowboat minus its oars (5th – 6th century CE).

Several unique panels produced in the glass *opus sectile* technique, dating from the mid– 4th century CE, with representations of harbors and Nitotic scenes, discovered at Kenchreai (the port of ancient Corinth) in the early 1960 are now displayed at the Isthmia Museum, Greece. Four of the surviving panels depict harbor scenes with several sailing ships shown with their rigging and steering gear.

One of the most complex scenes of mosaic manufacture is found at Palestrina (Praeneste), Italy. A large

1 Friedman, 1999, MA dissertation, p. viii, unpublished.
2 Dunbabin, 1978, p. 126.
3 BCE – Before Christian Era; CE – Christian Era.

variety of water vessels sailing in the Nile Delta during the flood period and the celebrations of the goddess Isis are depicted in this mosaic dated to the late 2nd century BCE. The vessels may be associated with river or coastal ships, a warship and small boats, all depicted with their means of propulsion. The mosaic formed the floor of a *nymphaeum* and it probably was covered with a thin layer of water to give the maritime design a more realistic appearance.

The ships depicted in the black-and-white mosaic floors of the offices at Piazzale delle Corporazioni at Ostia, Italy (dating from the 2nd – 3rd centuries CE), are unique and symbolic representations of shippers and traders from the Roman provinces, all gathered at the Imperial harbor at Ostia. Inscriptions indicating their homeland in the Roman provinces, and different harbor activities are appended to some of the ships.

The mosaics with depictions of vessels in the villa complex at Piazza Armerina, Sicily (early 4th centuries CE) are divided into two subjects:

1. Fishing scenes, in the *frigidarium*, Room 29 and the semi-circle *atrium*.
2. Three merchant vessels engaged in transporting African animals to Rome are depicted in the Great Hunt Hall (c.60 m long).

Not all the mosaics with ship depictions that have been found in the Mediterranean Basin could be incorporated in this monograph, although those considered represent them well. It is hoped that in the future all the remaining mosaics with ship representations will be catalogued and considered in detail, thus contributing greatly to our understanding of ship types and ancient shipbuilding techniques, both in the eastern and western Mediterranean.

Chapter 1
Introduction

1.1 Introduction

Mosaic settings are the most durable surfaces of decorative art since antiquity to the present days. From its earliest form of pebble inlay (8th – 4th centuries BCE) and then the later form of *tesserae* (4th – 3rd century BCE; small colored stone cubes), the decorated surfaces consisted of simple geometric patterns, which later were combined with figurative decorations, becoming more sophisticated throughout the centuries. Mosaic decorations are varied and also were influenced by the architectural setting (villas, baths, temples, synagogues, churches, etc.), and mythological, religion or symbolic themes.

Ships and boats in varied styles appeared from time to time in ancient mosaic surfaces. They are usually presented as part of a larger scheme. Mosaics with ship depictions decorated the floors and the walls of public structures, but seldom in the private domain. This motif was often used as an *apotropae*, to ward off evil, especially during the Roman period. From the Hellenistic period onwards with augmented interest in water-related cult practices symbolized by Nilotic scenes, more ships appear in such mosaics. Later, in the Byzantine period, Christian influence can be noted in the use of ship figures in allegories and symbols for Biblical illustrations. Ships depicted on the floors of baths, offices or private houses were also meant to be indicators of the owner's occupation.

1.2 Purpose of the Research

Ships depicted in mosaics, even those found in occasional publications on the subject of mosaics in general have not been paid enough attention nor have been researched in depth, especially from the point of view of marine archaeology. The present work began with the research for the author's MA dissertation, which dealt with ships depicted in mosaics in the Eastern Mediterranean, namely Israel and Jordan.[1] This research that extended through the author's PhD dissertation has been carried out with the purpose of understanding ancient ships and their construction, propulsion, and types of vessels, which may complement the long list of ship representations in other arts (frescoes, graffiti, reliefs, models, etc.). In general, illustrations of Greco-Roman ships are never at scale nor show their internal structure, but they permit us to observe and understand some of the most distinctive parts of such vessels. Ship depictions in mosaics provide us with a fair idea of the hull shape, and the rigging of ancient merchantmen or war-galleys, types and sailing environment.

When studying ship depictions in mosaics one should consider the following:

1. What can we learn from ship representations in mosaics?
2. Can these depictions be considered as relevant sources of information of ancient ships, types, rigging or steering gear?
3. Is it possible to relate these ships to references in ancient literature?
4. Are ship depictions in mosaics comparable to wrecks found in the Mediterranean?

In order to test these hypotheses, the present monograph includes studies of several mosaics from the Mediterranean: Egypt, Israel, Turkey, Greece, Italy and Sicily, with the purpose to answer the questions listed above. These mosaics with ship depiction were chosen because although some of them are known the ships have never been researched properly from the point of view of marine archaeology. These vessels will be studied in detail with emphasis on their propulsion (sail, oars, steering gear, etc.), building materials (papyrus/reed, skin boat and wooden planks), and identifying types mentioned in ancient literature, as well as comparing them to archaeological finds (shipwrecks). Specific gear will be studied in detail, taking in consideration the sailing environment, such as river, coast or open sea vessels. Another answer being sought is whether the ship depictions were made as standard units, and whether they indicate the size and convey the actual proportions of real Roman-Byzantine vessels? In order to answer this question data gathered from Roman-Byzantine wrecks will be utilized.

1.3 Methodology and Phases of the Work

The ships depicted in the mosaics studied in this monograph span from the late 3rd century BCE to the 6th century CE. The earliest representation of a part ship in mosaics appears as a headdress decoration in two panels depicting the bust of Berenike II (200 - 175 BCE) Egypt.

[1] Friedman, 1999, MA dissertation, unpublished.

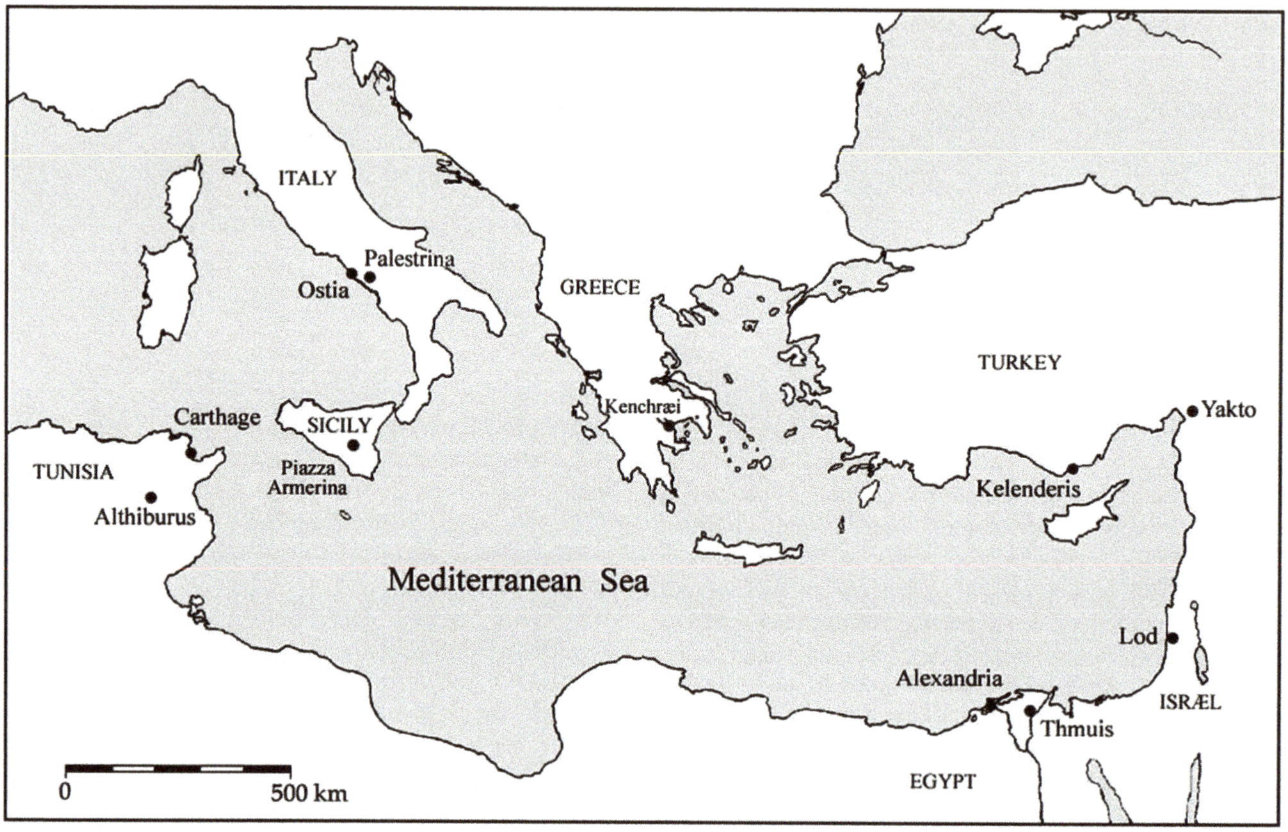

Fig. 1.1: Location map of the sites with the studied mosaics

The sites described below are according to their geographical location in the Mediterranean (fig. 1.1). The majority of mosaics with ship depictions were studied at their original location and in the museums where such mosaics are displayed. The Berenike mosaics from Thmuis (Tell Timai), Egypt, Yakto and Kelenderis, Turkey, were studied from partial publications of the sites. The mosaics from Lod in Israel, Ostia in Italy and Piazza Armerina in Sicily were studied in *situ*. The Nilotic mosaic from Palestrina, Italy was studied at the Museo Nationale Prenestino. Two of the Kenchreai glass *opus sectile* maritime panels with maritime scenes were studied at the Isthmia Museum, Greece, where they are displayed. The other two panels were studied from the Kenchreai archaeological report published in 1976. Due to their poor preserved condition these panels are stored in the storeroom at the Museum and therefore it was not possible to study them alive.

The research comprises:

1. Different techniques of mosaic making presented in the mosaic chapter and their application to the mosaics studied bellow will help us to understand the ship iconography.
2. A catalogue of the sites with mosaics containing depictions of ships.
3. Ship depictions will be referred to the general theme of the mosaic surface and its architectural setting.
4. Analyses of the propulsion and steering gear: sail, oars, steering oars, etc.
5. Historical reference for comparative substantiation of the types of vessels and their construction will be used.
6. Line drawings where necessary, when photos are insufficient to distinguish details of ship's elements.[2]
7. Classical wrecks found in the Mediterranean will be referred to when comparing the ship models in the surveyed mosaics. A table of wrecks dating from the 5th century BCE to the 7th century CE will list some ships at the sites where they were found, their full size reconstruction and cargo capacity, as a guides to deducing the actual size of the ships depicted in the mosaics presented in this monograph.

1.4 The Importance of the Research

The study is meant to help the viewers of the mosaics to estimate the validity of the ship depictions compared with ancient ships, shipbuilding traditions and their diffusion through the centuries in the ancient Mediterranean. Many ships depicting unique elements indicate that the mosaicist or the patron had some nautical knowledge, or probably that the models were produced from direct observations of real vessels anchored in harbors or from sketches circulated through sea-trade connections. The curiosity and observation of ancient ships, especially large merchantmen carrying grain, is beautifully described by Lucian with

[2] The illustrations used in this research were computer processed using PhotoShop 7.0 for PC. Most of the pictures were taken by the author at the sites where the mosaics with ship depictions are found.

reference to the large *Isis* ship that had anchored in Piraeus harbor, *en route* from Egypt to Rome:

> "*... hearing that such a huge boat, exceptionally large, had put into Piraeus, one of the Egyptian grain ships on its way to Italy?*[3] *... we stood a long time by the mast, looking up and counting the layers of hide, and marveling at the sailor going up among the shrouds and then running quite safely along the yardarm up there holding on to the ropes.*[4]
>
> *... what a huge ship! A hundred and twenty cubits long, the shipwright said, and well over a quarter as wide, and from deck to bottom, where it is deepest, in the bilge, twenty-nine. Then, what a tall mast, what a yard to carry! What a fore-stay to hold up! How gently the poop curved up, with a little golden goose below! And correspondingly, at the other end, the prow just right out in front, with figures of the goddess, Isis, after whom the ship is named, on either side. And the other decorations, the paintings, and the topsail blazing like fire, anchors in front of them ... and the cabins on the poop – all very wonderful to me. She was said to carry corn enough to feed all Attica for a year. And all this, a little old man has kept from harm by turning the huge rudders with a tiller.*"[5]

The combined information from mosaics with ship depictions and other sources, especially Roman-Byzantine shipwrecks, provides us with a better picture of a variety of vessels concurrent to the period of the mosaics presented in this monograph which also will serve to complement the gallery of ship iconography in other arts.

3 Lucian, *The Ship* or *The Wishes*: 1.
4 Ibid., 4.
5 Ibid., 5 - 6.

Chapter 2
Mosaic Production and its Application to the Ship Depictions

2.1 Introduction

Mosaics are among the most durable forms of decorative art that survived from antiquity to the present days. First they appeared in the ancient Greek floors made with natural pebbles and set into a mortar or plaster bedding. From the late 3rd century BCE the cube stones known as tesserae became the most common material for the mosaic production. This art form is very important because we can follow the evolution of pictorial, figurative and ornamental styles, and compositions over a time span of more than one millennium. The figurative scenes offer a wide range of information about the visual cultures of the period and the artists who commissioned them.[1] The majority of the mosaics were found in domestic contexts. The durability and water-resistance made them also suitable for baths and atrium. The mosaics with ship depictions studied in this monograph were chosen because they are less known, as well as that they permit us to follow the development of different mosaic technique that span a period from the late 3rd century BCE to the 6th century CE.

2.2 Definitions and Techniques

The name mosaic derives from the Greek word muse,[2] while the Hebrew and the Arabic term is psiphas stemming from the Greek word "strip"; it refers to the lead strip that marked the early Hellenistic mosaic work. Although, the term mosaic derived from the Greek word muse, no musa was appointed patroness for mosaic making.[3]

Mosaics are decorative surfaces made with small pieces of colored materials that are cemented together to form patterns or designs. The colored material consists of small pieces of stone, pottery fragments, shells or cubes of glass. The small square fragments are called tesserae (four in Greek), referring to the four sided small stone cube. When the mosaics are cemented into the floors they are called opus tessellatum and when placed on walls or vaults they are called opus musivum. The fine details made with tiny pieces of stone or glass and set wormlike (vermis in Latin) following the contour of the design is known as opus vermiculatum. Detailed figures and patterns known as emblema (emblemata) were made in the opus vermiculatum technique. Another mosaic technique made with pieces of marble or other materials (glass, pottery) cut to specific shapes of the design and fit together to form a smooth surface is known as opus sectile. In this technique, stones were used for floors, and glass, fine marble and even precious stones were utilized for walls. The earliest mosaics were made of pebbles readily available from riverbeds or the seashore; their range of hues was limited mostly to white, black, gray and some green. The introduction of the tesserae (end of the 4th – beginning of the 3rd century BCE) provided a wider range of colored stones that also permitted the making of more complex and detailed designs. During the Roman period the color range was widened by the introduction of glass tesserae which provided hues not found naturally, especially shades of green, red, yellow, blue, purple, etc. Glass tesserae became a more common material for wall and vault mosaics from the middle of the 1st century CE onwards. Human and animal figures, as well as different patterns in the Greek and Hellenistic mosaics were outlined with lead strips or later by fine strips of pottery shards.

The Roman mosaics were mainly classified into two categories:

1. The floor mosaics were known as lithostroton, vermiculatum or tessellatum.[4]
2. The wall and vault mosaics were distinguished by the intensive use of glass tesserae, as well as shells, pumice and other materials that were not used for floor mosaics. This type of decorations is known as *opus musivum*.

In the 1st century CE there was a clear distinction between different mosaic craftsmen: pavimentarii (making the floors), tesserarii (probably cutting the tesserae), tessellarii (mosaicists), all were related to the production of the floor mosaics, while musivarii and museiarii[5] were a distinct group of wall and vault mosaicists.[6] In the Edict of Diocletian (301 CE), the musearius and tessellarius were two separate crafts that also were differentiated by their payment; the musearius was paid 60 denarii a day, while tessellarius only 50 denarii.[7] The higher wage for the wall

1 Dunbabin, 1999, p. 1.
2 Ben-Dov and Rappel, 1987, p. 18.
3 Ibid.
4 Sear, 1977, p. 16.
5 Ibid., p. 17.
6 Ibid., p. 18.
7 Ibid.

mosaicist was recompensed for the difficulty and danger of his work on the high vaults and walls.[8]

The first written documented information on how a mosaic surface was prepared comes from the Roman architect Vitruvius that is mentioned again by Pliny the Elder. In his work "De Architectura", Vitruvius dedicated a whole chapter to the making of different kinds of floors. The relevant part for making mosaic floors is found in Book VII.I.3-4, 6:

> *3 "... A layer of stones is to be spread. After spreading the stones, the rubble, is to be mixed, three parts to one of lime; if it is of old material, five parts of rubble is to be mixed with two of lime. Let it then be laid on, and rammed down with wooden stamps. When the stamping is finished, it must be not less than nine inches thick. Upon this, a hard coat of powdered pottery is to be laid three parts to one of lime, forming a layer of six inches. On the finished coat, a pavement of marble slabs or mosaic is to be laid to rule and level.*
>
> *4 ... If it is mosaic, the edges of the tesserae are all to be leveled. For when the edges are not even, the rubbing down will be imperfect.*
>
> *6 ... Then the finishing coat is to be put on as already described, and the pavement is to be laid with tesserae about two inches thick, with a fall of two inches in ten feet. If this is well mixed and properly rubbed over, it will be safe against all damage. Then the mortar between the joints may not be affected by the frost, let it be soaked every year with oil lees before the winter."*

While the information given by Vitruvius is technical, in Pliny's work we find information about the origin of the mosaic floor and about the time when the tessellated floor was in common use in Italy:

> *"Paved floors originated among the Greeks and were skillfully embellished with a kind of paint-work until it was superseded by mosaics".*[9]
>
> *"... but tessellated pavements had already become common and extremely popular before the Cimbrian War (113 – 101 BCE) as is shown by the famous verse from Lucilius (180 – 102 BCE); 'Arte pavimenti antique emblemate vermiculato' (With paver skill and wavy inset stones)".*[10]

Later, Pliny mentions the time when mosaics came into use in Italy and the first use of glass mosaics on a vault in the baths built by Agrippa in Rome, in the 1st century BCE:

> *"Mosaics came in use as early as Sulla's reign (82 – 79 BCE). At all events, there exist even today one made of very small cubes, which he installed in the temple of Fortuna at Palestrina. After that, ordinary tessellated floors were driven from the ground level and found a new home in vaulted ceilings being now made of glass. Agrippa, in baths that he built at Rome ... he built vaults of glass".*[11]

2.3 Mosaic Materials

The colored stones used for the production of mosaic are mostly natural. The varied geomorphologic formations of the Mediterranean Basin provide rich colored material for mosaic making. The colors available from natural stones are: *black* – basalt, black marble and bitumen; *white, off-white* – limestone, white marble; *red* – natural reddish limestone, porphyry marble; *green* – Chipollino marble. The stones are easily cut into cubes (*tesserae*) and used for the work.[12] Cutting the *tesserae* into the required shape and sizes was done by hammer and chisel, where the stone was set into a wooden block as suggested by the depiction on a *stela* fragment at Ostia, dated to the beginning of the 4th century CE (fig. 2.1). Two seated men are hammering small objects whereas on the floor, near the foot of the man on the right, a fallen basket reveals thin stone slabs. Two figures instructed by a supervisor, carry sacks on their shoulders probably containing the *tesserae* that have to be taking to the site where the mosaics were laid.

The introduction of glass mosaics in Italy in the middle of the 1st century BCE expanded the variety of colors that could not be found naturally. The colored glass was obtained by introducing different oxides in the molten material which resulted in blue, green, brown, red and other hues. It is known that oxides (cobalt, iron, cupper, tin, etc.) were used from very early times to make opaque glass.[13] Gold and silver *tesserae*, used for wall mosaics, were obtained from glass cubes coated with a thin layer of gold or silver foil, over lapped by glass power and then heated at temperatures high enough to make solid and long lasting materials.[14] The materials used for *opus sectile* floors were colored stones and marbles, while for walls flat pieces of colored glass were used. There is no certain information on how and when flat surfaces of glass were made and also when they were used for windows. Pliny gives some hints for making flat glass when he describes the method of producing glass:

> *"After being reduced to lumps, the glass is again fused in the workshop and is tinted. Some of it is shaped by blowing, some machined on a lathe and some chased like silver".*[15]

[8] The difficulty of a wall mosaicist work is known from the case of a young man who fell to his death from a high scaffold; Sear, p. 18 and n. 45.
[9] Pliny, NH, XXXVI.60.184.
[10] Ibid., XXXVI.61.185.
[11] Ibid., XXXVI.64.189.
[12] Friedman, 1999, p. 207, unpublished.
[13] The white opaque glass is obtained by the addition of tin oxide; Lucas and Harris, 1989, p. 190.
[14] Friedman, p. 208.
[15] Pliny, XXXVI.66.193.

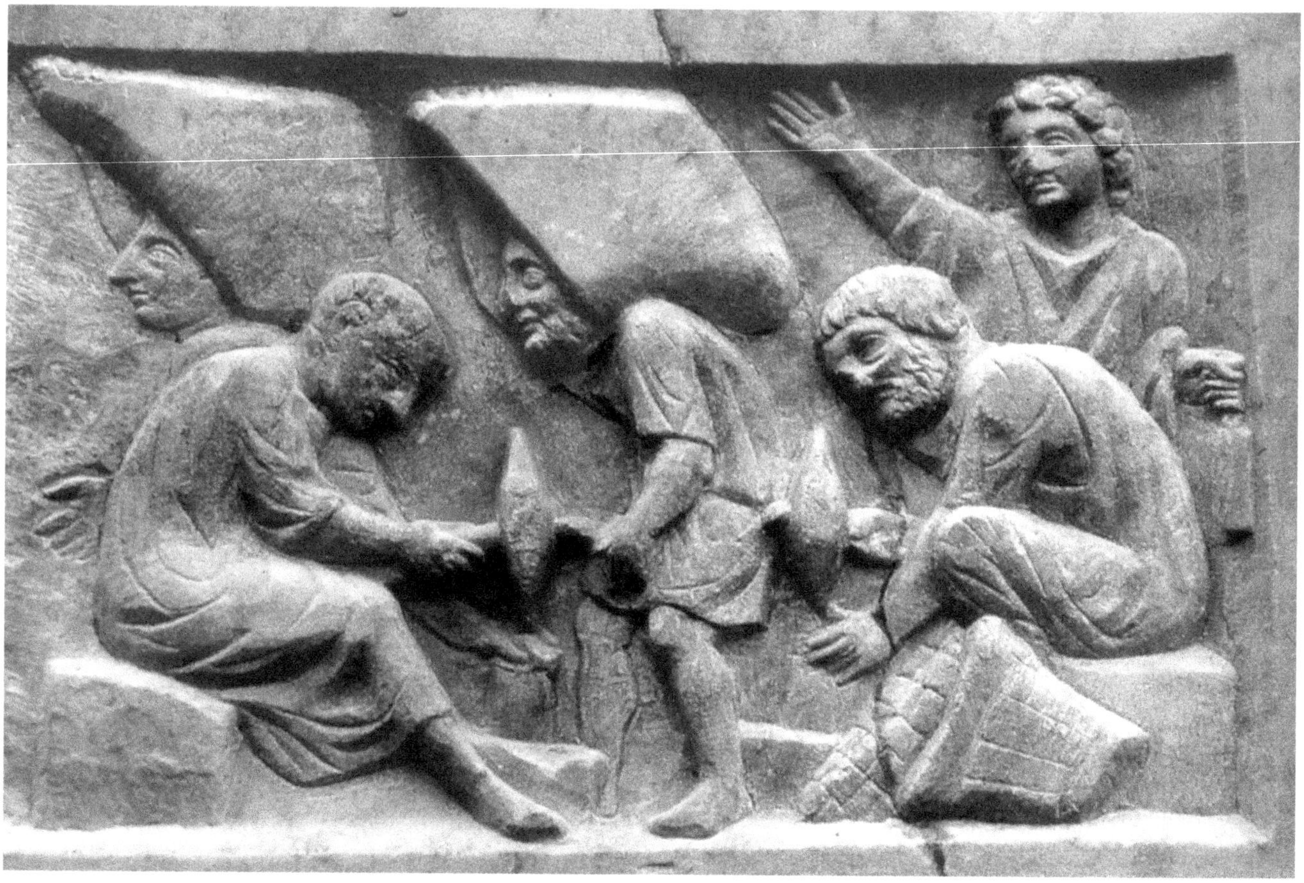

Fig. 2.1: Mosaicists cutting and preparing tesserae

This passage may suggest that pouring molten glass onto flat surfaces, forms of wet carbonized wood, or forms lined with sand, molded sheets of glass.[16] An alternative method was that large sections of multi-colored strips were assembled in flat rectangular forms and heated until the sections fused.[17] The upper surface of the panels was shiny due to the final polishing. Another method of obtaining flat surfaces of glass comes from glass blowing invented in the mid-1st century BCE. The glassmaker blew through a hollowed iron tube into a lump of molten glass attached at its other end. He blew until a bubble of glass emerged, which could be formed into any desired shape. In the 3rd century CE glass windows were made by the cylindrical method, in which an elongated bubble of glass was cut length-ways (when still hot) and opened to form a flat rectangular sheet.[18]

2.4 Layout and Compositions of the Mosaics

During the 2nd century BCE, *tesserae* mosaic became the common technique for making internal pavements in villas and public buildings. There are three methods of laying mosaics:

1. *Direct method*: the t*esserae* were individually set directly into the bedding mortar;
2. *Indirect method*: the *tesserae* were first embedded (face up) in sand and then strong paper or cloth was glued to their upper surface. Once the glue had set, the mosaic design could be removed from the sand and reset into the designated bedding mortar. When the mortar had dried, the glued materials were removed with hot water;
3. *Reversed method*: this method is similar to the indirect method except that the surface of the *tesserae*, instead of being laid face-up, were glued face down on a colored cartoon (paper or cloth) containing the design. After the section had been laid in the designated place and the cartoon had been removed, the picture was reversed (mirror image) to that which was shown in the workshop.[19]

Emblemae were produced in workshops either by the indirect or reversed techniques in their own tray and afterwards they were carried to the designated place within the mosaic surface. Such *emblema* could be set into frames or trays and also carried on journeys, as well as traded to different provinces. In a passage written by Suetonius, it is mentioned that transportable mosaic panels were carried during Caesar's military campaigns to adorn his tents.[20]

Mosaic making is not only a technical craft but also an art

[16] Grose, 1989, p. 357.
[17] Ibid.
[18] Adkins and Adkins, 1994, p. 326.
[19] The methods of the mosaics setting were summarized from Neal, 1976, p. 244.
[20] Daszewski, 1985, p. 16 and n. 16.

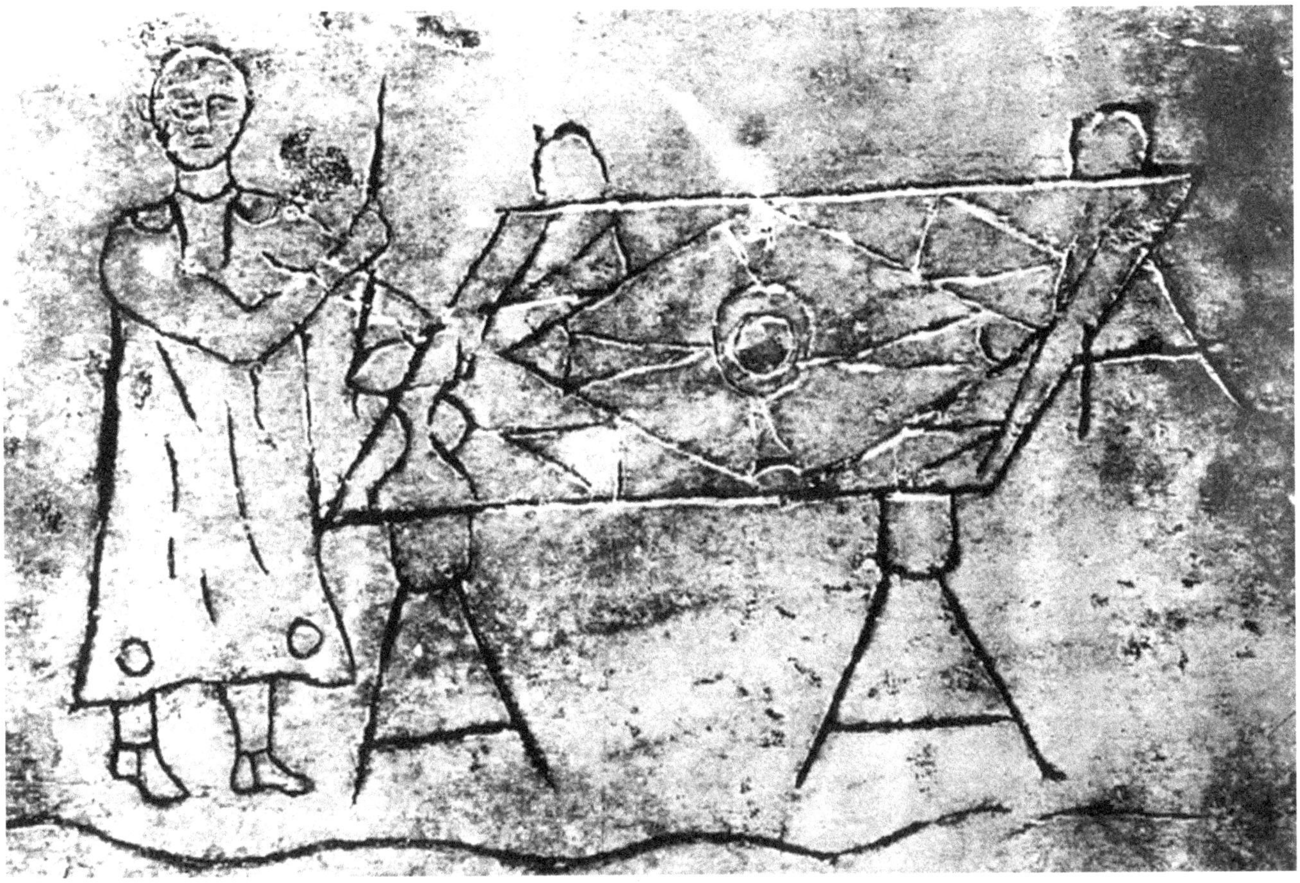

Fig. 2.2: Mosaicist working on opus sectile panel in his workshop

that needs extremely good planning, thus the final product will be an art in itself.[21] When the final bedding for the mosaic floor was finished, a very thin layer of fine mortar was smoothed over. When it dried guidelines were painted on this stratum with red or black color. A fine mortar layer was applied on the guidelines and the *tesserae* were then filled in, following the design. Pliny gives a hint on the making of mosaic floors by painting or drawing on the surface prior the laying of the tesserae:

> *"... paved floors originated among the Greeks, who embellished them with a kind of paint-work, until this was superseded by mosaics".*[22]

Guidelines were also scored into the moist mortar with pointed metal objects before the *tesserae* were set into the mortar bedding. The best remaining example of combined painted and scored guidelines was found on a mosaic floor in Villa Arianna at Stabiae, Italy,[23] or the mortar bedding of the mosaic floor in Herod's western palace at Masada, Israel (1st century BCE).[24]The use of lead strips as guidelines for geometric designs was a very common method used in Hellenistic mosaics and was especially suitable for complicated patterns like a meander, which required great accuracy in laying-out.[25] During the later part of the Hellenistic period the use of lead strips was abandoned and the mosaicists made preliminary drawings and incisions on the bedding. It has been mentioned above that *emblemae* with fine pictures were made in workshops in trays with a backing of pottery shards or marble strips which later were carried to their designated location on floors or walls. Evidence for prepared *emblemae* has been found at Pompeii, Kenchreai and Caesarea Maritima. A beautiful pictographic evidence of producing an *opus sectile* panel in a workshop is found on a funerary *stela* from Ostia, dated to the 4th century CE. A mosaicist is depicted standing next to his work-table and preparing a tray with a geometric *opus stectile* (fig. 2.2). Glass sheets were also used for *opus sectile* designs, mainly for producing wall mosaics. Pliny first mentions the use of glass tiles on walls in his description of a theatre built by Marcus Scaurus (58 BCE):

> *"This was his theatre, which had a stage arranged in three stories with 360 columns... The lowest stage was of marble and the middle one of glass (an extravagance unparalleled even in later times), while the top story was made of gilded planks".*[26]

Besides the polychrome mosaics, there is the black-and-

[21] Friedman, p. 213.
[22] Pliny, XXXVI.60.184.
[23] Robotti, 1973, figs. 1 – 4, pp. 42 – 44.
[24] Yadin, 1988 (reprint), p. 124 color, and p. 127 top.

[25] Dunbabin, 1999, p. 282.
[26] Pliny, XXXVI.24.114.

white style that emerged around 20 BCE, reaching its apogee in the 2nd and early 3rd centuries CE.[27] Economic factors may have dictated the use of black-and-white mosaics that were a substitute for the more costly polychrome mosaics. This new style took great pains to consider both the architectural setting and the viewer's movement around the spaces with such decorations. This new stile flourished in the 2nd century, coinciding with the period of prosperous economy and building development in the center of the Roman Empire. By the middle of the 3rd century, fewer black-and-white mosaics were produced and by the 4th century polychrome mosaics predominated. This black-and-white period was used during the reigns of Nero (54 – 68 CE) and Hadrian (117 – 138 CE) that is also known as the "Roman architectural revolution".[28] Black-and-white mosaics are found mostly in Italy (Rome, Ostia, Pompeii, Rimini, etc.) and only very few examples come from the eastern provinces (one example was found at Magdala or Migdal, on the NW coast of the Sea of Galilee, Israel[29]). Once the technique was established, the mosaic artists developed a unique visual effect by creating original and sophisticated compositions. The stylistic change in the black-and-white mosaics are often more visible than on any other media because such mosaic making required a simplification in modeling and use of *chiaroscuro* (light and shade).[30] The black-and-white mosaics were made in the *opus tessellatum* technique with *tesserae* of relatively equal size while the colored *emblema* was made in *opus vermiculatum* technique. The aim of this method was not to show the object as it actually was, rather to emphasize how it could be imagined in the mind of the viewer and more like a symbol. The decoration subject that is the focal point of the pavement occupies the field and the white background emphasizes this subject. The best known mosaics executed in this technique come from Piazzale delle Corporazioni at Ostia, Italy.

When mosaics became a common medium for interior decoration, artists who made these beautiful and complex works felt the desire to sign their names or to use alternative terms to be remembered. The earliest signature that appears in mosaics is found in two pebble mosaics dated to the late 4th century BCE, one is a fragment from Athens and the second is the Stag Hunt at Pella, both signed "*Gnosis epoesue*" (Gnosis made it). The first written evidence of an artist signing his name on a mosaic comes from Pliny:

> *"In this later field the most famous exponent was Sosos, who at Pergamu laid the floor of what is known in Greek as 'asarotos oikos' (unswept room), because by means of small cubes tinted in various shades, he represented on the floor refuse from a dinner table and other sweepings, making them appear as if they had been left there".*[31]

Only occasional Roman mosaics are found with the name of the artist inscribed into his work. In the east, especially Greece, such signed mosaics occur occasionally from the 2nd century CE.[32] The signature became more common on the floors of Byzantine churches from the 5th to the 7th – 8th centuries CE, in Israel, Jordan and Syria.[33] Other inscriptions in houses or ecclesiastic buildings indicate the date, the name of the bishops and priests responsible for the mosaic work, or of the donors who contributed. In many cases the name of the artist is not mentioned directly, rather the symbolic signature forms part of the prayer for the salvation of the mosaicists, or the allusion to be remembered for their devotion.

2.5 Subjects of Decorations

The designs of the early Greek pebble mosaic floors consisted of simple geometrical and floral patterns made with black and darker shades of white stones. The mosaics of the 4th century BCE from Gordion, Pella and Morgentina depict pictures of hunting and mythological scenes, as imitations of paintings on walls and pottery vessels. The mosaics made in the transitional phase from pebbles to *tesserae* (early 3rd century BCE), continued to use hunting and mythological scenes to decorate such surfaces. When the tessellated mosaic became a common interior decoration, rectangular or circular borders surrounded the central design on the floor. Within these borders were depicted geometrical patterns, meanders, waves, and scale patterns, either in two dimensions or in perspective.

When Alexander the Great conquered Egypt (332 – 331 BCE) and after Alexandria was founded, the Greeks were faced with a new and different culture, much older than their own. They were also influenced by the lifestyle of the Nile River and its Delta. These impressions were expressed in different arts that later influenced the whole Mediterranean Basin. One of the first mosaics with Egyptian influence is attested by the unique Nilotic mosaic from Palestrina, Italy (fig. 3.6.3). The Roman mosaics of the 1st – 2nd century CE depict scenes of everyday life, hunting, fishing, amphitheatre, circus, etc. During the 3rd – beginning of the 4th century, the artistic style of the Roman mosaics changed. The artists did not make them as realistic as the Hellenistic or early Roman mosaics, but they begun to use more sophisticated techniques. The designs became more symbolic and did not occupy the entire surface. When a mosaic was laid in the old style, the pictures were arranged in groups, realistic in appearance and depth. In pavements laid in the new style, the whole surface was divided up into medallions or geometric borders and the figurative decorations were inserted inside them, thus creating a series of individual *emblema*. This scheme of decoration continued to be used in the later Byzantine period.

The ship motifs depicted in the mosaics studied bellow are associated with varied themes: the Nile flood and the

27 Clarke, 1979, p. xix.
28 Ibid.
29 Friedman, pp. 5 – 19.
30 Ibid., p. 211.
31 Pliny, XXXVI.60.184.
32 Dunbabin, p. 273.
33 Ibid.

Goddess Isis festivities in the Nile Delta, hippopotami hunting in the Delta, fishing activities, harbor scenes and shippers offices. Such mosaics were found in a *nymphaeon* (Palestrina, Italy), baths, *atrium* and hall in a private villa (Piazza Armerina, Sicily), *triclinium* (Lod, Israel and probably both Berenike panels, Egypt), villa (Yakto, Turkey), a sanctuary (Kenchreai, Greece), commercial quarter (Kelenderis, Turkey), and the public Square of Shippers and Traders at Ostia, Italy.

2.6 Ships Depicted in Mosaics

Mosaic artists had to find ways around and solve the problems of ship representations, especially due to the limitations that the mosaic art provides. They did so in a variety of ways, mainly as to enchant the forms and details, which they chose to present. Laying the *tesserae* to render a design is not an easy task. The *tesserae* have to be cut into small cubes and chips for floor or wall decorations, or sheets for the *opus sectile* thus to fit within the mosaic design. Despite the fact that mosaic production is a difficult form of art it seems that rendering with *tesserae* of stone, glass, pottery, etc., it proved to be a flexible medium where the designs could be produce with quite accurate details following a previously made drawing or sketch. The depiction of ships in mosaics revealed that the mosaicists could produce such subjects with little difficulty. When a ship model was planned, it was likely that it first was painted schematically or outlined in detail on the mortar bedding and only afterwards the stones were set into the designated patterns. The vault mosaic in Tomb "M", beneath the St. Peter's Church, in Rome, substantiated that painted patterns were applied before the *tesserae* were laid. In places where stones were missing it was observed that some colors were used as guidelines rather than polychrome paintings for the colored stones that were placed over the drawings.[34] For example, a wash of grayish-blue was found underneath gray *tesserae*, ochre under lime-green *tesserae*, and pale yellow under yellow stone cubes.[35]

Mosaics with ships representations permit us to distinguish mosaic techniques from the later Hellenistic to the Byzantine periods. Ships depicted in Hellenistic mosaics (both Berenike panels and the Palestrina Nilotic mosaic) were rendered with rich and varied hues of polychrome *tesserae*, whereas the details suggest that they were copies made after paintings. The realistic touch of the vessels in the Palestrina Nilotic mosaic is given by perspective views, while the human figures are represented at some scale proportional to the entire scenes. The small size *tesserae* and chips of varied hues used for rendering the vessels, suggest that detailed drawings and paint samples may have been produced on the mortar bedding and then the laid stones followed these colorful details. The reefing of the sail from its middle foot, the brails sliding through brail rings/loops shown on the fore face of the bunt of both sailing ships in the Palestrina mosaic (figs. 3.6.4, 3.6.5) probably were produced from detailed drawings of such vessels observed *en route* on the Nile or along the Egyptian Mediterranean shores.

Although, the warship in the Palestrina mosaic (fig. 3.6.3) is depicted in a realistic way we may observe that the mosaicist had some difficulties when he used the perspective. The port side of the ship is well presented in perspective, as emphasized by the outrigger oarbox and the *en echelon* arrangement of two superimposed rows of the rowing oars. The fore face of the starboard oarbox with a "mass" of rowing oars projecting from beneath do not show clearly their arrangement as they appear on the port side. Probably the mosaicsit could not follow the detailed drawing on this part of the ship. Therefore this depiction appears as a frontal view with multitude of oars seen in the same plane. Besides this distorted depiction we may understand that the oars arrangement was identical on both sides and were true to the presentation of the port oars system. The *aphlaston* depicted in a two dimensional view indicates that in certain occasions the mosaicist was not able to use a three-dimensional representation (fig. 3.6.12). Despite the fact that the *aphlaston* when viewed as an individual element is flat (two-dimensional), it did not alter the three-dimensional rendering of the entire vessel.

The depiction of the *thalamegos* (cabin-carrier) engaged in hippopotami hunting in the Nile Delta shows that mosaicists did not have difficulties to render realistic views of vessels in mosaics (fig. 3.6.17). The perspective view of the vessel and three-dimensional representation of the crew give a vivid image to the scene. It appears that the mosaicist reproduced the vessel from a detailed drawing which may have been rendered with varied hues of paint in the places where the *tesserae* were laid atop. Even the lashing strops used to mount the rowing oars on the port gunwale indicated that such fine details could be produced in mosaics without difficulties (fig. 3.6.17).

Ships depicted in Roman – Byzantine mosaics are much simpler and mostly are shown in profile, in two-dimensional views. The vessels are less realistic and in certain examples they are quite distorted. Apparently the mosiacists were more concerned to show a vessel that is a symbol of a certain type and comprised it within the mosaic theme. The hulls reveal their general shape. The sails in certain examples are shown in some kind of perspective billowing over the gunwale or seen from the lee side (figs. 3.2.3, 3.4.7, 3.5.15, 3.515a, 3.7.4, 3.7.9, 3.7.14, 3.7.15, 3.7.27, 3.7.35, 3.7.36, 3.7.39, 3.7.44). Ships depicted in Roman – Byzantine mosaics mostly are symbolic representations of merchantmen as deduced from rendering of specific elements, such as the triangular topsail, the goose or swan head placed on the stern, or the lattice fenced hanging poop (figs. 3.2.3, 3.2.5, 3.5.7, 3.5.10, 3.5.11, 3.5.15, 3.5.15a, 3.5.19, 3.7.8, 3.7.19, 3.7.21, 3.7.22, 3.7.27, 3.7.29, 3.7.30, 3.7.32, 3.7.35, 3.7.36). When the hulls are rendered with polychrome *tesserae* they are rendered with wide strips of light and dark brown stones, thus indicating the strakes (figs. 3.2.3, 3.2.5, 3.4.7, 3.8.21, 3.8.25). In Roman-

[34] Sear, 1976, p. 238.
[35] Ibid.

Byzantine mosaics the *tesserae* are larger and the hulls are rendered with lesser details than the finer and painting-like representations in the Hellenistic mosaics. Although these ships are not represented in detail they are shown with specific elements that could not be produced in mosaics if the mosaicist did not possessed some nautical knowledge or the ship model was produced with such elements from sketches or pattern-books. Blocks and deadeyes that are very specific gear in sailing rigging, although depicted schematically, they still indicate their function and are properly placed on the tackle lines (figs. 3.2.4, 3.4.9, 3.8.21, 3.8.22, 3.8.23). It is not easy to render such elements with *tesserae*. The oval shaped blocks or deadeyes of Lod Ship 1 sail rigging are outlined with a strip of black *tesserae* and the field is made with ochre stones thus indicating the wooden material (fig. 3.2.4). They probably are shown from their lateral side, but not depicting the sheave. The rounded blocks or deadeyes with two short longitudinal black strips in the middle, depicted on the upper part of the brails and the masthead of the Sailing Ship 1 in the Great Hunt mosaic from Piazza Armerina indicate their real position and function (figs. 3.8.22, 3.8.23). Rendering of blocks and deadeyes in mosaics and their fairly accurate placement in the sailing rig augments that the mosaicists were familiar with ships and/or may have understood the function of these tackles.

In black-and-white technique, the mosaicists had to use more sophisticated ways to render ships. Usually, a strip of white *tesserae* is used to indicate the long seams between the strakes or outlining the wales on the black hull (figs. 3.7.7, 3.7.9, 3.7.11, 3.7.14, 3.7.16, 3.7.19, 3.7.21, 3.7.22, 3.7.24, 3.7.25, 3.7.27, 3.7.30, 3.7.32, 3.7.33, 3.7.35, 3.7.36, 3.7.38, 3.7.39, 3.7.40, 3.7.43, 3.7.44, 3.7.45). Some of the vessels seem to be shown in a three-dimensional view, as we may deduce from the cabin with the pitched roof in Ship 2, Station 18 (fig. 3.7.15), the capstan and the cabin with barrel roof depicted in Ship 1, Station 25 (fig. 3.7.24), or the merchantman in Station 51 (fig. 3.7.40). Apparently, when using the black-and-white technique, the mosaicists did not have any difficulties to use combined views, such as the bow of a ship seen from above (bird-eye view), the jars loaded on the deck and the seated *tabulator* who recorded the cargo of the ship on wax-tablet (bird-eye and frontal view) (fig. 3.7.40).

The Kenchreai glass *opus sectile* maritime panels indicate that the mosaicists were highly skilled in this art and were also able to depict ships with detailed sailing rigging and steering gear that were produced from flat pieces of glass cut to fit the desired shape within the design of the mosaic. These panels also reveal that the artists were not limited to render any details in the fragile glass *opus sectile* technique. Probably that the ship models were drawn in detail and then each element was cut out from molten glass to fit the design which afterwards was embedded in the mortar, previously prepared in trays. The hulls of the ships are made with strips of glass that were cut to follow these shapes (figs. 3.5.7, 3.5.10, 3.5.11, 3.5.12, 3.5.13, 3.5.18, 3.5.19, 3.5.20, 3.5.21). The lines rigging were made with twisted rods of glass, while the fully open sails were produced from a single piece of glass cut in the shape of a billowing bunt with an arching foot (figs. 3.5.7, 3.5.10, 3.5.11, 3.5.12, 3.5.13, 3.5.17, 3.5.18, 3.5.19, 3.5.20). Ship 4 in Harbor II rendered with a transom stern and the sail billowing forward with the starboard leech and side of the bunt turned backward suggest a three-dimensional view of the vessel (figs. 3.5.15, 3.5.15a). It also shows the high technological ability of the mosaicists to represent such a complex view in glass *opus sectile*. Ship 3 – Harbor IV is the most sophisticated depiction of a vessel seen from its prow through the entrance of a harbor (fig. 3.5.21). This view emphasizes a highly skilled mosaicist who was able to produce this complex scene of tiny cut pieces of glass to fit into the ship viewed from its prow. The artist probably followed a very detailed drawing of the vessel that may have been sketched by an artist who was familiar with ships and harbors.

2.7 Discussion

The mosaics with ship depictions surveyed in this monograph cover a period from the late 3rd century BCE to the late 6th century CE. The materials used in their making are polychrome stones, glass *tesserae*, and glass sheets for *opus sectile* panels. The decoration subjects range from portraits (both Berenike II mosaics), fishing scenes carried out from small boats, trade and shipping, and harbor activities that are found on the pavements or walls in private villas, public buildings and shippers offices. Each site will be discussed relatively to its mosaic technique and decoration theme. Rendering ships in mosaics indicate that the mosaicists were not limited in their art to produce any details and depiction of specific elements associated with the sailing rigging, steering gear, loading and unloading of cargo from a ship where the *artemon* also can be used as a crane, or the recording of the cargo in a ship before it is unloaded or shipped away.

The Berenike mosaics from Thmuis; Egypt

The first Berenike panel signed by Sophilos and the copy are the earliest mosaics that depict the fore-part of a Hellenistic warship, in the form of a headdress on the head of a female figure. In both works, wide borders decorated with meander, crenellated and scale patterns surround the central figurative panel. Both mosaics are made in *opus vermiculatum* technique using tiny polychrome stones and glass of varied hues. The first Berenike (210 – 200 BCE) is signed by Sophilos (fig. 3.1.1); it is the only example of a signed mosaic in this monograph. Sophilos probably produced the mosaic after a painting, while trying to be as accurate as possible to the source. The second Berenike (175 – 150 BCE) is a copy, probably produced as imitation of the Sophilos panel (fig. 3.1.6); it is less accurate and not as fine as the Sophilos mosaic. Both panels are *emblemae*, most probably made in the direct technique, produced in a workshop, and then were subsequently carried in trays to their location. In both mosaics the traditional lead strips of the Hellenistic technique were used to define the shaft of the mast and the yard, as well as the patterns within the geometric wide borders is noticeable.

Lod; Israel

The maritime panel within the middle of the mosaic pavement complements the entire floor decoration that is divided into two main carpets (northern and southern). All the figures in general were planned as being viewed from the same direction (looking from the south). The ships and the sea fauna (fig. 3.2.2) show strong links to the North African mosaics with similar scenes. The ships are depicted more like symbols of large merchantmen that sailed in open waters populated by very rich fishery. They may indicate that the patron of the villa owned the vessels or he traded fish, garum and grain from North Africa to Rome and in the eastern Mediterranean with North African ships. No human figure is associated with any scenes in the mosaics. The geometrical frame attached to the prow of each ship and the lack of human figures may indicate that the owner was Jewish and that he ordered the artists to make the design according to Jewish Law, which did not permit the use of the human figure in any kind of art depiction. The ships probably were also used as apotropaic signs associated with the safe return of the vessels, its owner and the thank-offering to celebrate such a return,[36] or they symbolically evoke his occupation.

We may suggest that the maritime panel with the ships and fishery were made in the direct technique on their designated site. The designs were probably laid out by using a square grid system for the entire designs, while guidelines were used to outline the patterns within each section. At least two mosaicists, who originated in North Africa, possibly designed the maritime panel and then the *tesserae* were laid by local artisans under their instructions. The artists possibly brought with them similar patterns used in the floors of their homeland and then adapted them to the taste of the villa's owner at Lod

Yakto Thetis mosaic; Turkey

This mosaic originates in Antioch and belonged to a villa complex; it was removed from its original setting and now it is displayed in the Hatay Archaeological Museum in Antakya. The mosaic is dated to the middle of the 5th century CE. The goddess Thetis and the fishing scenes within the surrounding border are made in the *opus vermiculatum* technique. The figures are quite realistic and were produced with colorful small *tesserae* (figs. 3.3.5, 3.3.6). The white background is made with bigger tesserae. Although Thetis is the main subject of the mosaic it does not compete with the fishing border; they complement one another. It appears that the mosaic was made by the direct technique, being laid directly at the site. The detailed work on the figures shows Hellenistic influence, but the design does not follow the Hellenistic tradition. The Thetis mosaic seems to fit into the scheme of the late Roman or early Byzantine period, where the mosaic is not a reproduction of a painting but a complex work of several scenes set together that were meant to be seen from different angles and therefore creating some dynamics to the room. The design of the mosaic is quite minimal while the details are found within Thetis, and the fishing scenes inhabiting the surrounding frame. When the visitor entered this space he could view the whole mosaic from a single place with the focal point on the Goddess and afterwards turning around on the fishing scenes.

Kelenderis mosaic; Turkey

The Kelenderis Harbor scene frame forms one third of the entire mosaic strip that measures 12m in length and 3m in width. This is the only figurative panel, while the other two are made with geometric patterns. The mosaic is dated to the 5th – 6th century. The harbor scene is depicted in combined frontal and bird's-eye views (fig. 3.4.6). The mosaicist illustrated the ship and the ship's boats as seen from astern, thus revealing the harbor from its entrance. It appears that his knowledge of perspective was not very deep which brought to the distorted view of the ship's stern and the close approach of the twisted prow to the shore (fig. 3.4.7). The buildings are spread flat on the side probably because he wanted to focus on the ship and show the harbor interior (he did not know how to present the complete scene in perspective).

We may suggest that the mosaic was probably made in the direct technique and at the site. The design is not particularly complicated and has modest dimensions (c.3 x 3m). A panel of this sort could be produced directly in its designated place without difficulty. It appears that the geometric panels beneath the harbor scene were also made directly on site. This mosaic fits into the Byzantine scheme where the field is divided into geometric frames with different designs in each one of them. This combined designs formed a whole unit of the surface, which probably gave some dynamics and a kinetic appearance to the surface. The harbor scene is realistic in its concept but abstract in its making, especially emphasized by the rendering of the wavy sea within the inner harbor (fig. 3.4.6).

Kenchreai glass opus sectile maritime panels; Greece

Mosaic panels made in the *opus sectile* technique are rare, whereas the Kenchreai panels are unique because they are made of glass. They were meant to decorate the walls of the complex known as the Temple of Isis (dated to mid-4th century CE). Of the twelve Nilotic panels that have survived, only four depict harbor scenes and ships. These glass *opus sectile* panels were produced in a workshop, in trays bedded with pottery shards, and then they were packed in wooden crates and shipped by sea to the port of Kenchreai.

Twisted rods and thin strips of glass were used for the rigging of the ships. The flat pieces of glass were cut into different shapes to fit the patterns. It appears that the pieces

[36] Dunbabin, 1978, p. 126.

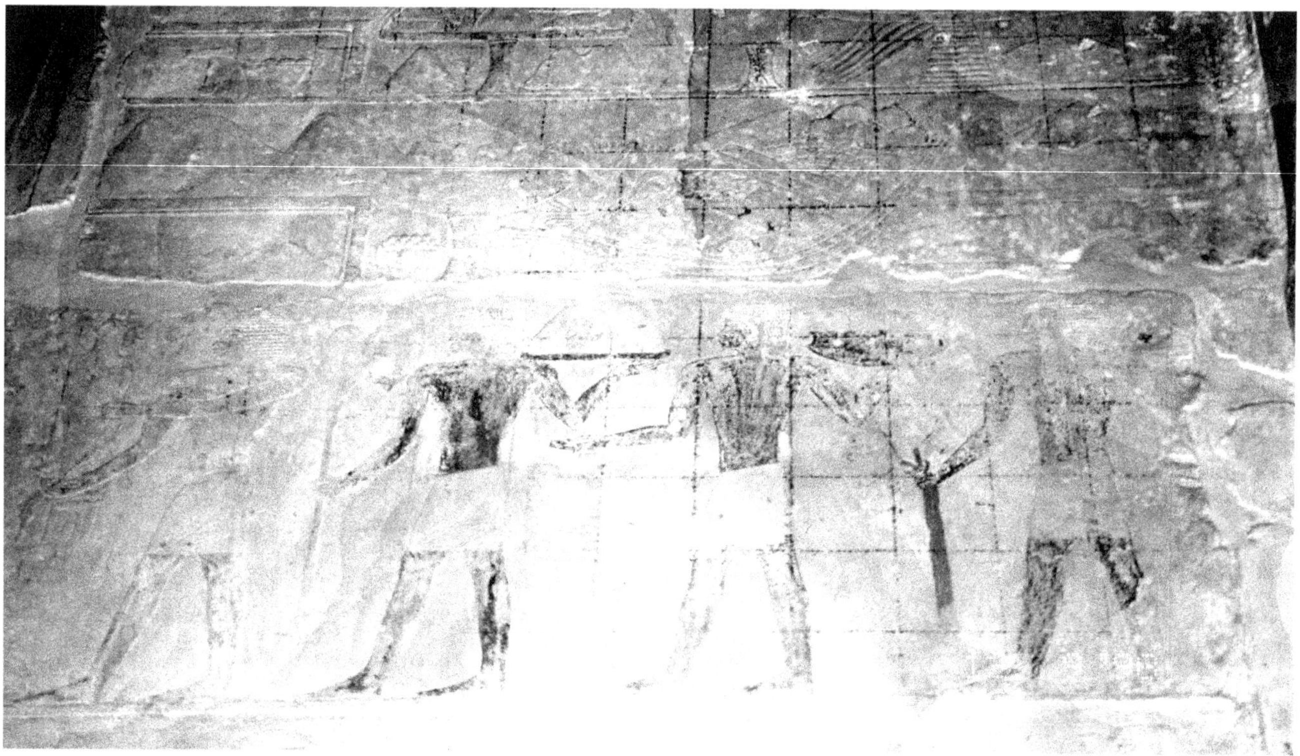

Fig. 2.3: Square grid system in the Mastaba of Akehetehep

were cut out when the glass was still close to its molten stage. The best indication can be observed on the sail of Ship 1 - Harbor I panel (fig. 3.5.7). The shape of the billowed bunt was cut from a single piece of white opaque glass and afterwards the brails and the reinforcing-bands were incised or cut into the face of the bunt with a sharp metal. These glass panels seem to resemble stained glass work that were not meant to be set against a light source for windows but to decorate the walls of a very elaborate structure. The panels were imported either from Italy or Egypt. Chemical analyses of the glass suggest Egyptian origin but there is no certain proof.[37] The theme decoration of these panels indicates their origin in Italy, especially in the area of Naples where seaside and similar Nilotic scenes depicted in frescoes were very common decorations in the villas around the Bay of Naples and Pompeii. Further investigations will be needed to verify this assumption.

Shipment of mosaic panels is attested by Suetonius when he noted that such luxurious items were carried along with Caesar to embellish the tents during his military campaigns. The Kenchreai panels present outstanding examples on the trade of finished mosaic panels *via* sea trade and how fragile art objects were circulating from one place to another. They also may be considered as examples of the high technology of ancient craftsmen who could produce and work flat glass surfaces.

The Palestrina Nile mosaic; Italy

This unique mosaic of the later Hellenistic period (110 – 100 BCE) may be considered as the earliest illustrated topography of Upper and Lower Egypt, especially the Nile Delta during the flood period and the dedicated festivities of the Goddess Isis (fig. 3.6.3). The scenes are depicted in combined frontal and bird's-eye views. The large figurative mosaic (5.85m at the base, and 4.31m deep) was made in the *opus vermiculatum* technique. Besides the polychrome stones, glass *tesserae* were also used, especially hues of blue and green. The figures, structures, animals and ships are quite proportional and realistic. Tiny chips of stones are used for the very fine details, while the *tesserae* in the entire mosaic are no larger than 5 – 6 mm.

The square grid is a basic tool used by the artists to convert a design that was first produced in a smaller scale.[38] The best examples of square grid usage are the Egyptian monuments, mainly found in tombs of the 18th Dynasty (New Kingdom). The earliest example for the use of the square grid system comes from the Mastaba of Akehetehep (5th Dynasty; 2400 BCE), now in the Louvre Museum (fig. 2.3). These grids of red painted or scored lines were used to obtain acceptable proportions of the human figures. We may assume that a square grid system was used in the Palestrina mosaic to outline the general decoration from detailed sketches, and then guidelines were either scored or drawn on the mortar surface to help in maintaining the same proportions as in the final design. Some color paint may have been applied on the mortar to indicate the exact location of the colored *tesserae* to be placed in the patterns. The layout of the Palestrina mosaic was likely produced by the direct technique executed at the site. Such a complex mosaic needed many

[37] Dunbabin, 1999, p. 126.

[38] Friedman, p. 215.

detailed sketches made by several mosaic masters probably on paper, cloth, *ostracas*, hides or even on wax-tablets, and also required many other assistants to lay the stones.

Piazzale delle Corporazioni, Ostia; Italy

The mosaics come from one place, Piazzale delle Corporazioni or the Square of the Shippers and Traders at Ostia. The floors of all shops were decorated with black-and-white mosaics, dated to the 2nd - 3rd century CE. The scenes are very minimal, depicting individual or groups of black objects on a white background. The ships either single or in pairs, are sometimes associated with a stepped structure assumed to be a lighthouse, and a *modium* for grain measure. The static designs were not meant to be viewed as artworks upon entering the shop rather they were messages and symbols to inform the viewer as to the function of the building. The inscriptions that follow some of the ships and the depiction of the exotic animals (elephant, tigers, boar, or deer) indicate the place of origin of the shippers or traders in the Roman provinces: Misua, Sylectum (North Africa), Turris, Karalitani (Sardinia), Narbo (Gaul), etc.

Although the ships are not depicted to scale, they show relevant details to indicate their type, assumed actual size, and their rigging.[39] A unique depiction appears in Station 51. The ship is seen reversed from the doorway. It shows a merchantman with a cargo of jars on the deck and a *tabulator*, taking records of the cargo by writing its account on wax-tablet (fig. 3.7.40). The horizontal white strips on the black tablet indicate the noted record, as we also understand from the man's stretched left arm counting the jars, and holding a *stylus* in the other hand. It appears that this is the only evidence in mosaics of recording sea cargo at the time when the ships arrived at a harbor or before leaving it. All the mosaics in the offices seem to be laid in the direct technique on the sites. There are many traces of repairs in the mosaics, which mainly replaced older ones, or probably the offices changed their function.

Piazza Armerina; Sicily

Ship depictions in the villa were found in areas with varied functions. All the mosaics, dated to the first quarter of the 4th century CE, were produced with polychrome stones. The composition of the maritime scenes is in registers (fig. 3.8.7; Room 29), a fishing scene following the layout of the structure (figs. 3.8.2, 3.8.13; *frigidarium, atrium)* and a long narrative strip (fig. 3.8.20; the Great Hunt Hall). There are two main subjects depicted in the mosaics: different fishing methods carried out from small boats (*frigidarium*, Room 29, *atrium*) and sea transportation of African animals to Rome for the games held in the Colosseum and Circus Maximus (Great Hunt Hall). It appears that the themes of the mosaics were suited to the function of their architectural setting.

The fishing mosaic in the *frigidarium* was set in an octagonal floor. The boats depicted in the middle and arranged in a circle, create some dynamics to the space and to the viewer who enters the room (fig. 3.8.2). This space is also the passage to the side apses and to other parts of the villa. The fishing scene in this cold room where the bathers came to entertain probably gave them the feeling of being on a lake or seashore. The scene in Room 29 is depicted in three horizontal registers: the waterfront colonnade is on top while the fishing boats form the second and third registers. The decorations arranged in one direction, can be seen by the viewer from the doorway without having to move around (fig. 3.8.7). To break the monotony of the scene, the artist set a seine net between Boat 1 and Boat 4, and the physical effort of the putti in both boats, straining to pull the net full of fish, bestows dynamism to the scene (fig. 3.8.12). The scene in the *atrium* is a very symmetrical setting (fig. 3.8.13). It appears to be a two registers layout: the waterfront colonnade and the structures behind form the upper part of the scene following the arc of the *atrium*, while the fishing boats form the second register. The boats are single features that occupy almost the whole width of the floor. The seine net pulled between Boat 3 and Boat 4 is the only connecting element between the scenes in the mosaic (fig. 3.8.13). It also emphasizes the center of the mosaic and provides some dynamics to the floor. Apparently the mosaic was produced in the direct technique.

The Great Hunt is the largest mosaic floor in the villa (60m long). The scenes are not depicted in registers but in a long continuous narrative strip. The middle section depicts two large merchant ships, showing animals being loaded in the harbor of Carthage and probably Alexandria, while on the other side, unloading them after having arrived at Ostia or Rome (figs. 3.8.21, 3.8.24, 3.8.25). The ships are not to scale with the human figures, but apparently some proportion is maintained between them and the animals. The ships represent very large *navis oneraria* that could carry cargoes of 400 –500 tons or even more. The mosaicists did not necessarily tried to depict the whole scene at scale rather they seemed to have been more interested in illustrating the message of the scenes. The mosaic was probably produced in the direct technique at the site and made by several artists with many other assistants.

Mosaics and patterns that were brought from North Africa were adapted to the taste of the villa's owner in Piazza Armerina. The African link is seen in the fishing scenes, especially the depiction of putti and the boats that are similar to those in a mosaic from Carthage. The middle ship in the Great Hunt mosaic, the crew and the loading of the animals, are very similar to the Dermech mosaic (fig. 3.8.28).

[39] See Chapter 4: Ship Archaeology.

Chapter 3
Catalogue of the Ships

3.1 The Berenike mosaics – Egypt

Introduction

Wall paintings, reliefs, sculptures and the monumental architecture of the Pharaonic Egypt had been studied in abundance. Less is known about the contribution and the influence of the Egyptian mosaic motifs from the Ptolemaic period upon different art representations in the Mediterranean. The mosaics found in Egypt (about 160 known mosaics) originate in four main areas: the Nile Delta (especially those from Alexandria and its vicinity), the Fayum Oasis, several mosaics from the Middle and Upper Egypt, and the Monastery of Santa Catherina, Sinai Peninsula.[1] Although different techniques of inlay or incrustations on sculptures, furniture, walls and sarcophagi were known in ancient Egypt since the Middle Kingdom, mosaic decorations were essentially foreign. Mosaics were Greek importation in the Nile Delta after the conquest of Egypt by Alexander the Great (332–331 BCE).[2] The early Egyptian mosaics display a Hellenistic influence, but within quite a short period they underwent changes in techniques and styles, adopting and reflecting the ancient Egyptian art tradition. Greeks who came to Egypt encountered new culture and artistic traditions much older than theirs. Representations of the fauna and the flora in the Nilotic landscapes, in the houses of the rich Greco-Egyptians in Alexandria or Canopus reminded their owners of the richness of the Delta and its influence on peoples' lives. These aspects of the Egyptian country houses became very popular outside Egypt due to commercial activities and cultural links with other countries around the Mediterranean.

This research will not deal with all the Egyptian mosaics but only with those with ship depictions that were found at Thmuis (Tell Timai) (fig. 1.1). Two mosaics depict similar theme of a female bust adorned with a headdress in the shape of a warship's prow. This model vessel (showing the prow and the fore section) seems to be the earliest ship representation in mosaics, c.200 – 175 BCE.[3]

[1] Daszewski, 1996, p. 145.
[2] Daszewski, 1985, p. 2.
[3] Ibid., p. 160.

Berenike of Sophilos

This rectangular mosaic frame (fig. 3.1.1) was discovered in 1918 and removed to the Alexandria Museum in 1924; Inv. Nr. 21739[4] (the dimensions of the entire panel are: H = 2.77m, W = 2.61m; the figurative inner frame is: H = 0.85m and W = 0.86m).[5] The mosaic is almost complete, but only the left-hand side and part of the outer border depicting a crenellated motif are missing. The border comprises three bands of different patterns:

1. The outer border depicts a crenellated motif in black and white tesserae.
2. The middle strip is a perspective meander made with white, yellow, ochre, dark brown, reddish-brown and black stones.
3. The inner band depicts a double braded guilloche made with similar hues as the middle band.

The *emblema* is outlined with white and black strips. The subject of the mosaic is a female figure crowned with a headdress shaped as the prow of a ship (fig. 3.1.1). The face is very realistic and made with stones of flesh hues (pink and white). Her large bulbous eyes stare into the open space without a focal point; the black eyelashes made with tiny bits of triangular tesserae emphasize the large eyes. She is dressed in military attire underneath which she wears a red tunic. Upon the tunic is a silver cuirass inlaid with gold. A chlamys, white underneath and purple outside, covers her left shoulder and part of the breast. The white ends of the chlamys are pinned together with a golden anchor-shaped fibula (fig. 3.1.2). She is also adorned with a golden necklace and earrings. The very large headdress shaped as the prow of a warship covers the top of her head (fig. 3.1.3). On her shoulder, she holds a long tapered mast with a horizontal yard that also may indicate a very large *stylis*. The mast and yard are outlined with thin strips of lead (fig. 3.1.1). The left side of the mast is made with dark brown tesserae and the right side with light brown-ochre stones. The yard is made with the same hues as the mast. The darker colors used on the lower and upper edges of the yard give it a three-dimensional appearance. A ribbon is attached to the tip of the masthead and both its ends undulate downwards, behind the head of the figure. The

[4] Daszewski, 1985, p. 142.
[5] Ibid.

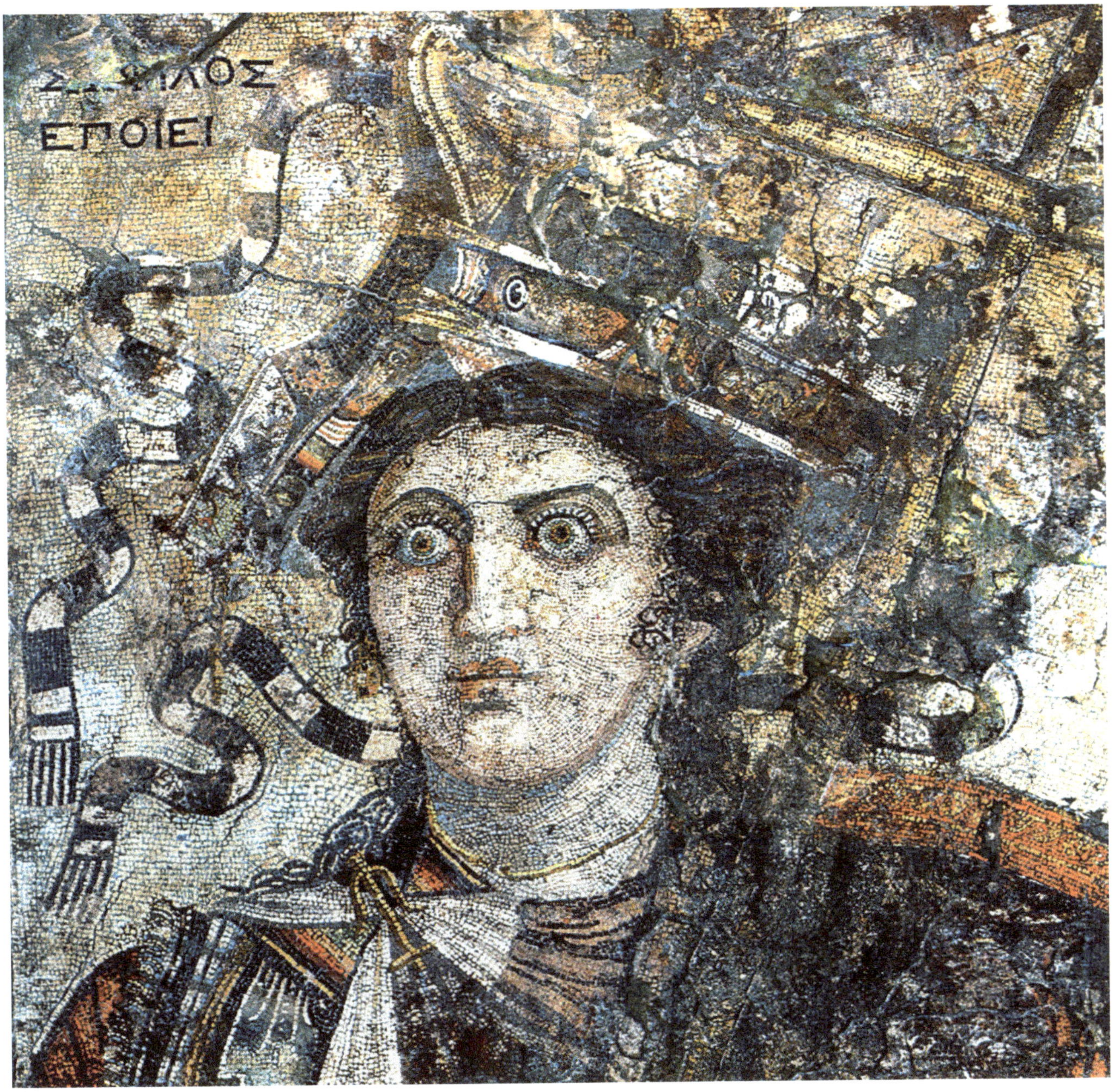

Fig. 3.1.1: Berenike of Sophilos

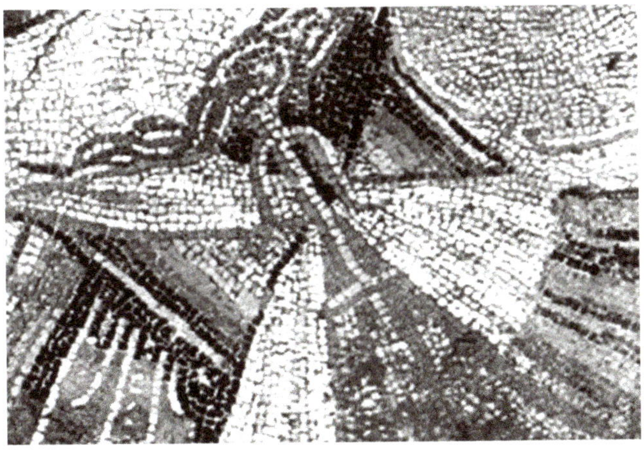

Fig. 3.1.2: Anchor-shaped fibula 1

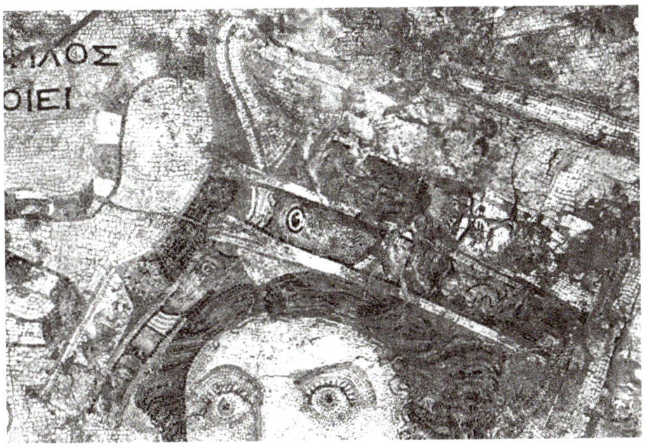

Fig. 3.1.3: Ship's prow headdress 1

ribbon is depicted with white and black intercalated strips and the lower ends finish in a fringe-like pattern, made with thin vertical strips of white and black tesserae. Two lines Greek inscription is found in the upper left corner of the panel; it inscribes the name of the mosaicist "*Sophilos Epoiei*", Sophilos made it (fig. 3.1.1).

The Ship

The ship seems to be seen from beneath the prow. The fore-sides of the hull appear to be open and project above the ears of the figure (fig. 3.1.3). Both, the starboard and port sides are seen from the front. The bottom of the ship is missing, being cut off for the opening of the headdress. The upraised, curved stempost is adorned with an inner turned small volute or a small globular tip (fig. 3.1.3). The central line of the stempost is made with a strip of dark brown tesserae while the sides are made with white, light brown and ochre stones, which was meant to give some perspective to the post. On the port side, between the stem and the stempost is a full bulwark. The below perspective view of the vessel does not reveal the upper part of the hull on the starboard side. Just beneath the gunwale on either side of the hull, outrigger oarbox are shown. The oarports are not depicted but two dolphins facing each other decorate their location. On the face of the starboard oarbox, a wreath with ribbons is depicted on a dark background. The same decoration is found on the face of the port oarbox, which is less preserved. On both sides of the upper prow, between the stempost and the oarbox, a light brown-ochre *caduceus* is depicted on a black background (fig. 3.1.3). The top wale is not visible on the starboard side, but it is emphasized on the port side. At their joint forward projection both wales are capped by a bronze *proembolion* (two blades top ram), shaped as a lion head (fig. 3.1.3). A bronze three-bladed ram capped the lower wales. This ram resembles the Atlit Ram[6] but in a more schematic depiction. Between the wales, on either side of the prow, is depicted a large *ophtalmos*, outlined with one row of black stones and a very large iris (fig. 3.1.3). A pair of dolphins with their heads pointing forward is depicted beneath the port outrigger oarbox. The same decoration was probably true on the starboard side.

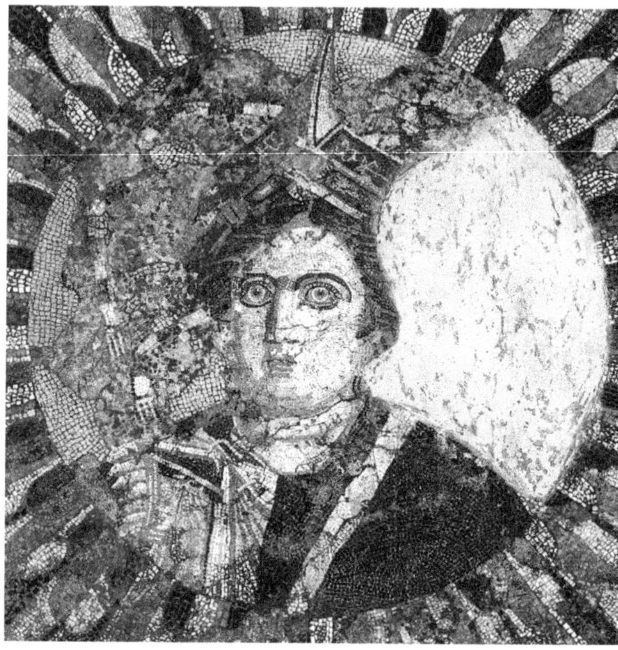

Fig. 3.1.4: Copy of the Berenike mosaic

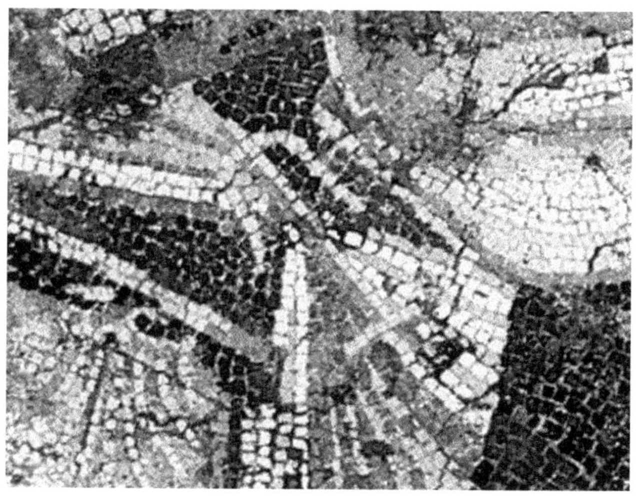

Fig. 3.1.5: Anchor-shaped fibula 2

Berenike Copy

The second Berenike mosaic was also found at Thumuis (Tell Timai), in 1923, or slightly earlier. As well this mosaic also was removed to the Alexandria Museum; Inv. Nr. 21736. The rectangular mosaic depicts a central, circular *emblema*, surrounded by geometrical borders. The entire panel measures: H = 1.44m, W = 1.44m and the diameter of the *emblema* is 0.79m.[7] Fragments are missing from the background and from the figurative panel. Other parts were broken and joined together. The left side of the head, part of the port side of the ship and the background are missing (fig. 3.1.4). The female figure adorned with the headdress shaped as the prow of a warship, apparently is a copy of the mosaic made by Sophilos (fig. 3.1.1). The face of this Berenike is less realistic, although it also was made with flashy hues tesserae. She has a small, partly open mouth. Her bulbous eyes emphasized by the eyelashes made with tiny black tesserae stare forward into the open space (fig. 3.1.4). The head is slightly turned to the right. She is also dressed in military attire. Underneath she wears a red tunic covered by a silver cuirass inlaid with gold. The left shoulder and breast are covered by a chlamys and the white edges are pinned together with a golden anchor-shaped fibula (fig. 3.1.5). The figure holds against her right shoulder a long mast with only visible left yardarm. The circular frame cuts off the right yardarm. The fingers of both hands are holding the mast. The band depicted with intercalated white and black strips is the ribbon that was attached to the masthead (not visible), similar to the mast in the Sophilos mosaic. The ends slide down above the female's right shoulder and

[6] Morrison, Coates and Rankov, 2000, fig. 33, p. 129.
[7] Daszewski, p. 185.

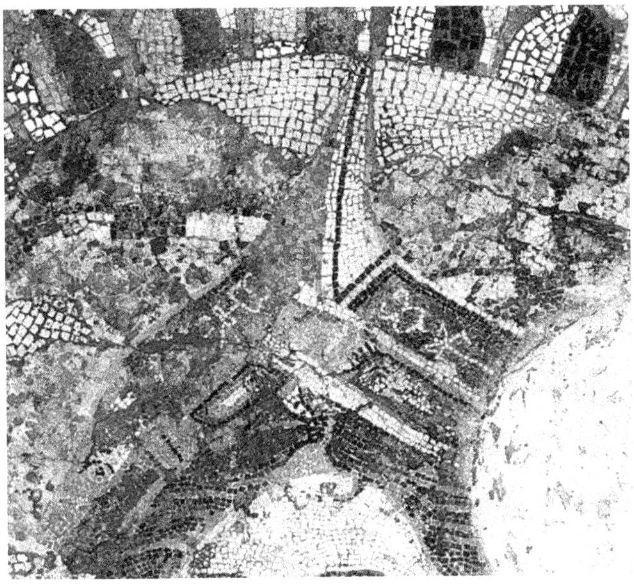

Fig. 3.1.6: Ship's prow headdress 2

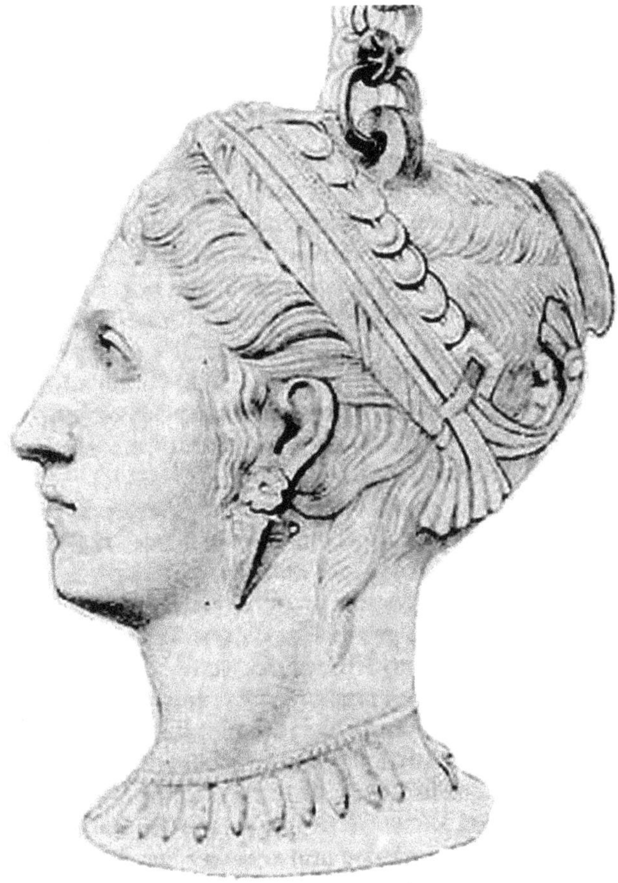

Fig. 3.1.7: Bronze head with ship headdress from the Tomb of Bruschi at Tarquinia

the tips are finished with a fringe-like pattern made with vertical white and black thin strips.

The Ship

The ship is viewed from beneath the prow and the headdress seems to be slightly offset from the central axis of the face (fig. 3.1.4). The curved, upraised stempost is finished with a small ball-like tip projecting above the top edge of the surrounding frame (fig. 3.1.6). The central line of the stempost is outlined with a row of dark brown tesserae. The left side is made with light brown-ochre and white stones, while the right side with dark yellowish-brown tesserae. The forward projection of the top wale, visible on the port side (made with white tesserae) is capped by a bronze two-bladed *proembolion* (top ram). The wales on the starboard side are not visible. The lower port wale, also depicted with white tesserae is finished with a bronze three-bladed ram (fig. 3.1.6). A horizontal golden *caduceus* is depicted on either side of the upper planks. Between the wales, on the port side, is depicted a dolphin with its head pointing forward. The outrigger oarbox is only visible on the starboard side. The port outrigger oarbox probably was depicted on the right-hand damaged mosaic. An elongated *ophtalmos* is rendered on the starboard stem in front of the oarbox (fig. 3.1.6). The oarports on the starboard oarbox are not indicated; however their place is decorated with the fragmentary head of a dolphin. Probably the same depiction was true on the port side, which was in the damaged section.

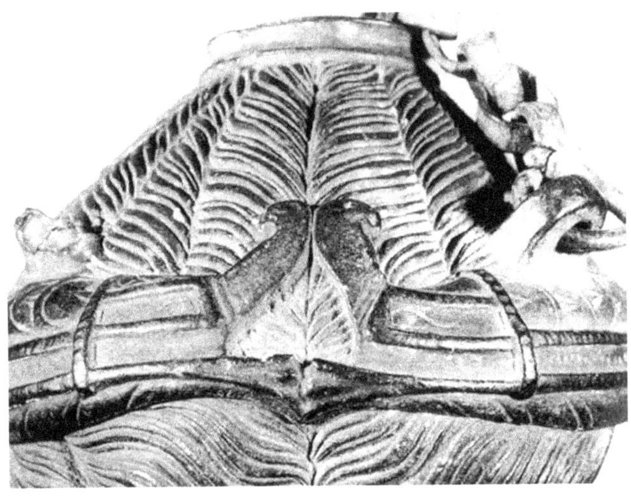

Fig. 3.1.7a: Close-up of the ships' prows

Discussion

The female figures in both mosaics from Thmuis (Tell Timai) were considered to represent the bust of the personified maritime Alexandria, the great Ptolemaic capital.[8] The attributes of both figures, indeed, emphasize the maritime aspects of the figures. The headdress representing a naval crown (*corona navalis*) resembles the turreted crown of a city goddess (*Tyche*),[9] but it depicts the prow of a warship. The figures, dressed with the red tunic covered by a silver cuirass inlaid with gold and a

[8] Daszewski, p. 146.

[9] Ibid, p. 147.

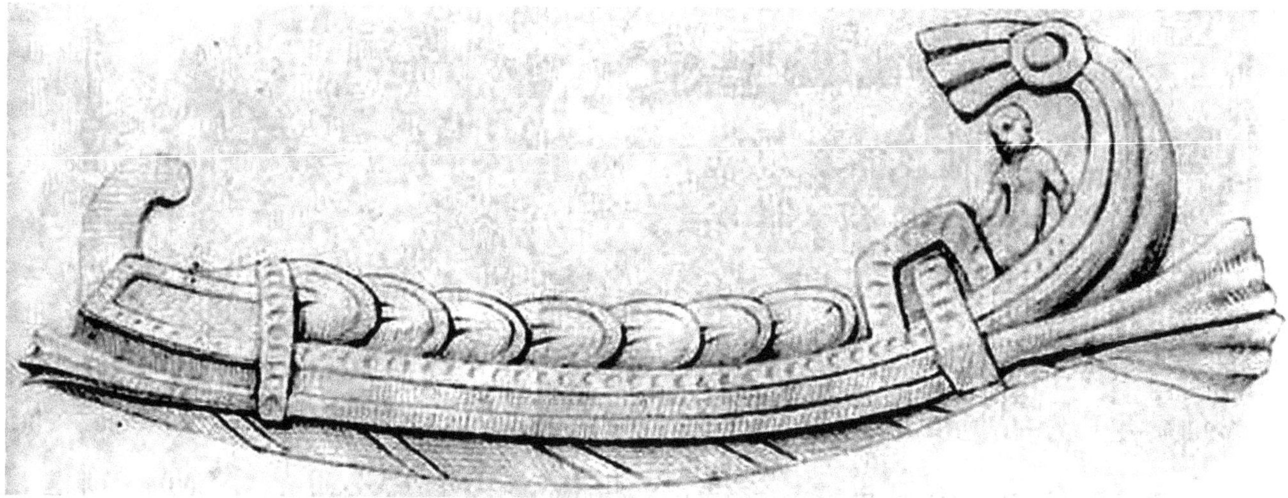

Fig. 3.1.7b: Drawing of one of the ships

purple chlamys, pinned by a golden anchor-shaped fibula, indicate the garments of a royal family.[10] In the abundant archaeological evidence, particularly coins, the Alexandria Tyche is represented as a woman wearing an elephant's skin and trunk on her head.[11] She wears a chiton and peplos, and sometimes holds ears of corn or a wreath scepter.[12] In Roman coins (especially Hadrianic), Alexandria is found with the attributes and dress of Isis.[13] The Sophilos mosaic (fig. 3.1.1) and the copy (fig. 3.1.4) do not represent an abstract personification of a maritime city, namely the personification of Alexandria rather they most probably depict one of the Ptolemaic queens. There were three significant queens in the Ptolemaic reign: Arsinoe II (278 – 270 BCE), Berenike II (246 – 222 BCE) and Arsinoe III (217 – 205/4 BCE).[14] Some ancient sources record the political activities of Berenike II. During the Syrian expeditions of her husband Ptolemy III, she fulfilled the role of Egypt's monarch. She was the first Ptolemaic queen who issued her own coins with the name and title "BERENIKES BASILISSES" by which she emphasized her supreme power.[15] The third queen, Arsinoe III, actively participated in the Battle of Raphiah, where she accompanied her husband Ptolemy IV.[16] The evidences presented above indicate that the female figure represents Berenike II.[17] Therefore the figures in both mosaics probably show an idealized representation of Berenike II, as a deified ruler.[18] The headdress in the form of a warship's prow probably is a symbolic reference to the Ptolemaic dominance of the sea (thalassocracy).[19] The Sophilos mosaic was dated to 200 BCE while the copy to the first quarter of the 2nd century BCE. Although, the portrayal of Berenike II in mosaic was made about 30-40 years after her reign, her strong figure survived as a motif used in paintings or other arts. Sophilos when made the mosaic, most probably relayed on a painting of the queen as his model.

Ships or parts of ships are well attested on coins from the 5th – 3rd centuries BCE. The Berenike ship prow as her headdress is the earliest ship depiction in mosaics. The ship represents a typical Hellenistic bireme or *dikrotos*, where the rowing oars protruded through oarports on the side of the outrigger oarbox (fig. 3.6.10). A three-bladed ram finishes the bow, while the *proembolion* (upper ram) is the forward extension of the top wales. The stempost has the characteristic raised inner curvature and finished with inner turned volute or a small rounded tip. The *stylis* held by each figure may have given Berenike II an extra meaning for the ruling power and the Ptolemaic thalassocracy. The *stylis* was a characteristic adornment at the sternpost of the Hellenistic warships, as we see mainly depicted on coins from the 4th centuries BCE.[20]

The Berenike mosaics are not the only female representations wearing a ship headdress. A similar headdress motif was found on two bronze female heads coming from Etruscan tombs in Italy. The first head was found in the tomb of Bruschi at Tarquinia, Italy (dating 3rd - 2nd century BCE) and now it is displayed in the Museo Nazionale di Tarquinia (fig. 3.1.7).[21] The headdress comprises two facing ships and their hulls extend around her head (fig. 3.1.7a). The ships are represented in profile. They are not shown with the propulsion gear (oars or sail) but only with two horizontal quarter rudders and a helmsman seated behind the quarter cabin (fig. 3.1.7 b).

10 Ibid.
11 Ibid.
12 Ibid.
13 Ibid.
14 The identification of Arsinoe II is well known and her association with the sea is demonstrated by the sanctuary dedicated to her by the admiral Kallikratos on cape Zephirion; Daszewski, p. 151.
15 Daszewski, p. 151.
16 She was the moral support to the Ptolemaic army under the threat of the Antioch forces; Daszewski, p. 151.
17 Daszewski, p. 153.
18 Ibid., p. 153.
19 Bowman, 1996, p. 218.
20 Basch, 1987: figs. 586-592, 680-683, 720-722.
21 The bronze head appear in the database of NAVIS II Project (http://www.2rgzm.de/Navis2/Home/).

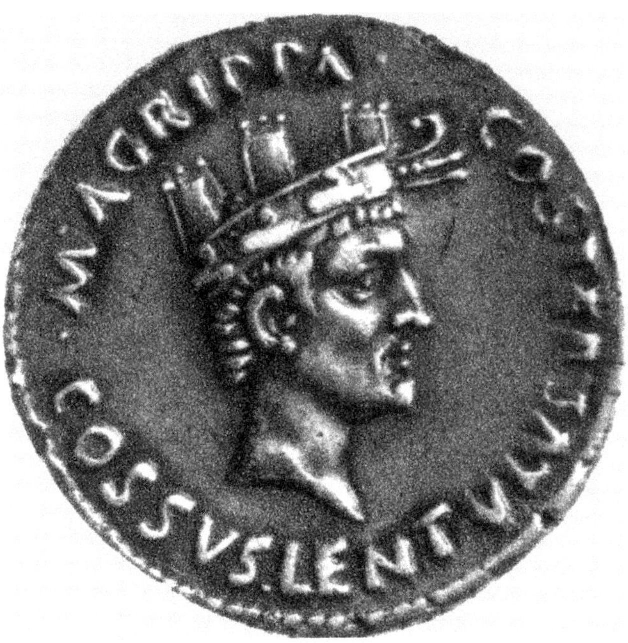

Fig. 3.1.8: The Cossus coin of Agrippa with corona rostra, 12 BCE

A projecting three-bladed ram surmounts the bow of both ships. The *proembolion* of both ships is adorned by the head of a lion. Above the gunwale are set seven shields forming the bulwark or the protecting strake. The origin of the second head is not known; it is also deposited at the Museo Nazionale di Tarquinia and dated to same period as the first statue. The ships on the second headdress are more schematic. On the projecting oarbox, the oarports are depicted as small dots. The ships on the headdresses from Tarquinia are not similar to the Berenike vessels. They probably represent *pentecontere* (50 oared), swift vessels and single banked. Such vessels were used as troop transporters.[22] *Pentecontere* originates since the 8th century BCE. They were still used by the Greek city-states of south Italy in the mid-3rd century BCE.[23] The Berenike ships are Hellenistic *dikrotos* (*bireme* in Latin), two-banked warships.

After the victory at Actium, the ship motif on headdresses was reused on the coins issued for the commemoration of Agrippa in 12 BCE, (fig. 3.1.8). On the reverse of such coin is depicted the head of Agrippa in profile adorned with a *corona muralis et rostrata*.[24]

22 Casson, 1971, p. 44 and n. 8.
23 Ibid., p. 124.
24 On the coin appears the inscription "M. AGRIPPA COS TER COSSUS" and on the obverse "AUGUSTUS COS XI"; the dimensions of the coin are D = 20 mm and weight 3.71 g; Sutherland, 1974, fig. 254, p. 141. Another appearance of Agrippa wearing the corona rostrata is found on a coin from Nemeusus (Nimes, France). Augustus and Agrippa are depicted back to back. Agrippa looks to the left; he wears a diadem of a warship, seen in profile with the forward projecting prow; Sutherland, fig. 231, p. 133. The latest appearance of Agrippa with a corona rostrata is on a coin of Cossos, dated 30 – 40 CE; ibid., fig. 280, p. 150. This coin was minted about 50 years after the death of Agrippa; ibid., p. 152.

3.2 Lod – Israel

Historical Notes

Lod is a town situated on the coastal plain of Israel, 16 km SE of Tel Aviv-Jaffa. The name first appears in Thutmosis III list of Canaan towns (1465 BCE). According to the Bible, Lydda/Lod was built by Shemed a Benjamite whose family settled in the northern Shephelah.[1] Lod, Ono and Hadid were resettled after the Jews return from the Babylon exile.[2] In 145 BCE Lod was included in the Hashmonean territory.[3] During the Hellenistic and Roman period, Lod is mentioned several times. Josephus mentions that Julius Caesar returned the privileges to the Jews of Lydda:

> *"... the Jews with regard to their high priest enjoy the same benefits which they have had formerly by the concession of the people and the senate, and let them enjoy the like privileges in Lydda".*[4]

In 43 BCE the inhabitants of Lydda/Lod were sold into slavery by Cassius, the governor of Syria. The Roman proconsul of Syria, Cestius Gallus, burnt Lydda on his way to Jerusalem in 66 CE. Between the First (66 – 70 CE) and the Second (132 – 135 CE) Jewish Wars, Lydda/ Lod flourished and was very prosperous. It had a large market, raised cattle, alongside dying and pottery industries that prospered. Vespasian occupied the town in 68 CE. A Christian community existed at Lydda in the time of St. Peter:

> *"... as Peter passed throughout all quarters, he came down also to the saints who dwelt at Lydda".*[5]

After the destruction of the Second temple in Jerusalem, Lydda became the seat of the Sanhedrin, where famous scholars like R. Eliezer ben Hyrcanus, R. Tarfon, R. Akiba, R. Joshua ben Levi, R. Eliezer ben Kappara or R. Hama, taught there. The Jewish community from Tarsus maintained one of the largest synagogues in the town. After the Second Jewish War or the Bar Kochba Revolt, the Jews remained in Lydda, though their agricultural plains were destroyed. In 200 CE, Emperor Septimius Severus established the Roman city called "*COLOLNIA LUCIA SEPTIMIA SEVERIA DIOSPOLIS*".[6] During this period the town remained partly Jewish. The population of Lydda took part in the revolt against Emperor Gallus in 351 CE, and therefore they were punished when the revolt failed. During the Byzantine era, Lydda was predominantly a Christian town. The town is the legendary birth place of St. George, later the patron saint of England.[7] Therefore Lod was also known as *Georgiopolis*, mentioned in late Byzantine and Crusader sources.

1 1 *Chron.* 8:12.
2 *Ezra* 2:33; *Nehemia* 7:37.
3 Negev and Gibson, 2001, p. 303.
4 Josephus, *Ant.* XIV, 10.6.
5 *Acts* 9:32.
6 Negev and Gibson, 2001, p. 303.
7 Ibid.

The pictographic representation of Lod is found for the first time in the Madaba Map mosaic, in the Church of St. George, Jordan (560 – 565 CE).[8] The Greek inscription depicted above the complex of two colonnaded streets, one straight running from east to west, and the second forming a semicircle around a prominent church, writes: "*Lod etoi Lyddea e k(ai) Diospolis*".[9] The second depiction of Lod appears within the topographic frame of the early Umayyad period mosaic floor in the Church of St. Stephen, at Umm al-Rasas (756 CE); the town is named "*Diospolis*" and shows a city without walls.[10] Lydda/Lod is never mentioned by Eusebius although he uses *Diospolis* as a reference point for various places.[11] Following the Arab conquest, the name *Diospolis* was changed to Ludd.

The Mosaic Floor

In 1996, a salvage excavation was carried out in Lod on the behalf of the Israel Antiquities Authority and the Lod Municipality, under the direction of Miriam Avissar. During this excavation, a beautiful and almost complete mosaic floor was revealed (17m long, and 9m wide, c.153 square meters). It is a unique archaeological find found in Israel that survived throughout the centuries in almost complete state, as being covered and protected only by 1 – 2.5 meters of soil and debris. The mosaic floor probably belonged to the dining room (*triclinium*) of a very opulent Roman villa. Within the covering soil were found colorful pieces of frescoes in a good state of preservation that decorated the wall of the room and other parts of the villa.[12] Among the numerous shards there were pieces of imported amphorae, dating from the late 3rd to the beginning of the 4th century CE. The majority of the coins found during the excavation date from the 3rd and no later than the 4th century CE.[13] The coins, the shards and the style of the mosaic indicate that the villa and its mosaic floor were constructed at the end of the 3rd or the beginning of the 4th century CE.

The designs of the mosaic are divided into two main carpets (fig. 3.2.1). Two rows of rounded *emblemae* inhabited with wild animals and birds are depicted in the upper part of the northern carpet. The main northern panel comprises a large circle with an inscribed octagon. The area between the outer edge of the circle and the octagon is divided into squares and triangles, which are inhabited by dolphins, birds and wild beasts. In the center of the octagon are depicted a lion and lioness confronting each other from two mountain peaks separated by a river from which a mythological sea creature emerges. Below this scene exotic African animals (an elephant, a giraffe, a rhinoceros, a tiger and a water buffalo) are revealed.

A long and narrow frame depicting an *akanthos,* whose leafy tendrils spread out forming spirals in different directions. This frame separates the northern and southern carpets, which also seems to be the dividing line of the room. The *krater* in the middle of the frame is flanked on its either side by a peacock. Beneath this frame is depicted a wide maritime panel comprising a sea inhabited by rich fishery and two sailing ships (the subject of this chapter) (fig. 3.2.2). The upper frame of the southern carpet depicts different birds standing on various branches ransomed on a white background. Guilloche borders forming hexagonal *emblemae* are depicted in the lower panel and a single wild beast, bird or fish inhabits these hexagons. The southern edge of the mosaic is missing. No human figure appears in the mosaic.

The Maritime Panel[14]

The marine panel is unique in the villa's floor decoration; it shows a variety of fish in different sizes. Many of them are depicted in realistic manner. A single dolphin is depicted beneath the bottom of Ship 2, while a large fish-monster with open jaw appears in the upper right corner as facing the ship (fig. 3.2.2). White tesserae are used for the watery background, while the marine fauna and the ships are depicted with a variety of polychrome hues of tesserae: blue, dark blue, yellow, ochre, light and dark brown, red, gray, green and black. Between the fish are depicted conical shells and some small conical features that may indicate fish or lobster traps, or some other objects. No shoreline or harbor installations are depicted, thus the white background seems to indicate the open sea.

In the lower left corner appears a ship rigged with fully open square sail. The second ship is depicted close to the top frame, almost in the middle of the panel (fig. 3.2.2). The static positions of the ships indicate that they are anchored, although they lack any mooring or anchoring device. The ships are not proportional to the scene, appearing quite small among the surface teaming with fish. The top ship sustained damage on its upper part by a cesspit dug during the Ottoman period.[15] Ship depictions, and especially elements of the ships, have to be studied with comparable materials, which in the end will result in a better understanding of the ancient vessels that may be considered concurrent to the period of the mosaic.

Ship 1

This ship survived completely and is shown with its full rigging (fig. 3.2.3). The bow points to the left, thus showing

[8] Piccirillo, 1993, fig. 63, p. 83.
[9] Piccirillo and Alliata, 1999, pp. 80-81, Section VII – The Sea – Coast.
[10] Ibid., p. 204, Pl. VII – Architectonic Representations; Piccirillo, 1993, fig. 345, p. 219.
[11] Di Segni, 1999, p. 117, in: Picirillo and Alliata (Eds.).
[12] Avissar, 1996, p. 157.
[13] Ibid.

[14] While being in the final stage of writing my PhD dissertation I received an offprint of the Ships depicted in the Lod mosaic, published by Haddad and Avissar in *IJNA* 32.1, 2003. The interpretation and the suggested reconstruction of Ship 2 (damaged by the Ottoman cesspit) are problematic. Haddad and Avissar referred to the damaged Ship 2 as representing a vessel that suffered a "marine trauma", which is the opposite. The reconstructed rigging of Ship 2 proposed by Haddad and Avissar was not properly understood and therefore they showed the mast as broken. This interpretation resulted from the lack of nautical archaeology of both authors. I published a detailed article with a counter interpretation of both ships in the mosaic; in: *IJNA* 33.1, 2004, pp. 164 – 168.
[15] Avissar, 2001, p. 48.

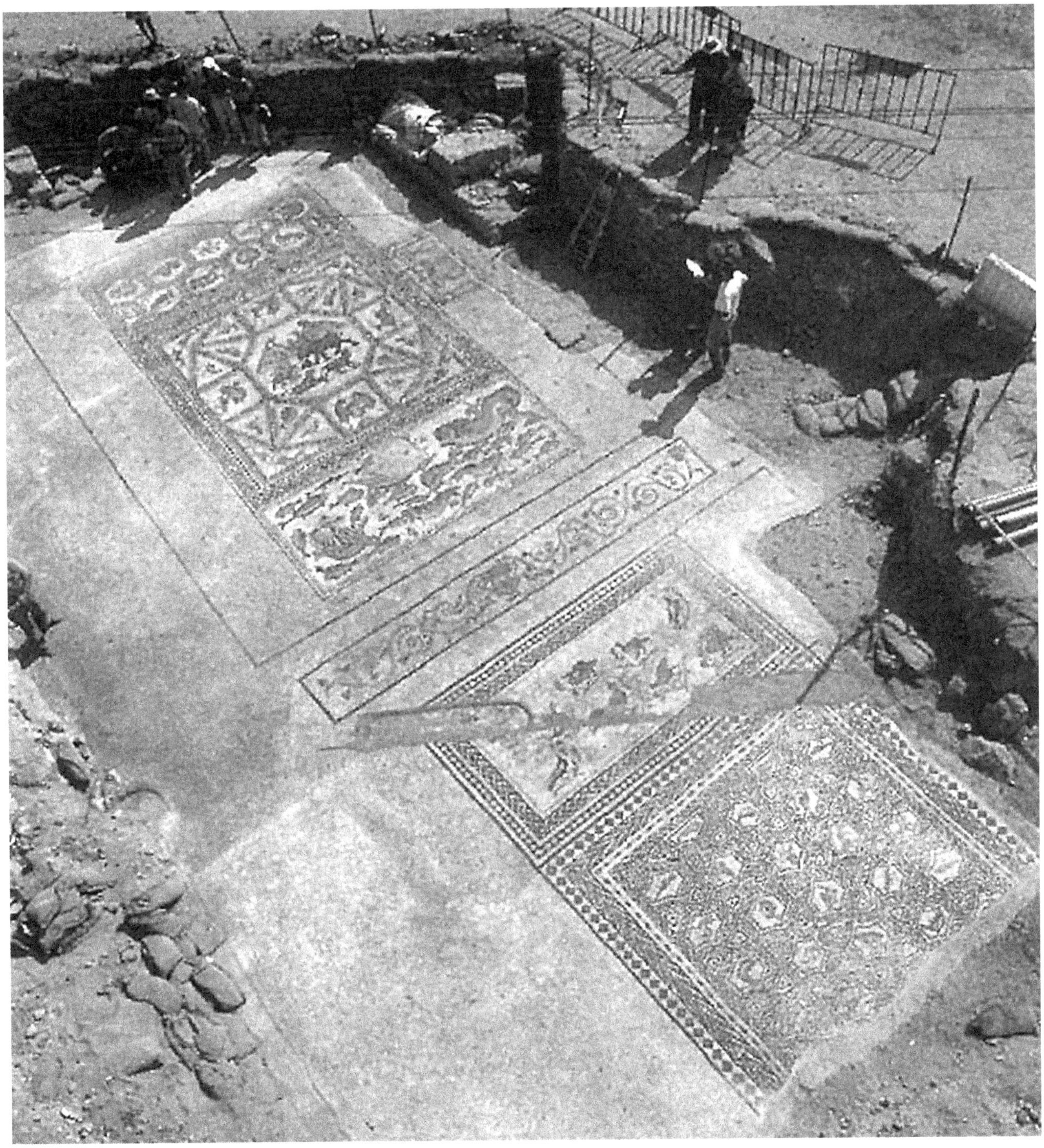

Fig. 3.2.1: General view of the Lod mosaic; looking to NE

its entire port side. The ship is depicted with elements that rarely are represented in ship iconography in any arts, and especially in mosaics. The hull has a spoon-like shape. The planks are made with ochre and brown tesserae. The lower part of the hull is rendered with three rows of black tesserae, probably indicating the pitch/bitumen coating of the bottom to make the hull watertight (fig. 3.2.3). No fence or bulwark is shown amidships. Neither a wale reinforcing the hull longitudinally is shown. The stempost has a forward extension with a slight downward arching. On top of the prow and stempost is place a partial bulwark rendered with three rows of brownish-red tesserae. A square frame outlined with black tesserae is set above the stempost. A circle with an inscribed cross-like pattern is depicted within the purple background. The rear extended hanging poop is fenced with a lattice screen. Above the fence projects the head of a duck looking forward (fig. 3.2.3).

The rigging comprises a broad tapered mast, stepped amidships, a horizontal yard and a fully open square sail that billows forward the mast and over the starboard gunwale. The black and yellow intercalated bands of both posts

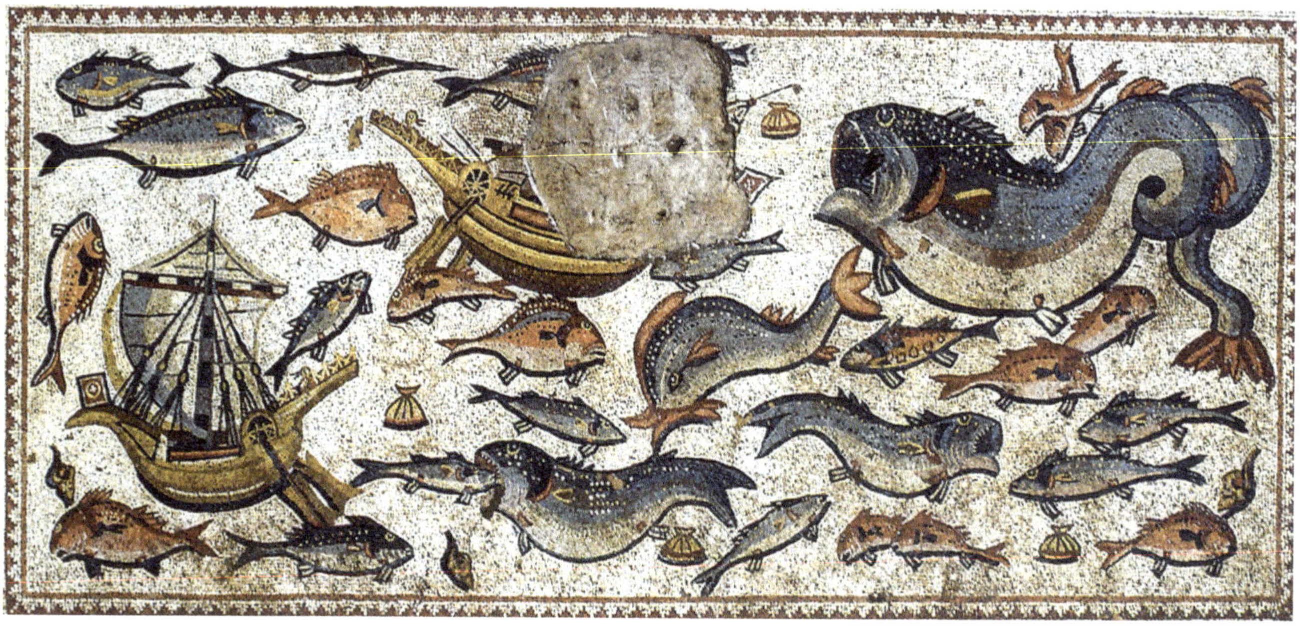

Fig. 3.2.2: The maritime frame with both ships

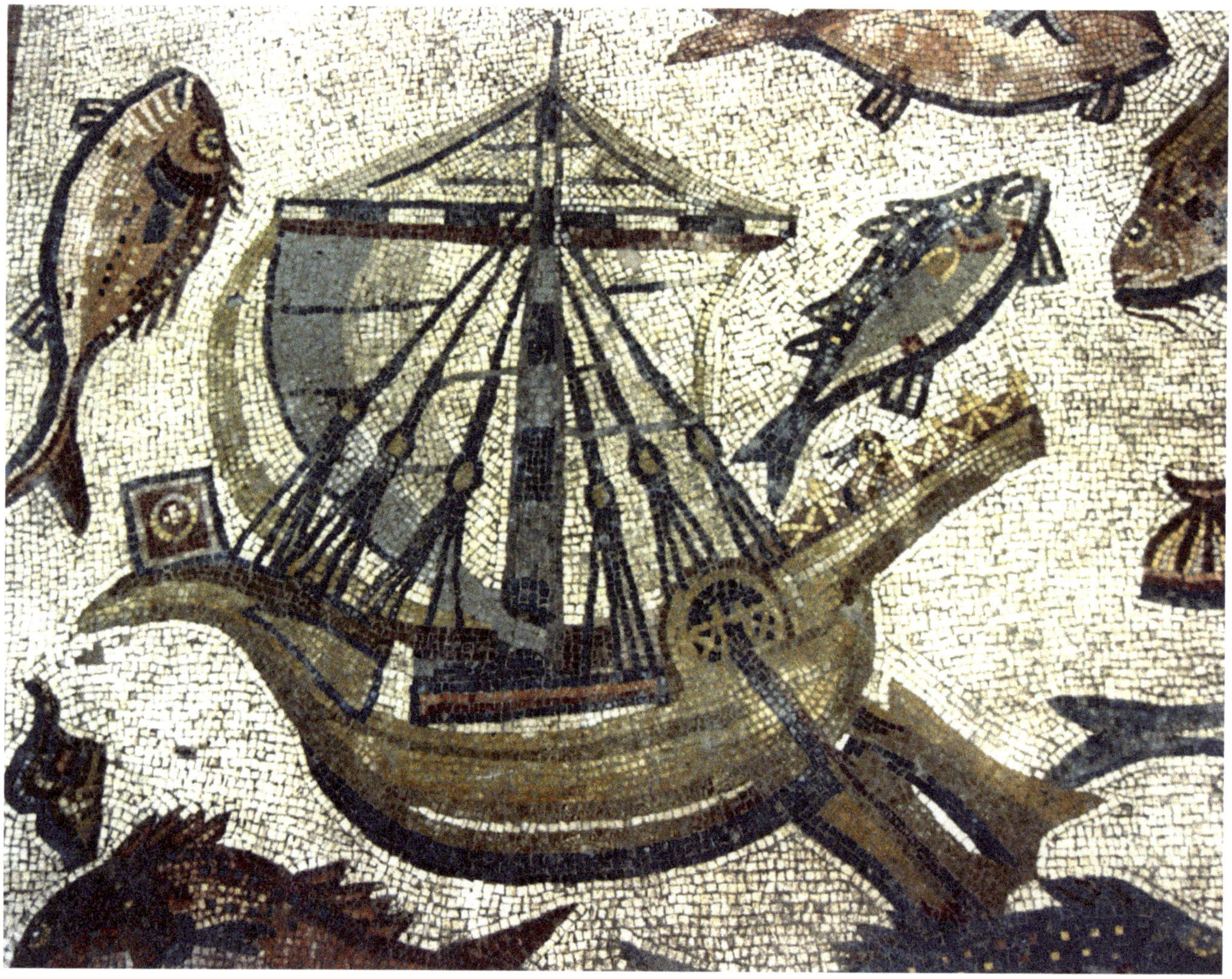

Fig. 3.2.3: Ship 1

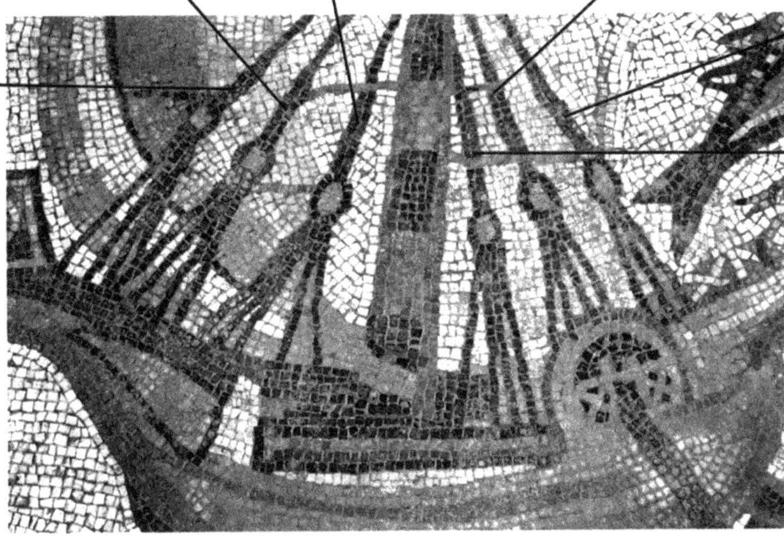

Fig. 3.2.4: Ship 1 – The tackle lines with the deadeyes

symbolically indicate a composite mast and yard girdled with wooldings at evenly spaced intervals (fig. 3.2.3).[16] The mast is secured in place by two forestays, two backstays and one shroud on either side. The lower end of each line splits into three to five shorter arms forming the connecting elements on the gunwale. Between the lower ends of the main tackle lines and the splitting arms is found a wooden block or deadeye. This gear is outlined with a rounded strip of black tesserae, while the field is made with ochre or yellowish-brown tesserae thus emphasizing the wood (fig. 3.2.4). The lower ends of each line are tight to the gunwale (figs. 3.2.3, 3.2.4). All the lines are rendered with black tesserae and clearly indicate their individual function and position. The left-hand line stretching from the masthead (above the yard) to the top of the bow represents the *port forestay*; the four arms of the lower end are attached to the port fore-gunwale, behind the square frame set on the bow; the middle line stretching from the joint of the yard and the mast to the fore gunwale is the *starboard forestay*; the four arms are attached to the starboard fore-gunwale; the line closer to the left side of the mast indicates the *port shroud*; the five arms are attached to the starboard gunwale, a bit fore-amidships (fig. 3.2.4). The line closer to the right side of the mast, as stretching from behind the mast, is the *starboard shroud*; the four arms are attached to the port gunwale just before the rudder-cabin. This depiction is a mistake made by the mosaicist who did not understand the proper setting of the lines. These arms should have been attached to the starboard gunwale and not as they appear in the mosaic. The middle line indicates the *starboard backstay* with the four arms attached to the quarter gunwale behind the rudder-cabin; the right-hand line (extreme) stretching from the joint of the yard and the mast to the quarter is the *port backstay*; the three arms are attached to the quarter gunwale aft the rudder-cabin (fig. 3.2.4). The masthead projecting above the yard also supports the triangular topsail. One lift stretching from either side of the masthead to the tips of the yardarms holds the yard; they also indicate the leeches of the triangular topsail. The large square sail made with white, ochre and gray tesserae billows before the mast and over the starboard gunwale. The port leech outlined with black tesserae and the side of the bunt turned backwards on the port side, apparently giving perspective view to the sail , it also may symbolically indicate that the wind blows from astern or the port quarter (fig. 3.2.3). An arching strip of ochre stones indicates the starboard leech. The port clew seems to be attached to the port gunwale at amidships. The halyard, brails, braces and the sheets are not shown.

One rudder is mounted on either quarter. The upper shaft projecting between two vertical stanchions of the fenced rudder-cabin with an arched roof is the port rudder (figs. 3.2.3, 3.2.4). The left side of the shaft is rendered with black stones and the right side with dark-brown tesserae, thus giving the shaft a three dimensional view. The lower shafts of both rudders transverse the blades longitudinally, hence indicating that they were made with two wings. The shoulders and the lower edges of the wings angle towards the shafts. Both elongated edges of the rectangular wings have shallow concave cuts. The left side of either blade is made with a strip of black tesserae giving them some perspective, or it may indicate copper or lead sheathing to protect the blades from any damage.

Ship 2

The ship is located close to the top frame. It is quite damaged, but bears enough fragments to allow us to distinguish its hull shape and the rigging (fig. 3.2.5). The

[16] Casson, 1971, p. 69, n. 123; p. 232, n. 31.

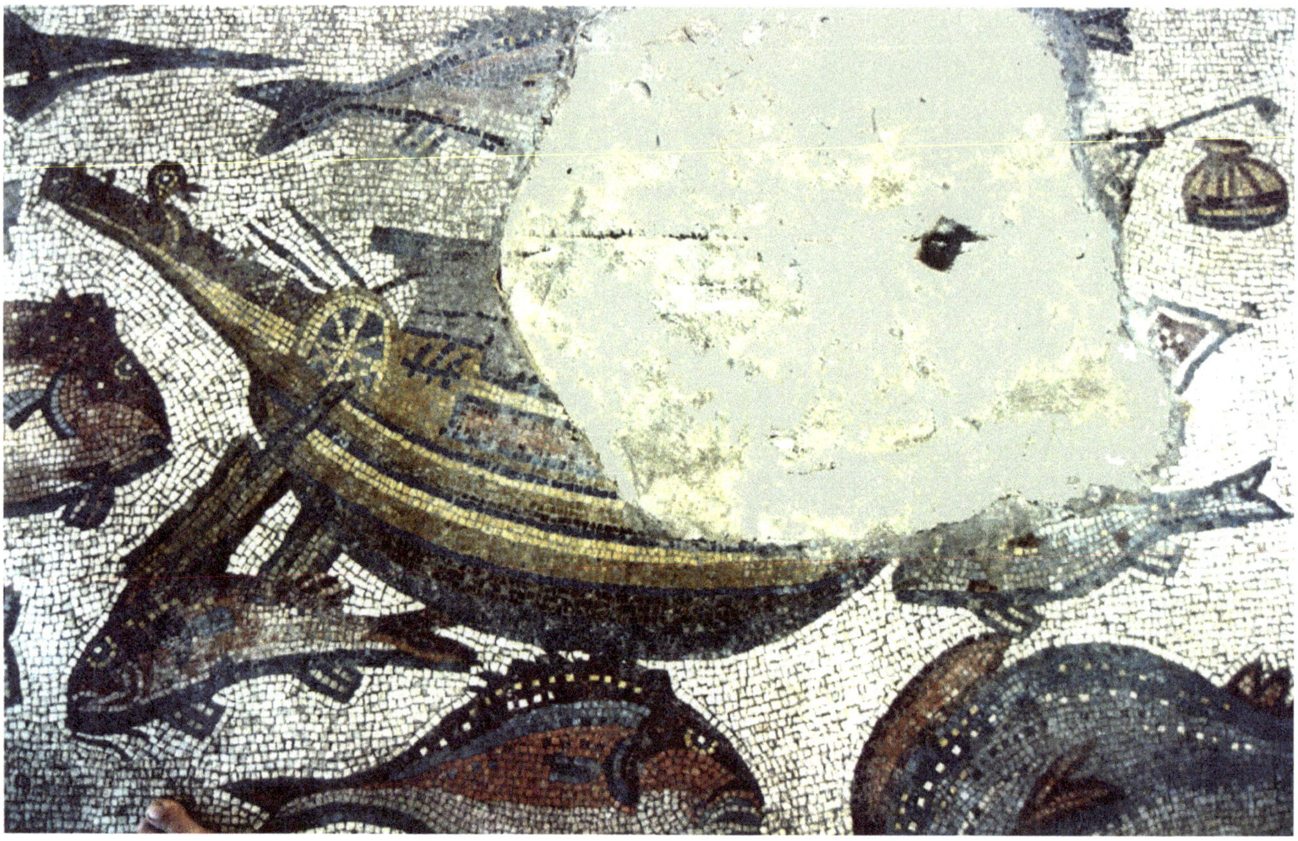

Fig. 3.2.5: Ship 2

bow points to the right. The hull is very rounded, spoon-shaped with both ends extending outward, similar to that of Ship 1 but a bit longer. The seams between the planks are indicated by one row of black tesserae. The upper part of the starboard strakes is made with ochre stones and the lower part with brown tesserae. A wide plank that probably forms the bulwark around the deck is placed above the gunwale. It is outlined with a black strip and the field is made with reddish-brown or purple tesserae. The projecting wale reinforcing the vessel longitudinally is depicted at the middle of the starboard hull. Its upper and lower edges are outlined with one strip of black tesserae. To give a three-dimensional appearance to this element the mosaicist rendered the upper part of the beam with two rows of light yellowish-ochre stones while the middle and the lower part are rendered with three rows of dark ochre tesserae (fig. 3.2.5). The bow, the stempost and the rigging are damaged. The preserved top and right-side of the rhomboid frame outlined with black stones is set above the forward extension of the stempost, similar to Ship 1. Individual white cubes are placed within the purple background, creating a checkerboard pattern. A lattice screen surrounds the hanging poop with a rear extension. Above the screen rises the head of a duck, looking forward (fig. 3.2.5).

The Ottoman cesspit cut thought the mosaic affected the rigging and the fore part of the vessel. The remains of this rigging provide us significant information on how it was positioned in the ship. The mast did not survive the damage; only the masthead with a small flag attached to its tip, the port upper corner of the sail and part of the yardarm were preserved. The small ball-like element at the base of the flagstick indicates the *parrel*. The outer lines stretching from either side of the *parrel* towards the yardarms indicate the lifts that support the yard (fig. 3.2.5). The inclined position of the masthead and the flag projecting over the bow, indicate that the mast was retracted from the mast-step and lowered on the deck. The yard and the furled sail were set above the lowered mast on the deck. The preserved left-corner of the sail is made with grey tesserae. The black horizontal line (beneath the yard) depicted on the bunt represents a reinforcing-band. The segmented black line stretching from the yardarm to the top of the arched roof of the rudder-cabin, most probably indicates the brace or the lower end of the backstay. The function of both black parallel lines behind the rudder-cabin is not exactly understood.

The steering gear comprises a pair of rudders, one oar mounted on either quarter. The port rudder is indicated only by a small part of the rectangular blade, visible between the bottom of the ship and the fin of the fish below. The blade is outlined with one row of black tesserae. The starboard shaft outlined with one strip of black stones, projects through the fenced rudder-cabin with an arched roof (figs. 3.2.5, 3.2.6). It seems to be mounted forward the central stanchion of the fence. The lower ends of both shafts transverse the blades longitudinally, thus emphasizing the wings. The shoulders of the wings are slightly angled towards the shaft. The longitudinal outer edges of the blade have a

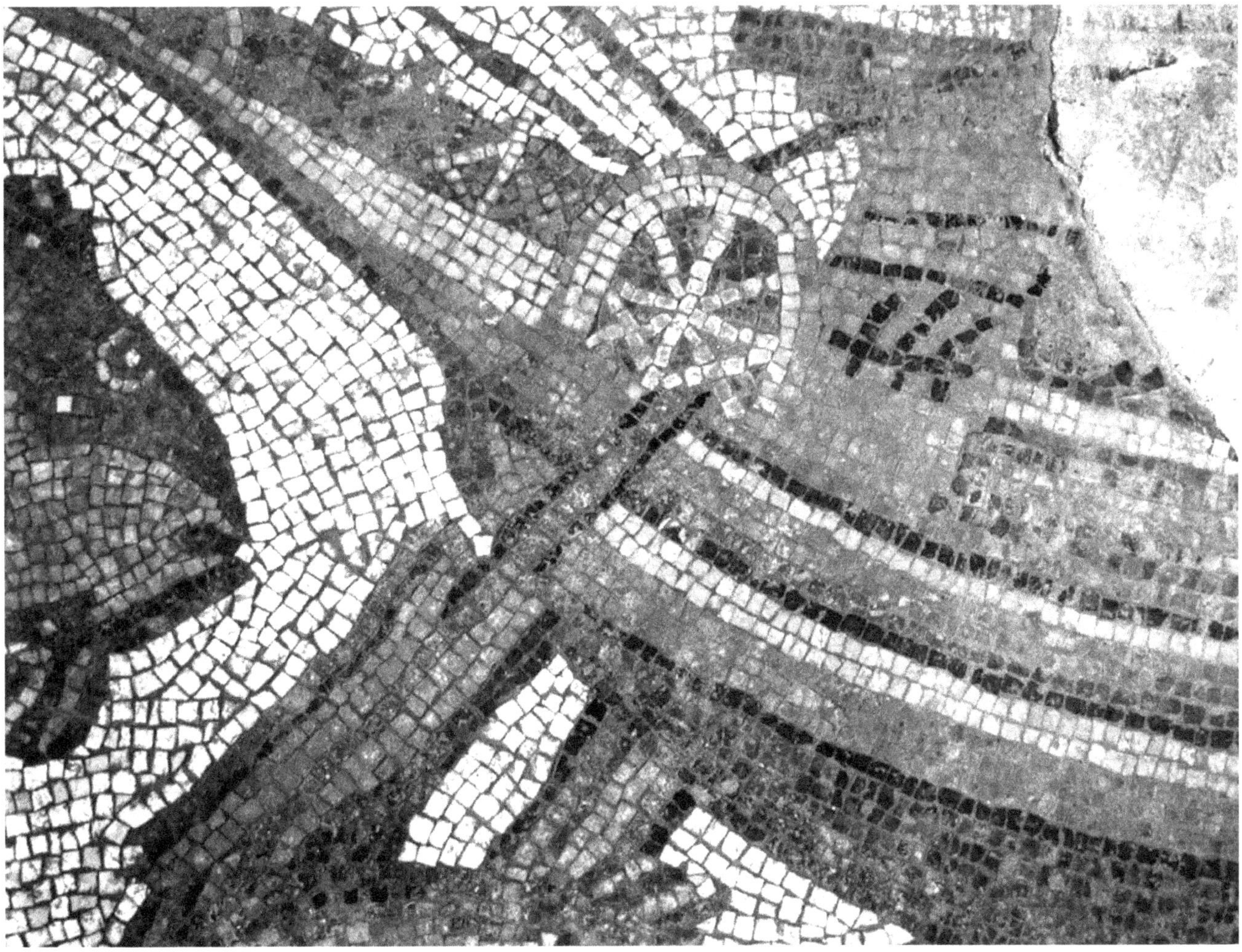

Fig. 3.2.6: Ship 2 – Steering gear

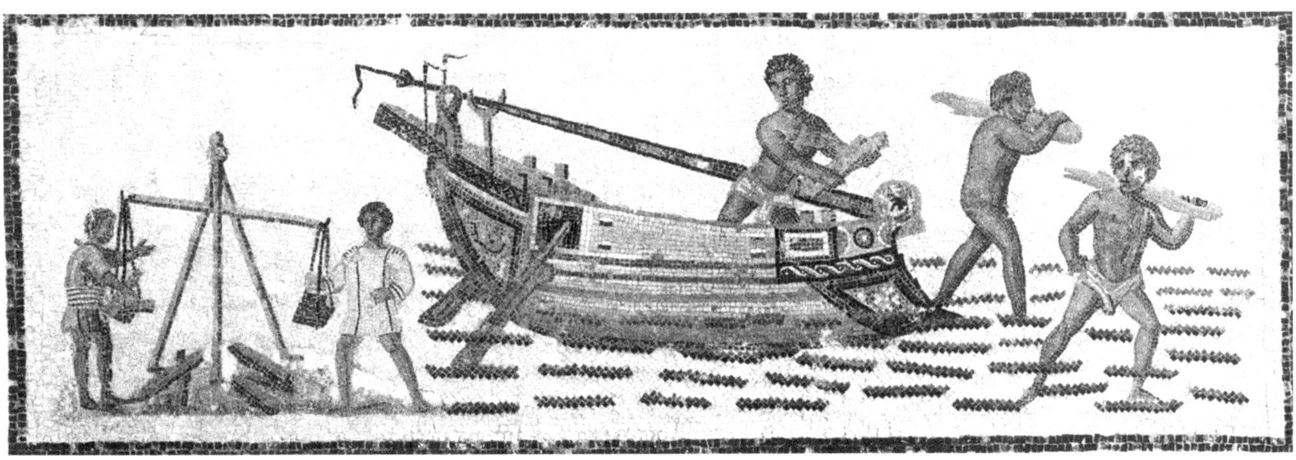

Fig. 3.2.7: Merchant ship with lowered mast on the deck; Sousse, Tunisia

slightly concave cut and are rendered with dark brownish-red tesserae. This coloring may indicate the copper or lead sheathing to protect the blades, or the mosaicist intended to give a perspective view to the blades.

Discussion

Although the Lod ships are not shown with an anchor or mooring lines, it does not mean that they were not anchored. Depictions of ships in any arts and especially

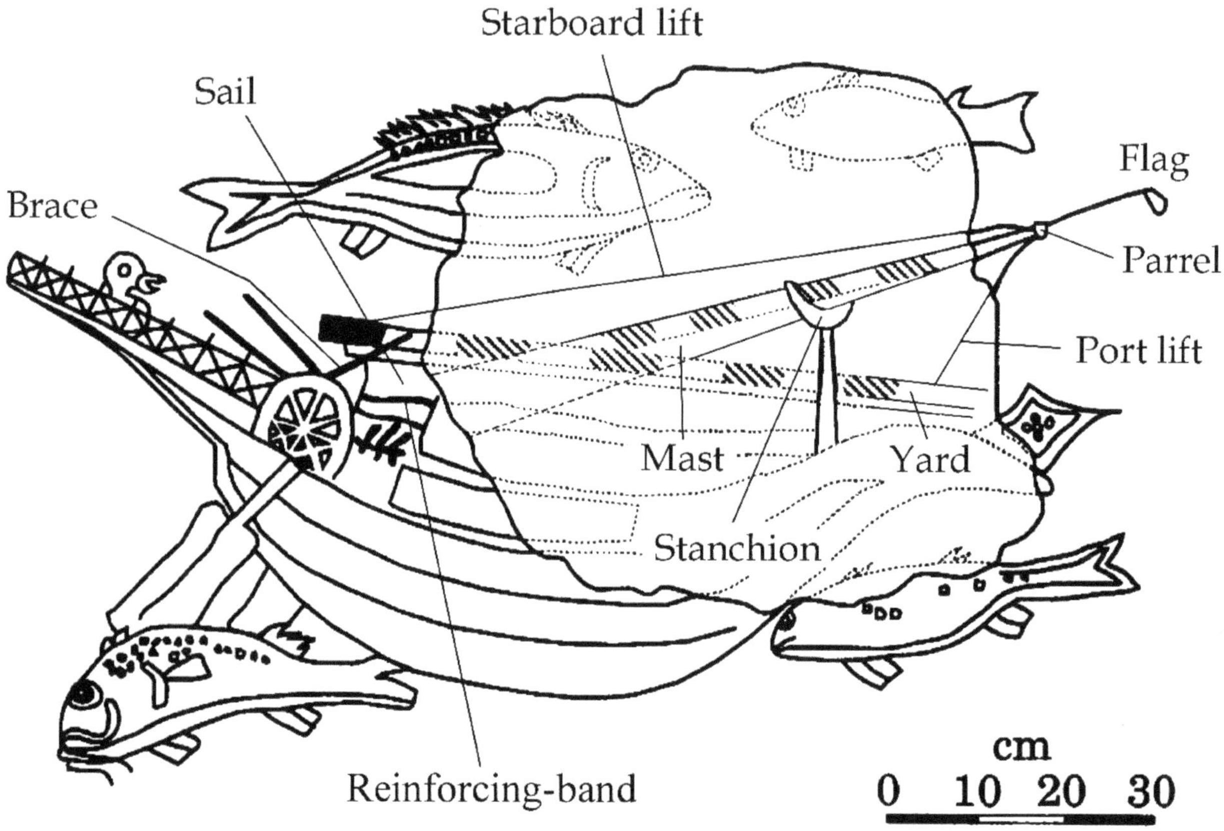

Fig. 3.2.8: Ship 2 – Reconstructed sailing gear

in mosaics (Piazzale delle Corporazioni, Ostia and the Catalogue of ships in the Althiburus mosaic, Tunisia) are not shown with anchors, or the sail, but are static. Though that the Lod ships are not depicted at scale, they present us with enough details to indicate seagoing merchantmen. The broad tapered composite mast and yard, the billowing square sail, the triangular topsails and the projecting head of a water bird above the fenced poop are distinctive features of Roman seagoing merchantmen. The vessels probably symbolize medium size merchantman that could carry a cargo of 80 - 150 tons. Such vessels also were used to transport passengers overseas, as known from some references in the ancient literature. When Josephus describes one of his journeys from Caesarea Maritima to Rome, he wrote that the ship (*phasolos*) in which he sailed wrecked in the Adriatic Sea.[17] He mentions that in the ship were six hundred passengers. After the wreckage the crew and the passengers had to swim for their lives all the night. Josephus and other seventy nine people were lucky to be rescued by a ship from Cyrene.[18] Synesius, later known as the Bishop of Ptolemais (404 CE), describes in an amusing letter (*Epist.* 4) his voyage from Alexandria to Cyrene in a *phasolos* that nearly ended with the wreckage with its passengers, in the proximity of the Libyan shores. We learn that the vessel had a crew of thirteen and carried on board fifty passengers:

> *"The crew of twelve with the skipper made it thirteen. More than half of them were Jews... The rest were a collection of peasants, who even as recently as last year had never handled an oar... We had taken on board, more than fifty passengers, about a third of them women..."*[19]

Whenever studying ship depictions in any arts and especially in mosaics, we have to be bear in mind to avoid mistakes and misinterpretations. The lowered mast of Ship 2 (fig. 3.2.5) indicates that when a vessel anchored for a period of time in a harbor or an anchorage, the mast was not left in its vertical position but was lowered to the deck or even removed from the ship. A similar example with a lowered mast on the deck of a cargo ship while it was engaged in unloading lead or gold ingots in the vicinity of the shore appears in a mosaic that from Sousse, Tunisia, now displayed in the Bardo Museum (fig. 3.2.7). The heel of the lowered mast projects forward above the prow, while the masthead rests on a forked stanchion placed on the aft-deck. The preserved sections of the rigging of Lod Ship 2 allow us to reconstruct the original gearing and the position in which mast was lowered on the deck. We may suggest the raised position of the masthead of Lod Ship 2 probably rested on a forked stanchion, similar to that of Sousse Ship (fig. 3.2.8). Both examples show that when a mast was lowered on the deck, it could be laid on either position,

17 Josephus, *Life*, 3.
18 Ibid.

19 White, 1986, p. 156; Casson, 1995, p. 268 and n. 1.

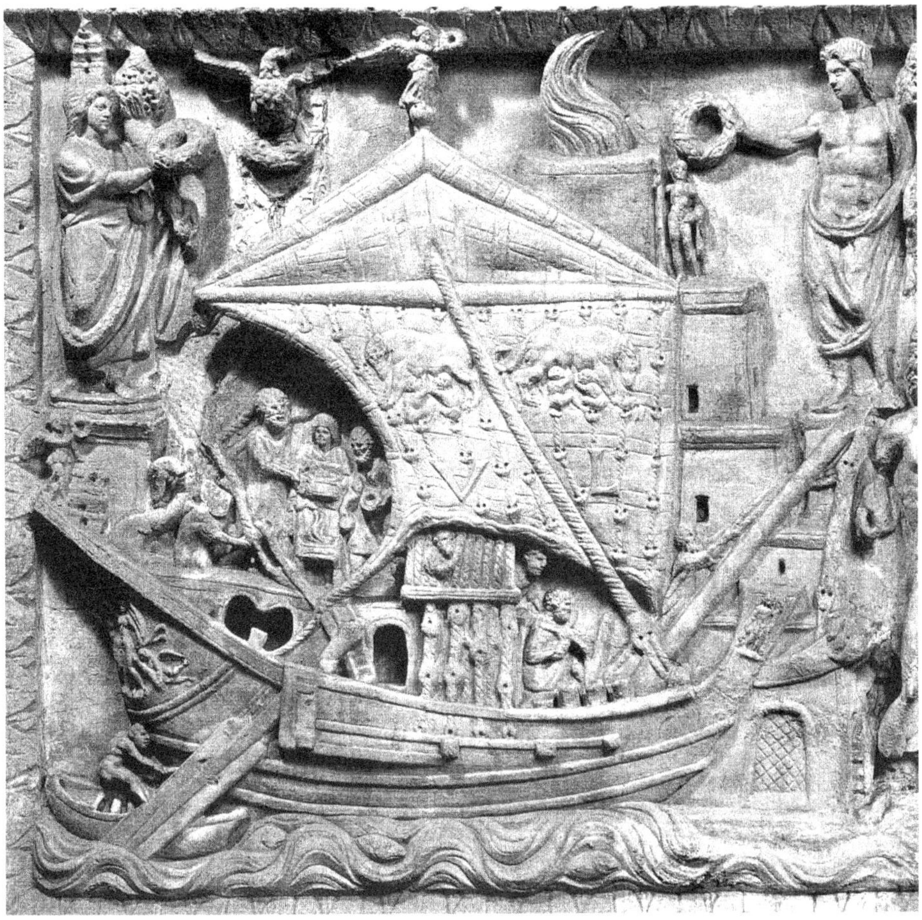

Fig. 3.2.9: The Torlonia relief

when the vessel was anchored. The blocks or deadeyes used for stretching the tackle lines are not unique representations in Ship 1 from Lod. One outstanding example of such tackles comes from a merchantman depicted on a relief from Torlonia, dated to the early 3rd century CE (fig. 3.2.9).

The rudder-cabins with lattice-fence as they appear on the Lod ships are unique representation in any ship iconography (figs. 3.2.3, 3.2.5). The helmsman probably sat inside the cabin and worked the rudders by long tillers (not visible in the mosaic) as we can see in other representations. The rounded and irregular openings on the starboard side of the quarter-cabin in the Torlonia relief are assumed to show the window and the side door. The helmsman probably sat inside the cabin and worked the rudders by a long tiller inserted perpendicularly into the head of each loom (fig. 3.2.9). We may assume that the rudder-cabin of the Lod ships resemble a similar quarter-cabin as depicted in the Torlonia relief. Generally in representations of merchantmen and warships, the rudder shaft projects from a rounded or square oarport at the end of an oarbox[20] (fig. 3.2.8) and not from a rudder-cabin as evidenced by both Lod ships (figs. 3.2.3, 3.2.5).

The rhomboid and square frames attached above the prow of the Lod vessels are quite unique elements used to indicate their trademark. On Mediterranean seagoing merchant ships, the most common decoration was an *oculus* or a dolphin painted or applied on either side of the lower stem just above the waterline. With careful consideration, we may suggest that the owner of the ships was probably Jewish and therefore he followed the Jewish law that prohibited the use of any human or animal figure, or even an *oculus* decoration on the stem, and for that reason he chose geometric frames to mark his ships.

The Lod vessels are typical Roman *navis oneraria*, probably of medium size (80 to 150 tons) that may be associated with *kerkouros* or *corbita*[21] type, carrying *garum* and grain from North Africa to Rome and the eastern Mediterranean. The ships also may indicate *phasolos* (*phaselis* in Latin) vessels suited for carrying passengers and cargo.[22] The North African origin of the Lod ships is suggested by a sailing ship that has a similar rounded spoon-shape hull which is depicted in a mosaic floor from Djemila (Cuicul), dated 4th – 5th century CE (fig. 3.2.10). The Djemila ship does not have the outer extended stempost as the Lod ships, but the rounded spoon-shape and the coloring of the hull, as well as the open square sail with its tackles, and the rear extended fenced poop are very similar. The lack of *artemon* rigging (fore-sail) in the Lod Ship 1 indicates that such

[20] Casson, figs. 129, 131, 140.

[21] Casson, fig. 137/*1*, Althiburus Catalogue of Ships.

[22] *Phasolos* (*phaselis* in Latin): this vessel was suited for carrying passengers and cargo. They were in use in the Mediterranean in the 1st centuries BCE - CE. It relied on sails and had no oars; Torr, 1964, p. 120. The larger version served as a man-of-war; Casson, 1971, p. 168 and notes 58 - 59. There are no references to indicate when such vessel ceased to sail in antiquity.

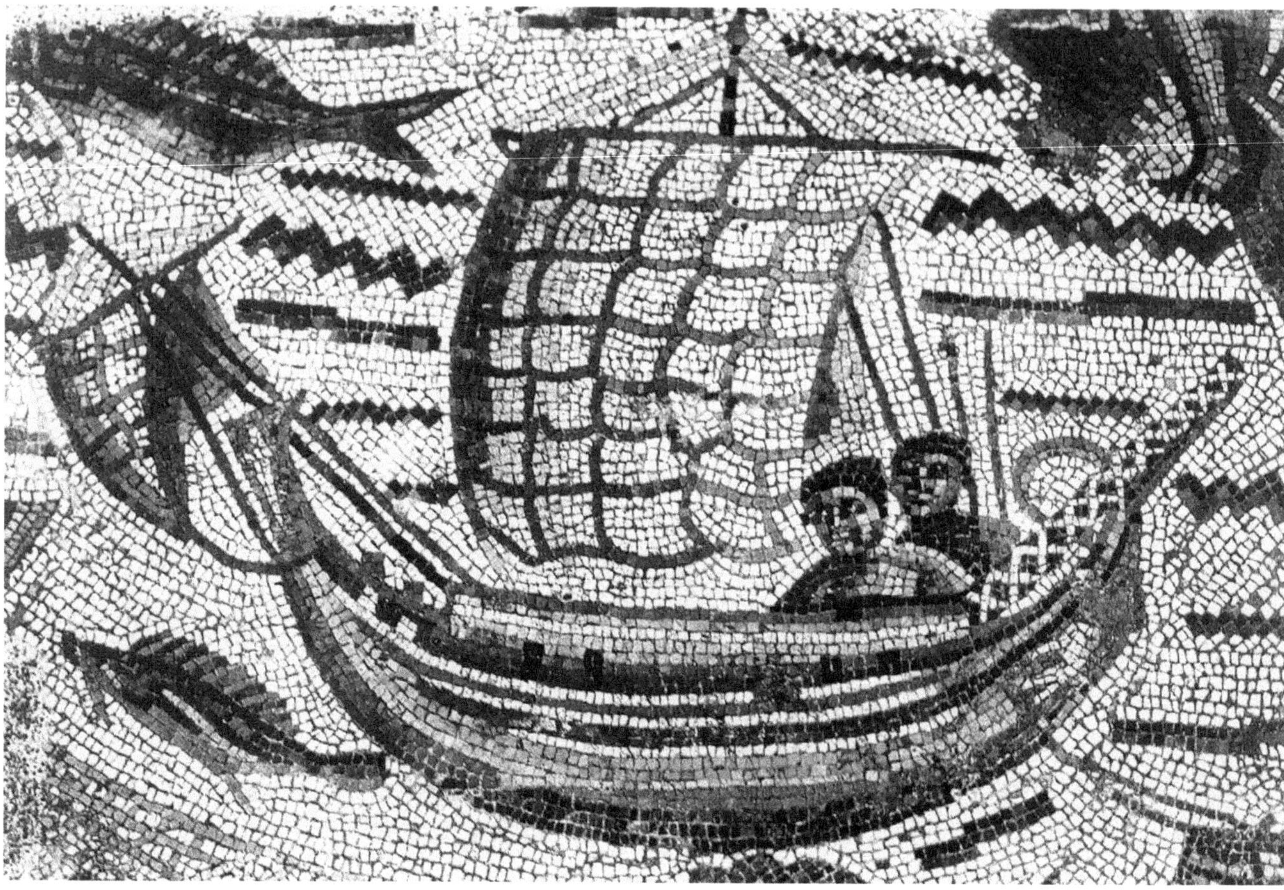

Fig. 3.2.10: Sailing Ship in the mosaic from Djemila, Algeria

vessels were generally rigged with one main mast and sail, or they could have both rigging, as indicated by the Djemila ship. The ship is sailing as indicated by its open sail and both men seated on the quarter. This ship is not static as the Lod Ship 1 (fig. 3.2.3).

The style and subject decoration in the Lod fishery frame indicates a strong North African connection. A rectangular mosaic frame inhabited by very rich fishery from Cap Matifou, Algeria, dated 4th – 5th century CE,[23] does not show ships, dolphins or fish-monsters, but various types of fish in realistic appearance and similar to the Lod fish. The composition of the Lod maritime frame may have been produced from similar patterns that were common decoration themes in the North African mosaics. The Lod ships probably are symbols to the profession of the villa's owner, who owned or traded in North African ships, or else also were used as *apotropaic* signs associated with the safe return of the vessels and the thank-offering to celebrate such a return.[24] We may assume that the mosaicists came from North Africa and brought the patterns which were used as guidelines for the overall designs of the Lod frames, while local artisans were used to carry out the laying of the mosaics. The ships with the detailed gears could not have been produced by the mosaicists if they were not guided by the patron who possessed nautical knowledge or was familiar with similar ships. A closer inspection of the mosaics reveals mistakes made by the workers, especially in the geometric borders, thus indicating that the work was carried out at the site, and the panels were not prefabricated in a workshop and then carried in trays to the designated site within the mosaic.[25]

The sailing ship from Djemila and the fishery frame from Cap Matifou augment the North African influence in the subject decoration and themes of the Lod mosaic. The complete preservation of the Lod mosaic indicates that the villa did not have a long life and probably was destroyed and abandoned at the end of the 4th or the beginning of the 5th century CE. There are no indications of "iconoclastic adjustments", which augment that the mosaic was already covered and protected by soil and debris at the time of the iconoclasm (8th century CE).

[23] Ferdi, 2000, pp. 166 - 167.
[24] Dunbabin, 1978, p. 126.

[25] Personal observations of the mosaic while studying the maritime scene in *situ*. Being a marine archaeologist, specialized in ship iconography in mosaics, I was fortunate to receive the permission from Mrs. Avissar to carry out the study the ships during the excavation in 1996. After the excavation the mosaic was recovered. It was uncovered in the spring of 2010, and removed from its site for restorations. Within two years the mosaic will return to its new home at the mosaic museum to be built on the site.

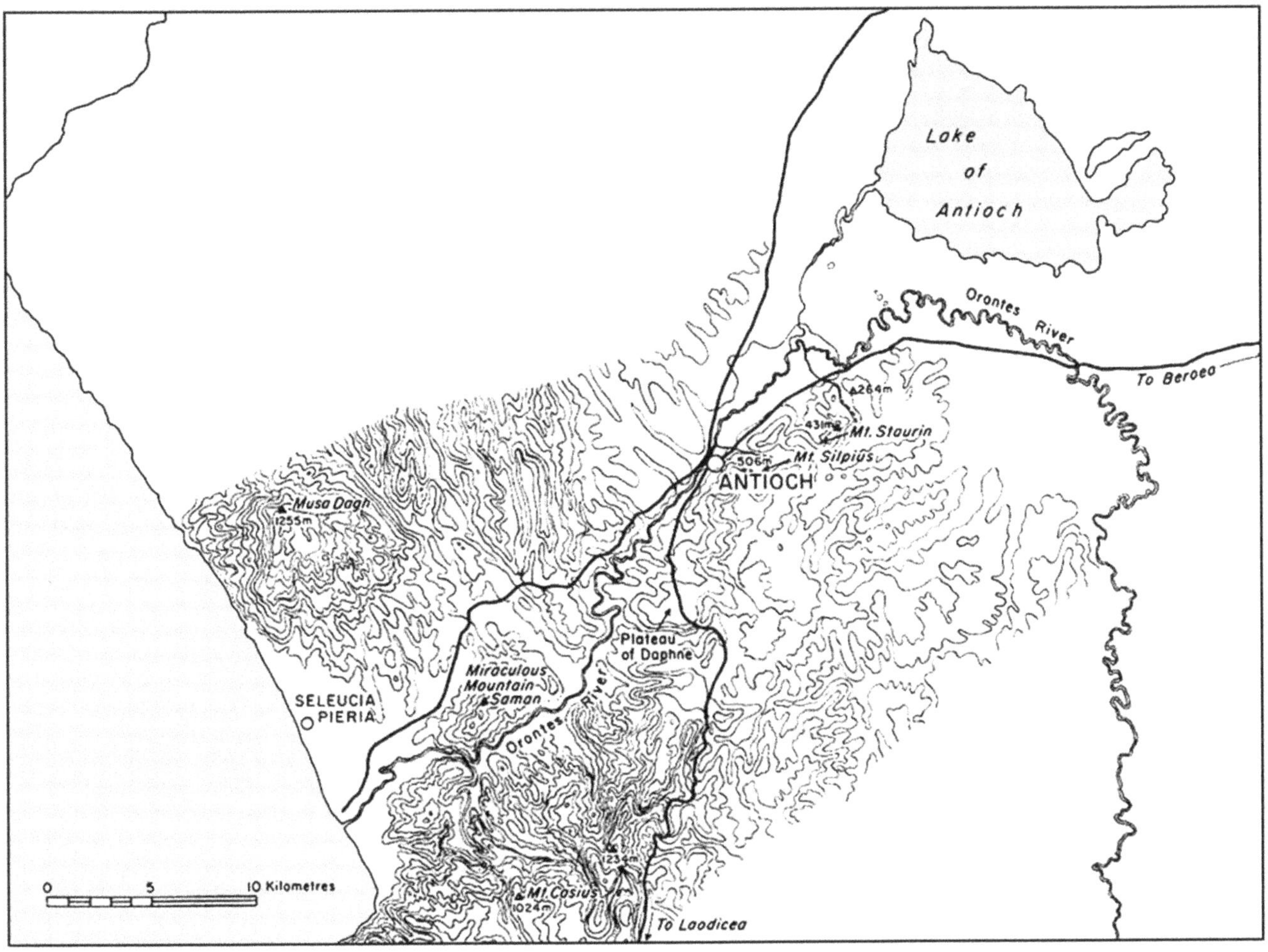

Fig. 3.3.1: Location map of Antioch-on-the-Orontes

3.3 Yakto mosaic (Antioch) - Turkey

Introduction

Alexandria, Rome and Constantinople were important cities during the Roman – Byzantine periods but quite little is known about the wealth of Antioch in the same periods. The rich and expansive life of 4th century CE Antioch is mostly known from the detailed description provided by Libanius, a native of the city, in his autobiography, in the *Letters* and *Orations*.[1]

Antioch-on-the-Orontes and the Excavations

Antioch is situated on the southwest corner of the Amuq plain. The Orontes River flows along the southern edge of the plain, cuts through the mountains and flows into the sea on the northeast coast of the Mediterranean (fig. 3.3.1). The Silpius Mountain Range stretches along the river's right bank and rises to a height of 506m above sea level.[2] Facing the city, the lower slopes of the mountain are gradual and in antiquity, terraces were cut into these slopes, providing building ground for villas and baths with a magnificent view of the Orontes River and the city below. At the NE part of the city, cutting into the mountain, is a ravine through which ran the road to Apamea, as well as to the eastern and to southern Syria.[3]

Although the Orontes River is not navigable today between Antioch (now Antakya) and the sea,[4] several sources mention that during certain periods in antiquity the river had very busy water traffic.[5] Antioch was supplied with products from the hinterland, the lake and the sea by small vessels that sailed on the Orontes. Products like timber and fish came up the river from the seaport of Seleucia.[6] When Libanius speaks about the outlet of the river he refers to the merchants brought to Antioch from its harbor:

> *"I am impelled to mention the harbor (Seleucia), that among other harbors, this is the most sailed that are spread over the seas".*[7]

1 Libanius, *Autobiography* 4.4, p. 1.
2 Downey, 1961, p. 15.

3 Ibid., p. 17.
4 In addition to the silt deposits from Mt. Silpius, since the Middle Ages, earthquakes, which are quite common in the Antioch region, caused debris from the city wall to topple into the stream of the river, filling it up. The arm of the river that ran between the island and the city itself had filled up since Crusader times; Downey, p. 18.
5 Ibid., p. 8.
6 Leibeschuetz, 1972, p. 75 and notes 2 - 3.
7 Libanius, XI.264.

Libanius in his Autobiography gives a nice summary of Antioch in the 4th century CE:

> *"As having the good fortune to be a citizen of a great and famous city, let us consider the size and character of the city of Antioch, the extent of its territory, the streams which water it, and the breezes which it backs. Even without seeing it, one can have full knowledge of it from hearsays, for there is no other corner of land or sea to which the fame of the city has not spread".*[8]

Antioch has been visited by a number of European and American travelers since the Middle Ages who wrote about what they saw during those periods. There are evidences of monuments that no longer exist or that have deteriorated since they were recorded.[9] Some of these travelers copied ancient inscriptions or drew sketches, and made photographs of these monuments (especially in the 19th century).

On occasional surveys carried out at Antioch, it was observed that the ancient city walls on top of the mountain were in fairly good condition and they survived in several parts. In aerial photos taken by the French military authorities in 1932, are seen distinct traces of the ancient streets and of some buildings.[10] In order to stop the looting of antiquities and treasure hunting excavations in the area of Antioch, the Committee for the Excavations of Antioch and its vicinity was formed in 1930, under the chairmanship of Prof. C. R. Morey of the Princeton University. The Committee also included the Museés Nationaux of France. The University of Princeton directed the excavations that began in the spring of 1932 and continued until 1939, when the outbreak of the Second World War prevented further works to be carried out.[11] Excavations carried out at Antioch and its immediate vicinity (at Daphne, suburb of Antioch) and at the seaport Seleucia-Pieria, hinted at the prosperity and luxuries during the 4th – 5th centuries CE. Many mosaics covering the floors of private houses, public structures, pools and baths were discovered during these investigations. The mosaic themes are varied and rendered with wide ranges of hues. The mosaics showed that their decorations were similar to other Roman cities of the same period. At Antioch, even in houses with modest facilities, had floors made of mosaics, thus emphasizing the richness of the city.[12] The excavations revealed that figural mosaics since the 2nd century CE continued to be used until the 6th century, the end of classical Antioch.[13]

The Yakto Complex - Upper Level

In November 1932, the Committee for the Excavations of Antioch, while excavating at Daphne (the suburb of Antioch) revealed an interesting mosaic whereas in the center is depicted a large medallion of a woman's bust and labeled "*Megalopsychia*".[14] The boarder surrounding the panel illustrates varied structures assumed to represent the topography of ancient Antioch and Daphne, along with scenes of everyday life in the 4th – 5th century (fig. 3.3.2).[15] The style of the mosaic and some artifacts related to this floor dated it to the middle of the 5th century.[16] Contrary to what other scholars believed, Doro Levi maintained that there was no orderly progression of the buildings and no certainty that this border shows the architectures outside Daphne.[17] Although the excavators identified the colonnaded streets and the imposing buildings in the city, the mosaic border is not a cartographic representation of Antioch. A precise date of the "*Megalopsychia*" mosaic is given by the name of "PRIBATON ARDABURIUS" depicted atop of one building within the border; he was a *magister militum per Orientem* from 450 to 457 CE, with his headquarters in Antioch-on-the-Orontes, where he resided as late as 459 CE.[18]

To the NW of the Yakto villa complex are found two adjacent rooms: Room A, with the figurative mosaic panel of *Megalopsychia*, and a smaller room, Room B, paved with a beautiful mosaic depicting in the center the bust of the goddess Thetis, while several putti are engaged in fishing activities, and riding on dolphins (fig. 3.3.3). The walls of these rooms, as well as many parts from the edges of both mosaics, had completely disappeared. No datable material was collected during the excavation of Thetis mosaic floor. It has been mentioned that the *Megalopsychia* mosaic dates to the middle of the 5th century, according to the inscription within the topographical border and some related artifacts. The Thetis mosaic most probably was made during the same period which may be deduced from the fact that both rooms are adjacent to each other (fig. 3.3.3).

The Thetis Mosaic

The main theme of the mosaic comprises the figure of the Goddess rising from the sea, holding a rudder in her right hand and laid on the shoulder, while in her left palm she holds a small dolphin (figs. 3.3.4, 3.3.5). Two claws of a lobster or crab rising above the forehead of the Goddess are made with light brown and reddish-brown-ochre tesserae. A monstrous snake, with a dog or wolf's head coils around the bust of Thetis. The rudder has a short shaft made with ochre and dark brown stones. The blade comprises two long and quite narrow wings, each inserted on either side of the lower shaft (fig. 3.3.5). Both their ends are angled toward

8 Ibid., *Autobiography* 2.
9 Downey, p. 26.
10 Ibid., fig. 6, p. 26.
11 Ibid., p. 28.
12 Ibid., p. 33; Levi, 1947, I, p. 1.
13 Lassus, 1976, p. 63.

14 Campbell, 1934, p. 201.
15 Downey, p. 31; Levi, I, fig. 136, p. 324; II, pls. LXXVII b – c, LXXVIII a – c, LXXIX a – c, LXXX a – d; Campbell, p. 202. According to the reference of Evagrius (*Hist. Eccl.* III, 28), Daphne received its status of town under the Emperor Zeno (474 – 491 CE), and a certain Mammianos erected an agora at his own expanse; Levi, 1947, p. 326.
16 This mosaic along with others, were removed from their original location and now they are displayed in the Hatay Archaeological Museum, at Antakya (Antioch).
17 Downey, p. 31.
18 Levi, p. 323.

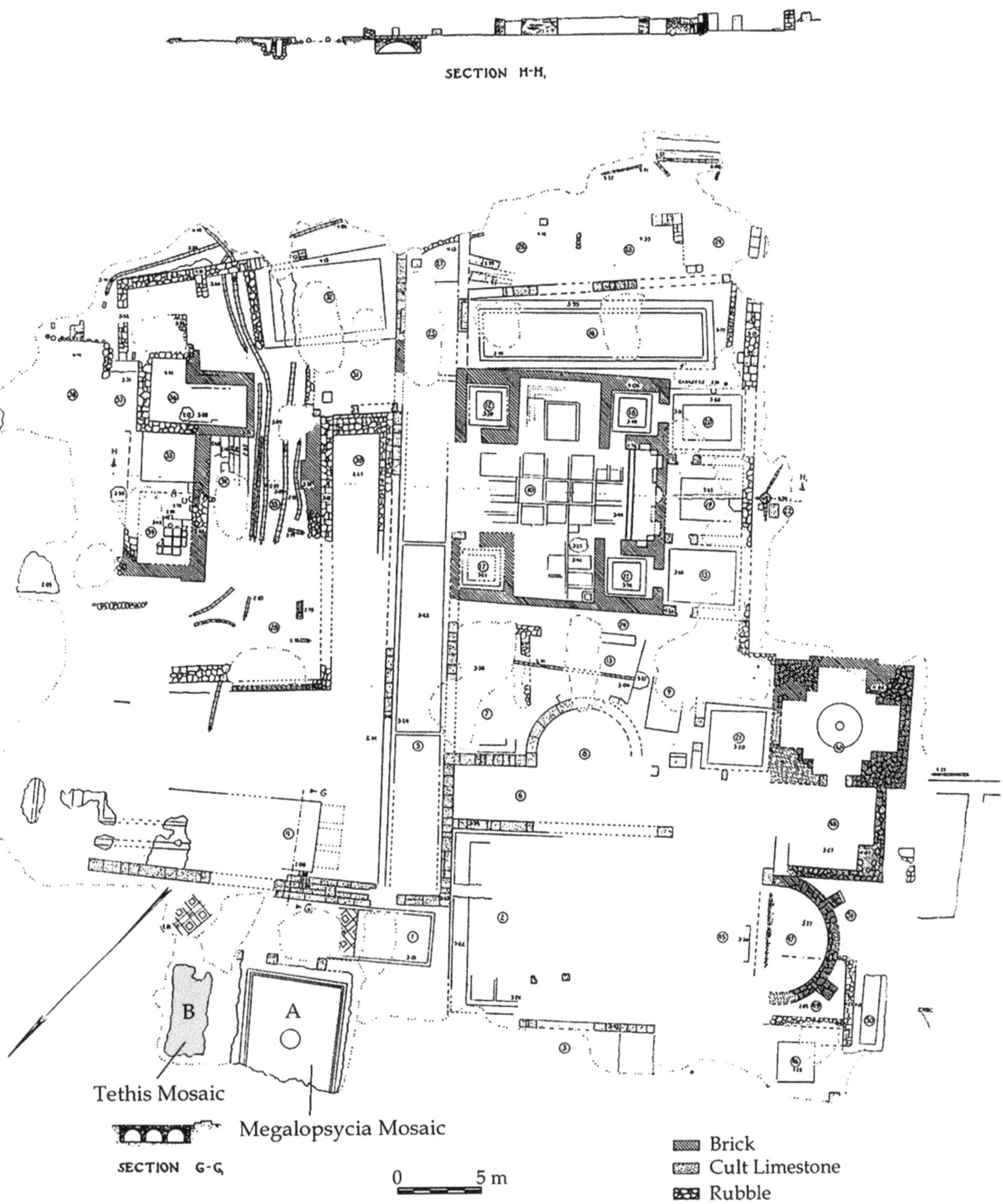

Fig. 3.3.2: Plan of the Yakto Complex

the shaft. The left wing is depicted with ochre tesserae and the right one with dark brown stones. It appears that each wing formed an individual unit inserted in a long groove cut into the lower shaft. They seem to be locked by wooden treenails or pegs, or even bronze nails, indicated by the darker dots along the visible face of the shaft (fig. 3.3.5). Two putti riding on the back of a dolphin approaches the goddess from the right and the left.

A wide border, only fragmented parts of which have survived, surrounds the panel. Within the border are depicted varied maritime activities of putti who are pulling

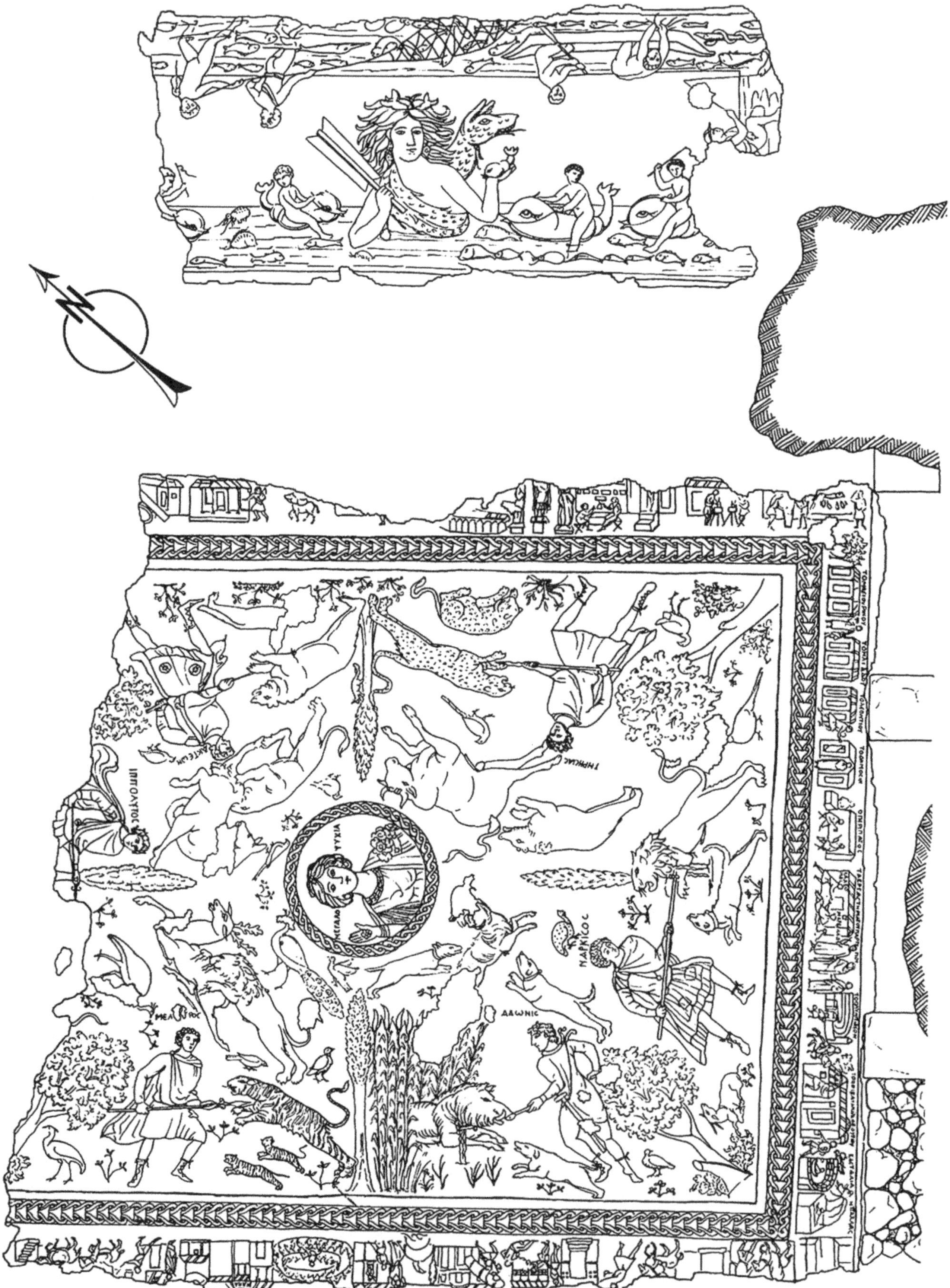

Fig. 3.3.3: Thetis and "Megalopsychia" mosaics

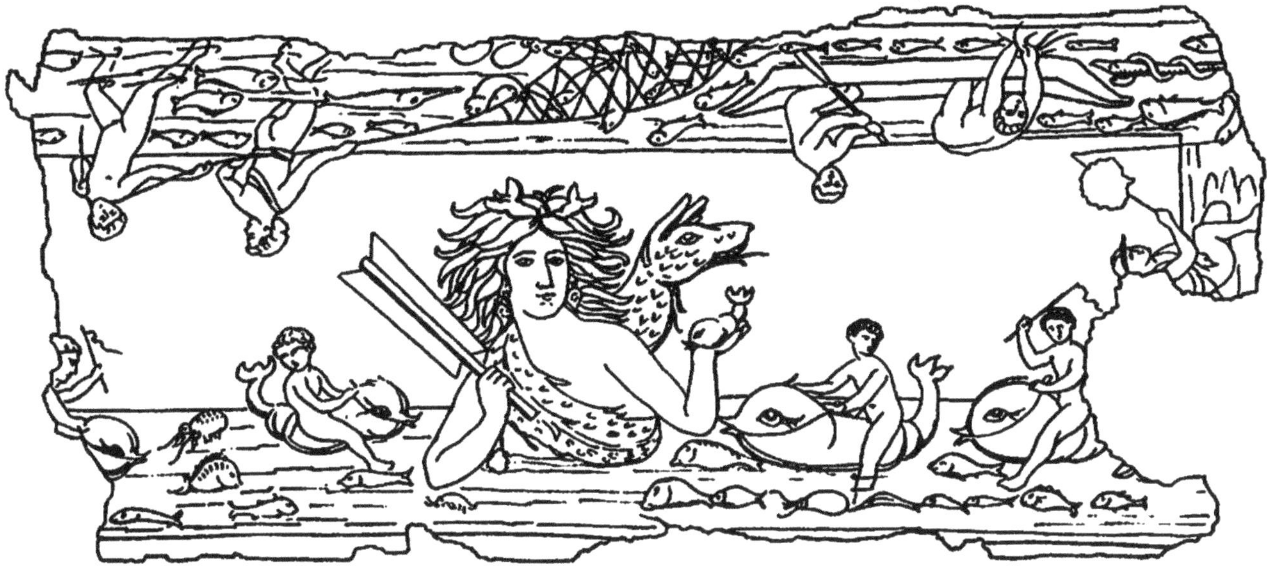

Fig. 3.3.4: Thetis mosaic panel

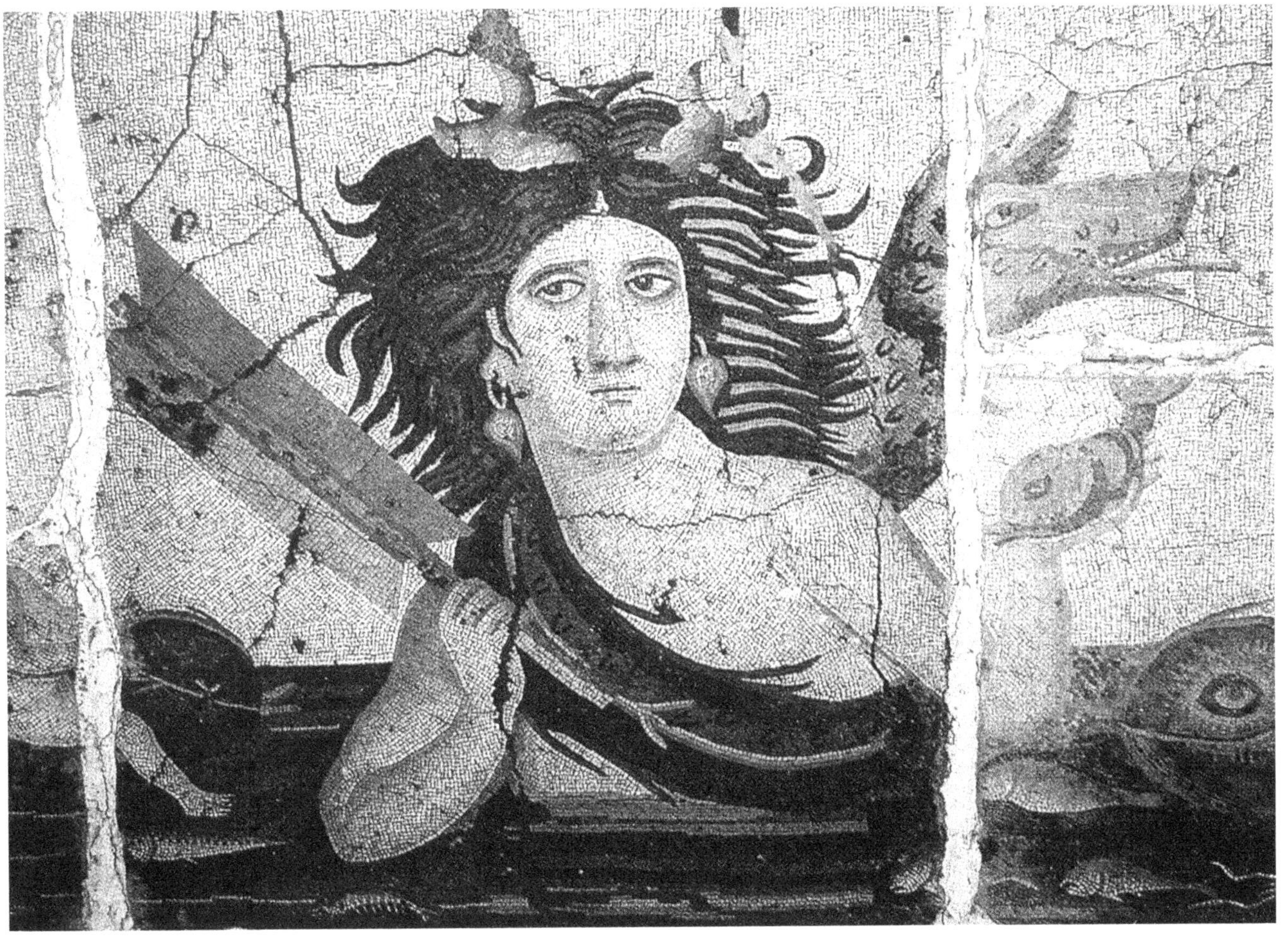

Fig. 3.3.5: Thetis rising from the sea

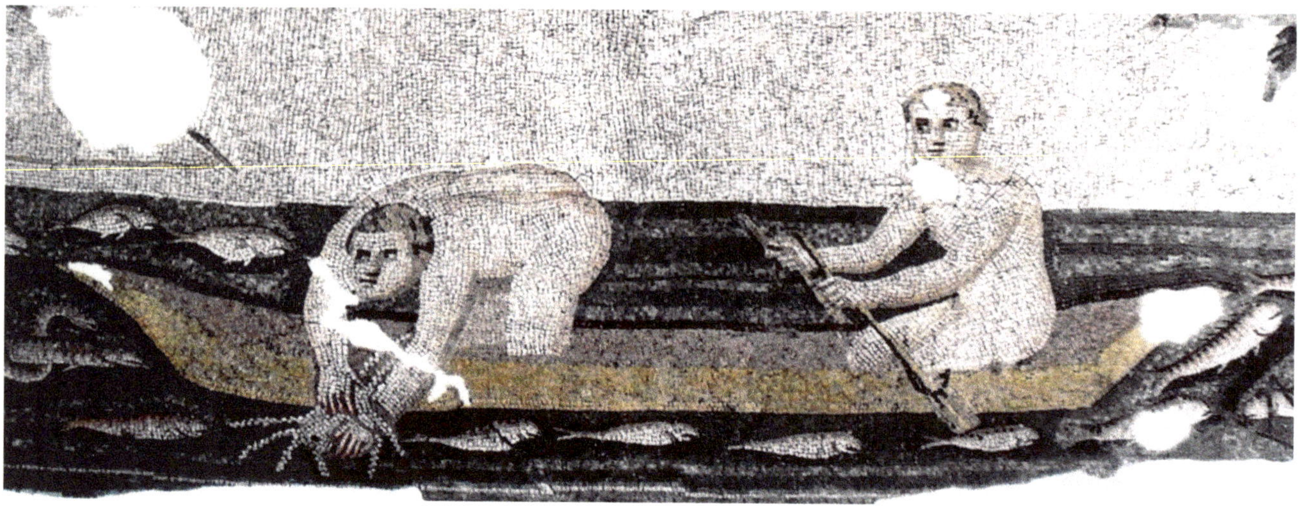

Fig. 3.3.6: The fishing boat in the Thetis mosaic

a heavy net with fish. Their legs are seen up to the knees through the transparent water. Near this scene is depicted a boat inhabited by two men, one bent over the port side probably pulling a net out of the water while the second one works the steering oars to stabilize the boat (fig. 3.3.6). In the right-hand border is depicted a fisherman seated on a rock who fishes with a fishing rod and a fish is just caught in the hook. The water is rendered with dark green tesserae, hues of lighter green, while the dark gray and white stripes represent the waves, as well as the transparency of the water. The background is made with small, white, fine tesserae. The sea is inhabited with different kinds of fish, eels and shells.

The Boat

The only boat in the mosaic is depicted in the border with the fishing scenes, above the head of Thetis in reverse (figs. 3.3.4, 3.3.5). The boat is seen in some perspective, revealing part of its starboard inner hull. It is engaged in fishing with a net that seems to be lowered into the water or the bent figure over the side of the boat is taking an octopus out of the water (fig. 3.3.6). The prow of the boat is pointed to the left. The lower part of the hull is outlined by two rows of light brown-ochre tesserae, thus indicating its keel, stem and stern. The port hull is rendered with dark brownish-yellow tesserae. The inner starboard hull is depicted with dark brown stones. The stern and stem are angle with outward horizontal extension and finished by straight cuts. The hold of the boat is quite shallow, as deduced from the figure bent over the port side of the boat. The gunwale reaches to his knees (fig. 3.3.6). The feet of the seated helmsman rest on the flat floor of the boat. His knees also reach the height of the gunwale. A pair of oar, one mounted on either quarter, comprises the rowing and/or the steering gear. The port oar is depicted with a long and thin shaft that its lower end transverses the blade longitudinally. The oar is secured to the port quarter by a metal ring or a leather strop (fig. 3.3.7). The short rectangular blade comprises two blades that seem to be inserted into a groove cut along the lower part of the shaft and then secured by wooden treenails or bronze nails. The starboard oar is indicated only by the upper end of the shaft which most probably was secured in the same way as the port oar.

Type of Boat

The Yakto boat with the slim hull, flat bottom and even-colored shear, most probably represents a logboat fashioned from a single log of wood. Such vessels were quite easily made from either an entire log or half log that was split longitudinally. First the bottom of the boat was flattened. The hold was hollowed along almost the entire length of the log. The ends with enough thickness, that also formed the thwarts, were left as a solid block and only leveled horizontally.[19] The bow and stern were shaped with an outward angle to provide the hydrodynamics of the boat. The tools used to make such boats comprise adze, ax and chisel. The timber for a logboat naturally depended on what was available locally. Ethnographic evidences suggest that the tree chosen for such boat would be in the vicinity of water, would be tall and strong, and not split or have spiral grain, which would create difficulty in working the log.[20] The log was hollow-out by an adze and other carpenter's tools, or by fire. When fire was used, it was normally lit on top of the horizontal log and constrained by spraying water or with wet clay.[21] The charred wood was removed by adze or scraped away, thus leaving a smooth surface (fig. 3.3.8). The final shape of the boat's hold most probably was done with conventional tools (adze, chisel) and not fire. For a better understanding of how a hollowed logboat was made we must rely on a 20th century examples from Sweden,

[19] There are many hollowed logs or half logs that survived in land excavations, especially in northern Europe, dating from pre-historic times until the 19th century; McGrail, 1998, pp. 57 – 9, figs. 6.3, 6.9, 6.13, 6.14, 6.15, 6.17.

[20] McGrail, p. 59.

[21] Ibid., p. 62

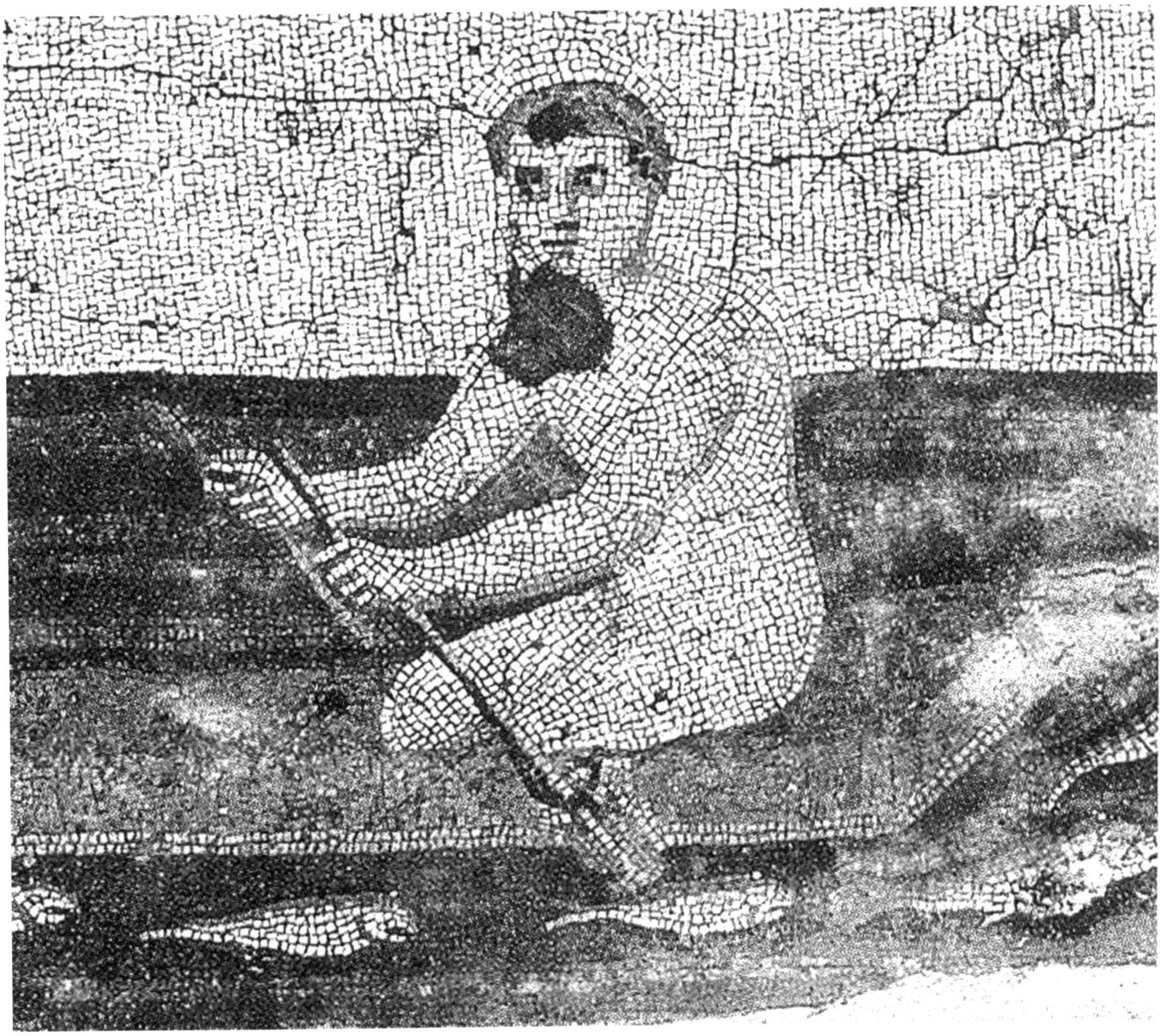

Fig. 3.3.7: Close-up of the rower

Finland, Hawaii, Austria, and North or South America.[22] The boat-builder used a line smeared with charcoal to mark out the edges of the hold of the boat, with transverse lines every 0.30m. A section between two lines was then hollowed down to the thickness gauge hole; then the next section was hollowed in the same way.[23] The ends were hollowed last.[24]

[22] Ibid.

[23] Ibid.

[24] Evidence from north Europe shows that a logboat was not finished in one season, but stored, often underwater, until the following year, when the work was finished. In Denmark for example, logboats from Hasseled were sunk unfinished and held underwater for two years. It was claimed that the logboat timber became softer and more workable; McGrail, p. 63. In the Society Islands, Hornell reported that after the first hollowing, the boats were filled with water or submerged for several days. The wood absorbed the water and probably by this method the timber kept fresh and moist above the fibers' saturation point, so that the boat would not split while hollowing it; ibid, p. 63. It seems that the method of unfinished logboats being stored underwater was and still is popular in places were timber is found for building such boats.

The Marina Mosaic in the Nile Villa at Leptis Magana

The Villa with the Nile mosaic is situated near the port of the ancient city of Leptis Magna. Its name comes from three excavated mosaics depicting Nilotic scenes. One of the boats depicted in the Nile Villa that decorated the floor of a *tepidarium* may be compared to the Yakto boat (fig. 3.3.6). The excavations of the site that began in 1916 and continued until 1930 indicate a possible date of the mosaics around the 2nd century CE.[25] On the left hand side of the mosaic are depicted two boats (fig. 3.3.9). They are part of a rectangular panel illustrating several fishermen engaged in varied fishing methods. The top boat with an elongated and slim hull, with outward extended ends and a flat bottom (fig. 3.3.10), is similar to the Yakto Boat (fig. 3.3.6). The static position of the boat and the five sitting figures, who probably are engaged in a party, suggest that

[25] Aurigemma, 1960, p. 49; the size of the mosaic: L = 3.78m; W = 1.2m.

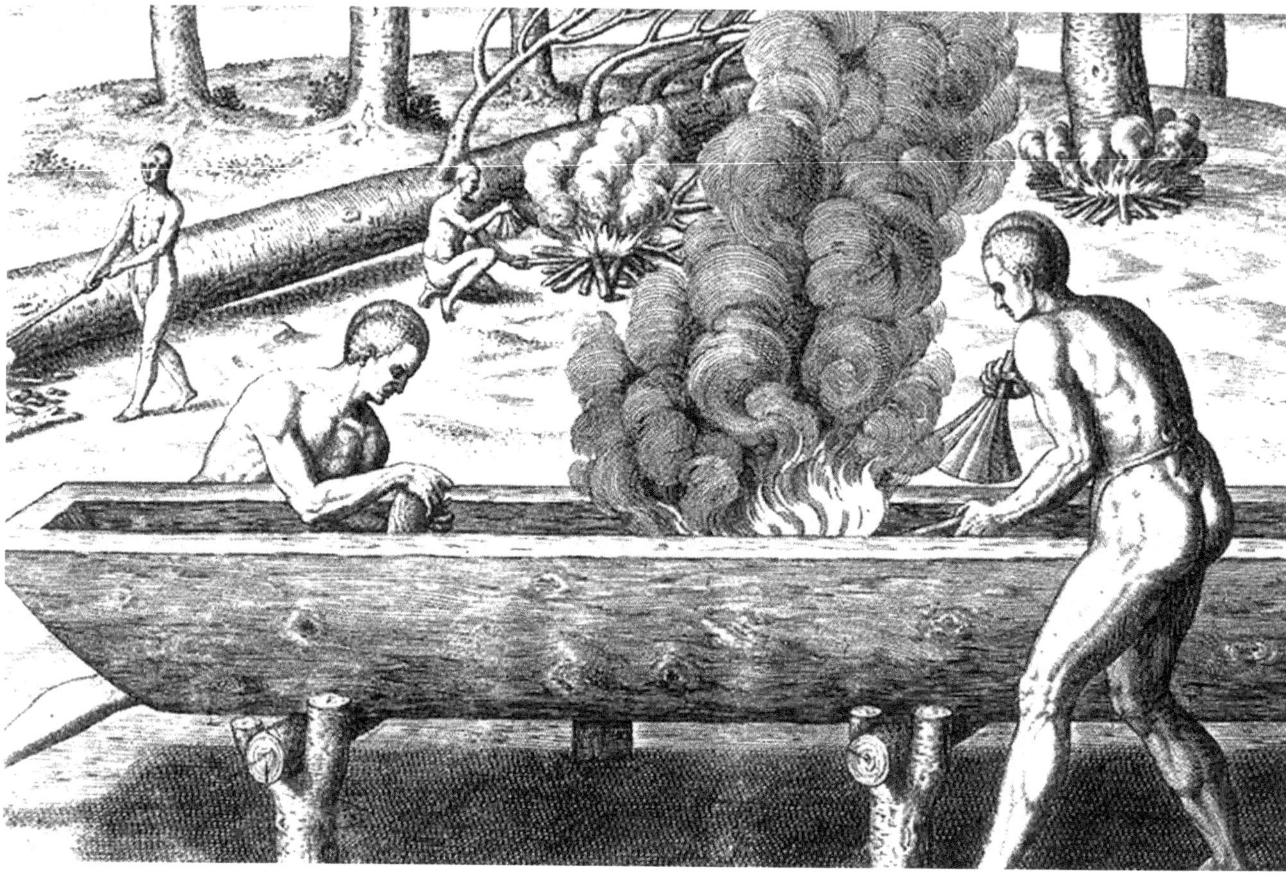

Fig. 3.3.8: A late 16th century drawing of a logboat hollowed by the fire method

the boat is anchored, though no anchor or mooring device is shown. The prow pointing to the left has a similar shape to the Yakto vessel. The stern, now damaged, probably had a similar shape to the bow. The steering gear of the boat shows one oar indicated by the angled shaft laid against the port quarter (fig. 3.3.10). The figure sited on the quarter near the oar, presumably is the helmsman. His right arm stretches forward while the left hand probably held the loom (now damaged). The shape of the boat and the uniform coloring of the hull rendered with dark brown tesserae, indicates a logboat of about 5 – 7m long and *c.*1m abeam. Its shallow hold is deduced from the seated figures on the starboard and port gunwales, with their feet resting on the floor while their knees reach above the height of the gunwale.

The second boat found in the middle section (fig. 3.3.9), also has an elongated slim hull. The raised rounded sternpost ends with a pointed tip. The bow is damaged. Four figures inhabit this boat. Two are engaged in rowing; one at the quarter with the back turned to the viewer (probably a female) holds the looms of each oar mounted on the quarter. The second figure is identified only by his head, and presumably sitting in front of the helmsman, the third sits amidships with his head turned to the quarter and works the rowing oars mounted on either side of the hull. The fourth figure suffered some damage and may be identified only by his right shoulder and part of his back.

Discussion

The Yakto Thetis mosaic (460 – 465 CE) is not the only one depicting this sea goddess in Turkey. It appears that this theme was quite a common motif for the eastern mosaic decorations in Turkey and Syria, especially in Antioch. The Thetis motif is noticed from the 2nd century onwards.[26] The Yakto Thetis however is the only example where the Goddess is surrounded by putti engaged in various fishing activities and water games. It has been mentioned above that the boat might represent a logboat, which is a rare representation in any art form. The Yakto Boat (fig. 3.3.6) is comparable to the top boat (fig. 3.3.10) depicted in the Nile mosaic from Leptis Magna (2nd century CE). If the boat in the Leptis Magna mosaic indeed represents a logboat then we have the earliest representation of such a vessel, pre-dating the Yakto Boat by about 300 years. It is well known that pattern decorations in the North African mosaics and those from the NE Mediterranean are very similar. Pattern books and other sources of decoration circulated

[26] The earliest sea goddess was found in the House of the Calendars in Antioch (Levi, pl. VI a), dated by the archaeological evidence, shortly after the earthquake in 115 BCE; Wages, 1986, p. 124. A later representation of the sea personification –*Thalassa*, appears in a circular medallion centered in the mosaic floor in the Church of the Apostles at Madaba. The mosaics and the church date from 578 CE. In the inscription around the medallion is mentioned that it was dedicated to Anastasius, Thomas and Theodorus, and the mosaic was made by Salaman the mosaicist; Piccirillo, 1993, fig. 78, pp. 96, 106.

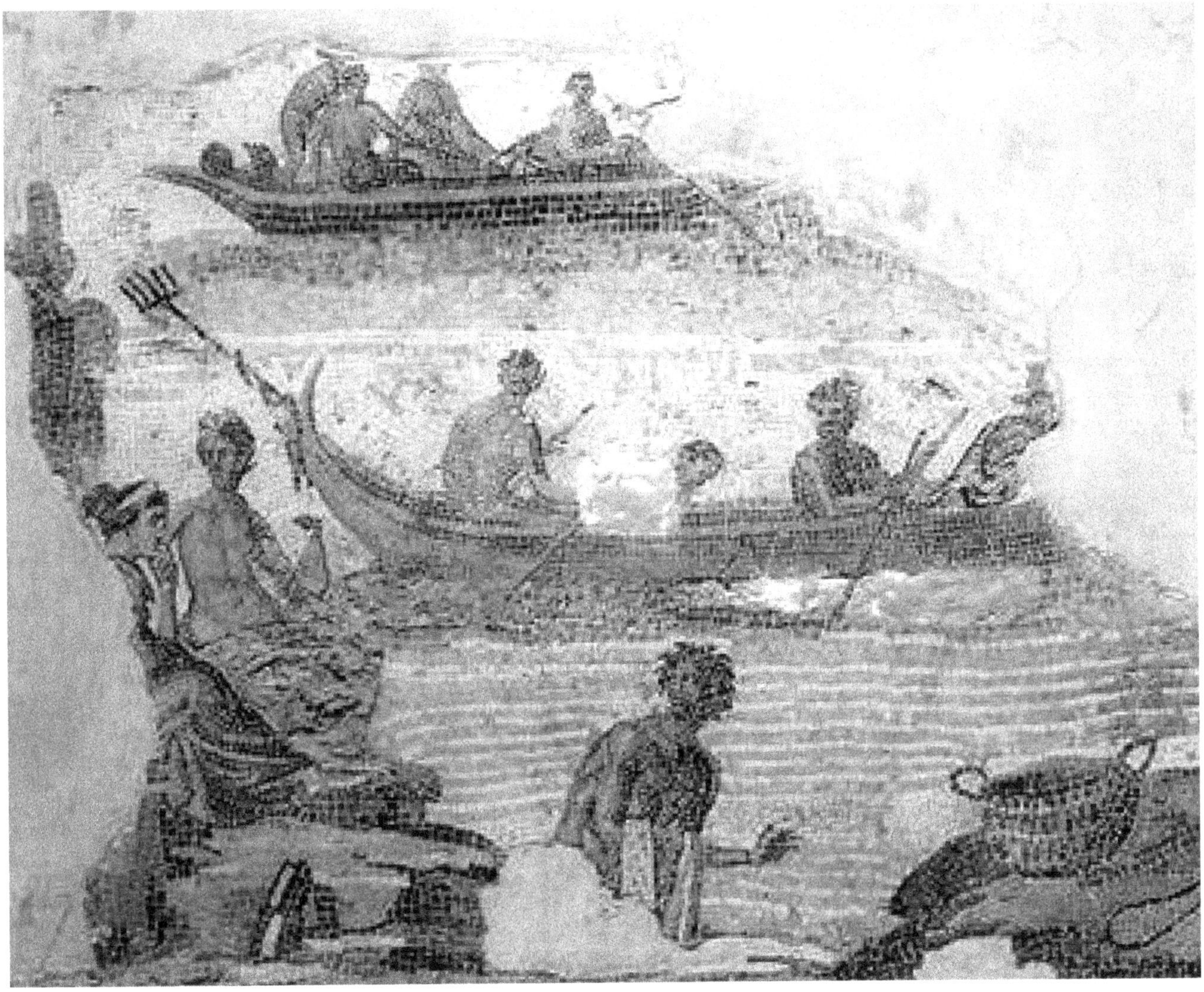

Fig. 3.3.9: The "Marina" mosaic in the Nile Villa at Leptis Magna

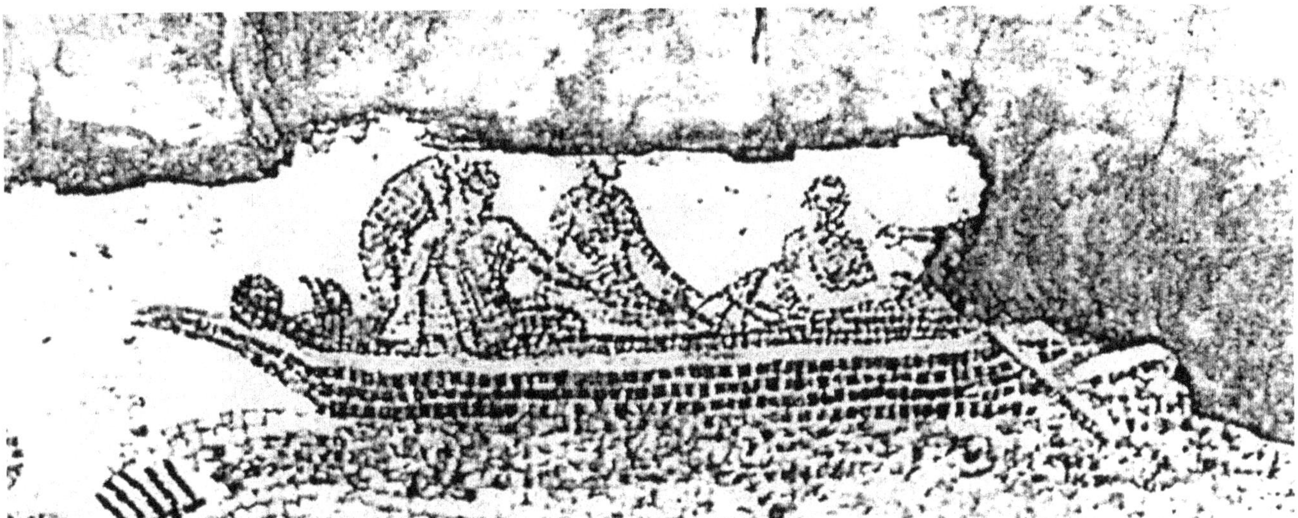

Fig. 3.3.10: The top boat in the "Marina" mosaic

around the Mediterranean through sea trade connections and movement of goods and people.

The Yakto Boat may be considered a relevant representation of a logboat considering the fact that the site is found in close proximity to the Orontes River and the mountain ranges of Amanus and Silpius. Although, the Orontes was not navigable in late antiquity and neither today (between the sea and Antioch), ancient sources record that at certain periods the river had busy water traffic.[27] The *Gourub papyrus* mentions that Orontes was navigable in 246 BCE.[28] In a commentary written by Libanius concerning Antioch, he said that the city profited greatly from the river because "*many things were brought to the city by the vessels sailing on the river*".[29] Debris from the city walls that toppled into the river since the Middle Ages have blocked the passage for sailing boats. It also appears that the riverbed rose as a result of earthquakes, which are frequent in the area.[30]

The economy of Antioch in the 4th century was very prosperous and the city enjoyed great luxuries and opulent buildings. Libanius mentioned that the rich forests on Mts. Silpius and Amanus provided the wood needed for the construction of roofs, and fuel for bakeries and baths:

> *"... forests furnished the roofs, and assist in other ways, and in addition provide fire for the bakers and the baths by means which we not only live but have to live in comfort".* [31]

Though Libanius does not mention anything about boat building near the Orontes, we may assume that the wood from the mountain ranges and overseas trade provided the raw material for their construction. This assumption is suggested by Libanius: "*the river was a way of transportation for variety of woods from everywhere*".[32] The logboat was the simplest vessel to build that neither required much space nor a special shipyard or a large team of shipwrights. The elementary tools were adze, axes, chisels and fire.

The mosaicist who made the Yakto Thetis mosaic may have seen or been familiar with such simple river craft that probably still sailed on the Orontes in the 4th century. He also may have seen such boats engaged in fishing on the river, the Orontes Lake or nearby the seashore, and so such scenes were a good source for decorations associated with mythological stories.

27 Downey, p. 18 and n. 12.
28 Ibid., p. 18.
29 Libanius, XI.260.
30 Downey, p. 18.
31 Libanius, XI.25.
32 Ibid., XI.262.

3.4 Kelenderis (Aydincik) Harbor mosaic - Turkey

Location and a Short History of Kelenderis

Kelenderis known today by the name Aydincikis is located in the province of Mersin, on the Mersin-Antalya D 400 highway and also it is a port of the southern Asia Minor coast on the Mediterranean (fig. 3.4.1). The location of the site belongs to the central part of the Rough Cilicia which is called Tracheia or Aspera.[1] The natural bay and the coast provide a significant anchorage in the eastern Mediterranean since antiquity. The Kelenderis was connected with the Central Anatolian Plateau through passages in the valleys but mainly it was the port which linked the southern Cilicia with Cyprus and other countries lying in the Mediterranean coasts. The land around the ancient city was not adequate for farming, or other agricultural activities. On the plateau behind the hills there are vineyards, olive plantations and stockbreeding.[2] Woods, mainly cedar and pine, which were essential materials for shipbuilding in antiquity, provided one of the important traded products. The topography of the city was not suited to extend the settlement in antiquity and thus developing land communication. Therefore the connection with other countries had been mostly made by sea way. The port being a transit site between the eastern and western Mediterranean, formed an important overseas trade lane. Later in the Roman period the port became insufficient for big fleets and the city lost of its importance. Nowadays Kelenderis bestows a safe shelter for yachts sailing in the eastern Mediterranean.

Kelenderis is mentioned by few ancient writers. Pliny referred to Celendiritis as the district, including "*a town, and harbor, and cave of the same name*".[3] Pomponius Mela, when describing the southern coast of Cilicia, wrote that Celenderis and Nagidos were colonies of the Samians:

> *"... there are two promontories: Sarpedon, once the boundary of the kingdom of Sarpedon and Anemurium, which separates Cilicia from Pamphylia. Between them lie Celenderis and Nagidos colonies of the Samians, but Celenderis is the one nearer to Sarpedon".* [4]

Kelenderis area was colonized by the Samians at the end of the 8th century BCE and later by the Ionians.[5] The first settlement layers of the ancient city date from the 7th century BCE. This was substantiated by the late Geometric pottery found at the site during the excavations carried out from 1987 to 1992, by the Selçuk University, Konya, under the direction of Prof. Levent Zoroglu.[6] The city was most prosperous during the 5th – 4th century BCE. Kelenderis became a member of the Delian League between 460

1 Zoroglu, 2007, p. 68.
2 Ibid.
3 Pliny, *NH*, V.xxii.192.
4 In: Romer, 1998, p. 58 - Pomponius Mela I.77.
5 Mitford, 1976, p. 445.
6 Evrin et al., 2005-3, ANMED.

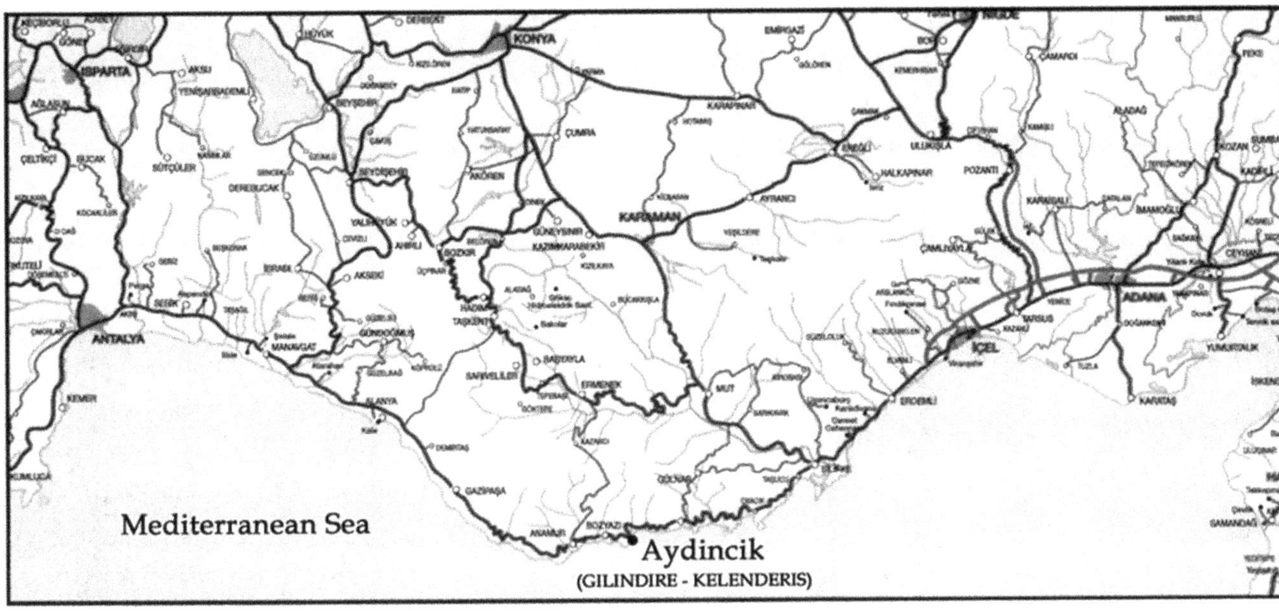

Fig. 3.4.1: Location map of Kelenderis

and 456 BCE, being a station on the route to Egypt.[7] The Persians defeat by Alexander the Great brought Kelenderis to be a free city and therefore it established relations with the Ptolemaic Egypt. When the Ptolemaic patronage withdrew from this area in 100 BCE, Kelenderis allied with the Romans against the Isaurian pirates. During the 2nd century CE, the city became the metropolis of the *regio Cilinderitis*, an area stretching between Anemurium and Seleuceia. It had struck its own coins from the 5th century BCE to the time of Decius.[8] The Parthians destroyed the city around 260 CE. In late antiquity, Kelenderis became smaller but the harbor did not lose its importance up to modern times. Gilindiri, derived from the ancient name Kelenderis; it was declared a town in 1973, and Aydincik became a district in 1987.

The Cilician Harbor

Since its early days, Kelenderis was the harbor of the mountainous area of Cilicia that connected with Cyprus and other harbors in the Mediterranean. Voyagers who preferred the sea route to reach the southern coast of Asia Minor sailed to Kelenderis. They described some of the ruined fortifications, towers, graves, the theatron, the harbor-bath, aqueducts and few cisterns.[9] Captain Beaufort visited the harbors of Cilicia in 1811 – 1812 and prepared a map which also shows the location of Kelenderis (fig. 3.4.2). Barlett drew a panorama of the harbor with a huge tower on the peninsula (fig. 3.4.3). The ruins today are overlaid by the expanding modern town. Ancient fortifications may still be traced around the modern lighthouse, on the promontory which forms the harbor.[10]

The Harbor Mosaic

The earliest settlements were probably located on the peninsula. Later, the city extended to the west of the harbor. The excavations carried out since 1987 near the harbor revealed a continuous area occupied from the archaic period to late antiquity. Additional excavations were made in 1992 and on this occasion a beautiful mosaic was revealed and unearthed in *Square* K *I*.111 (fig. 3.4.4). Indicative historical layers of Kelenderis were revealed in the stratigraphy of this square.[11] The mosaic pavement measuring nearly 12m in length and 3m in width was set above the top layer of *Square* K *I*.111. A floral border surrounding the entire mosaic comprises lotus buds. The field is divided into two parts, one-third (the top, towards the harbor) comprises the harbor scene (3.1 x 3.1m)[12] and the other two-thirds are decorated with geometric patterns (fig. 3.4.5). The harbor panel seems to form the main decoration of the mosaic. The scene is depicted in a combined view of low perspective and bird's-eye view with the main focus on the large sailing ship anchored within the port (fig. 3.4.6).The harbor comprises a long and straight quay (on the lower panel), and an arched colonnade (*stoa*) following the shape of the shore (upper part). On the lower quay are shown different buildings, enclosed by crenellated city-walls. To provide the panoramic view of the harbor, the mosaicist laid down the buildings on the quay bellow. The structure with a tall arched façade, in the lower left corner probably indicates a ship-shed (fig. 3.4.6). Other buildings include a tavern with propylon façade and pitched tile roof, two structures with crenellated tops, a double arched gate, a square tower with a crenellated top, and an additional structure with a pitched roof. The colonnade (*stoa*) is reached from the quay,

7 Mitford, p. 445.
8 Ibid.
9 Zoroglu, p. 71.
10 Mitford, p. 445.

11 Personal communication with Prof. Zoroglu; September 2000.
12 The study and measurements of the harbor frame were undertaken by the author in *situ*, in August 2007, while I was invited by Prof. Levent Zoroglu to join the excavation season.

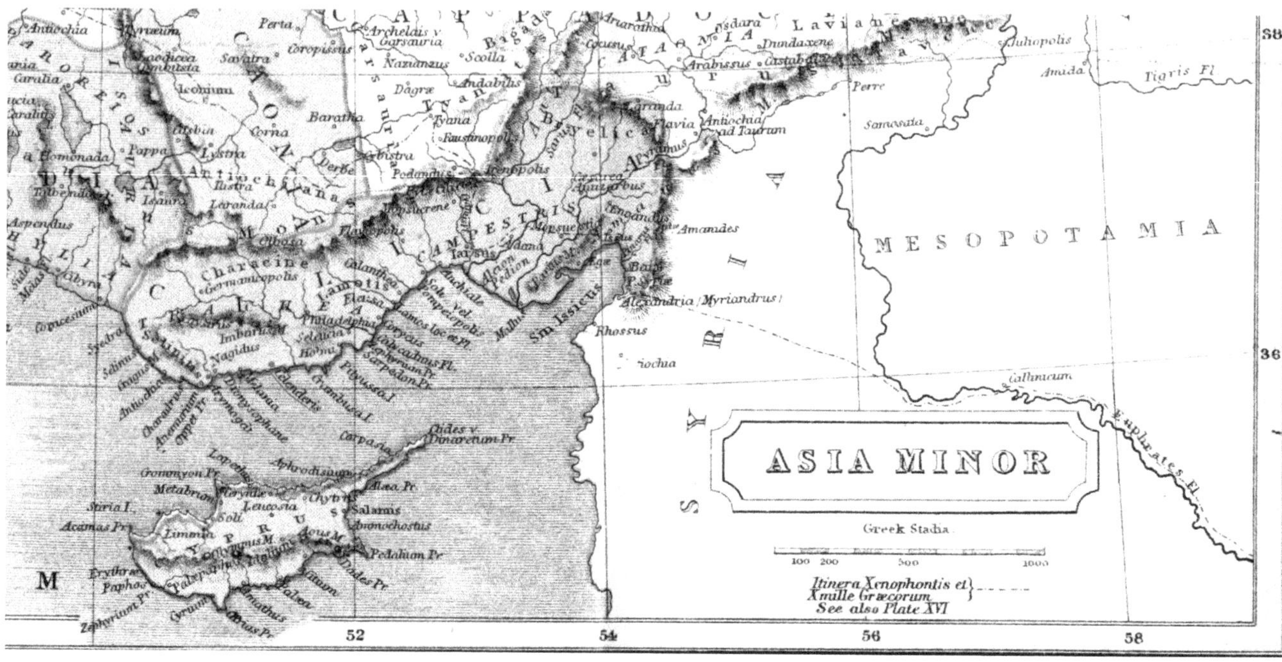

Fig. 3.4.2: Beaufort Map of the Cilician harbors with the location of Kelenderis

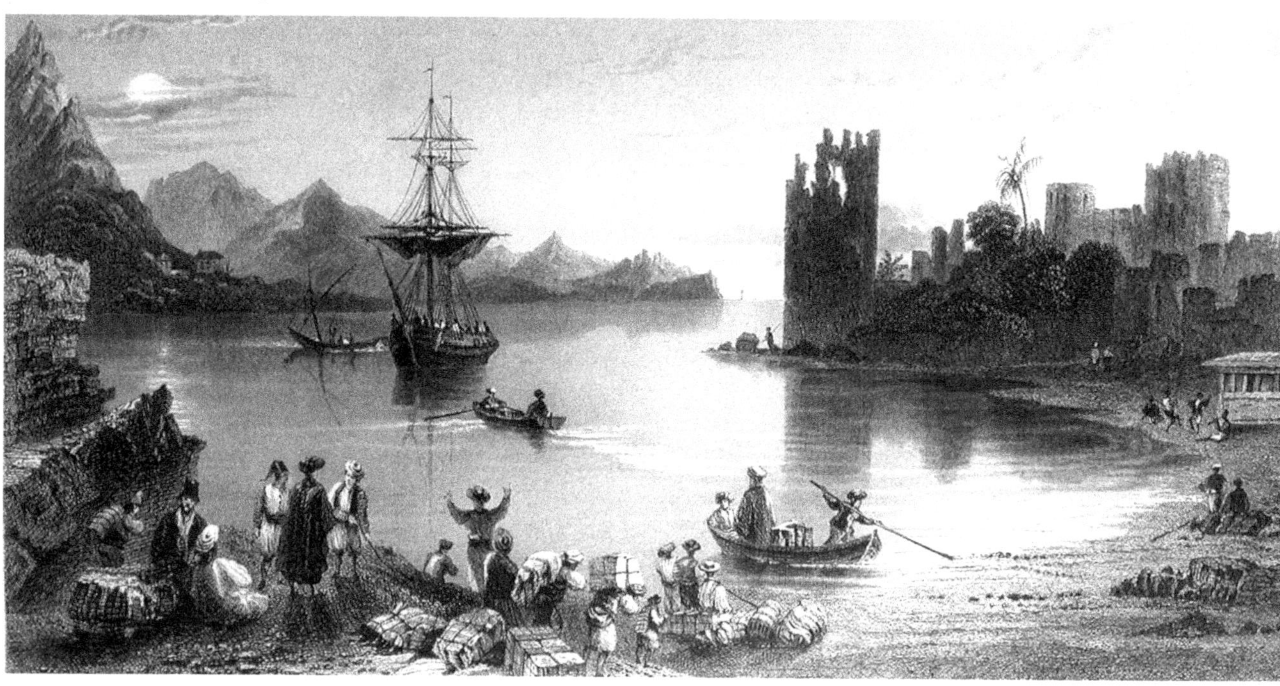

Fig. 3.4.3: Engrave of Kelenderis Harbor by W. H. Barlett, 19th century

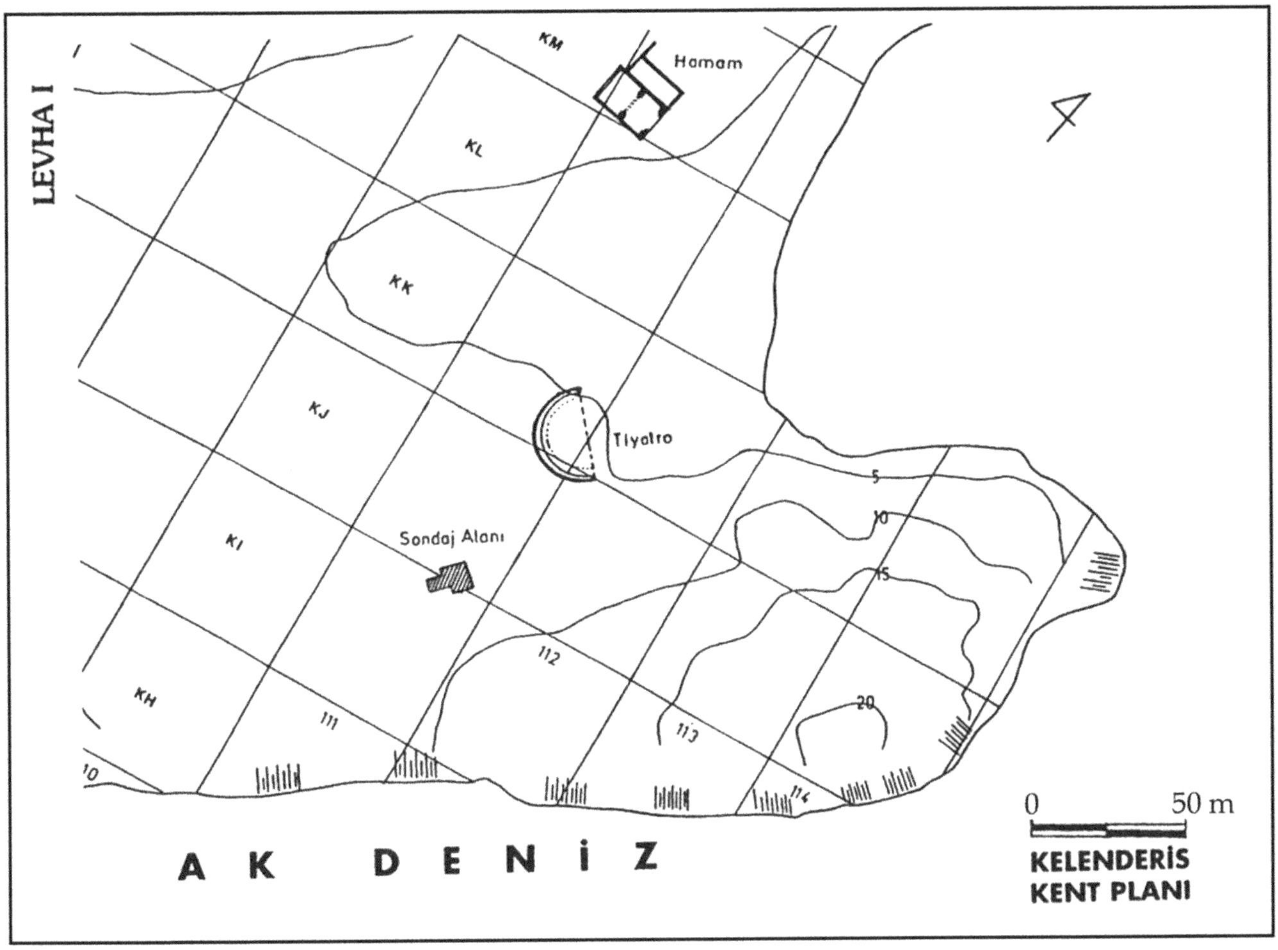

Fig. 3.4.4: Plan of the excavation site

to the left, by five steps and an arched entrance (fig. 3.4.6). Behind the *stoa*, in the left top corner, there is a three-arched structure with large windows, probably a church or the harbor's bath.

The function of the building complex paved with the almost-intact mosaic is not clear and Prof. Zoroglu has been unable so far to identify it. Zoroglu suggested that the mosaic probably dates to the 5th – 6th century CE, based on comparison of the Kelenderis harbor scene with other mosaics depicting similar themes, and the archaeological stratigraphy.[13] He also assumed that the mosaic probably depicts the real harbor of Kelenderis. The baths located to the north-west of the harbor, with almost intact remains, resemble the structure with three arched windows depicted in the top left of the mosaic. The topography of the harbor seems to be the same with the illustration within the mosaic.[14]

The Vessels

Within the semi-circular harbor is depicted a very large sailing ship seen from its stern and towing two smaller vessels, one of which is rigged with a fully open quadrilateral sail, while the other is a rowboat, minus the oars (fig. 3.4.6). The scene is viewed from the harbor's entrance. The towing or mooring lines are looped on either side of the large ship's stern. Both boats are seen from the port side with their bows pointing to the stern of the sailing ship.

The Sailing Ship

The ship has a long and broad hull, as evidenced by the full stern (figs. 3.4.7, 3.4.7 a). It is depicted in a distorted perspective, seen from below the port quarter. The bow is facing the shore and appears close to it. The mosaicist turned the bow towards the port side, thus viewing it in the same plane with the port hull. The raised stempost is outlined with one row of black tesserae and the field is made with ochre stones. A short vertical thin black spar topped by a small flag or bird's head adornment is attached to the tip of the stempost. On the port side of the hull are depicted four parallel long strakes made with white, ochre and dark brown cubes, while the seams are made with one row of black tesserae. On either side of the stempost there is a white trapezoidal plank that emphasis the sides of the

[13] Zoroglu, 2000.
[14] Ibid.

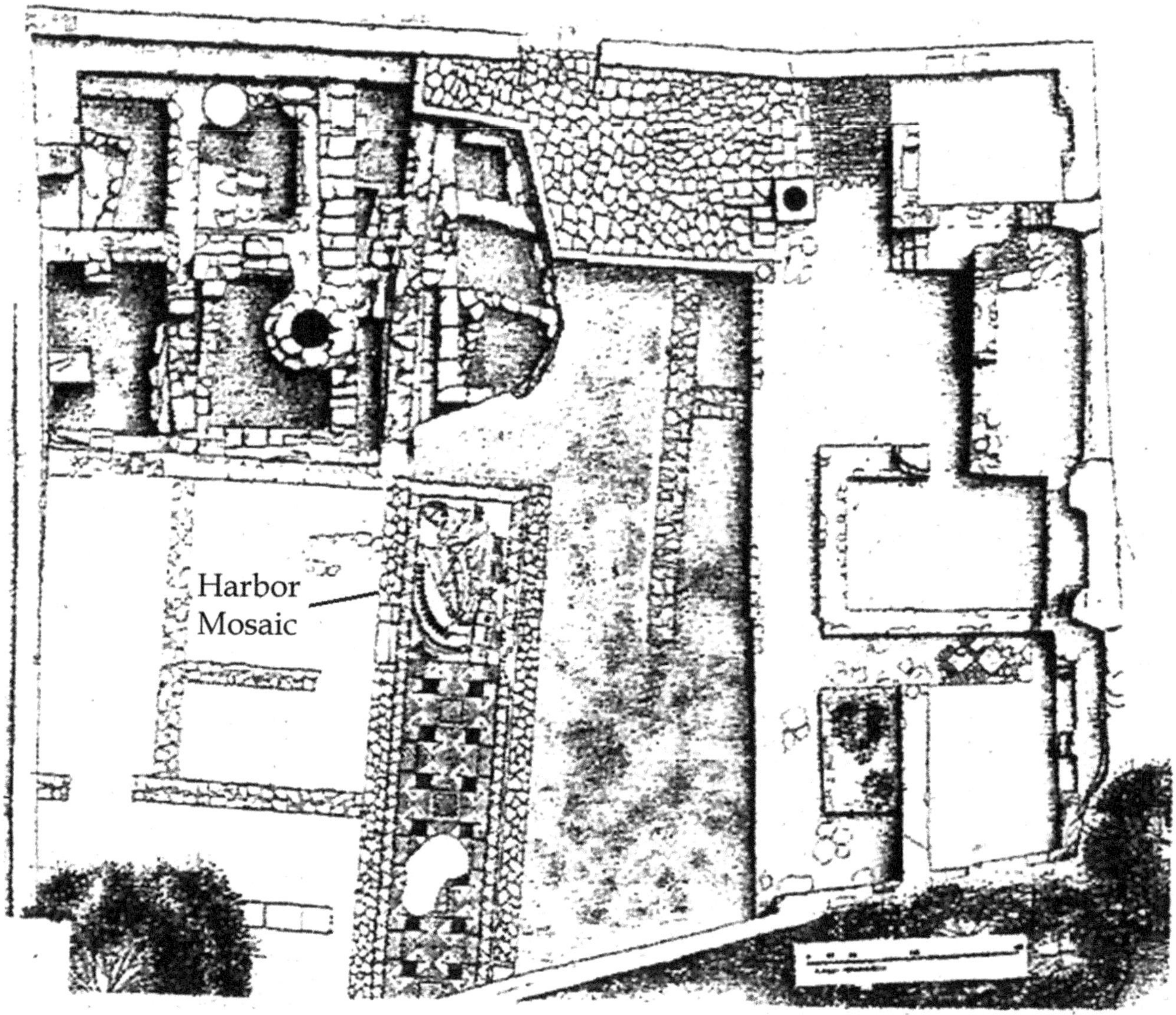

Fig. 3.4.5: Plan of the complex with the mosaic floor

stem. A strip of dark brown tesserae indicates the gunwale. The dark-brown coloring the lower part of the hull may indicate the pitch coating to make the vessel watertight and also protect it from *teredo navalis* (ship worms). The port aft strake seems to widen to form a wing-like aft extension that houses the port steering-oar (figs. 3.4.7, 3.4.7 a). The top edge of the strake suffered some damage. The distorted broad rounded stern is depicted with yellow, dark-brown and white tesserae. The wide sternpost seems to divide the stern into two parts, each with an almost straight board, thus resembling transom boards. The post-like border was probably meant to indicate the edge of the starboard aft-wing extension. The tow or mooring line of the sailing-boat is looped around the tip of the starboard aft- wing while the line of the rowboat is looped around the loom of the port steering-oar.[15] A large rectangular cabin is placed on the mid-deck; it is made with yellowish, dark brown, black, white and pink tesserae. Five bitts, each made with two vertical rows of dark brown-ochre tesserae are depicted on the port fore-gunwale (fig. 3.4.8).

The ship is rigged with a very large fully open quadrilateral sail, attached by brail-fairleads to a long thin yard. The yard is outlined by one row of black tesserae on its upper part, while the lower part is made with light brown stones. Its length appears to equal the length of the ship. The port yardarm is lowered over the fore deck and the starboard one is raised above the stern. The yard is attached to the fore side of the masthead by a *parrel* but misses the lifts (figs. 3.4.7, 3.4.7a). Usually at the tip of the masthead of ancient ships a small flag or banner was attached to indicate the wind direction, or for festive occasions. The tip of the Kelenderis masthead is adorned with a black conical cap with a forward bent tip. Many examples of ship iconography show small banners of flags attached to the tip of the masthead; e.g. several of the vessels depicted on the floor of the shippers' offices at Piazzale delle Corporazioni, at Ostia, Italy.

The tapered vertical mast of the Kelenderis ship is stepped a bit forward, with a slight backward rake. Only the lower part of the pole is visible between the port gunwale and the middle part of the mid-cabin. The rest of the mast seen as

[15] Friedman and Zoroglu, 2006, p. 110.

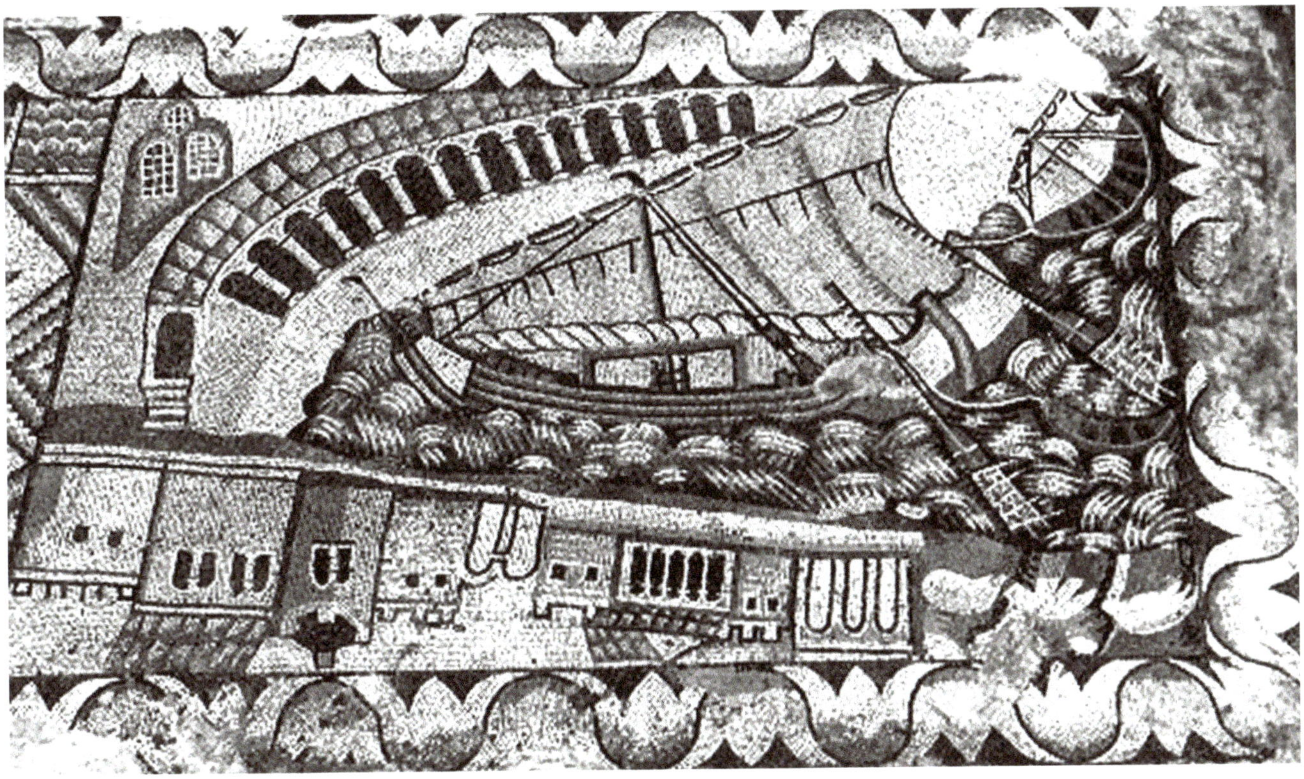

Fig. 3.4.6: The harbor scene

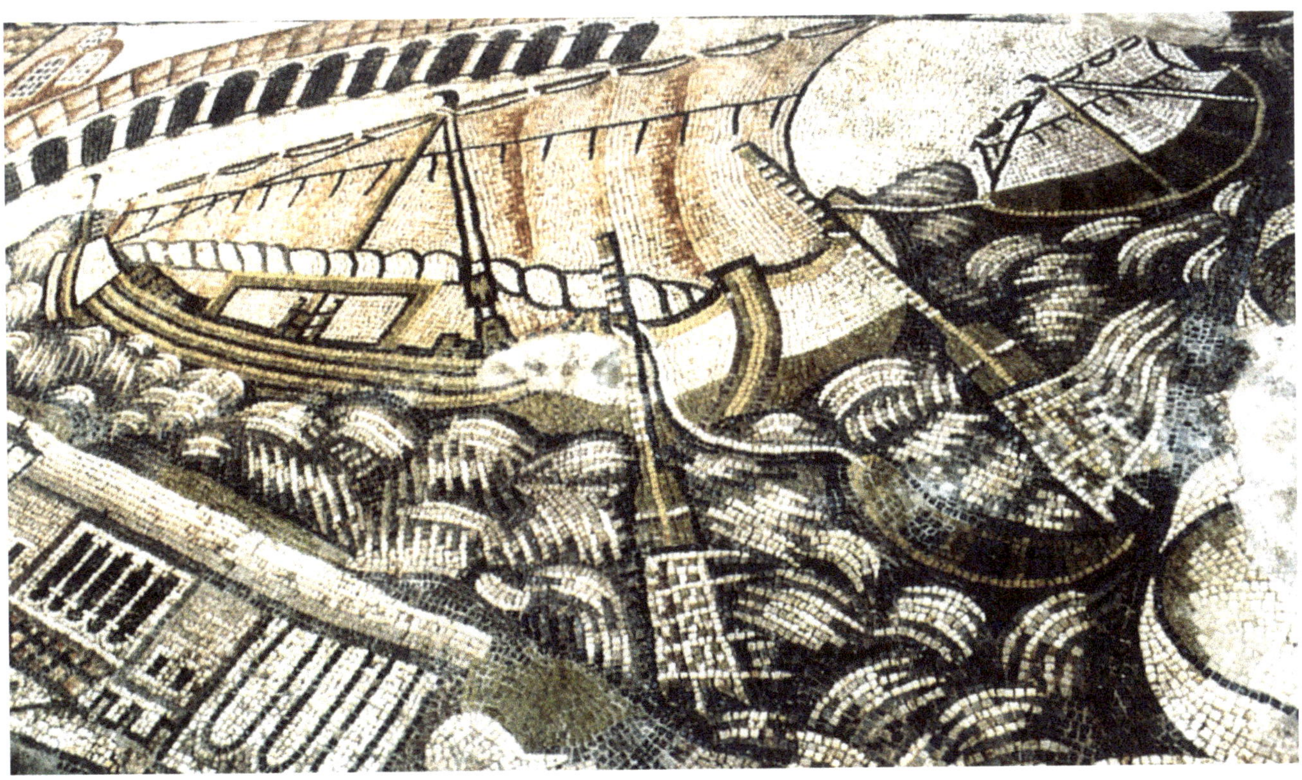

Fig. 3.4.7: The vessels

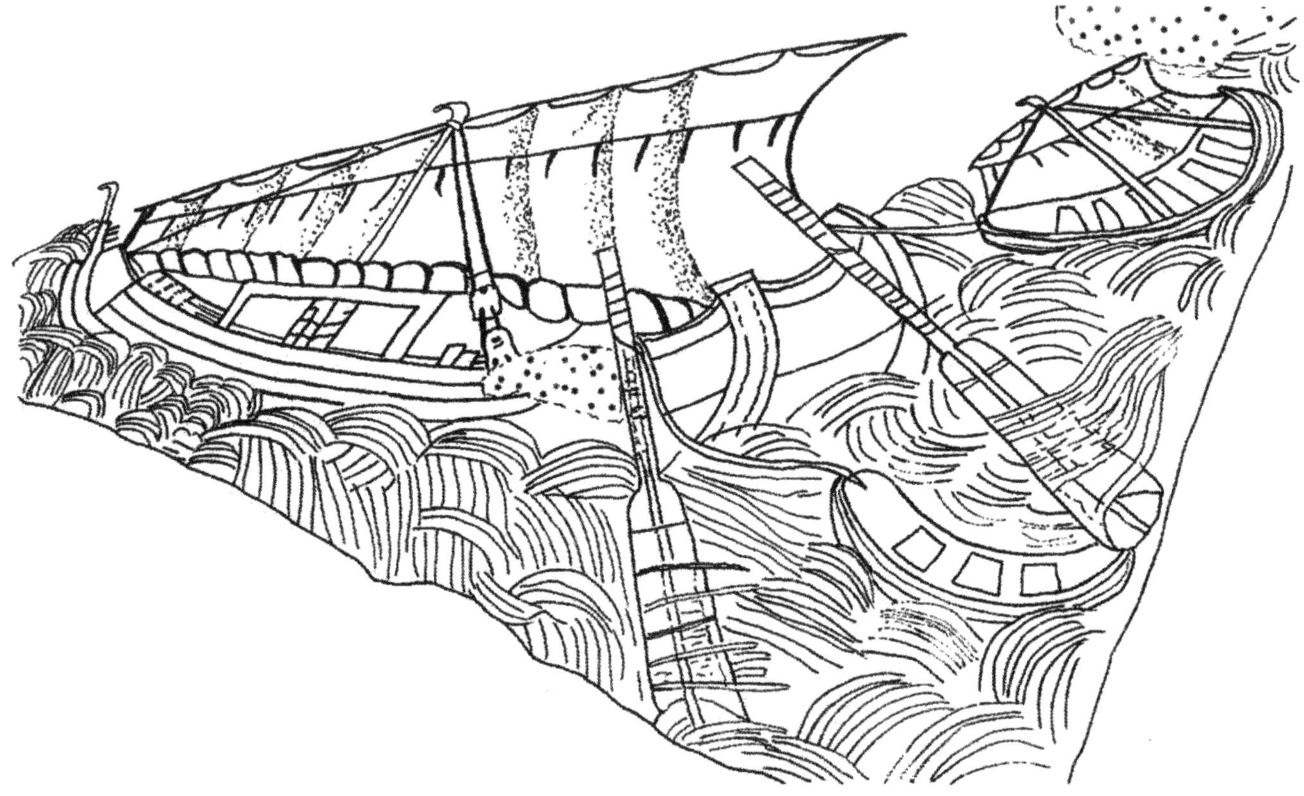

Fig. 3.4.7a: Drawing of the vessels

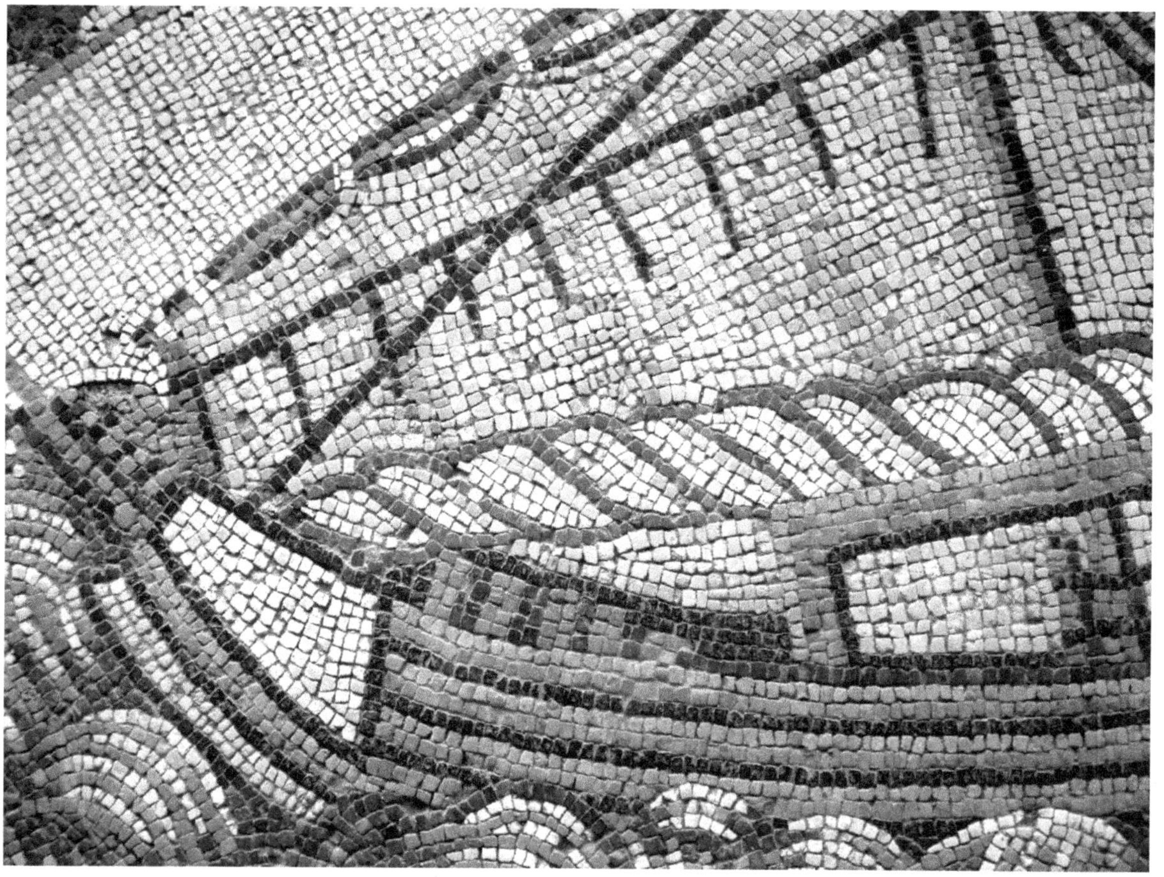

Fig. 3.4.8: Close-up of the bitts

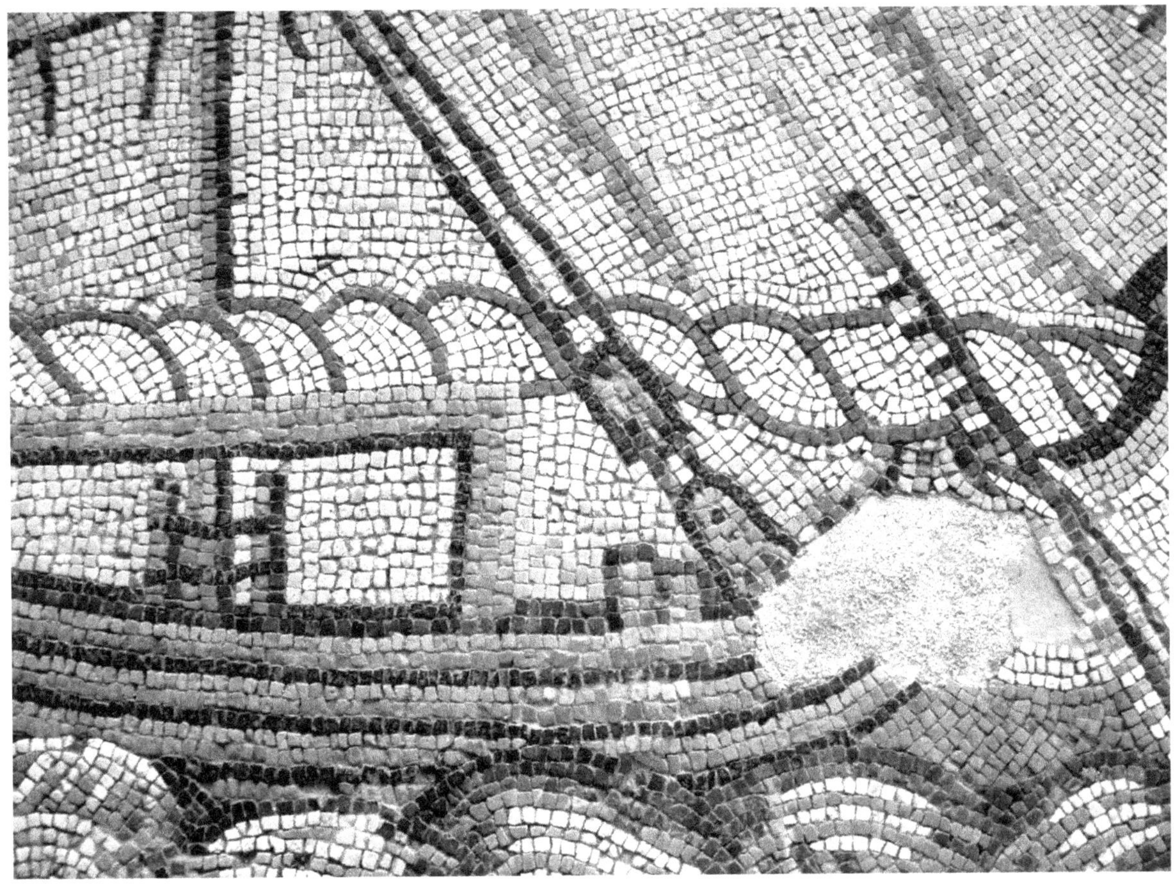

Fig. 3.4.9: Detail of the mast bracing-timbers, the deadeye and the block-sheave

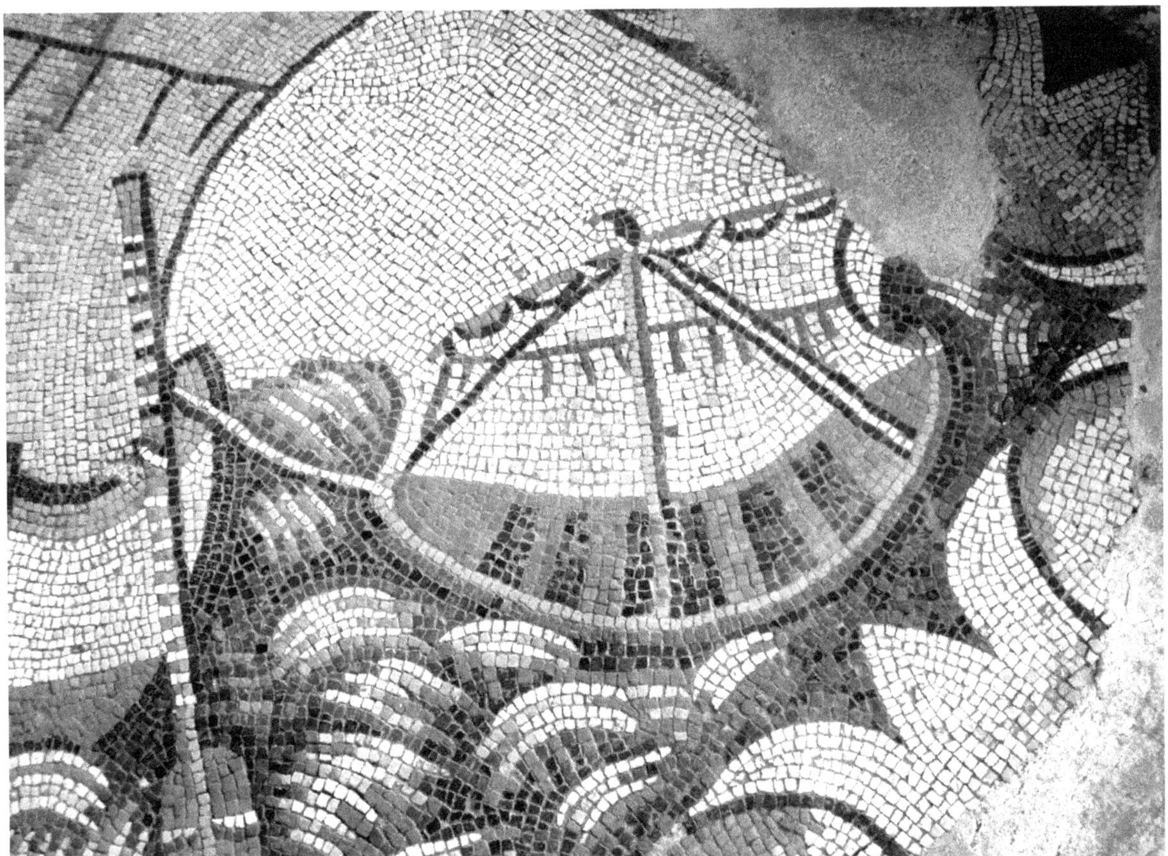

Fig. 3.4.10: Kelenderis sailing-boat

a shade on the lee side of the bunt (a distorted depiction) is rendered by hues of white and light brown tesserae, while the right side is outlined by one row of black stones. On the lower part of the mast there are two black vertical and horizontal lines, probably indicating two lateral stanchions and ropes for the bracing timbers to give extra support to the lower mast (fig. 3.4.9). Two vertical parallel lines made with black tesserae, stretching between the masthead and the port quarter indicates the double-lines backstay. Their lower ends go through two blocks or deadeyes, one attached to the ends of the lines and the second is mounted above the quarter gunwale (fig. 3.4.9). Due to the distorted depiction of the stern, it appears that the backstay represents the shrouds, but the opposite is the case. The angled line stretching from the masthead to the tip of the port stem is the forestay. Both stays form the standing rigging to secure the mast in place.

The quadrilateral sail billowing forward is viewed from its lee face, thus indicating that the wind blows from astern or the port quarter; it has the same width as the yard. The head is attached to the yard by brail-fairleads on top of the spar. The port leech is made with brownish and black stones and the starboard one with black tesserae. The bunt is made with white, pink and light orange hues. Several longitudinal lines of light, reddish-brown tesserae are depicted on the lee face of the sail, probably indicating the shadows of the brails running vertically on the fore face of the bunt. On the upper part of the sail, beneath the yard, is depicted a long black angled line with sixteen short vertical ones attached beneath it. This line indicates a reef-band with reef-points sown on the lee face of the sail. The seventeen short black reef-points were used to hoist the sail when it was furled to the yard. Such a depiction is a rare representation in any art.

The steering gear consists of two large steering oars, one mounted on either quarter. The right side of both shafts is outlined with a row of black tesserae while the field is made with dark brown stones. The shaft of the port-oar seems to go behind the aft-wing extension, while starboard shaft appears to be laid against the inner side of the starboard aft-wing (figs. 3.4.7, 3.4.7a). Their lower ends transverse the blades longitudinally, thus indicating that the blade comprises two wings probably inserted into a groove cut in the lower shaft. The elongated blades have rounded shoulders, while their lower ends angle towards the shafts. Beneath the shoulder of both blades is depicted a single horizontal white strip, probably indicating some reinforcement, or a decoration resembling the fashion on the oars of the Pharaonic Egyptian ships since the Middle Kingdom. Undulating white strips crossing the face of the submerged blades indicate a rough sea. The white single tesserae depicted within these white strips may indicate the foam of the waves breaking on the submerged oars.

The Boats

Astern of the large sailing ship two small boats are towed or moored (figs. 3.4.7, 3.4.7a). The upper craft is a sailing-boat, as indicated by its fully open quadrilateral sail, while the lower one is a rowboat missing its oars. A tow or mooring line secures each boat to the stern of the sailing ship. The sailing-boat is seen from above in bird eye's-view, while its open sail is viewed from the lee side. The elongated banana shaped hull has a pointed stem and a rounded raised stern. The gunwale is outlined with one row of light yellowish tesserae. The fore and aft ends of the vessel are partly decked at the gunwale level. Thwarts rendered with very dark brown tesserae are set above the gunwale. Due to the top view of the hull, not much is visible of its sides. The top strake, visible on the port side, is depicted with dark reddish-brown tesserae and the lower strake with black stones (fig. 3.4.10), which may indicate the pitch or bitumen coating to make the hull watertight and also protect it from the *teredo navalis* (ship worms). The sailing rig is very similar to that of the large sailing ship, but smaller. It comprises a fully open square sail viewed from the lee side. The rounded vertical mast stepped amidships is made with light and dark brown tesserae; it is secured by the forestay stretching from the masthead to the tip of the bow, and double-lines backstay, stretching from beneath the yard to the port quarter gunwale (fig. 3.4.10). The tip of the masthead projecting above yard seems to be covered by a black conical cap with a forward bent rounded tip similar to the masthead decoration of the large sailing-ship. A slightly downward curving yard is secured by a *parrel* indicated by the white loop surrounding the masthead just above the yard (fig. 3.4.10). The lifts are missing. The length of the yard and the hull are almost the same. The tip of the starboard yardarm and the upper corner of the sail are damaged. The sail is made with white and light brown tesserae. The head of the sail is attached to the yard by brail-fairleads. The port leech is outlined with black tesserae. The tack of the starboard sheet, pulled backwards, seems to coil around the tip of the sternpost (fig. 3.4.10). The foot of the sail is hidden by the starboard gunwale, due to the billowing of the sail on that side. This depiction symbolically indicates that the wind blows from port quarter. The boat is anchored, although that it does not show any such device, deduced from its static position. On the upper part of the sail is depicted a slightly angled black line with nine short vertical reef-points attached beneath it and similar to the sail of the sailing ship (figs. 3.4.7, 3.4.7a). These reef-points were used to hoist the furled sail to the lowered yard.

The tow or mooring line of the small rowboat is looped around the upper end of the port oar's loom (fig. 3.4.7). The hull is very similar to that of the sailing-boat. The boat is seen from above in a bird-eye view. It appears that the rounded end of this boat, facing the stern of the sailing ship, may indicate an angled transom bow, while the pointed end is the stern. The port gunwale is outlined with one row of dark ochre tesserae. The bow and the stern are partly decked above the gunwale. There are two thwarts in the boat, placed across the hull. The part-decks and the thwarts are made with dark brown stones. This boat is slightly shorter than the sailing-boat. Both boats appear to be the ship's boats towed astern, whereas they lack their steering and rowing gear. The water within the harbor is not calm, as indicated by the short curving strips depicted in a crisscross

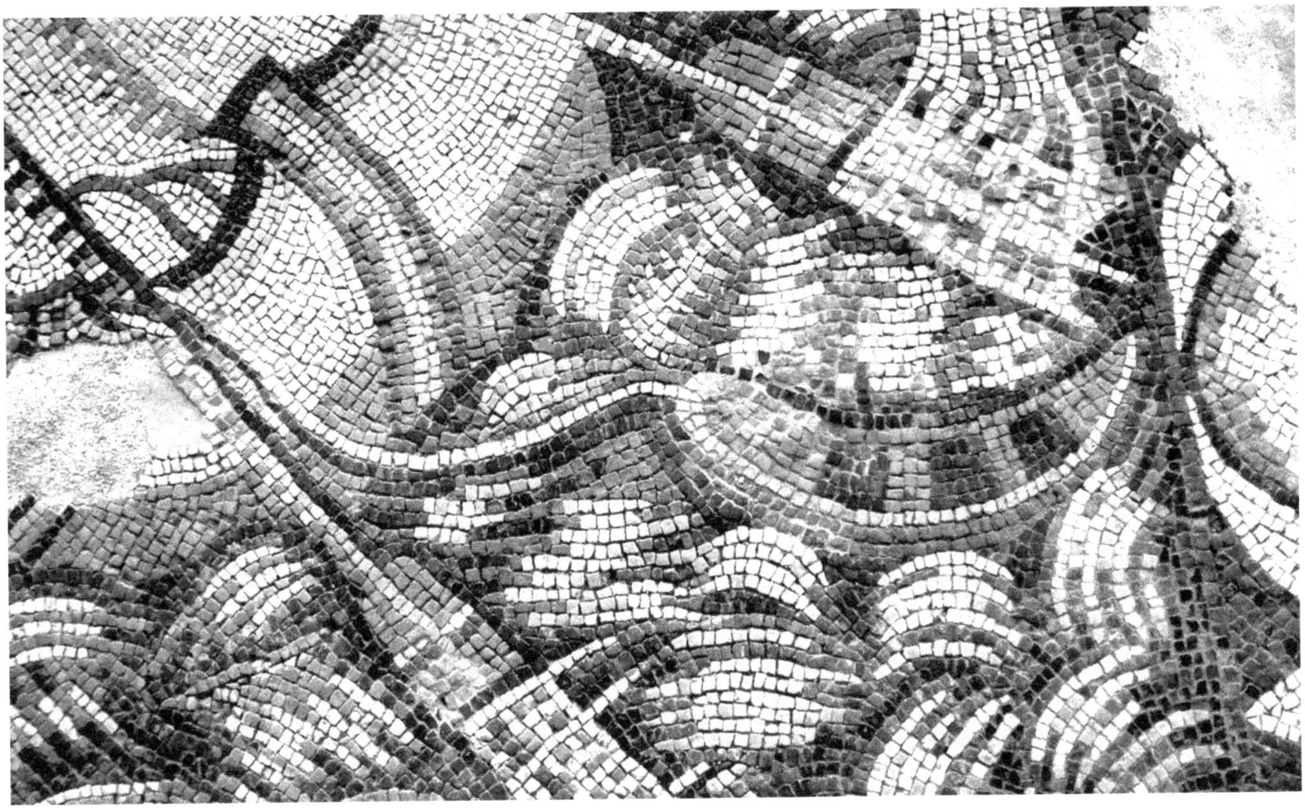

Fig. 3.4.11: Kelenderis rowboat

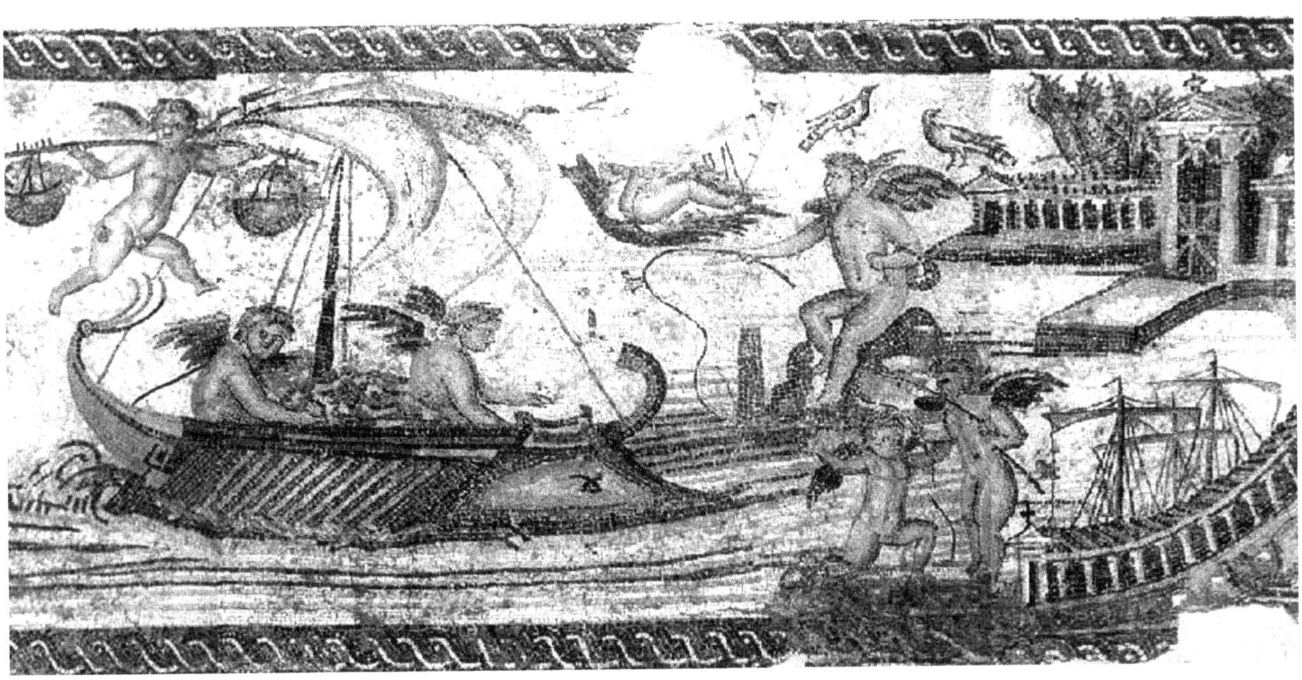

Fig. 3.4.12: Harbor scene in the Nile Villa at Leptis Magna

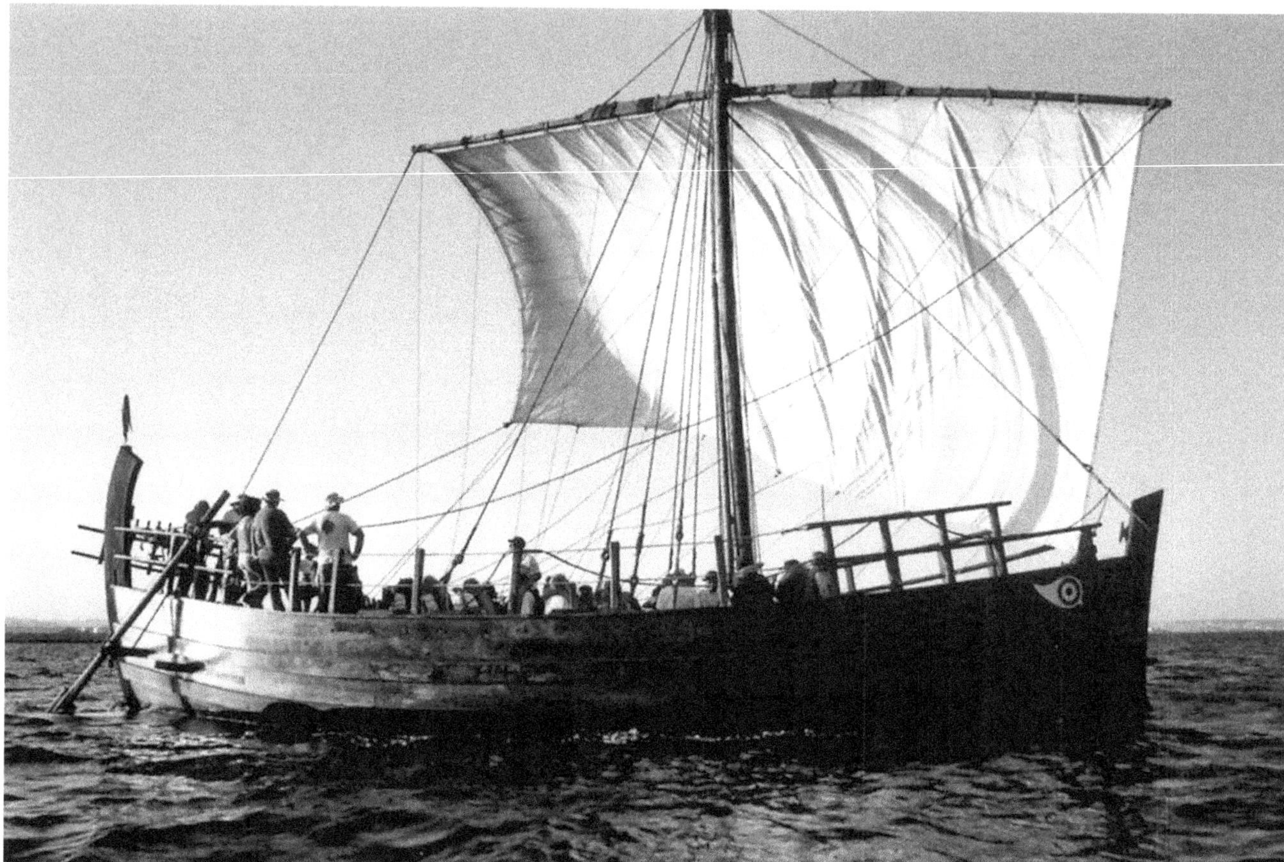

Fig. 2.4.13: Kyrenia Liberty trial sailing

pattern. The white strips crossing the blades of the rudders also emphasize a rough sea.

Discussion

The Kelenderis harbor panel described above is a rare representation of such scenes, especially in mosaics from Turkey. Depictions of harbor scenes with waterfront colonnades surmounted by pitched tile roofs and boats engaged in fishing activities were common decorations in Nilotic scenes. They were popular themes on wall paintings at Pompeii and the area around the Bay of Naples. One of the earliest depictions of a harbor scene in mosaics with an arching colonnade following the shape of the shore is found in the mosaic floor of the *tepidarium* (2.93 x 1.2m), in the Nile Villa at Leptis Magna, dated to the 2nd century CE (fig. 3.4.12). In this depiction, the harbor is seen from behind the colonnade, as looking from inter-land towards the sea. Within the harbor, facing the inner colonnade there are three masts, probably indicating three ships anchored along-side the quay. Only the top part of the masts and yards are visible above the roof of the outer colonnade. The sails are furled beneath the yard and the masts are still stepped in their vertical position. The standing rig of each vessel comprises the fore-and-back stays and the shrouds. The yards are attached to the yards by lifts stretching from each masthead to the tips of the yardarms. The braces or sheets of each sail hang vertically to the deck. At the tip of two mastheads is attached a small flag that indicate the wind direction. The maritime scene is inhabited with many putti engaged in fishing activities, and two putti in the larger sailing ship with a projecting cutwater sail into the harbor (fig. 3.4.12). The forward billowing sail appears to be made of two sections whereas the forestay stretching from the masthead to the stempost passes between these sections. The sailing rig also comprises the triangular topsail set above the yard. The figures in the vessel are much larger and are not proportional to the harbor structures and the anchored ships facing the inner colonnade. This depiction may suggest that the mosaicist emphasized the focal point of the scene on this ship.

This mosaic dated to the 2nd century CE was discovered between 1916 and 1939.[16] Other examples, illustrating a harbor scene with waterfront colonnade and adjacent buildings are found in the maritime panels of glass *opus sectile* from Kenchreai, in Greece[17] (figs. 3.5.5 a, 3.5.5b, 3.5.6a, 3.5.6b), in Room 29[18] (fig. 3.8.7) and in the semi-circular atrium (fig. 3.8.13), at Piazza Armerina, Sicily. The waterfront colonnades and some of the structures in both

[16] Aurigemma, 1969, p. 48.

[17] Ibrahim L., Scranton R., Brill R., 1976, figs. XVI, XVII, XVII and XIX. The panels dated to 350 – 375 CE, were found underwater in the building where they were meant to decorate the walls. A severe earthquake struck the site during the restoration of the building and the great destruction led to its abandonment.

[18] Carandini, Ricci, de Vos, 1982, pl. XXXVII.

Kelenderis and Kenchreai mosaics have some similarities. To give some depth to the Kelenderis harbor scene, the mosaicist made some attempts at foreshortening that resulted in quite distorted representations. The distortion is emphasized by the frontal view of the prow seen almost in the same plane with the port hull and the broad stern. The misinterpretation of the starboard aft-wing edge and the distorted representation of the full stern resulted in a depiction that any viewer would see it as a transom stern between two sternposts.

The large sailing ship is depicted with an intricate sail (figs. 3.4.7, 3.4.7a). The forward billowing bunt may indicate that the wind is blowing from astern or from the port quarter. The shortened port side and the fully open starboard side of the large sail apparently points to the mosaicist's intention to show the scene in perspective. The pronounced lower rake of the port yardarm to the bow may also suggest a lateen sail, but the opposite is the case. The angled reef-band with the sixteen reef-points is thought to be a lateen rig, especially as not much is known about this type of rigging in the 5th – 6th century CE.[19] The sail of the Kelenderis large sailing-ship is an elongated quadrilateral bunt apparently set diagonally to the hull (indicating a fore-and-aft sail) and not parallel as would be expected in a vessel rigged with a lateen sail; the starboard sheet stretching backward is attached to the starboard quarter gunwale which also indicates a square sail rigging. The angled reef-band was intentionally made by the mosaicist thus to emphasis the perspective of the scene and giving a three-dimensional appearance to the billowing sail. The free foot of sail is in the process of furling whereas the reef-points already hoist it. Bowen LeBaron mentioned that when dipping the lug with short luff, the tack is set forward and then the sail appears to be triangular from a distance. Such a sail could be copied by any artist as being almost triangular, thus an indicative of a lateen sail.[20] The sail of the Kelenderis Ship apparently is similar to Bowen LeBaron's observation. Such an image is augmented by the Kyrenia Liberty's trail sailing in the harbor of Larnaka, Cyprus (fig. 3.4.13). In square rig the mast is generally stepped amidships. The Kelenderis ship shows clearly the function of the standing rigging and the bracing timber at the lower end of the mast. The mast-partners (bracing-timbers) of the Kelenderis ship is confirmed by the Wreck D of the Black Sea Ship, which still carries the intact vertical mast stepped amidships (fig. 3.4.14). Near the lower part of the mast was revealed the upper part of the bracing-timber, with probably some rope coiled around.[21]

The static view of the Kelenderis ship suggests that the vessel is anchored rather than sailing, although no anchor or mooring line is shown. Apparently the vessel is anchored in the outer harbor's basin where the water was deep enough to accommodate the ship's draught. This observation is also emphasized by both steering oars hanging freely on either quarter with a backward rake; in shallow waters the steering oars would break and therefore they had to be lifted or removed. Both sailing-and-rowing boats towed astern the sailing ship indicate their function, ordinarily used for communicating with the shore when the large vessel is anchored off shore.[22] They were probably used to transport passengers from the large ship to the shore or vice-versa. The thwarts in both vessels indicate that only passengers could be transported in them with their personal belongings. Carrying cargo in such vessels was inefficient because the thwarts limited the hold capacity of the boats. The harbor scene and the vessels are shown in a combined lower perspective and bird eye's view. The fore-shortening technique used by the mosaicist, showing the bow close to the quay, is inaccurate because it would be impossible for such a ship to anchor so close to the quay; the water would be to shallow for its draught.

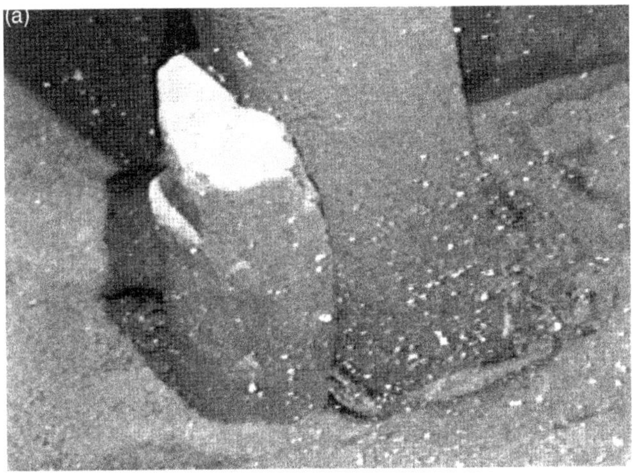

Fig. 3.4.14: Mast bracing-timber; Wreck D, Black Sea Shipwreck

There are many ancient written sources referring to large merchant vessels towing a small boat astern. Apparently the Kelenderis merchantman could carry cargo and passengers as well. The large cabin on the mid-deck could accommodate passengers, the crew and some cargo as well. The sailing ship may be associated with a more specific type known as *phasolos*, especially suited to carry travelers as well as cargo.[23] They were in use in the Mediterranean in the 1st century BCE. The Kelenderis mosaic was dated to the 5th - 6th century CE. There is no indication or any written references of *phasoloi* ships used in later periods. Traditional construction techniques of ancient ships were maintained for very long periods without major changes. We may assume that the Kelenderis ship represents a *phasolos* type that continued to be built and used in the 5th – 6th centuries CE in the eastern Mediterranean.

19 Friedman and Zoroglu, p. 114, p. 115 and n. 1.
20 Bowen LeBaron, 1956, p. 241.
21 Friedman and Zorogle, p. 111, fig. 5; Ward and Ballard, 2004, fig. 12a, p. 10.

22 Torr, 1964, p. 103 and n. 226.
23 Ibid., p. 120.

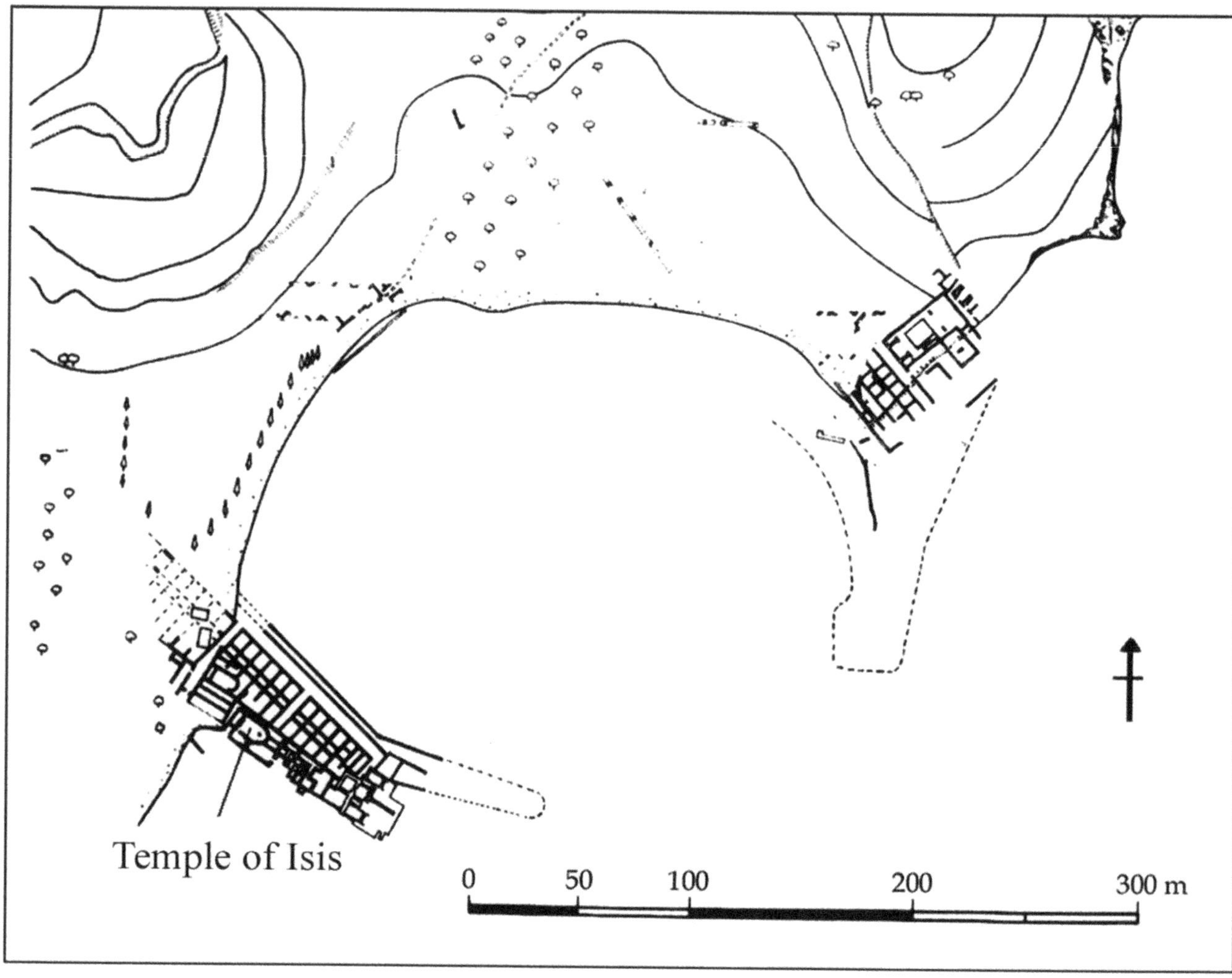

Fig. 3.5.1: Location map of Kenchreai

3.5 The maritime panels of glass *Opus Sectile* from Kenchreai - Greece

Location and Excavations

Kenchreai is situated on the NW coast of the Saronic Golf of the Corinth Isthmus, and formed the eastern port of the Corinth in antiquity. The topography of Kenchreai comprises an alluvial plain facing a straight beach about 500m long.[1] Along the north side of the valley there is a steep ridge, which extends eastward beyond the beach. On this ridge are the remains of the ancient moles that formed a small harbor of 250m in diameter (fig. 3.5.1). Today, a small fishing village with the same name uses the bay that provides good sheltering and anchorage for fishing boats and sailing yachts in the Aegean Sea. Lechaeon, on the Golf of Corinth was the western port of the ancient Corinth. With these two ports, the ancient cosmopolitan metropolis of Corinth was able to control the east-west traffic by sea, while the north-south traffic was controlled by land. Kenchreai is mentioned by few ancient writers. In Book 11 of Apuleius's novel "*The Golden Ass*" is referred that Lucius' transformation from man to ass occurred at Kenchreai during the Isis festival.[2] Pausanias in the "*Description of Greece*" wrote that Kenchreai as being the port of Corinth derives its name from Cenchrias, one of Poseidon and Preirene children. The second harbor was named Leches, after the second son of the gods:

> *"The name of the Corinthian harbors were given them by Leches and Cenchrias, said to be the children of Poseidon and Preirene ... "*[3]

The port of Kenchreai provided a communicating seaway with the Aegean, the Black Sea, the eastern Mediterranean and the countries that bordered them. The site is hardly mentioned in Greek literature, especially when related to the armies that controlled the Corinth Isthmus.[4] In the New Testament, Kenchreai is mentioned twice in relation to St. Paul: once when speaking of Phoebe,[5] who was the bearer of his letter to the Church of Rome, and the second

[1] Scranton, 1976, p. 446.

[2] Hohlfelder, 1976, p. 219.

[3] Pausanias, II.II.3.

[4] Scranton, 1967, p. 163.

[5] She was a deaconess or church worker in the port of *Cenchreae*; Comay and Brownrigg, 1980, p. 366.

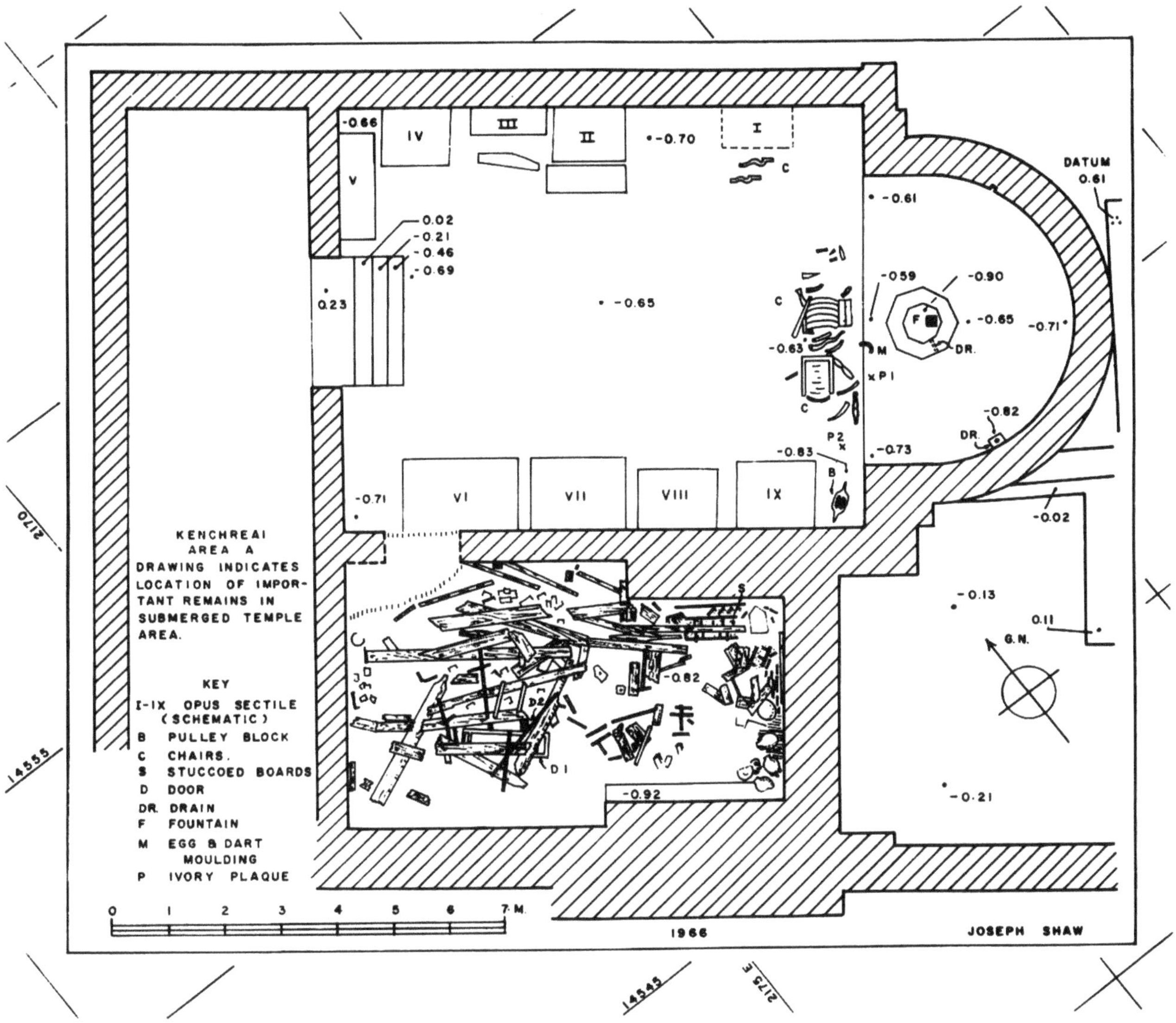

Fig. 3.5.2: The Isis Complex with the location of the glass opus sectile panels

time, while he was pursuing a mission in Greek lands.[6] Specifically the name or pictorial representations of the Kenchreai appears on coins issued under Nero, Hadrian and Antonius Pius.[7] The port depicted on the reverse of one of Antonius Pius coins, apparently is the pictorial representation of Pausanias's description of Kenchreai:

> *"... on the road leading from the Isthmus to Cenchreae there is a temple and an ancient wooden image of Artemis. In Cenchreae are a temple and a stone statue of Aphrodite, after it on the mole running into the sea a bronze image of Poseidon and at the other end of the harbor sanctuaries of Asclepius and of Isis".*[8]

The description given by Pausanias seems to suggest that the port of Kenchreai reached the height of its development by the reign of Antonius Pius.[9] This statement is supported by the coins series struck at Corinth by the same emperor, and depicting the harbor of Kenchreai.

Although, Kenchreai is mentioned in ancient literature, little was known about this site prior to the excavations conducted under the aegis of the American School of Classical Studies at Athens and carried out by the Universities of Chicago and Indiana. Occasional exposures of artifacts led to the investigation of the site in the summer of 1963.[10] Major excavations were undertaken on the submerged structures and those near the waterline, in 1964. Trial trenches were carried out at various points along the waterfront and on the submerged moles. These trials indicated that the Roman harbor was constructed in a natural bay. The protection from storms and prevailing sea conditions were afforded by two promontories and augmented by the artificially

[6] Scranton, p. 163.
[7] Hohlfelder, p. 223.
[8] Pausanias, II.II.3.
[9] Hohlfelder, p. 222.
[10] The excavations were carried out between 1963 and 1966; Scranton and Ramages, 1967, p. 124.

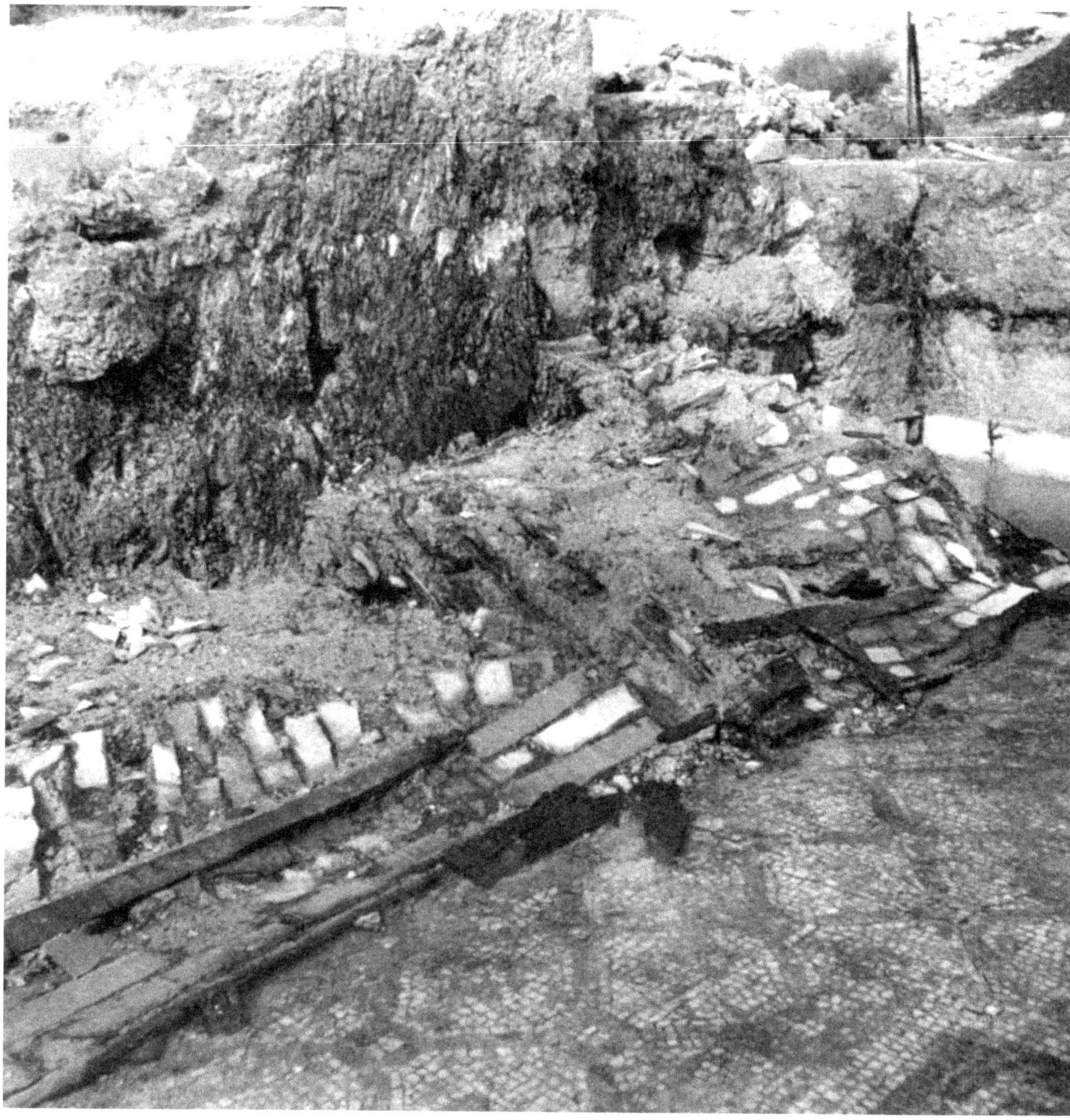

Fig. 3.5.3: S-Wall with the damaged panels as they were found during the excavations

construction of two large breakwaters (fig. 3.5.1).[11] The southwest breakwater comprised a pier area with numerous structures. It projected into the sea for a distance of 100 to 150 meters. The northeastern breakwater served as an extension of the promontory formed by the headland. It ran into the sea in a southward direction for about 110 meters. The breakwaters were linked by seawalls, which formed the base of the quays and roads. Adjacent to these seawalls various structures comprised the harbor offices and storages. The structure known as the early Christian Church, at the SW end of the harbor was largely excavated, as well as the Roman buildings found in the area.[12] The building in which unique panels of glass *opus sectile* (118 – 126) were discovered comprised an apsidal structure with a small octagonal fountain, a rectangular room paved with simple geometric mosaics and an adjacent cellar (fig. 3.5.2). These panels embedded a unique "cultural time capsule". The shipping wooden crates containing the glass *opus sectile* panels were laid against the walls in the room where these panels were meant to decorate the walls (fig. 3.5.3). From 1965 until early 1970's efforts were made to salvage and preserve all the panels. Four of these panels depict rare and almost complete harbor scenes with ships sailing in or out of the harbors. The uniqueness of these four maritime panels is invaluable for the way of combining both Roman ships and harbor iconography along with the

[11] Hohlfelder, p. 221.
[12] Scranton, 1967, p. 164.

intricate technique of glass *opus sectile*. They also bring evidence of cultural heritage and diffusion of ships and harbor construction in antiquity.

The structure with the glass *opus sectile* panels was associated with the temple of Isis, as mentioned by Pausanias in his "*Description of Greece*". The building is submerged and originally it occupied an area of 20 x 15m, with a series of adjacent rectangular rooms. The apsidal unit consists of a rectangular space, 7.7m long, 9.9m wide, and the apse slightly larger than a semi-circle is 5.2m wide (fig. 3.5.2). The walls were built of local sandstone (*kurkar*) blocks 0.6m thick; some in secondary use. In most part they are preserved to a height of 1.2m (one or two courses) above the floor.[13] The floor was paved with geometric mosaics. Now it lies at 0.75m below sea level. Apparently, this structure was in the process of remodeling or reconstruction when the entire area was hit by earthquake in 365 CE and then again in 375 CE. These earthquakes had a violent impact on the site and caused extensive damages. The epicenter apparently laid in the Aegean Arc, whereas the following tidal waves (*tsunami*) devastated the Kenchreai harbor and the nearby settlements, which also had a great impact on the Eastern Mediterranean. Probably shortly after this event, the floor of the court and the temple sank to sea level or below, together with the crates containing the glass *opus sectile* panels. It seems that immediately after the earthquakes the site was abandoned and the building filled in with rubble and debris that partly covered the stacks of the crates and thus preserved them. The final destruction apparently occurred at the end of the 4th century and afterwards the whole site fell into ruin. It seems that the Avar invaders completed the final destruction of Kenchreai at the end of the 6th century.[14] The excavated land and submerged ruins attest that the port functioning as an important harbor in classical periods, at least until the end of the 4th century CE. There are several indications of the rise and fall of land that occurred during the long history of Kenchreai. It is clear today that the present shore is c.2m lower in relation to the sea than it was at the beginning of the 1st century CE.[15]

Some of the ancient literature describing the "universal" earthquakes and followed by *tsunami* waves, in the 4th century CE, are augmented by several seismic events in the Eastern Mediterranean (Crete, Sicily, Libya, Nile Delta, Cyprus, Antioch and the Levant coats). Few ancient authors, e.g. Ammianus Marcellinus (26.10.15-19), Jerome/ St. Hieronymous (*Chron.* 244c; PL27.693-694), Athanasios of Alexandria (*Life of Athansius*, PG 25, *ccx*), Libanius (*Or.* 18.292) and Zosimus (5.6.2), recorded such earthquake events. Recent geological studies indicate a cluster of seismic activities around the Eastern Mediterranean occurring between the middle of the 4th and the middle of the 6th centuries CE, with severe impact on the region. One comparable site that suffered from the same earthquake event as Kenchreai (365 CE) is the city of Kourion on the southern coast of Cyprus. The Eustolios Complex on the Akropolis at Kourion, with its rich mosaics and *opus sectile* floors was sternly damaged.[16]

The Glass Opus Sectile Panels

In the central room of the Isis temple were found dozens of shipping wooden crates which contained the most unique panels of glass *opus sectile*. The archeologists deduced that there might have been 59 to 63 complete crates of which only 50 were preserved in a fair state, and about 9 – 13 crates with the panels were entirely lost.[17] In some of the preserved crates it was discovered that two panels were set face to face. It was therefore deduced that originally there were between 118 to 126 panels and among the lost ones were 18 – 26 panels.[18] The crates were stored on the floor of the apsidal room in nine stacks (fig. 3.5.2). In each stack, the first crate leaned against the wall. The rest of the stacks were arranged similarly, each crate leaning on the preceding one.[19] The crates that were closer to vertical position were preserved better than those, which had slumped (fig. 3.5.3). In the 50 surviving crates, it was found that only one whole panel was preserved and the second survived only fragmentarily or was completely lost.[20] It was assumed that both panels in each crate were identical or similar in decoration. The surviving pieces of wood from the crates permitted to reconstruct their original construction (fig. 3.5.4).

The *opus sectile* decorations were mainly made of flat pieces of glass with varying thickness from 3 to 4 mm. Glass rods were used to indicate the ropes rigging of the ships, parts of birds and plants in the Nilotic panels. Many glass sheets weathered very badly and therefore the thickness varies. Only few of them preserved their original thickness of 2 – 5 mm.[21] In the process of the consolidation, the panels were impregnated with *polyvinyl acetate* (PVA), which preserved the shape of the individual pieces.[22] When the glass *opus sectile* panels were originally made, they were set in trays, which were backed with potshards and glued with a resin mixture acting as a binding agent for the mortar, the shards and the glass.[23] It seems that the top surface of the adhesion was a mud-resin component.[24] Originally the glass appeared in hues of blue-green, turquoise, purple, red, yellow, greenish-white, etc. Due to the seawater, algae and other marine aggregates that have had an effect on the panels, most of the glass lost its original coloring, but where it has survived its remain was a grayish-white, with very

[13] Scranton and Ramages, p. 140.
[14] Scranton, p. 163.
[15] Ibid.

[16] David W. Rupp: "*The Polychrome Floor Mosaics, Opus Sectile Floors and Architecture of the Eustolios Complex at Kourion, Cyprus*"; Shelby White – Leon Levy Grantee 2004 - 2005.
[17] Ibrahim, Scranton and Brill, 1976, p. 11.
[18] Ibid.
[19] Scranton, 1967, p. 166.
[20] Ibrahim, Scranton and Brill, p. 11
[21] Ibrahim, 1976, p. 17.
[22] Ibid., p. 16.
[23] Ibid., p. 17.
[24] Ibid., p. 18.

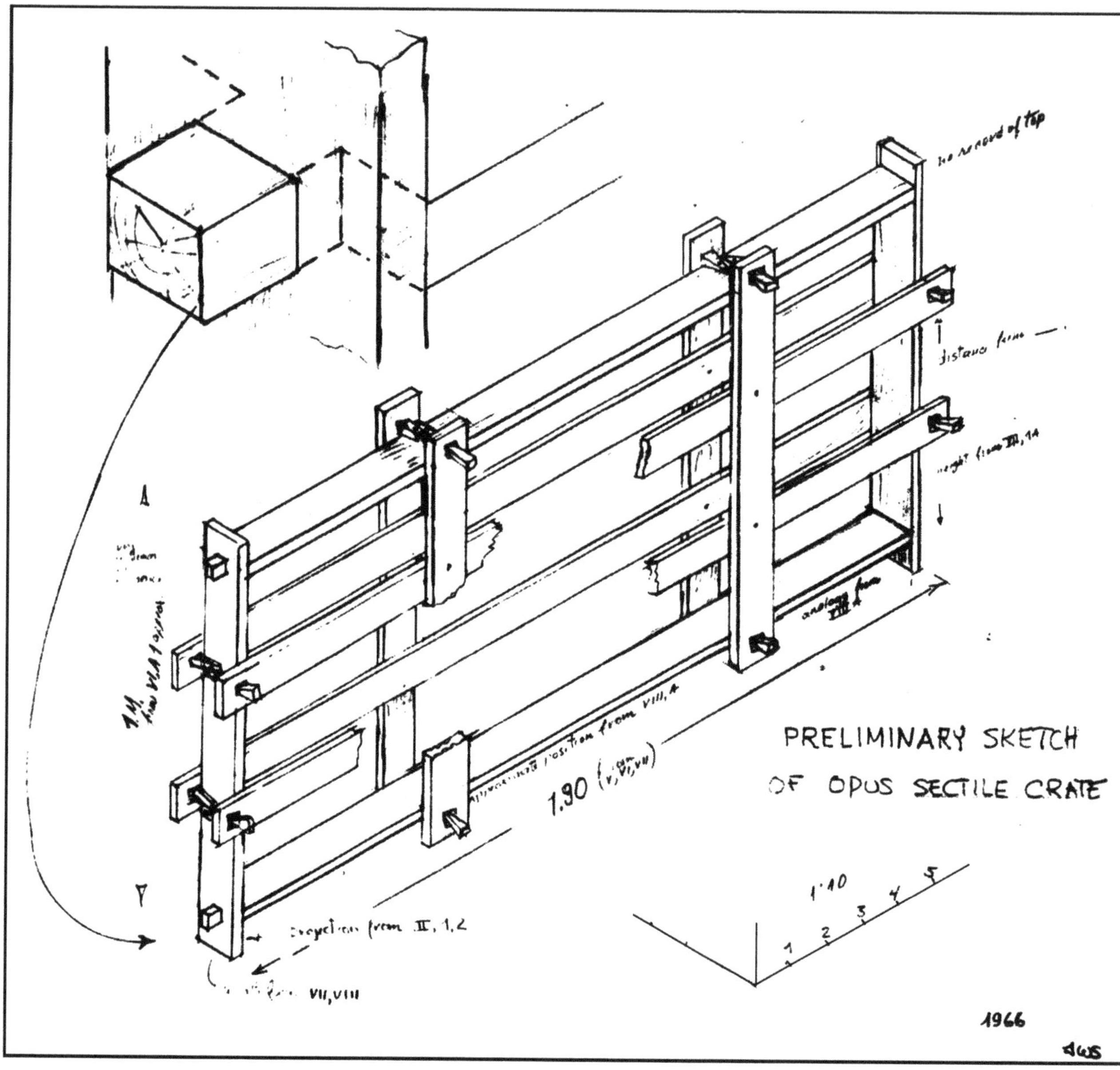

Fig. 3.5.4: Schematic reconstruction of a shipping wooden crate

few weathered blue, some yellow and very rare red, purple and pink.

The Maritime Panels

The panels recovered during the excavations may be classified by the subject matter. This classification will not be discussed here because it is not relevant to the theme of the present study; only the maritime panels will be described below. The overall sizes of the panels were 0.80 – 0.85m in width and 1.85 – 1.90m in length.[25] There were twelve panels depicting Nilotic scenes, which have survived nearly completely or in fragments. The crates deposited in the Isis temple room were laid in stacks against the wall, apparently placed according to their location on the wall decoration above; the archeologists numbered them with Roman letters from I to IX in counter-clockwise order (fig. 3.5.2). The descriptive panels comprise of buildings along a coast with ships sailing inside or outside the harbor, swimming fish in the water and men fishing with a line or net. The fishermen are not proportional or in scale with the buildings, some are smaller and others taller than the structures. They are not placed in the town but peripheral to it, as at the entrance of the harbor or outside the port area (fig. 3.5.5a).[26] The smaller figures follow the Hellenistic tradition of realistic impression, while the larger ones are typical of late Roman tradition in which the figures in the landscape call for the viewers' attention.[27]

Four of the surviving panels depict complex maritime scenes comprising waterfront buildings of porticoes with

[25] Ibid., p. 58.

[26] Ibrahim, Scranton and Brill, Pls. XX B, XXI, XX A.

[27] Ibrahim, p. 58.

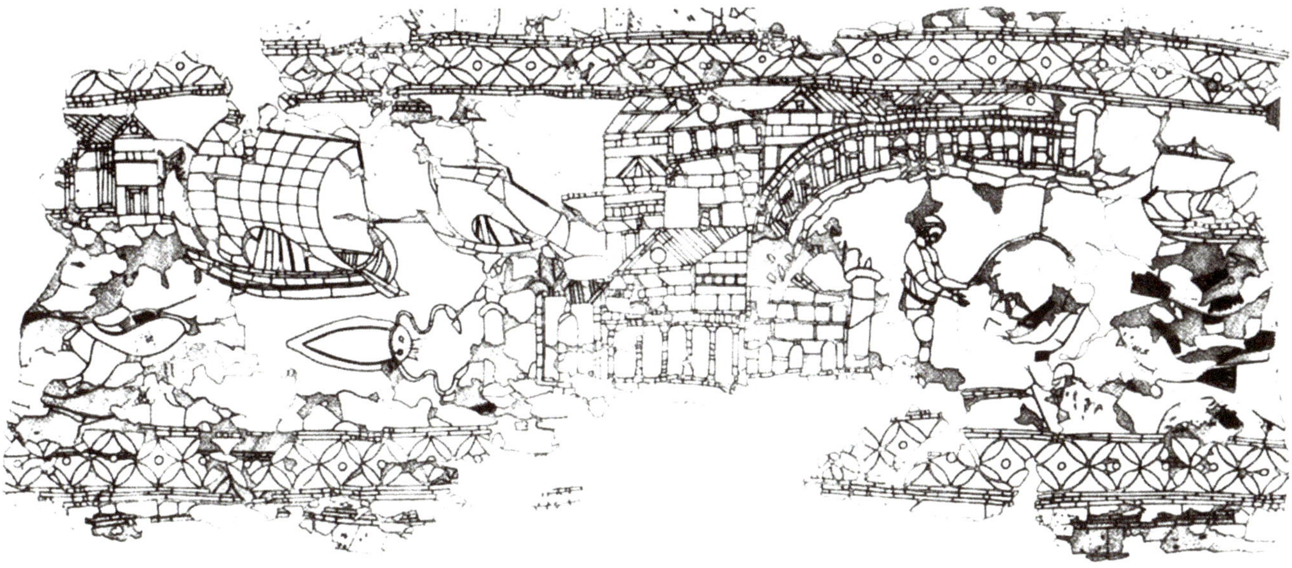

Fig. 3.5.5a: Harbor Scene I and the ships

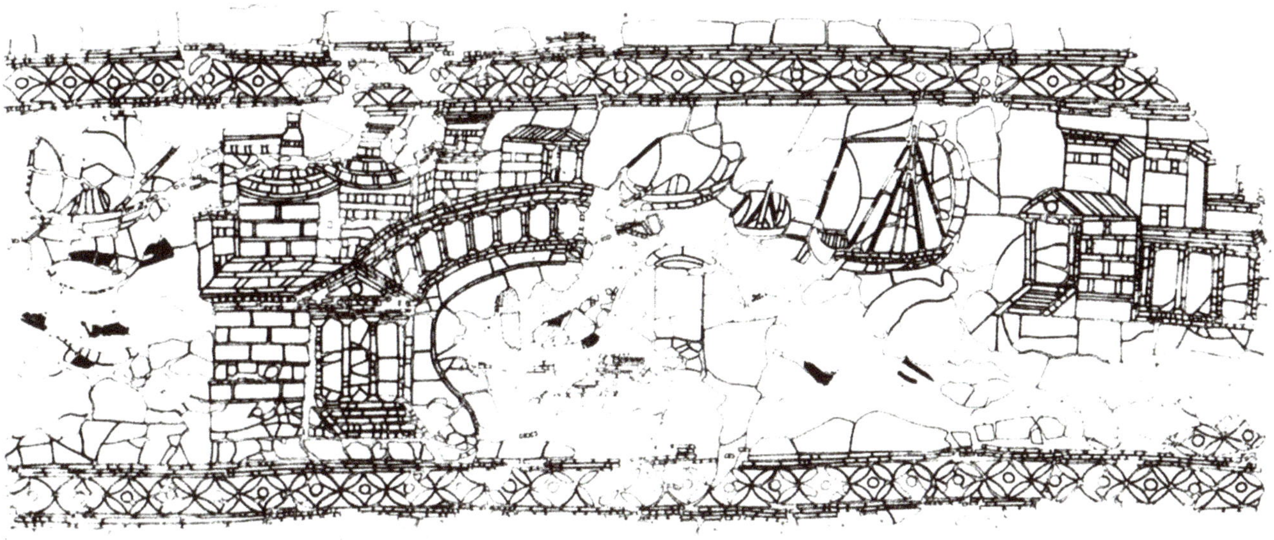

Fig. 3.5.5b: Harbor Scene II and the ships

colonnades, temples topped with pitched tile roofs, walled structures and ships sailing in or out of the harbor. The colonnade porticoes are depicted in two distinct shapes: one is semi-circular (figs. 3.5.5 a, b) and the second is hexagonal (figs. 3.5.6 a, b).[28] The buildings are seen from a bird-eye view, while the ships and the human figures from eye-level or frontal view. These scenes related to the harbor area are quite static; the ships and the buildings are not associated with any human figure. The ships, in or out of the harbor, are depicted with fully open sails, but no movement is indicated. They probably are anchored, although no anchor or a mooring line is shown. Neither the water line nor the waves are represented. Almost all the vessels are shown from the side, in profile, with two exceptions, in fig. 3.5.5b, where Ship 3 (figs. 3.5.15, 3.5.15a) is depicted in a three-quarter view, and in fig. 3.5.6b, Ship 3 is viewed from the prow, thus indicating that it leaves the harbor or it is anchored within the harbor basin (fig. 3.5.21).

Harbor Scene I and the Ships

When the panel was discovered, it measured 0.85m in length and 0.77m in height.[29] The central part of the panel comprises an arching colonnade following the shape of the shore that defines the harbor basin and several buildings topped by pitched roofs, which are located behind the colonnade and to the right facing the harbor entrance (fig.

[28] The numbering and order of the maritime scenes described bellow do not follow the same sequence as the illustrations in: Ibrahim, Scranton and Brill, 1976. Detailed pictures of the ships in figs. 3.5.5a, 3.5.6a, were taken by the author at Isthmia Museum while studying the panels in September 2002 for my PhD dissertation.

[29] Ibrahim, p. 97.

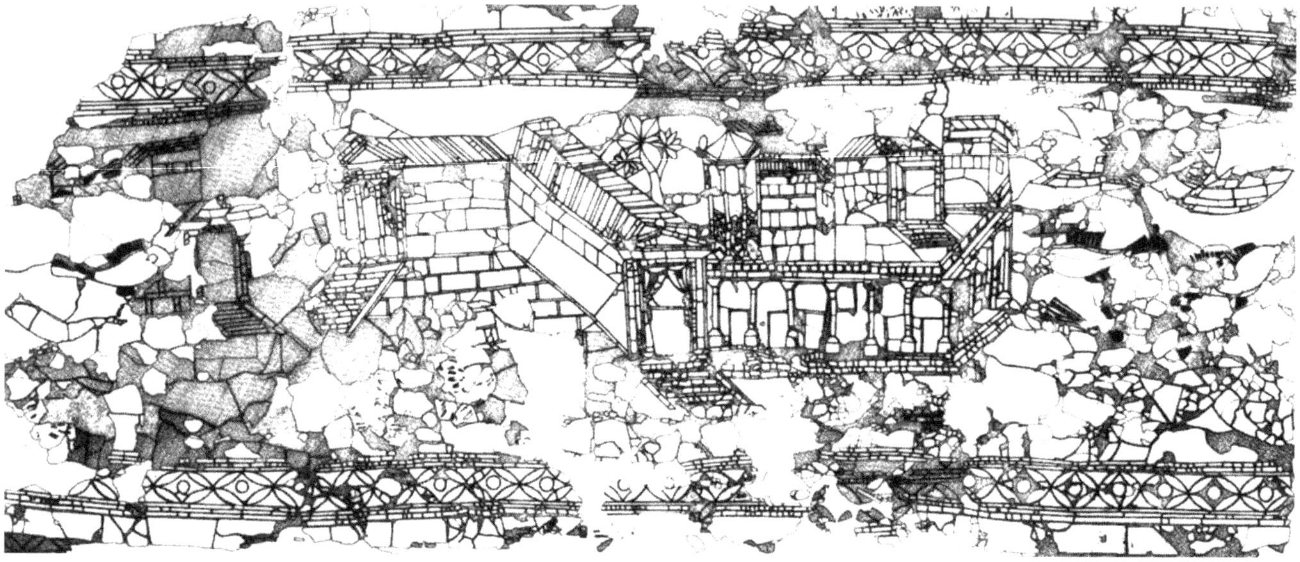

Fig. 3.5.6a: Harbor Scene III and the ships

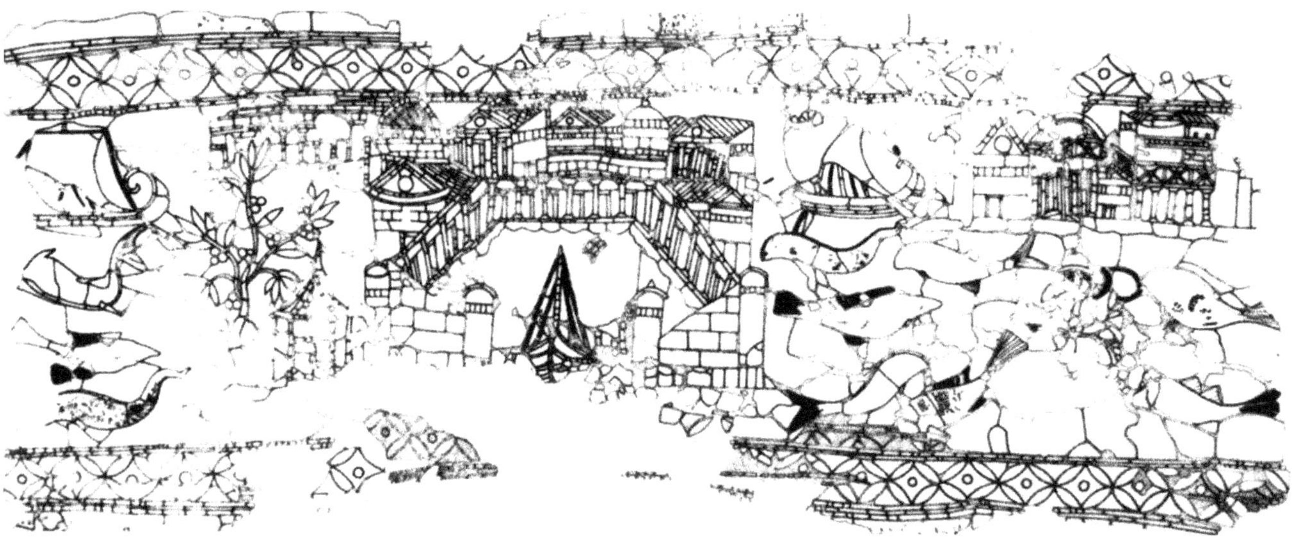

Fig. 3.5.6b: Harbor Scene IV and the ships

3.5.5a). The colonnaded portico describes an arc from left to right that is enclosed by a cylindrical tower on either side (fig. 3.5.5a). To the right of the frontal building adjacent to the lower part of the colonnade, there is a high stone-wall indicated by five courses of rectangular blocks. This wall enclosed the harbor basin. On the lower part of this wall there are three arched passages, probably for flushing the seawater into the harbor's inner basin to prevent its silting (fig. 3.5.5a). This wall is joined on its right by a cylindrical tower topped by a domed roof with a conical cap on top. To the left side of the colonnade (upper part) is found a similar tower. These structures may indicate lighthouses whereas the conical cap on the domed tops may represent the beacons. A very large fisherman whose back is turned to the left-hand tower fishes with a fishing rod bending from the weight of the fish just caught into the hook at the end of the line. To the left side of the harbor complex are two sailing vessels with their prows pointed to the left. They appear to be sailing away from the harbor. At the top right corner of the panel is found another sailing ship with its prow pointed to the left, probably sailing into the harbor.[30]

Sailing Ship 1: This is the largest ship of the three rendered in the Harbor Scene I panel. The elongated hull is made of four strakes (fig. 3.5.7). The upper plank comprising five very thin strips of glass indicates the gunwale or bulwark. The second strake is a wider strip of glass (made with several segments; originally it may have been a single piece), thus following the shape of the hull. The third strake is made of three long superimposed thin strips of glass that probably indicates the wale just above the waterline. The lower strake that forms the flat bottom of the ship is made

[30] The vessels in this panel are numbered from left to right.

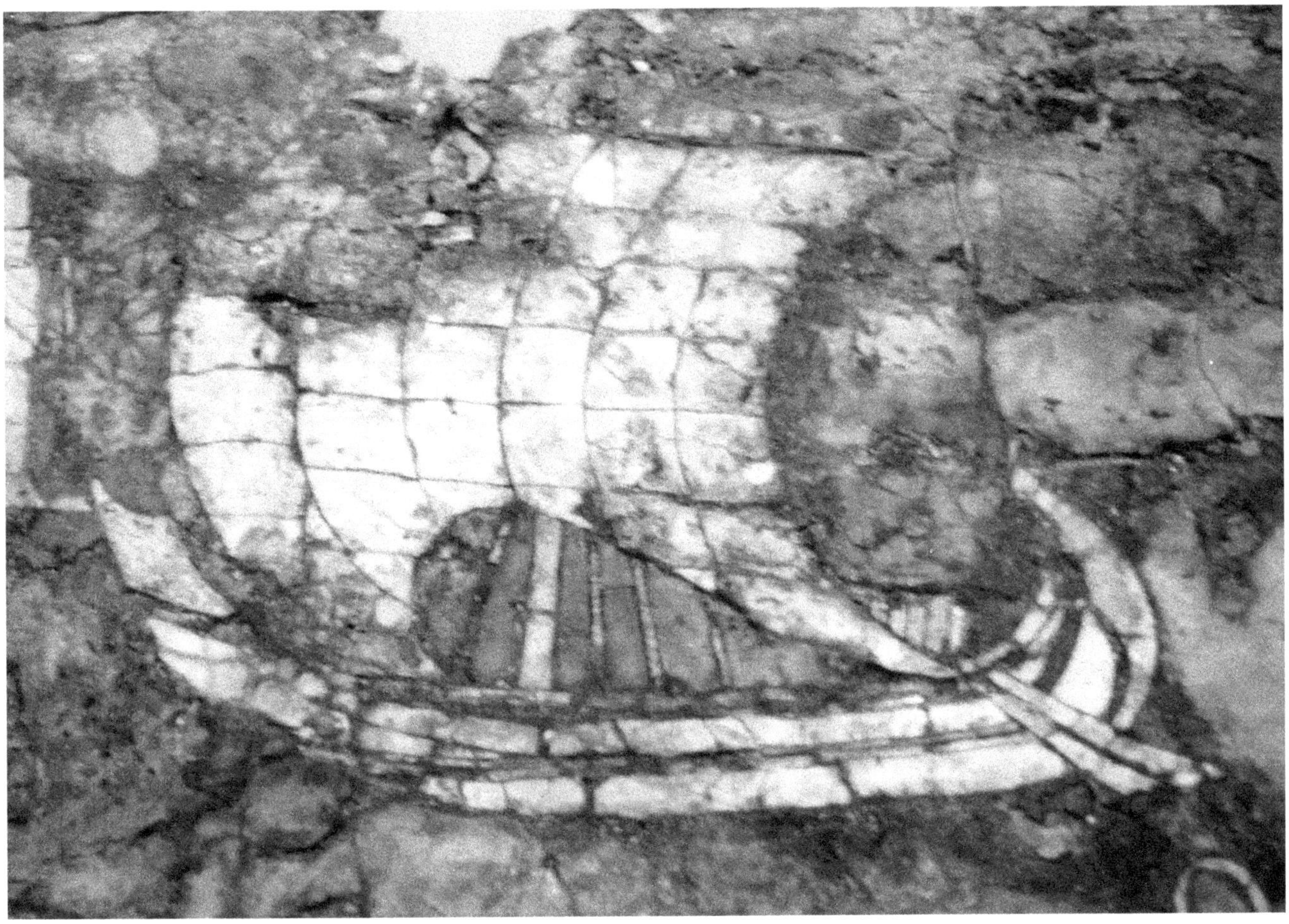

Fig. 3.5.7: Sailing Ship 1 – Harbor I

of four tapered segments (originally it was a single piece). Two pieces of glass that completed the stern are missing (fig. 3.5.7). A wide curving sheet of glass designates the sternpost. The strake atop the port quarter may indicate the bulwark. The raised rounded stem is surmounted by a short vertical stempost with a concave cut tip. Some of the original reddish-purple color is still preserved on this piece (fig. 3.5.7). The highly curved stern is surmounted by an upraised sternpost slightly curving above the quarter. The stern plank is formed by the triangular piece. A rectangular structure made of five vertical thin strips topped by a horizontal strip forming a flat roof, indicates the quarter cabin (for the helmsman or the crew).

The rigging consists of a fully open square sail set on a horizontal yard, a tapered mast, several ropes and a steering-oar or rudder. The forward billowing square sail seen from its fore face is made of a single piece of white opaque glass. The upper starboard part of the sail and the yardarm are missing. Five vertical lines following the curvature of the bunt indicate the brails, while the six horizontal ones represent the reinforcing-bands. Apparently the lines were incised with a sharp blade or knife onto the surface of the glass when it was still in its molten stage, after the shape of the sail was cut first. The foot of the sail is highly arched, as if reefed at the center or resulting from the wind blowing on the lee side of the sail, from astern or the starboard quarter (fig. 3.5.7). The yard is made of a very thin strip of glass (1 mm thick) and is equal in length to the width of the sail's head. The sail was attached to the yard by brail-fairleads set on the yard, whereas the brails also passed through. On either side of the masthead above the yard there is a triangular white topsail each cut from a single piece of white glass. The port sheet is attached to the gunwale, just near the rudder's shaft and covering the lower part of the quarter cabin. The starboard sheet is attached to the quarter gunwale. The forward billowing of the sail was intentionally made to give some perspective to the scene, thus suggesting that the mosaicist wanted to show an actual vessel sailing into the wind. No seamen are depicted in any of the vessels within all four maritime panels.

The mast is revealed between the arching foot of the sail and the gunwale. It seems to be a tapered pole made of two elongated pieces of white opaque glass that originally may have been yellow and probably comprising a single piece. The mast is stepped forward with a slight backward inclination. Between the right side of the mast and the port sheet there are three almost vertical lines, all made of twisted rods of glass (figs. 3.5.7, 3.5.8). The middle thicker line may indicate the port shroud. The right-hand line closer to the port tack is the backstay (misplaced by the mosaicist who did not understand the proper placing of the line). The line closer to the left side of the mast probably represents

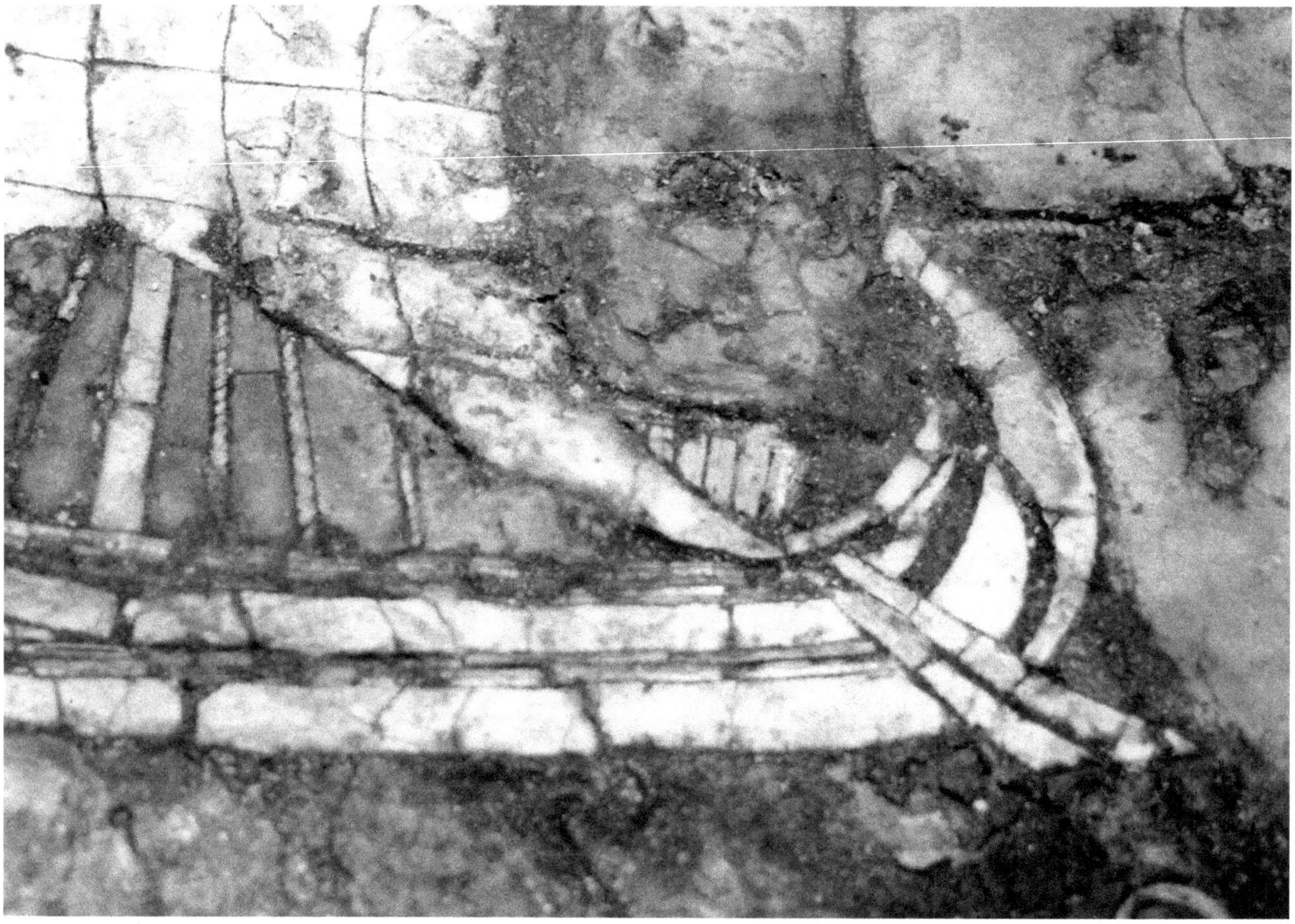

Fig. 3.5.8: Close-up of the quarter cabin, stern and rudder

the starboard shroud. The negative imprint of the left-hand line, closer to the starboard clew, indicates the forestay (fig. 3.5.7).

The steering gear is indicated by the tapered blade of the steering-oar or rudder mounted on the port quarter. The lower part of the shaft transverses the blade longitudinally, thus defining both elongated triangular wings, with their lower edge cut horizontally (fig. 3.5.8). The wings probably were inserted into a groove cut on the lower shaft and afterwards they were reinforced by wooden treenails and bronze pins. The shaft of the port rudder seems to be housed behind the aft top quarter plank (fig. 3.5.8). The loom is hidden by the port sheet. A similar arrangement may have been used on the starboard quarter.

Sailing Boat 1: The vessel is towed astern of the Sailing Ship 1 by a towline made of a twisted rod of glass that stretches between the tip of its stempost and the sternpost of the Ship (fig. 3.5.9). Most of the hull is missing. The preserved upper strake, the prow and the stern which survived the damage indicate a banana shape hull (fig. 3.5.10). The top strake is made of five segments that probably broke during the panel's destruction. The width of the strake apparently was cut in a single elongated strip. The tip of the bow has a horizontal cut. The triangular broad piece of glass with its straight cut base placed above the oar indicates the broad rounded stern (fig. 3.5.10). The boat is rigged with a fully open square sail, two triangular topsails, one mast, some ropes and one steering oar. The sail, seen from its fore face, billows forward; originally it was made of a single piece of white opaque glass (now it is fragmented and poorly restored). The foot has a very high arch on its middle section (fig. 3.5.10), thus indicating that the wind blows from the starboard quarter or astern. The starboard sheet is attached to the gunwale and the port sheet stretches to the port quarter gunwale, near the loom of the oar. The horizontal yard is made of a very thin grayish strip of glass. The head of the sail equals the length of the yard. Above the yard, on either side of the masthead, is set a triangular topsail each made of white opaque glass. Between the vertical sides of the topsails can be seen the tapered masthead (made of two segments) projecting above the yard (fig. 3.5.10).

The fragmented mast pole is visible between the arched foot of the sail and the gunwale. Originally it may have been made of a single piece of glass. It has a slight forward inclination, being stepped slightly fore-amidships. The almost vertical line between the starboard sheet and the left side of the mast may indicate the forestay (misplaced by the mosaicist, who did not understand the exact function of this line), and therefore it appear to indicate the port shroud. The line closer to the right side of the must with a

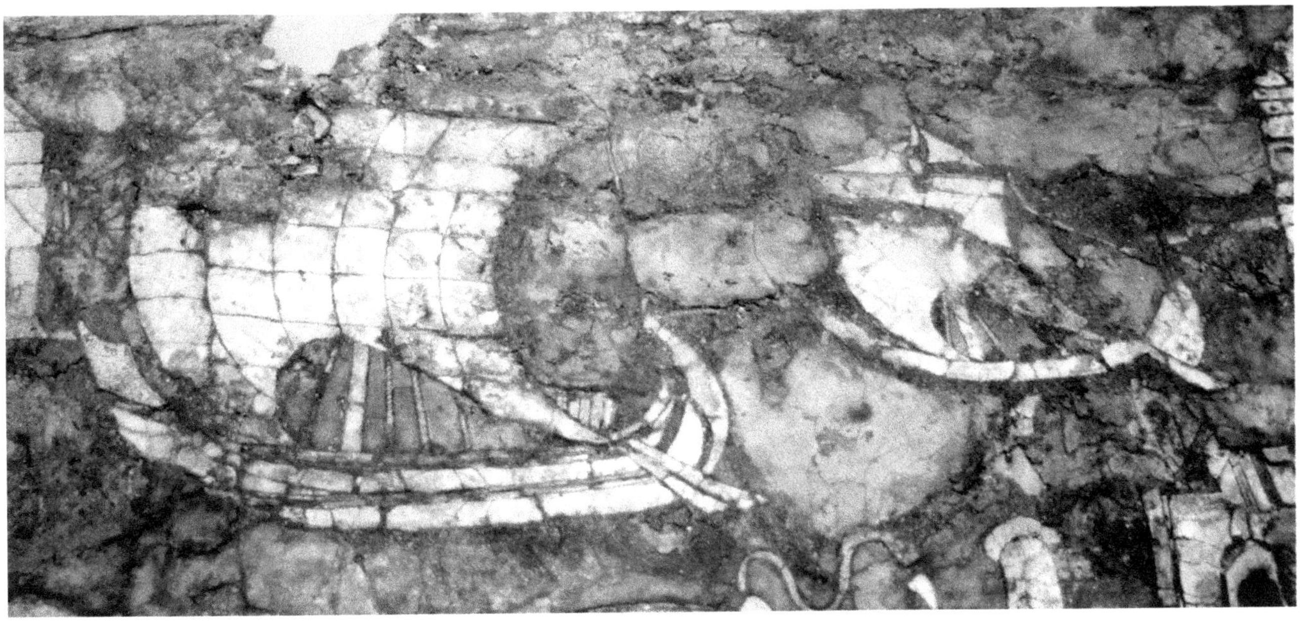

Fig. 3.5.9: Towing ship – Harbor I

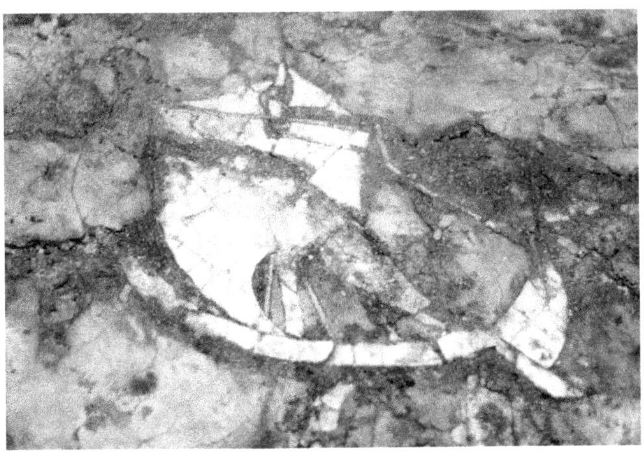

Fig. 3.5.10: Sailing Boat 1 – Harbor I

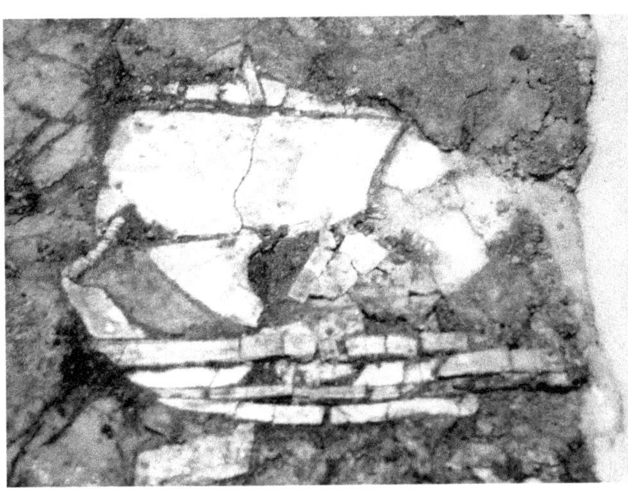

Fig. 3.5.11: Sailing Boat 2 – Harbor I

backward rake may indicate the backstay (fig. 3.5.10). Its position was misplaced by the mosaicist or the restorer who did not posses nautical knowledge and therefore he lacked to understand the function of this line. Thus, one looking at this rig line may interpret it as the port shroud, which is the opposite. The line that stretches between the tip of the port yardarm and the tip of the sternpost (now angled, not a straight line) indicates the port brace. This angle is the result of the restoration made in the 1970's.

The steering gear oar mounted on the port quarter has a backward slanting (fig. 3.5.10). The surviving oar is made of a single tapered piece of white glass that may indicate the wing of the blade. The second wing is missing, as well as the lower part of the hull. A similar oar may have been mounted on the starboard quarter

Sailing Boat 2: The vessel found on the top right corner of the panel is sailing into the harbor (fig. 3.5.5a). The short and broad hull is indicated by four horizontal strakes (almost the same width). The third one is thinner and it consists of three very thin superimposed horizontal strips which may indicate the lower wale, just above the waterline. It is assumed that originally each strake was made of a single strip and not fragmented as they appear now in the depiction, which resulted from the destruction of the panel. The boat has a rounded stem but the stern is missing (fig. 3.5.11). An unsuccessful attempt was made to reconstruct the stern during the restoration of this boat with unsatisfied results. The stempost is made of a short vertical piece with a concave cut tip; it is very similar to the stempost of the Sailing Ship 1 (fig. 3.5.7). Remains of the original yellow color are still found on this piece, thus indicating the wooden built boat (fig. 3.5.11).

The vessel is rigged with a fully open square sail billowing forward and seen from its fore face. Originally, the sail was made of a single piece of white opaque glass. As a result of the catastrophe, the lower right corner of the sail was damaged and an attempt at its reconstruction resulted in a very crude depiction (fig. 3.5.11). Near the port leech there is a mass of glass whose function is not clear. The crude restoration resulted from the lack of understanding of the ship's rigging. The head of the sail equals the length of the yard. The masthead projects above the yard with a slight forward inclination; it probably resulted from the reconstruction. The foot of the sail has a high arch, deduced from the preserved lower starboard sheet. This sail is very similar to that of the Sailing Boat 1 (fig. 3.5.10). On either side of the masthead is set a triangular topsail, each made of a single piece of white opaque glass (fig. 3.5.11). The only rope associated with the rigging is depicted between the tip of the stempost and the middle starboard leech. The line, made of a glass twisted rod, probably indicates the forestay stretching between the junction of the yard and the mast to the stempost. The steering gear is missing.

Harbor Scene II and the Ships[31]

This panel when discovered measured 1.86m in length and 0.82m in height (fig. 3.5.5b).[32] Since its discovery and restoration it is stored in the storeroom at the Isthmia Museum, in very poor preserved condition. The scene is very similar to that of Harbor I (fig. 3.5.5a). The harbor area in this panel occupies a much larger part, revealing almost its entire inner basin. In the foreground there is a temple with a propylon façade set on a podium with five steps leading to its entrance. Behind the temple, the colonnade stretches in a semi-circle following the shape of the shore (fig. 3.5.5b). Several rounded and rectangular buildings are found behind the colonnade. In the middle of the harbor basin there is a broad, tall cylindrical tower. The lower part of the tower is damaged making it impossible to see if it was placed on a podium. To the right side of the panel there is a structure that probably was a temple or it formed part of the harbor's administrative offices; it is placed on a podium with four steps leading from the shore to its entrance. To the right side of the structure, and slightly lower, there is a portico faced by four columns; the right side of the panel is damaged and therefore it is not possible to see the end of the colonnade (fig. 3.5.5b). This complex probably indicates the agora. Unlike the previous panel, there are no human figures to be found within this scene.

On the top left side of the panel a ship with its prow turned to the right seems to sail towards the port. Three vessels are depicted within the harbor basin. The prows of all ships point to the right, thus indicating their sailing direction, probably leaving the harbor (fig. 3.5.5b).

Sailing Boat 1: This vessel is found in the upper left corner of the panel, just to the left of the harbor complex (fig. 3.5.5b). The broad hull, made with three strakes has a spoon-shape, resembling vessels of the Pharaonic Egypt. The prow has a slender raised curve with a pointed tip. The top strake seems to be missing (fig. 3.5.12). The rigging consist the main fully open square sail, a mast, several ropes, and a second mast with a forward rake above the stem, and one steering oar. The stern and the sternpost are damaged and therefore it is not possible to deduce their original shape. The main square sail, billowing forward, viewed from its fore face was cut from a single piece of glass. The head of the sail equals the length of the yard. The foot has a high arch, thus indicating that the wind blows from the starboard quarter or astern. The tapered mast, visible between the arched foot of the sail and the gunwale shows that it is stepped amidships. There are three ropes associated with the main rigging. The line to the right side of the mast that its lower part stretches towards the stem and hidden by the port sheet probably indicates the forestay. The second line to the left side of the mast, with a backward rake, visible between the foot of the sail and the aft gunwale, probably indicates the backstay. The vertical line stretching between the tip of the starboard yardarm and the quarter gunwale indicates the starboard brace. All the lines are rendered with glass twisted rods. The forward angled mast above the prow indicates the *artemon* (foremast); its yard, sail and the lines rigging are missing (fig. 3.5.12). The vessel lacks its top sail or probably it was destroyed during the catastrophe.

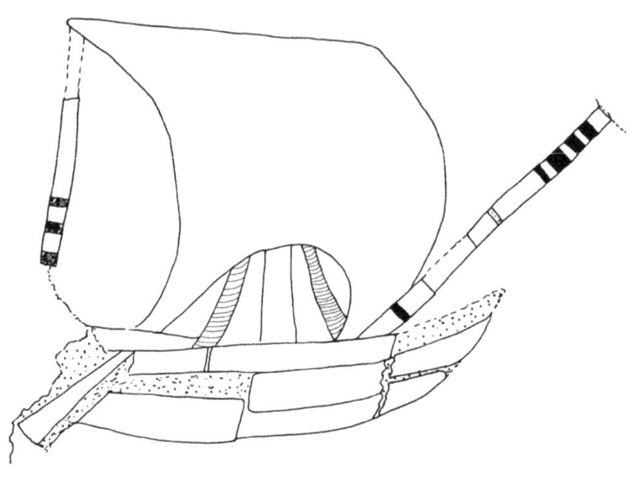

Fig. 3.5.12: Sailing Boat 1 – Harbor II

The negative imprints in the hull indicate the missing glass planks. The steering gear is indicated by a single oar mounted on the starboard quarter with a backward rake. The oar made by a tapered elongated piece of glass indicates the top wing whereas the lower wing is missing. A similar oar may have been true on the port quarter. Beneath the bottom of the vessel are depicted two fish that are not in scale with the vessel, appearing much larger (fig. 3.5.5b). This disproportion was probably made intentionally by the mosaicist to create some perspective to the panel. The large fish shown on the foreground were probably

[31] This panel was described according to the illustrations from the excavations report. Panels II and IV are stored at the Isthmia Museum and due to their bad state of preservation, no permission was granted to research them alive.
[32] Ibrahim, p. 91.

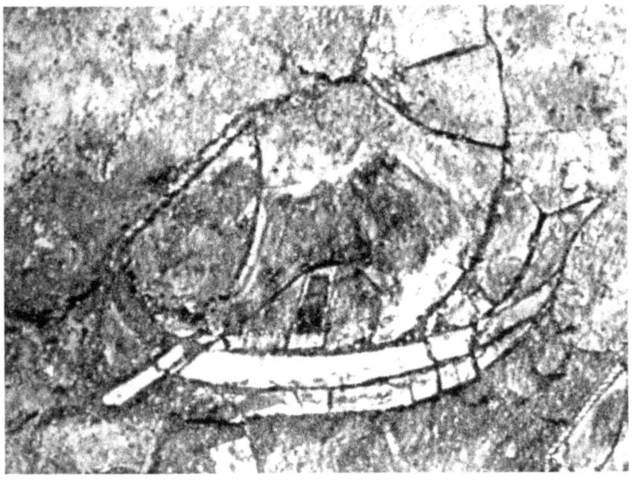

Fig. 3.5.13: Sailing Boat 2 – Harbor II

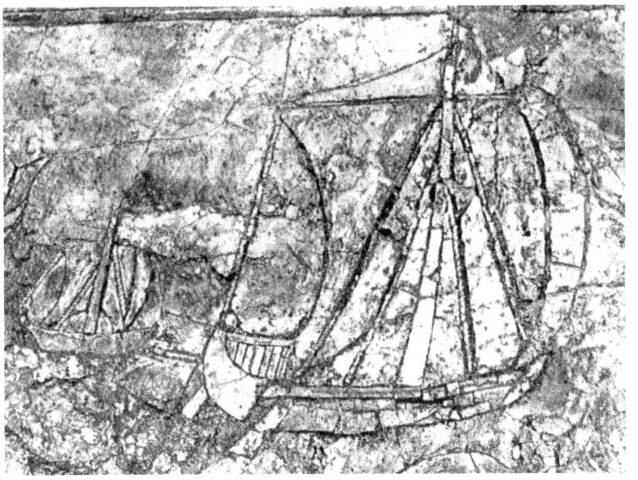

Fig. 3.5.14: Sailing Boat 3 towed by Sailing Ship 1 – Harbor II

used to emphasize the vessel sailing in the open sea in the background.

Sailing Boat 2: This vessel is found within the harbor basin close to the upper end of the colonnade. The hull has a banana shape and is very similar to that of Sailing Boat 1. Two strakes indicate the hull; the top one is quite wide, probably indicating a wide wale alongside the starboard hull or a bulwark (fig. 3.5.13). The prow has a slender curve. It is topped by a vertical stempost that its outer edge has a slight forward extension. The stern is broad and rounded. The stern top strake indicated by a thin strip, seems to follow the shape of the starboard quarter that its aft-end projects over the sternpost. The rigging consists of a fully open square sail, the tapered mast, one rope and one steering oar. The sail, seen from its fore face, is made of a single piece of white opaque glass cut in the shape as billowing forward (fig. 3.5.13). The highly arched foot of the sail indicates that the wind blows from the port quarter or astern. Each sheet stretching backwards probably was tight around a bitt attached to either quarter gunwale. The visible mast between the highly arched foot and the gunwale indicates that it is stepped amidships with a slight forward rake. The topsail and the masthead are missing. The angled line stretching between the tip of the starboard yardarm and the head of the sternpost represents the starboard brace. The single steering oar is mounted on the starboard quarter with a forward rake. It is represented by a tapered blade and the upper part of the shaft which seems to extend forward the starboard sheet (fig. 3.5.13).

Sailing Boat 3: This is the smallest of all the vessels represented in the panel. The narrow hull with angled ends is made of a single trapezoidal strip of glass (figs. 3.5.14, 3.5.14a). The stern and the stem with pointed ends finish at the gunwale level. The stern is topped by the sternpost with a backward rake. The sailing rigging is shown from the lee side. The mast is stepped amidships and has a slight forward inclination. The masthead projects above the yard. The topsails are missing or were not to be set originally. A fully open square sail bellies over the port gunwale. The

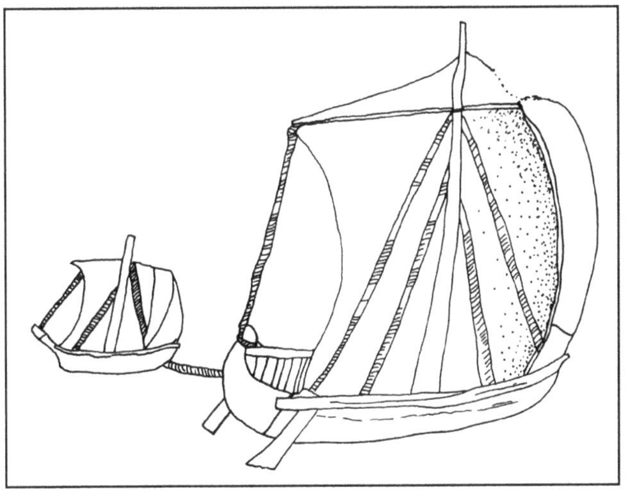

Fig. 3.5.14a: Drawing of Sailing Boat 3 and Sailing Ship 1 – Harbor II

starboard leech and edge of the sail are turned backwards, thus revealing part of the fore bunt. The angled port brace stretches between the tip of the port yardarm and the sternpost. Its upper end was misplaced, thus appearing to be attached to the upper leech. The line that stretches from the left side of the mast to the starboard quarter indicates the backstay, while the angled line to the right side of the mast is the forestay. The upper ends of the fore-and-backs stays are not placed in their right position at the joint of the yard with the masthead. This depiction resulted from the mosaicist lack of understanding of this rigging. The steering gear is missing. A short horizontal line that stretches from beneath the stem of the Sailing Boat 3 to the port quarter of the Sailing Ship 1 indicates the towline. We assume that this vessel represents the ship's boat and apparently it used a sail when the wind conditions were favorable (figs. 3.5.14, 3.5.14a).

Sailing Ship1: This is the largest vessel in the panel, being depicted in a three-quarter view. The ship is viewed slightly from above the stern, thus revealing the port inner side

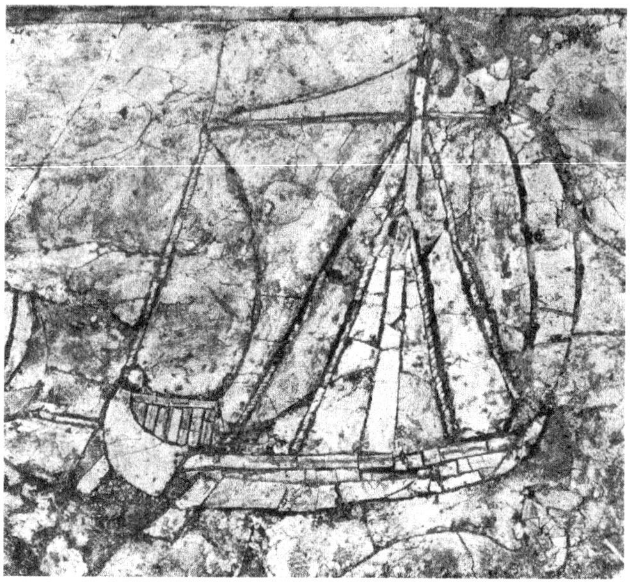

Fig. 3.5.15: Sailing Ship 1 – Harbor II

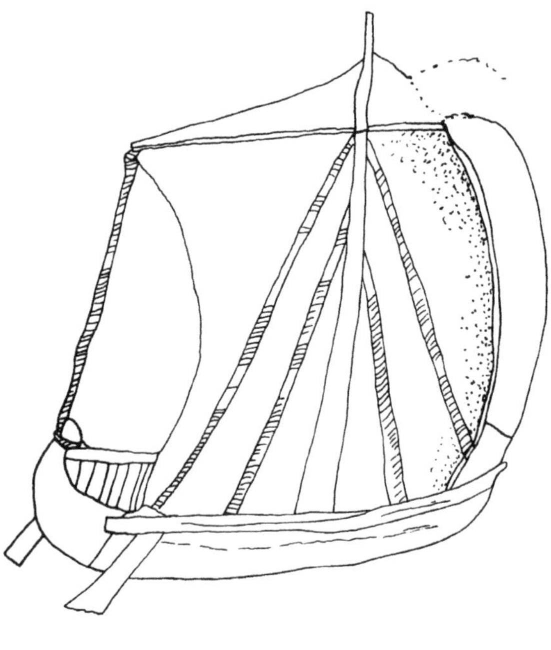

Fig. 3.5.15a: Drawing of Sailing Ship 1 – Harbor II

of the hull (figs. 3.5.15, 3.5.15a). The hull of this ship is different from the vessels described above. It has a rounded shape with an upraised curved prow finished with a pointed tip and flat bottom. A straight cut board made of a single piece of glass indicates the *transom* stern (figs. 3.5.15, 3.5.15a, 3.5.16). The vertical lines depicted on the visible inner port hull may indicate the upper part of three frames and the inner side of planks attached to the frames. The thin horizontal strip topping the frames delegates a caprail beam. The thin strake alongside the upper edge of the starboard hull may represent the top wale; its aft-end projects over the starboard edge of the transom board (figs. 3.5.15a, 3.5.16). The wale reinforced the hull longitudinally. The aft-wing houses the loom of the steering oar.

The sailing rigging comprises a tapered mast, sail and ropes, all seen from the lee side. The fully open square sail billows forward. Originally it may have been made of a single piece of glass cut in the shape of the billowing sail; now it is fragmented due to the traumatic damage of the earthquake. The starboard side of the sail turned backwards reveals the leech and part of the fore bunt (figs. 3.5.15, 3.5.15a). Five lines associated with the standing and running rig are made of twisted rods of glass. The tapered mast is stepped forward, with a slight forward rake. On either side of the mast there are two angled lines. The left-hand line stretching from beneath the joint with the yard towards the tip of the starboard edge of the transom board probably is the backstay. The closer one to the left side of the mast may represent the port shroud. The right-hand line stretching from beneath the joint with the yard towards the stem indicates the forestay, while the closer one to the right side of the mast may represent the starboard shroud. All the lines of the standing rig (fore-and-back stays, and shrouds) seem to be depicted on the same side, as their lower ends being attached to the starboard gunwale. The proper depiction of the port shroud should have been attached to the port gunwale. The yard probably had the same length as the width of the sail's head. The tip of the starboard yardarm, the leech, and the right side of the topsail are damaged. The vertical line stretching from the tip of the port yardarm to the quarter bitt (at the end of the transom's port side) represents the port brace. Its lower end is looped around the bitt. Two triangular topsails were set above the yard, on either side of the masthead. The starboard topsail is preserved near the masthead, while its lower leech and side are missing (fig. 3.5.15a). The depiction of the lines and the square sail viewed from the lee side, created some problems for the mosaicist. The upper ends of the shrouds are misplaced. In a proper representation they had to be set on either side of the mast just beneath the joint of the yard with the masthead. The decision of the mosaicist to show the rigging lines on the same side probably was done intentionally by trying to present a complete picture of the rigging. An additional explanation for the representation of the shrouds may be attributed to the artist's lack of understanding the meaning of such rigging lines in a ship. The back turned leeches and the edges of both sails (Boat 3 and Ship 1) probably were made to give some perspective and dynamics to the vessels.

Two oars with elongated blades mounted on either quarter comprise the steering gear. Only the short thin rectangular blade projecting beneath the hull indicates the port steering-oar or rudder. An elongated tapered blade made of a single piece of glass represents the starboard steering-oar that its shaft probably was laid on the cross-beam supporting the aft-wing (fig. 3.5.16).

Harbor Scene III and the Ships

The dimensions of this panel when found were 1.87m in

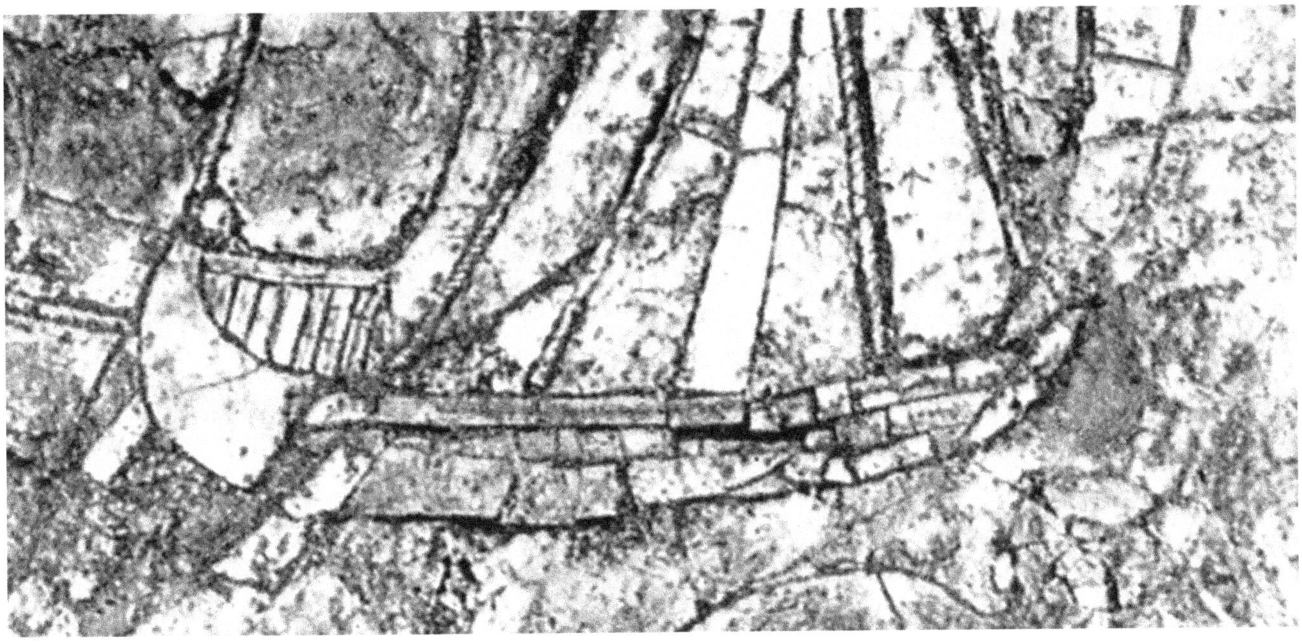

Fig. 3.5.16: Close-up of the transom stern – Ship 1

length, and 0.81m high.[33] The left hand building shown in combined frontal and side views probably represents a temple set on a podium. The propylon façade is reached by six steps from the ground. An elongated structure joins the building at right angles, though extending towards the lower frame. Its entrance facing the viewer can be reached from the ground by six steps. Adjacent to the right side of the entrance there is a colonnade set on a small podium reached by three steps (fig. 3.5.6a). The right side the colonnade turns upwards, making an angle, and reaches the top background square tower, thus forming an enclosure. The entrance of this harbor probably was place on the top of the panel (not visible). It may be indicated by the *tholos* with three visible columns and topped by a dome-like terra-cotta roof. It is placed to the right side of the harbor entrance. This complex may indicate that the harbor basin is enclosed by the colonnades built on top of breakwaters that were constructed in open waters, probably in a hexagonal shape, therefore resembling the Trajan's Harbor at Ostia. The lack of perspective to the scene brought to a crumble of all the buildings as being depicted in a two-dimensional view. Two sailing vessels are depicted in the upper right corner of the panel, with their prows pointed to the left, as they seem to leave the harbor.

Sailing Boat 1: The vessel is shown in profile, revealing its starboard sides; the hull is damaged, especially at its bow (fig. 3.5.17). A thin fragmented strip indicates the gunwale or the top wale. The bottom and the stem are not distinct, due to the damaged panel, as well as from the misinterpreted restoration. The stern is rounded and broad, as indicated by the triangular piece of glass. The starboard quarter gunwale is made with a thin piece of glass; its aft-end extends above the sternpost. Remains of reddish-

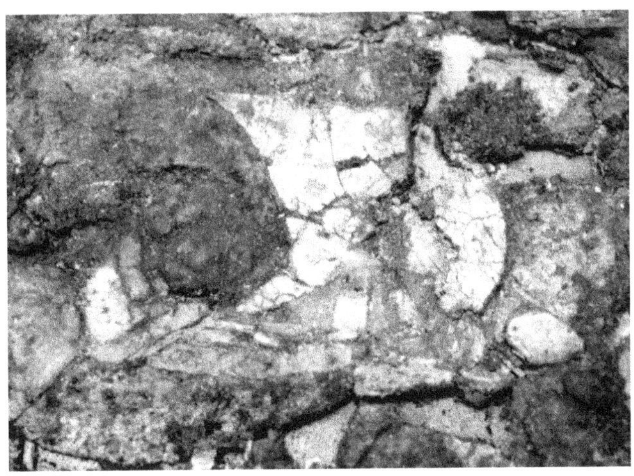

Fig. 3.5.17: Sailing Boat 1 – Harbor III

purple color are preserved on the starboard middle strake that probably indicates the original coloring of the wooden hull (fig. 3.5.17).

The sailing rig comprises a rectangular, fully open sail billowing forward which is seen from its fore face. The arching foot of the sail and the starboard leech emphasize the billowing of the bunt, thus indicating that the wind blows from astern or the port quarter. The starboard sheet stretching backwards is attached to the gunwale just near the shaft of the steering oar. The port sheet is damaged; it may be concluded that it was attached to the port gunwale in the same fashion as the starboard sheet. The tapered mast visible between the arched foot of the sail and the gunwale is stepped forward with a slight forward rake (fig. 3.5.17). The tip of the masthead projects above the yard. It still bears some original reddish-purple coloring (fig. 3.5.17).

[33] Ibrahim, p. 67.

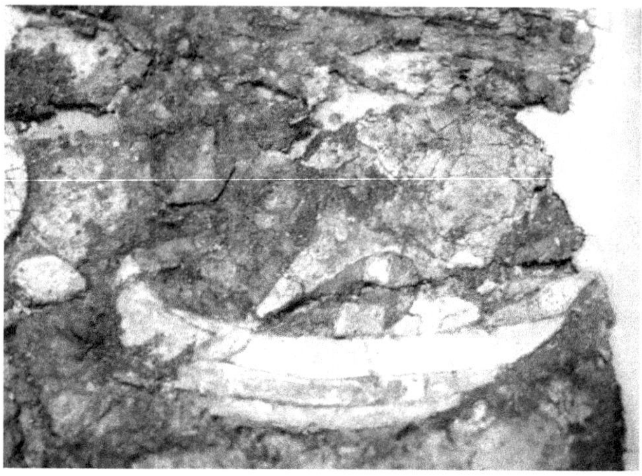

Fig. 3.5.18: Sailing Boat 2 – Harbor III

The topsails are missing. The steering gear is indicated by a fragmentary trapezoidal piece of white opaque glass set on the starboard quarter with a backward slanting. Due to the damage, the shape of the entire oar is not very clear. A similar steering oar may have been mounted on the port quarter.

Sailing Boat 2: The hull of this vessel is much better preserved that that of Boat 1. It is very rounded, resembling a spoon-shape (fig. 3.5.18). The hull is made with three wide strips: the first strake is indicated by a strip of white opaque glass, the middle one is made of a grayish-white glass and seems to have a backward extension. It probably indicates the wale just above the water line. The third strake forming the bottom of the vessel seems to have preserved the pinkish hue from the original reddish or purple coloring of the hull. The stem has a slender upraised curvature. The short vertical stempost with a shallow concave cut tip is made of a single piece of glass that still bears the original purple coloring (fig. 3.5.18). The curving stern is finished with a raised pointed tip.

The rigging consists of a tapered mast, a square sail and a steering oar. The lower part of the tapered mast (indicated by a small trapezoidal piece of glass), visible beneath the highly arched foot of the sail appears to be stepped amidships (fig. 3.5.18). The fully open square sail shown from its fore face is made of a single piece of white opaque glass (now fragmented from the damaged panel). The highly arched foot of the sail indicates that the wind blows from aster or the starboard quarter. The starboard sheet is attached to the gunwale behind the shaft of the steering oar. The port sheet is attached to the port aft gunwale. The fragmented shaft mounted on the starboard quarter with a backward rake represents the steering gear. The blade is missing. The negative print of the blade indicates that the oar was housed behind the aft-extension of the middle strake (fig. 3.5.18). A similar steering oar was mounted on the port side.

Harbor Scene IV and the Ships[34]

When the panel was found, it measured 1.83m in length and 0.78m in height.[35] Since its discovery and restoration this panel also is also stored in the storeroom at Isthmia Museum in very poorly preserved condition. The center of the scene focuses on the colonnade bordering the inner harbor (fig. 3.5.6b). A three-sided colonnade seen from above in "bird-eye" view defines the harbor basin, which seems to be built on breakwaters constructed in open sea. It is similar to the Harbor scene III. The entrance of Harbor IV is facing the viewer though placed in the open sea. The numerous buildings that surround the harbor are very closely spaced, due to the scene being shown in combined frontal and "bird-eye" views, and also the lack of representation to scale and perspective. Two frontal cylindrical towers define the entrance of the harbor: one tower at either end of the colonnades, while other two towers are in the back at the end of each colonnade (fig. 3.5.6b). All the towers are topped by domed roofs. On the upper part of all four towers there is a fenced balcony probably indicating a passage. Each tower at the entrance may indicate a lighthouse, deduced from the short conical device on tip of the domes which may represent a beacon. The back towers probably are watchtowers. All the towers provided landmarks to guide the ships entering into harbor. A stonewall made of large blocks is found between the towers and the lighthouses in the foreground that define the harbor basin (fig. 3.5.6b). The lower part of the wall and the towers to the left side are missing. The right-hand wall is better preserved. Close to the base are two arched openings that may indicate flushing channels to circulate the sea water between the inner harbor and the open sea, to prevent the silting of the harbor basin by continuous circulation of the water. We may assume that the left-hand wall also was built atop the flushing arched channels.

There are three ships associated with this scene, one on either side of the harbor, and the third one is viewed from its bow, visible through the entrance to the harbor, though it is sailing out or probably being anchored in close proximity to the harbor entrance (fig. 3.5.6b).

Sailing Boat 1: This vessel is found on the left upper side to the harbor complex. Apparently it leaves the harbor though sailing out into the open sea. The prow and the stempost are missing. The elongated hull is made of three thin strips of glass and the bottom is flat. The broad and up-curved stern is surmounted by an inner turned volute sternpost (fig. 3.5.19). The fully open square sail billowing forward is seen from its fore face. Originally it was made of a single piece of glass molded or cut in the shape of a billowing square sail with a highly arched foot; now it is fragmented. The mast is missing, probably occurring at

[34] The description of the scene and the ships is based upon the illustration in the excavations report. The ships were drawn to emphasis their construction and rigging. The drawings of all three vessels were made from enlargements of Pl. XVII; Panel no. 7 (VI5.A) – Back, from the excavations report.

[35] Ibrahim, p. 81.

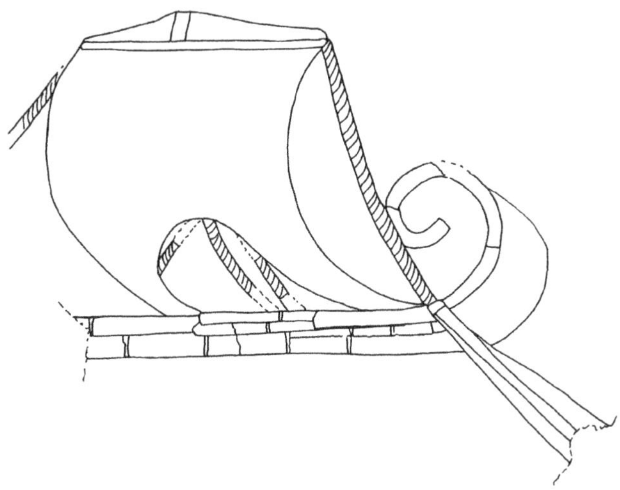

Fig. 3.5.19: Sailing Boat 1 – Harbor IV

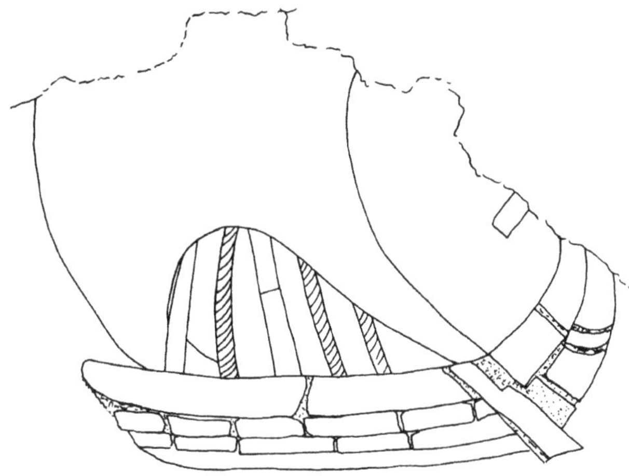

Fig. 3.5.20: Sailing Boat 2 – Harbor IV

the time when the panel was rescued from under the sea or during the restoration process. The masthead projects above the yard and triangular topsails are set on either side. The yard is made of a very thin strip of glass that equals the width of the sail's head. The port sheet with a backward stretch is attached to the port quarter gunwale, near the oar's shaft. The starboard sheet is attached to the aft quarter gunwale. The highly arched foot of the sail resulting from the billowing of the sail indicates that the wind blows from the starboard quarter or astern. Three ropes are revealed beneath the arched foot; they indicate the standing rigging used to secure the mast in place. The left-hand line with a forward extension, just beneath the starboard arching foot of the sail, may represent the forestay. The right-hand one, closer to the port sheet, probably indicates the backstay, and the middle line may be the halyard (fig. 3.5.19). Two additional lines are associated with the sail's rigging. The line that stretches forward from the tip of the starboard yardarm indicates the starboard brace. The port brace stretches between the tip of the port yardarm and the quarter gunwale. All the lines are made with twisted rods of glass.

The oar mounted on the port quarter with a backward slanting represents the steering gear. The upper shaft seems to be secured to the gunwale by a metal rowlock or a leather strop. The shaft transverses the blade longitudinally, thus dividing it into two wings; probably each wing was fastened into a groove cut onto the lower shaft, then locked by wooden treenails or bronze nails. A similar oar may have been mounted on the port quarter as well.

Sailing Boat 2: The vessel is found on the upper right side of the harbor complex and its prow point to the left. Its hull is similar to that of Boat 1, rather slightly broader; it is indicated by four strakes whereas the top one is the thickest (fig. 3.5.20). This strake may indicate the bulwark or the top wale. The stern is similar to that of the previous vessel; the top part of the arching sternpost is missing, where it finishes with a shallow inner turned volute. The bow is rounded and its top misses, or it may have been shaped as shown in the panel.

The sailing rig comprises a square. The upper part of the sail, the yard, the masthead and the topsail, are missing that probably resulted from the damage of the panel. The fully open sail billowing forward is seen from its fore face (fig. 3.5.20). The sail apparently was cut out of a single piece of glass shaped as a billowing bunt with a highly arched foot. The port sheet stretching backwards is attached to the gunwale behind the shaft of the steering oar. The starboard sheet is attached to the starboard aft gunwale. The lower part of the mast and four lines are revealed beneath the arched foot of the sail. The mast is stepped slightly fore-amidships with a slight forward inclination. The fore-line found closer to the arched starboard foot of the sail probably indicates the forestay. It appears to be made of smooth strip of glass. The lines found on either side of the mast may indicate the starboard and port shrouds (fig. 3.5.20). The right-hand line closer to the port sheet probably indicates the backstay. The only visible oar attached to the port quarter with a slight forward rake represents the steering gear. Only one wing of the blade made of a tapered piece of glass survived the damage. The second wing and the shaft are missing. A similar steering was mounted on the starboard quarter.

Ship 1: The ship viewed from its prow provides a rare depiction in mosaics or any other arts. It seems to be leaving the harbor entrance, or it seems to be anchored in the inner basin and facing the entrance (fig. 3.5.21). On either side of the stem are shown three wales and the strakes (not seen). The lower wale is just above the water line. These wales may indicate a sea going merchantman, whereas they reinforced the hull longitudinally. The projecting gunwales are seen on either side of the hull. If it would have been depicted from its stern then the steering oars would be still mounted on the quarters or at least the edges of the aft-wings could be visible. Therefore we may deduce that the ship is viewed from the prow. The lower right side of the hull is missing.

The rigging comprises the vertical mast made of several small segments (fig. 3.5.21). This depiction may indicate a

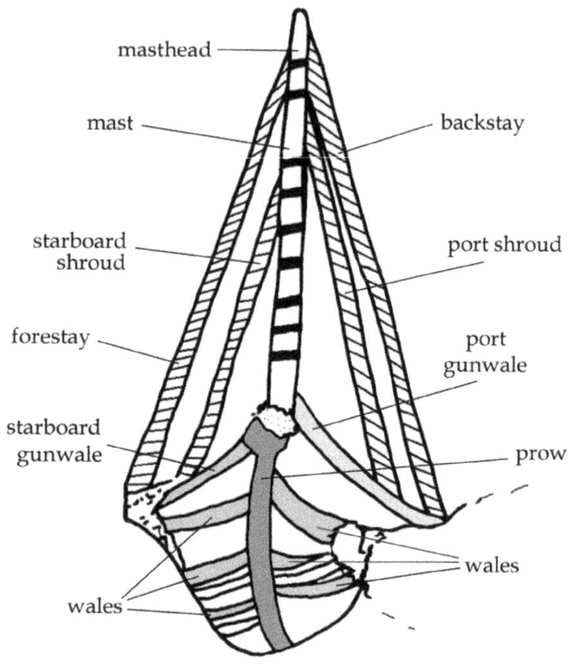

Fig. 3.5.21: Ship 1 – Harbor IV

composite mast, girdled with wooldings at fixed intervals,[36] or it was restored from the recovered broken fragments of glass. On either side of the mast are two lines: the extreme ones may represent the fore-and-back stays, while the lines closer to the mast probably represent the starboard and port shrouds. The yard, the sail and the steering gear are missing.

Discussion

All four maritime panels rendered with harbor scenes basically represent two types of harbor construction: one with a semi-circular basin and the second is an enclosed hexagonal basin. Colonnaded porticos facing both semi-circular basins and following the shape of the shore, indicates that it was built in a natural bay or cove. The hexagonal type appears to be built on constructed breakwaters in the open sea. Within all these scenes are depicted twelve vessels of different sizes. The hulls of the vessels are rounded or elongated with rounded ends. There is a rare representation of a hull with a transom stern (figs. 3.5.15, 3.5.15a, 3.5.16). All the ships are merchantmen (*navis oneraria*) that also were probably engaged in transporting cargo and/or passengers. None of the vessels are associated with human figures. Besides their fully open sails billowing forward, the ships are static, probably anchored but not sailing, although no anchor or mooring lines are shown. All the sails have the same shape with a highly arched foot in the middle (figs. 3.5.7, 3.5.10 – 3.5.13, 3.5.17 – 3.5.20). Apparently such depictions may indicate a standard convention of sailing vessels, not necessarily sailing, rather they are anchored and lack any human figures associated with the crew.

[36] Casson, 1971, p. 231, and n. 31.

The brails and the horizontal reinforcing-bands depicted on the fore bunt and the high arched foot was mainly a common depiction of vessels in Roman mosaics dated to the 3rd century CE, at Ostia and in North Africa.[37] The sail of the **Sailing Ship 1** depicted in the Harbor Scene I (fig. 3.5.7) being similar to the square sail of the ships depicted in the Nilotic mosaic from Palestrina, dated the late 2nd century BCE (figs. 3.6.4, 3.6.5). These riggings indicate that not much change was done on the sails throughout the 300 – 500 years. The sailing vessels depicted in the Kenchreai maritime panels (mid-4th century CE) augment that sails and their method of working when sailing into the wind (blown from astern or the quarter) did not change through centuries, and were still preserved around the Mediterranean in later periods.

Sailing Boat 1, in the Harbor Scene II, is the only vessel depicted with two masts, the main and the *artemon*/fore-mast (fig. 3.5.12). All the other vessels in the maritime panels are shown only with the main mast and square sail. The **Sailing Ship 1** in this scene is the only vessel depicted with a transom stern (figs. 3.5.15, 3.5.15a, 3.5.16). Such vessels are rarely represented in any arts. The earliest appearance of transom sterns actually originates on cargo ships of the Old Kingdom in the Pharaonic Egypt.[38] It appears that transom sterns were also constructed on small Roman boats as we can see in a Nilotic fresco from Pompeii (1st century CE), now displayed at the Museo Nazionale di Napoli (fig. 3.5.22). In the *Catalog of Ships* mosaic from Althiburus in Tunisia (3rd century), such a rowboat is flowed by its name *horia*.[39] Transom sterns became a common feature of construction in wooden vessels from the 14th century onwards. Today, some small boats are built with a transom stern because it provides a good support for the outer attached engine.

Sailing Boats 1 and **2** in the Harbor Scene IV probably represent wooden *papyriform* vessels (figs. 3.5.19, 3.5.20). The thin strips used to make their hulls and the broad rounded stern surmounted by an inner turned volute sternpost are characteristic features of such vessels that apparently still sailed on the Nile and the Delta in the Hellenistic and the Roman periods. A relevant comparison for such a type rigged with a fully open square sail appears in the Palestrina mosaic (fig. 3.6.5).

Depictions of vessels seen from the prow are rare, especially in mosaics. A similar view of a sailing ship seen from its prow appears in a mosaic from Dougga, Tunisia (now in the Bardo Museum). The vessel engaged in pulling a large fishing net full with fish belongs to the panel of "*Dionyssos and the Tyrrhenian pirates*" (fig. 3.5.23). The Dougga ship reveals the full view of the prow and the fore part of the port hull. The vessel is rigged with a mast, an arched yard, and a furled sail beneath the yard and tight with robands, the port and starboard shrouds, the port and starboard braces,

[37] Casson, figs. 145, 154; Basch, 1987, figs. 1049, 1050, 1955, 1056, 1058, 1059, 1109.
[38] Landstrom, 1971, figs. 177, 179, 183.
[39] Casson, fig. 137/*20*.

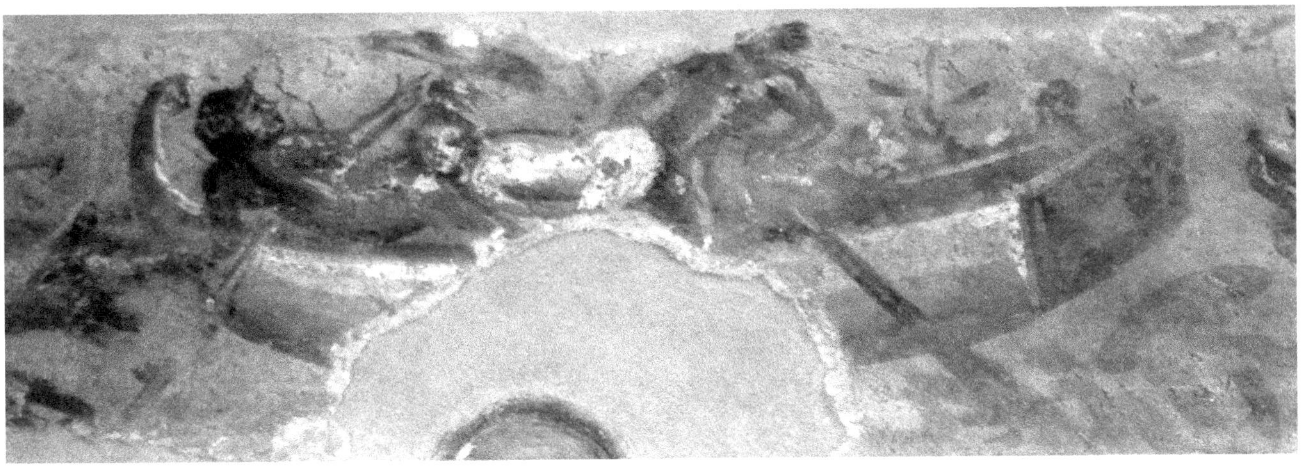

Fig. 3.5.22: Boat with transom stern in a Nilotic scene from Pompeii

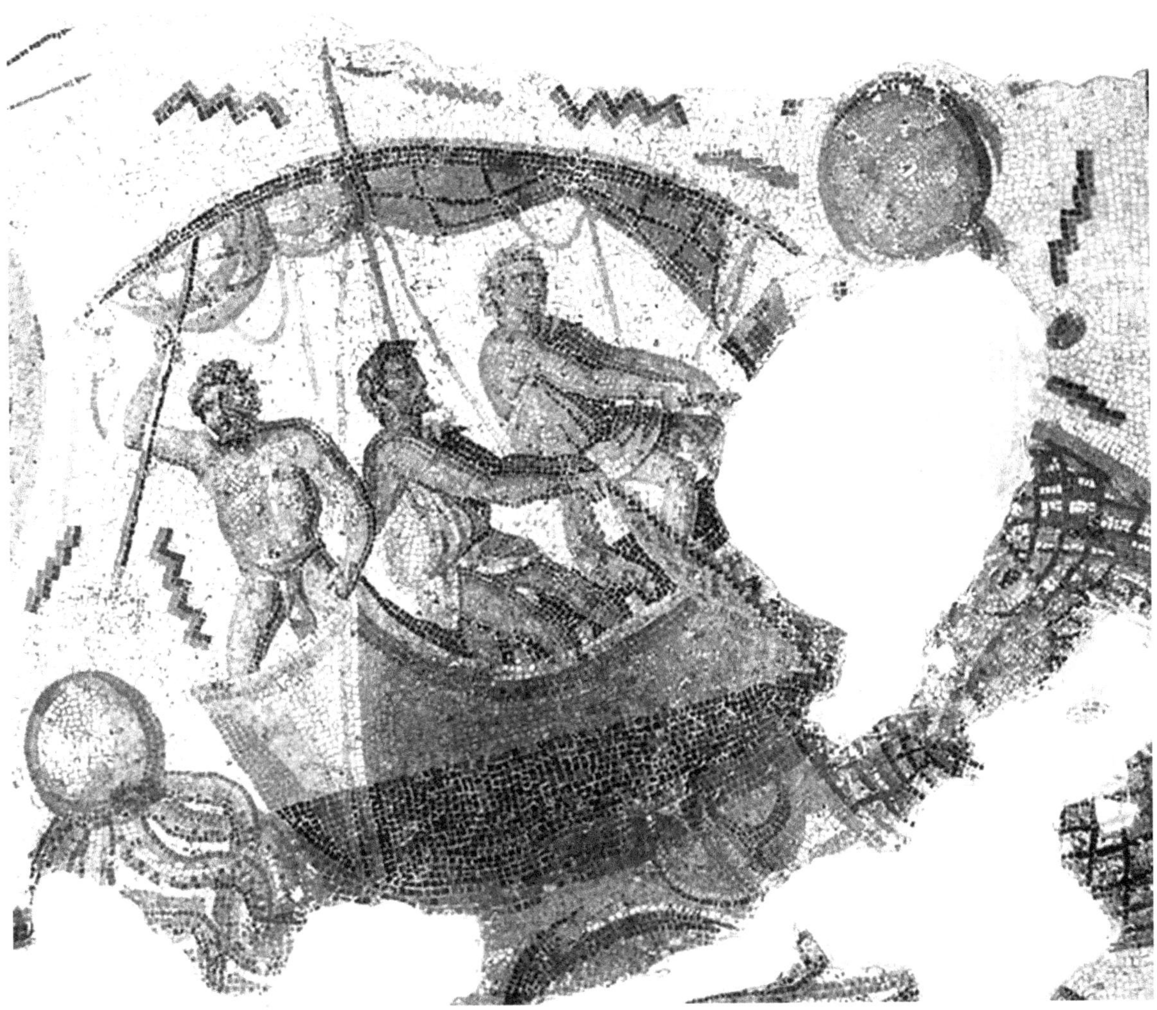

Fig. 3.5.23: Fishing ship in the "Dionyssos and the Tyrrhenian Pirates" mosaic; Dougga, Tunisia

and some arched brails looped around the yard (fig. 3.5.23). The black painted lower hull indicates the pitch/bitumen coating to make the hull watertight, as well as to protect it from the *teredo navalis* (ship worms) or wood rot (when the vessel was left for long periods in water). In the Kenchreai Ship 1, the yard and the sail are missing (fig. 3.5.21). The wales shown in the Kenchraei vessel are not depicted in the Dougga ship. They indicate that the Kenchreai ship is a seagoing vessel; the wales gave an extra strength alongside the hull exposed to the great stresses and forces of the sea waves and currents. The Dougga ship may represent a small vessel that mostly sailed on a river or closer to the coast and was not engaged in long voyages, rather in fishing.

Two of the Kenchreai panels (Harbor I and Harbor II) depict ships with boats towed astern (figs. 3.5.9, 3.5.14, 3.5.14a). There are several representations in any arts of large ships towing smaller boats astern. Such boats were used for harbor work, carrying cargo and passengers from the large freighters anchored in open waters that could not anchor near the quay where the water depth was to shallow. The depiction of a towed boat with fully open sail indicates that when not towed such vessels could use wind the propulsion, which was more efficient than rowing them. A similar depiction of a large sailing ship towing a small sailing boat appears in the Kelenderis harbor mosaic (figs. 3.4.6, 3.4.7).

The Kenchreai and Kelenderis maritime mosaics may be considered as representing typical Mediterranean harbors in Turkey, Greece, Italy and North Africa, concurrent to the period of the mosaics, the 4th – 6th centuries CE. Apparently harbors faced by waterfront colonnades were mostly used as themes in the frescoes from Pompeii and Herculaneum. These scenes were popular depictions in North African mosaics from the 2nd century onward. The Kenchreai maritime panels of glass *opus sectile* show that the traditional maritime panorama mostly found in frescoes in Roman villas around the Bay of Naples were also used as mosaic decorations, especially on walls. The ships depicted in these panels have rounded hulls and lack the forward projection of the cutwater, a typical device of vessels in Roman frescoes, mosaics, or coins. The only oars associated with the Kenchreai vessels are the steering-oars or rudders. These oars are indicated by elongated trapezoidal blades and short shafts housed behind the quarter aft-wings. The Kenchreai vessels are merchantmen (*navis oneraria*) of small and medium size burden (4 to 200 tons) that can be associated with a certain type known from ancient literature and the Althiburus *Catalog of Ships*, which indicates their character and size:

corbita (figs. 3.5.11, 3.5.12, 3.5.17, 3.5.18, 3.5.21), a sea going merchantman with broad hull and both rounded ends;

uperetika, scaphae: merchantman of a great size with rounded and broad hull (figs. 3.5.10, 3.5.15, 3.5.15a);

phasolos: vessel in a fleet or any vessel towed astern by a larger ship; the name refers to a ship's boat[40] (figs. 3.5.10, 3.5.14, 3.5.14a), especially suited for carrying passengers, sailing in the Mediterranean in the 1st centuries BCE – CE.[41]Such vessels varied from modest size, a little more than a skiff, to crafts capable to travel throughout the Mediterranean and carrying hundreds of men.[42] There are no records mentioning when *phasoloi* ships ceased to be used in later periods.

[40] Torr, 1964, pp. 114 – 16.
[41] Ibid., p. 120.
[42] Casson, p. 167 and notes 54, 55.

3.6 Palestrina Nile mosaic – Italy[1]

History of the Nile Mosaic

The large Nilotic landscape depicted in the mosaic known as Palestrina or Barberini has been the object of many studies. Yet no proper or detailed research had been carried out or any publications of the vessels depicted within this mosaic and especially researched from the point of view of marine archaeology.

Originally, the mosaic formed the floor of an apsidal recess which was attached to the eastern end of a long basilica building at the north side of the Praeneste's ancient forum (fig. 3.6.1).[2] The apse in which the original mosaic was set apparently was cut like a grotto into the rock of the hill behind. The water that seeped through the rock over the mosaic surface gave it a sparkling effect and brought the scenes to the realm of the Nile in flood. The length of the grotto's base was 6.87m, 4.35m deep and about 10m high.[3] The semi-circular apse also had three niches adjacent to the arc of the structure (fig. 3.6.2). The mosaic, as it appears today, measures 5.85m at the base and 4.31m deep (the original mosaic was about 1m wider at its base). Since 1954 the mosaic is set on the wall of the main room in the Palazzo Barberini the formal Museo Nazionale Prenestino. As it appears today, the mosaic is a combination between the original sections and other reconstructions made during its long history of restoration.

The mosaic was discovered between 1588 and 1607 in the place that served as a cellar of the building, which had taken the shape of the ancient apsidal structure, then the Bishop's Palace adjacent to the Cathedral of S. Agapino.[4] The mosaic was never excavated, though that it was partially visible in the dump and the basement of the episcopal palace.[5] The first antiquarian interest in the mosaic dates to 1614, when Prince Federico Cesi, one of the four founding members of *Accademia dei Lincei*, came to Palestrina to marry his future wife, Artemisia Colonna. The domain was part of her family's property. The considerable remains of the temple, the terraced structure and the mosaics aroused Cesi's enthusiasm. He ordered that drawings should be made of these remains and architects were employed to carry out the restoration of the structure as well.[6] The account of Cesi's observations was incorporated into the history of Praeneste written by Giuseppe Maria Suares.[7] His observations and comments on the mosaic points out on the difficulties of viewing it at its original location:

[1] A detailed article about the preserved shipbuilding traditions as depicted in the Nile mosaic was published by the author in: *Archaeologia Maritima Mediterranea*, vol.3, 2007, pp. 115-145.
[2] Whitehouse, 2001, p. 71.
[3] Meyboom, 1995, p. 8.
[4] Whitehouse, 1976, p. 5.
[5] Whitehouse, 2001, p. 72.
[6] Ibid.
[7] Suares was the librarian of Cardinal Francesco Barberini. He also wrote the most detailed contemporary information about the drawings of the mosaic in Cassiano dal Pozzo's possessions; Whitehouse, p. 5.

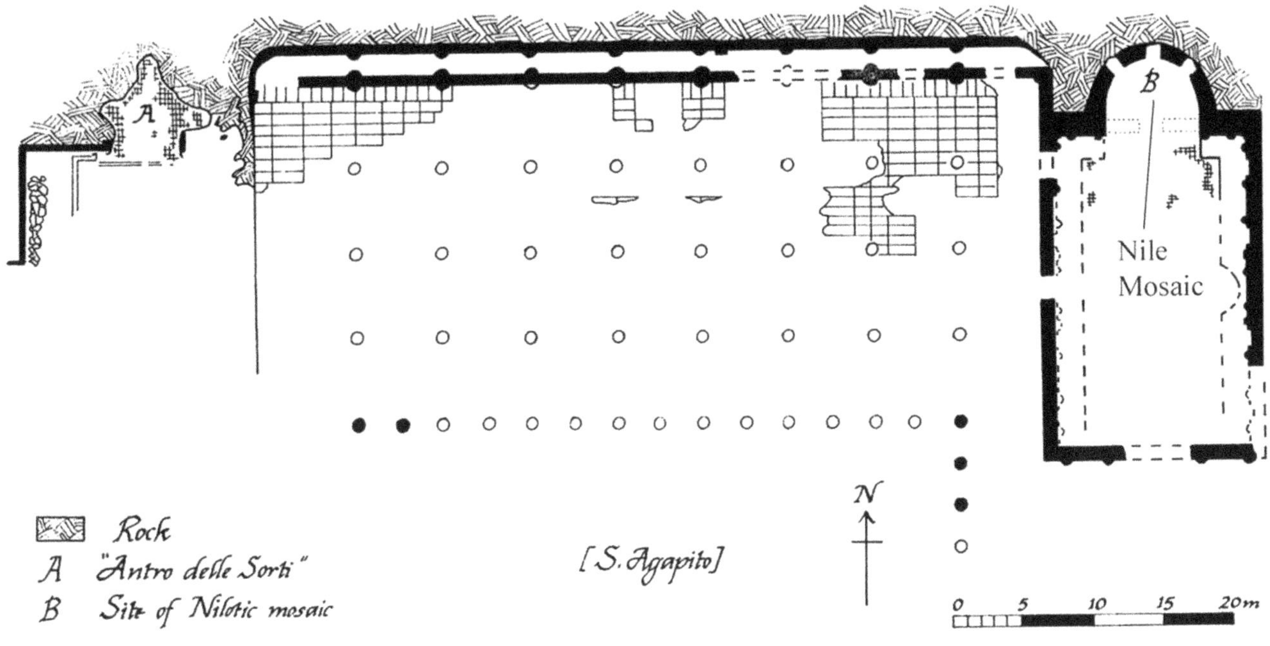

Fig. 3.6.1: Plan of the Nile mosaic complex at Palestrina

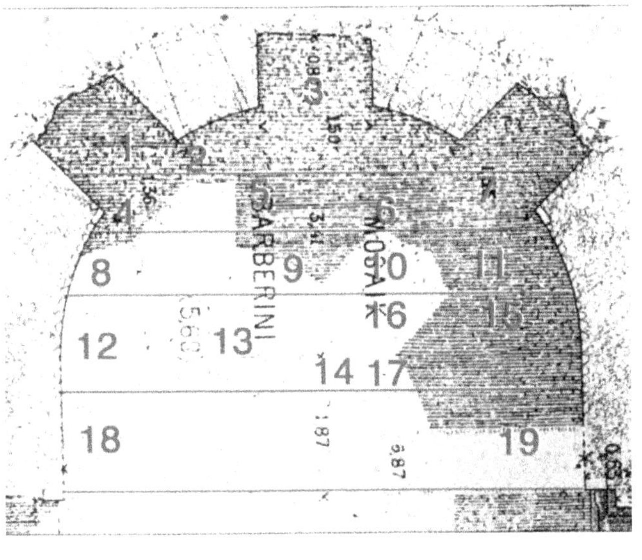

Fig. 3.6.2: Plan of the apse where originally the Nile mosaic was set

"Many indeed are the things to be seen in it (the mosaic), composed apparently of little chips of stone and cut fragments with assiduous attention to detail and painstaking care. At the site itself, it was possible to see or draw only by torchlight and with water pouring over (the surface of the mosaic). There are various representations of men and beasts, Elephants, Rhinoceros, the name written in some special letters; and in the water there are ships. You may behold these things meticulously portrayed, one by one, at the home of our fellow Lynceus and colleague, D. Cavaliere Cassiano dal Pozzo, noted alike for his erudition and excellence".[8]

This note should have been written by Cesi before his death in 1630, but after August 1622, when Cassiano became a *Lynceus*.[9] The drawings of dal Pozzo probably were in existence at the time of Cesi's writing.[10] The mosaic is mentioned again during the years 1624 – 1626, when the archbishop of Palestrina, Andrea Peretti, began the process of removing it in sections from the basement of the palace and sending them to Rome for restoration. The removal was done without a plan or drawings of the original mosaic *in situ*. Probably some parts of the original mosaic still remained *in situ*. In 1630, the Colonna family sold the domain of Palestrina to the Barberini family. The pieces of mosaic in Rome were passed from Peretti's heir to Cardinal Lorenzzo Magalotti who then gave them to his nephew Cardinal Francesco Barberini.[11] When the mosaic was returned to Palestrina in 1640, a fatal accident occurred:

"... having ordered that the mosaic should be packed in boxes and replaced in the Palazzo at Palestrina; the boxes had been propped up the wrong way round with the result that they had crushed and dislodged the whole mosaic; but with the drawings already executed by the inestimable Cavalier Cassiano

[8] Whitehouse, 1976, p. 6 and n. 31.

[9] Whitehouse, 2001, p. 75.

[10] Ibid.

[11] Megalotti retained one piece for himself, which later came to Berlin and it is still there. This section presents a banquet of several people underneath an arched canopy. The section in the mosaic was entirely reconstructed from the dal Pozzo watercolor drawings; for more details on this section see Whitehouse, 1976, pp. 23-24. Cardinal Francesco Barberini was the director of the library at Vatican (1627 – 1636); in 1632 he also was vice-council at Vatican; Bernard Andreae, 2002, p. 18.

dal Pozzo, and the lengthy care and great skill of ... Calandra (a mosaicist), it was all put together again; and Taddeo Barberini having appropriated the hemycircle ... for the consideration of a room, there being a large niche, platform or apse ... he put the said lithostroton back there so well restored that not even the tiniest stone is missing from it, nor might there be one dislodged or put out of the original order... "[12]

The mosaic underwent considerable restoration on its return. Cassiano dal Pozzo had commissioned artists on behalf of Cardinal Barberini to make watercolor drawings of the pieces that were sent to Rome by Peretti before 1640.[13] Giovan Battista Calandra, an expert mosaicist who came to Palestrina, was able to put the pieces together using the dal Pozzo drawings. After its restoration the mosaic was not set in its original location, but in a room with an apsidal niche (fig. 3.6.1; apse B). The original fragments which remained in *situ* were also inserted into the reconstructed mosaic. The mosaic remained in place until 1853, when it was again taken to Rome for restoration by the Vatican mosaicists. The restored work was returned to Palestrina in 1855.[14] During the Second World War, in December 1943 (just before the allies bombardment of Palestrina in 1944), the mosaic was again removed to Rome for safety and stored in the Museo Nazionale Romano. Giorgio Gullini and Salvatore Aurigemma made the first scientific research of the mosaic in early 1952, when carrying out the restoration work. On this occasion the *tufa* backing of the mosaic was removed and the reverse sides of the tesserae were revealed. They were able to distinguish between the original sections of the mosaic and the restorations made in 1640 and 1853 – 1855, by variations of the tesserae thickness and the type of cement used to fix them. The operation was fully documented by drawing and photographing the sections that appeared to correspond to the original work. When the mosaic returned to Palestrina in 1953, it was set in the wall of the main room at Palazzo Barberini, Museo Nazionale Prenestino, where it may be seen today.

After dal Pozzo's death in 1657, his great collection of drawings of Roman antiquities and mosaics passed to the Albani family in 1703, whereas in 1762 a large number of them came into the possession of King George II of England; afterwards they entered the Royal Library at Windsor.[15] Among these drawings, were found the nineteen copies of the Palestrina mosaic, which had been thought to be lost. The copies are still to be seen at Windsor. Hellen Whitehouse published beautiful colored reproductions of these drawings in 2001 in the book "*Ancient Mosaics and Wallpaintings*".

Date of the Nile Mosaic

In spite of many contradictory opinions on the date of the "Barberini Complex", it is generally accepted that the sanctuary of Fortuna, and especially the lower complex, belong to the time of Sulla (82 BCE). This date is deduced by Pliny's reference that Sulla ordered the rebuilding of the sanctuary by Varro Luculus and paved it with floors of *lithostroton*.[16] The remains of the upper complex, which were badly destroyed by the Allied bombardments in 1944, were researched and published by Gullini and Fasola in 1953. Their conclusion that this grandiose complex was built shortly after the middle of the 2nd century BCE was based on epigraphic evidence,[17] which mostly dates from the period of the foundation of the Roman colony at Palestrina in 82 BCE, and on the character of the plan, the style and the construction of the sanctuary in relation to Late Hellenistic architecture in Latium.[18] Degrassi's study of the epigraphic evidence showed that the upper sanctuary must have been completed in 110 – 100 BCE.[19] Many inscriptions found in the area indicate magistrate titles of independent *civitas Praenestina* and names of Praeneste nobles, which may date between 110 – 100 and 82 BCE.[20] The upper complex was most probably built between 120 and 110 BCE. The wall and floor decorations in the sanctuary are presumed to pre-date Sulla's colony. The Nile mosaic could also be dated to the period of the complex's construction between 120 and not later than 100 BCE. The study of the vessels also will emphasize that the date of the mosaic is not later than the end of the 2nd century BCE.

The Theme of the Nile Mosaic

Besides the Nile mosaic, another floor depicting varieties of aquatic fauna was found within the Barberini complex (fig. 3.6.1; apse A). It is assumed that both surfaces were not meant to walk on, but decorated the floors of the *nymphaea* and were covered with a thin layer of water to emphasize their aquatic character, thus giving a more realistic appearance to the scenes. The Nile mosaic may be interpreted as illustrating the combined topography of the Ethiopian Heights and the Nile Delta at the period of the inundations, the festivities of the flood and the celebration of the goddess Isis (the Roman goddess Fortuna). The top of the mosaic depicts the rocky landscape of the cataracts region of the Upper Nile. Dark-skinned hunters are engaged in hunting games with bows and arrows of exotic animals, most of them identified by Greek names (fig. 3.6.3). Strabo describes a similar scene of hunters using bows:

"The Aethiopians also use bows, which are four cubits long, are made of wood and are hardened by

[12] Whitehouse, p. 7 and n. 38.

[13] The nineteen watercolor drawings in the possession of dal Pozzo were probably made between 1626 and 1630, after the pieces that were sent to Rome by Peretti before 1626.

[14] Whitehouse, p. 8.

[15] Ibid., p. 59.

[16] Pliny, *NH*, XXXVI.54.189. Researches carried out on mosaic floors in recent years suggest that *lithostrotos* is not a mosaic floor but a pavement made of marble or other stone slabs; Meyboom, p. 15.

[17] Meyboom, p. 10.

[18] Ibid., p. 11.

[19] Ibid.

[20] Ibid.

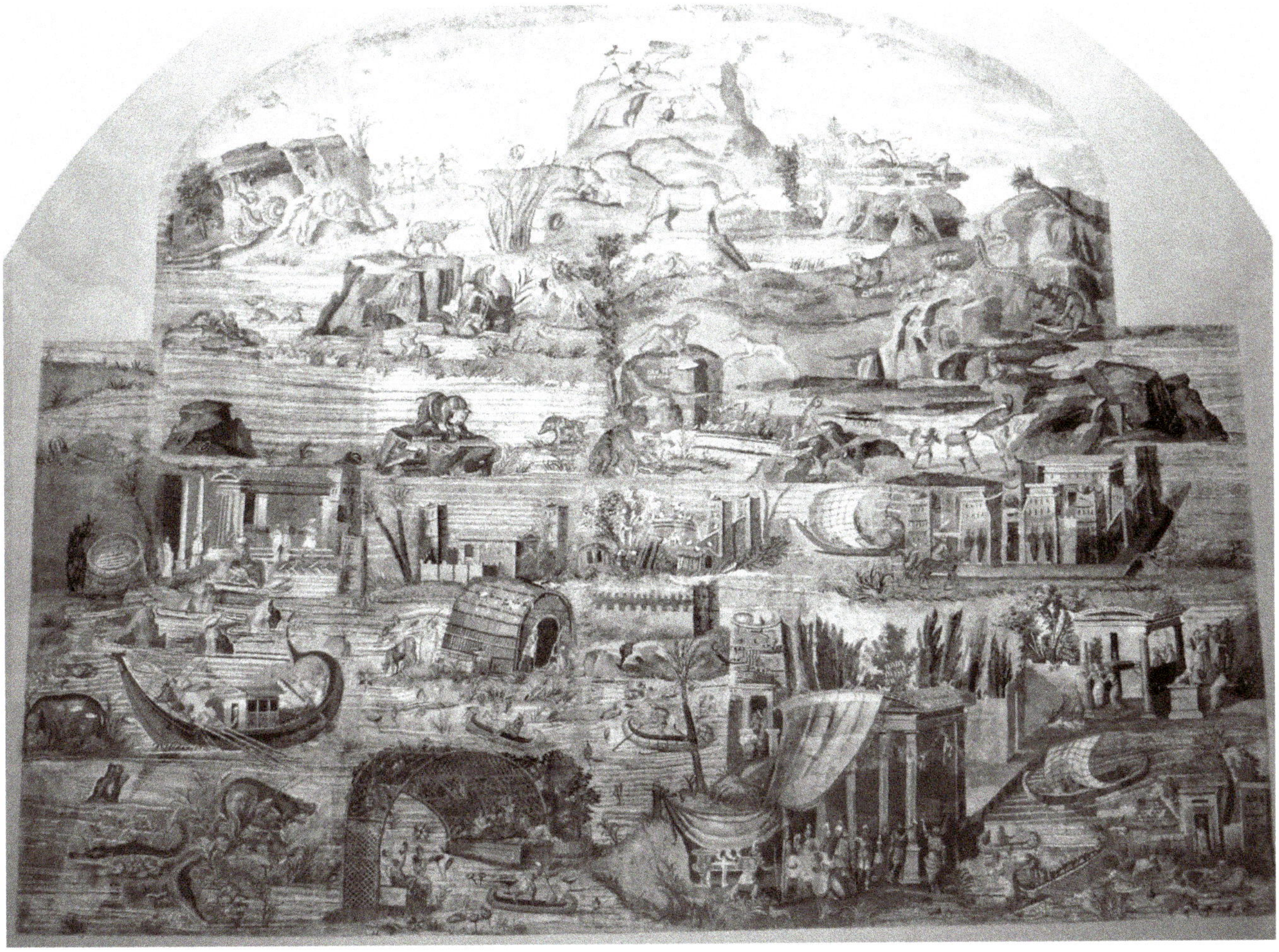

Fig. 3.6.3: General view of the Nile mosaic

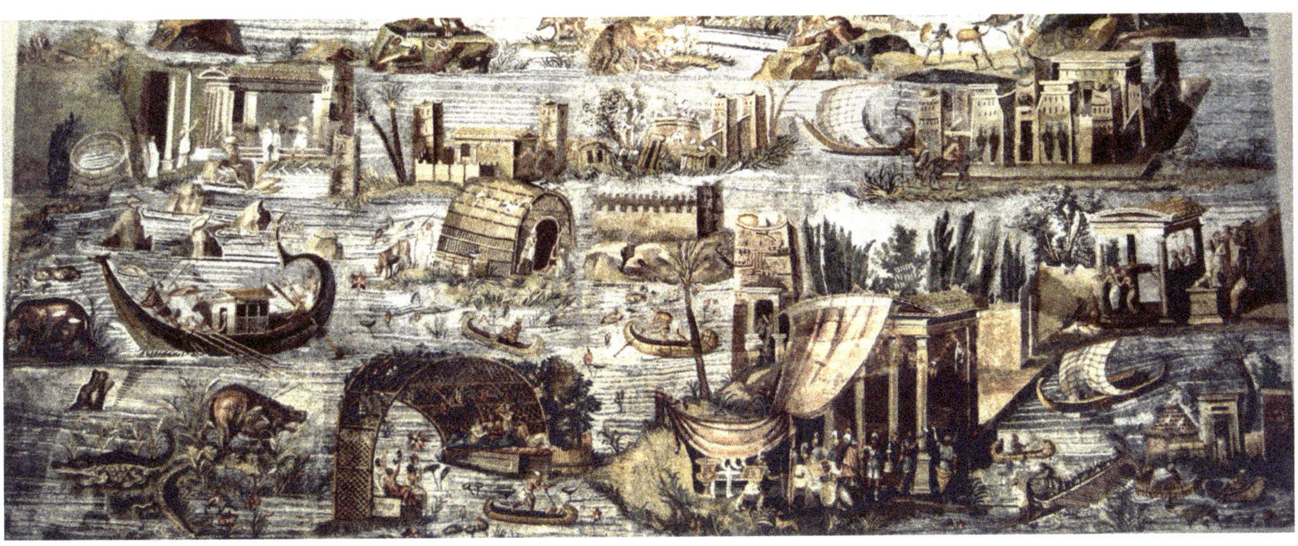

Fig. 3.6.3a: Close-up of the maritime scenes

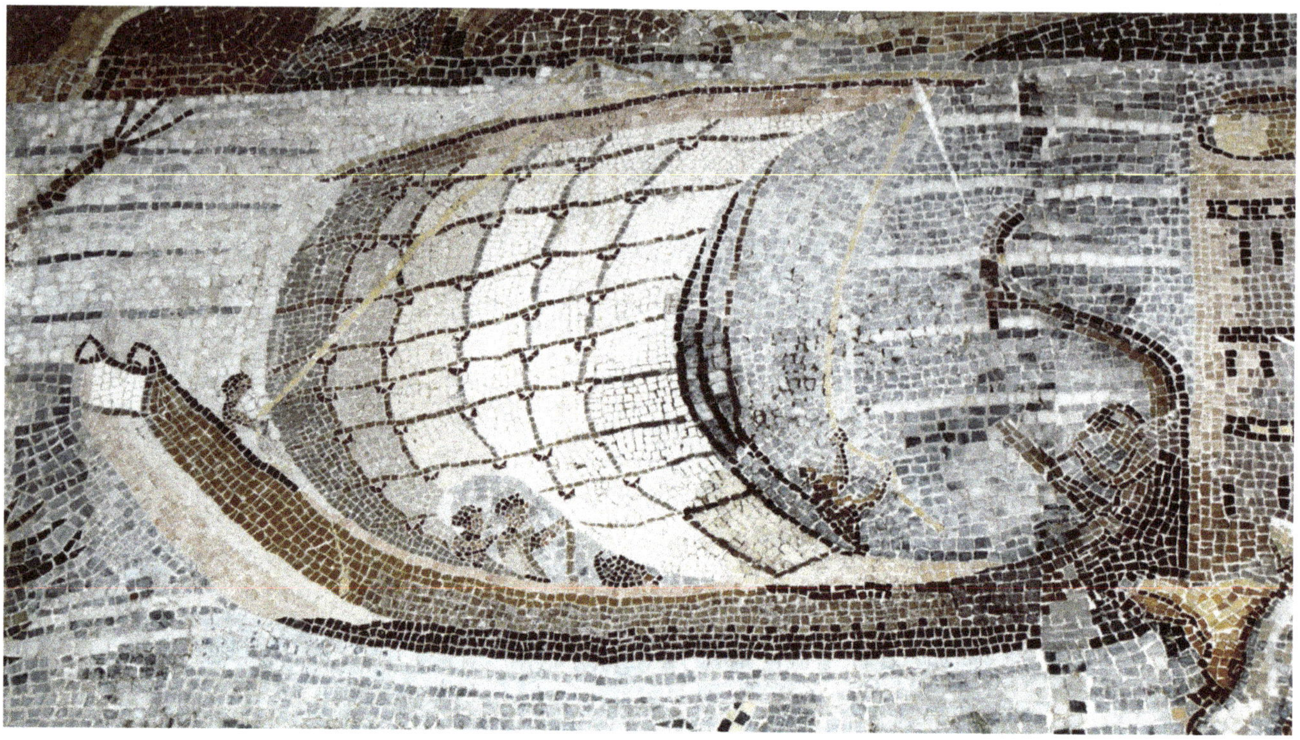

Fig. 3.6.4: Sailing Ship 1

fire; and they wear sheep skins, since they have no wool, their sheep having hair like that of goats".[21]

The lower part of the mosaic, which occupies a bit more than half of the entire surface, illustrates the northern part of the Nile, specifically the Delta at the inundation period, when the Nile flooded the fields and villages that became like islands. During this period the population living along the Nile rested from their labors and participated in hippopotami hunting, flood ceremonies and festivities of the goddess Isis. Diodorus gave a vivid description of the inundation period that augments the depiction in the Palestrina mosaic:

> *"The rise of the Nile is a phenomenon which appears wonderful enough to those who have witnessed it, but to those who have only heard of it, quite incredible. And since the land is a level plain, while the cities and villages, as well as the farmhouses, lie on artificial mounds, the scene comes to resemble the Cyclades Islands. The wild land animals for the larger part are cut off by the river and perish in its waters, but a few escape by fleeing to higher ground; the herds and flocks, however, are maintained at the time of the flood in the villages and farm-houses, where fodder is stored up for them in advance. The masses of people, being relieved from their labours during the entire time of the inundation, turn to recreation, feasting all the while and enjoying without hindrance every device pleasure".*[22]

Philostratus also gives a nice summary of the Nile's flooding period:

> *"At its flood the Nile makes Egypt open to the boats; then when it has been drunk up by the fields, it gives the people a fertile land ..."*[23]

The Vessels

Within the Nilotic scenes are depicted nine vessels of different constructions, types and size. These vessels were not studied in detail yet, except for only few of them which were referred in a general way or related to other contexts by some scholars. They will be described in clockwise order. There are two types of sailing ships, one large luxurious cabin-carrier, one warship, four papyrus boats, and a single coracle/hide-boat (fig. 3.6.3a). Each vessel will be described with its means of propulsion, type, function and size.

Sailing Ship 1: The ship sails from behind a rounded towered structure. The bow is pointed to the left. It is depicted in combined views of side and frontal perspective (fig. 3.6.4). The long, slim hull is made with white, pink, yellowish-brown and dark brown tesserae. The vessel has a flat bottom and a raised stem (c.45 degree) ending with a transom board, made with pink and white tesserae. The tips of the caprails (outlined with one black strip of tesserae) are visible above the upper edge of the transom board (fig. 3.6.4). The gunwale is indicated by one row of dark brown stones in the fore part, amidships is yellow, and on the aft

---

21 Strabo, *Geography*, 17.2.2.
22 Diodorus I.36.7 - 10.
23 Philostratus, Book I.6.

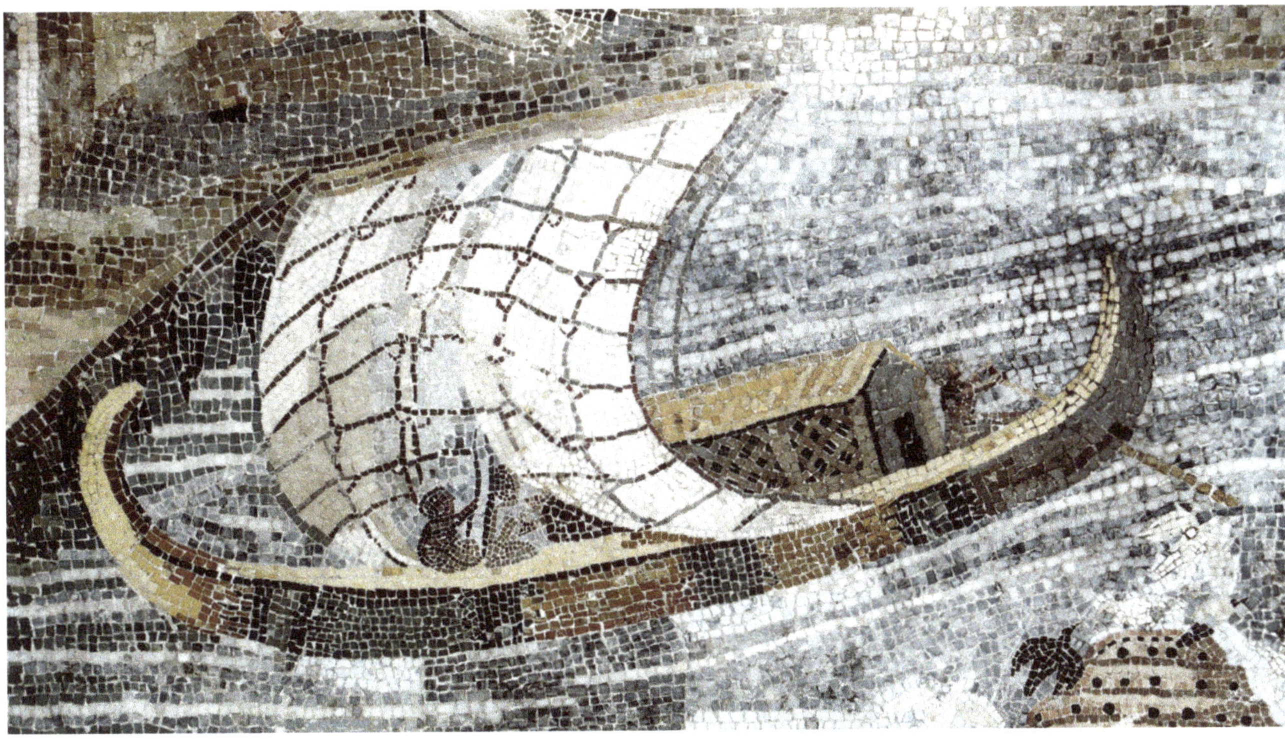

Fig. 3.6.5: Wooden Papyriform Ship

side is made with dark brown tesserae. The port quarter and stern are made with dark-brown tesserae, thus indicating the bitumen/pitch coating to make the hull waterproof, as well as protecting it from the *teredo navalis* (ship worms). Although the tower hides part of the stern, its shape may be deduced. The broad rounded stern with an upraised curve is surmounted by an inner curved sternpost above the head of the helmsman. The tip of the sternpost is adorned with a fishtail decoration.

The ship is rigged with a fully open, large square sail, seen from its fore face billowing forward. The yard has a slight downward curvature and is set perpendicularly to the hull, thus indicating that the wind blows from astern and the ship sails into the wind. The upper part of the yard is made with one row of light brown tesserae and its lower edge with one row of dark brown tesserae. Three lifts on the port side of the masthead and two on the starboard side support the yard (fig. 3.6.4). The mast is hidden by the billowing bunt. The head of the sail is made with three rows of pinkish-brown tesserae, while the bunt is rendered with dark-and-light gray and white tesserae. The middle and lower port leech is depicted with one row of dark brown stones. On the fore bunt are shown five brails and ten reinforcing-bands. At the intersection of the lines are depicted the brail-rings/loops through which run the brails (fig. 3.6.4). The crew comprises five men that each one of them has a distinct job in the propulsion of the ship. The figure seated on the bow facing the sail pulls a line that represents the foresail, and the he is in the process of securing the line to the foredeck or to the stempost. The foot of the sail is highly arched in the middle. Two seated figures on the deck are revealed beneath the arched foot. They work the lower end of a line (made with ochre-brown tesserae) that probably indicates the halyard or the middle brail (fig. 3.6.4). The port sheet stretches backwards and appears to be attached to the quarter gunwale. The starboard sheet is not visible, but we may deduce to be fastened in the same way as the port sheet. A figure standing behind the sail (his back turned to the lee face of the sail) pulls the port brace stretching from the tip of the port yardarm. Between the sail and the figure there are two lines, each made with one row of dark brown cubes. They represent the brails hanging on the lee side of the sail. The helmsman sits on the raised quarterdeck platform and works the port rudder by a short tiller inserted perpendicularly beneath the loom's head; this gear is rendered with dark brown tesserae (fig. 3.6.4). We may assume that a similar steering arrangement was true on the starboard quarter. The helmsman is rendered with a lighter pink hue, while the crew has dark-skinned, thus suggesting that they probably are Egyptian.

***Papyriform* Sailing Ship:** This vessel is found within an anchorage together with other crafts that will be described below. The bow of the ship points to the left towards the quay, as revealing its port side hull (fig. 3.6.5). The elongated, slim hull is finished with an upraised curved stem and stern. The stempost made with ochre tesserae has a slight inner curve and finished with a rounded tip. The sternpost is raised vertically and finished with a pointed tip; it is made with dark brown stones. The hull is decorated with dark green, black, dark and lighter shades of brown strips. The shape and coloring scheme of the wooden hull indicate a *papyriform* vessel.[24] The dark green and the

[24] Landstrom, 1971, p. 57.

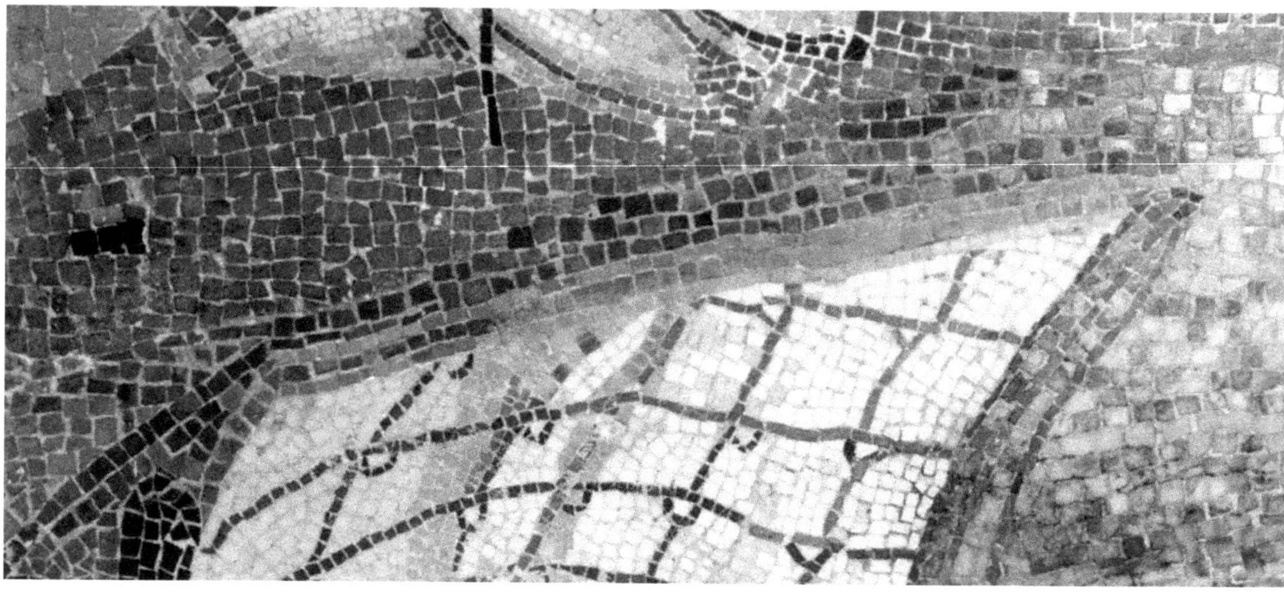

Fig. 3.6.6: Close-up of the yard

brown shades probably symbolically indicate the papyrus bundles, while the vertical black strips resemble the cords used to tie the bundles (fig. 3.6.5). The deck platform at the gunwale level is depicted with light ochre stones. Between amidships and the aft deck there is an elongated rectangular cabin, topped with a pitched tile terracotta roof made with light brown and ochre tesserae. The port side of the cabin is covered with lattice screens. The door is located in the back wall.

The ship is rigged with a fully open large square sail seen from its fore face billowing forward, probably indicating that the wind blows from astern, thus the vessel sails into the wind. The yard, set perpendicularly to the hull, is depicted with one row of ochre cubes and one row of light brown tesserae. Originally it was attached to the masthead by lifts. Probably in one of the restorations (long history) these lines were erased and replaced by dark-brown tesserae to make the shore's background (fig. 3.6.6). It also is distinguished by the size of the tesserae that are different from the original ones and the mosaic setting. The dal Pozzo watercolor reproduction of the ship shows that formerly the mast was secured by the forestay stretching from the intersection of the lifts and the masthead to the starboard fore-gunwale (fig. 3.6.7). The sail is made with white and pink tesserae. The port leech is depicted with one row of brownish-red stones. On the fore bunt are represented nine reinforcing-bands and five brails; all the lines are made with one row of dark brown tesserae. At their intersection are depicted the brail-rings/loops that the brails run through (fig. 3.6.5). In the dal Pozzo drawing seven brails are depicted on the lee side of the sail with their lower ends attached to the starboard quarter (fig. 3.6.7). In the mosaic the brails are missing. The ship as appears in the mosaic now is inhabited by three men engaged in varied propulsion jobs. In the dal Pozzo drawing there are only two men working the sail from the middle of the sail's foot. Two seated men amidships with their backs turned to the prow are revealed beneath the arch foot of the sail. They work two vertical lines, probably the double lines halyard, to lower the sail when sailing into the harbor (fig. 3.6.5). The port sheet stretches backward and its end is attached to the aft port gunwale. The starboard sheet has the same method of attachment. Two curved lines hanging from the tip of the port yardarm may indicate the port braces (fig. 3.6.5).

The vessel is steered by one steering oar mounted on the starboard quarter, slanting to the rear. A standing figure on the quarter, facing the stern, works the loom of the long tapered shaft, depicted with ochre-brown stones (fig. 3.6.8). The blade is not visible being submerged into the water. In the dal Pozzo drawing the helmsman is missing (fig. 3.6.7). Probably in one of the mosaic restorations, the helmsman maneuvering the steering oar was added and replaced the brails shown in the drawing. Apparently, this depiction augments that the restorers did not use the drawing as a reference.

Coracle: On the lower right corner of the mosaic, there is a small rowboat with an elliptical hull, depicted with dark and lighter shades of brown, pink and black tesserae (fig. 3.6.9). The boat is seen from the bow in a three-quarter view. The rim/gunwale is emphasized by one row of dark brown stones. A seated man with its back turned to the bow, rows the boat from amidships using a pair of oars. A long shaft represents the starboard oar with its blade (not visible) submerged in the water. The lower side of the shaft is made from one row of dark brown tesserae and the upper part with light brown stones. The shaft is mounted to the tholepin inserted amidships into the gunwale (fig. 3.6.9). From the position of the figure, we may deduce that he works a similar oar (not visible) mounted on the other side of the boat. He is in the process of pulling the oars, as deduced from his backwards lying (fig. 3.6.9).

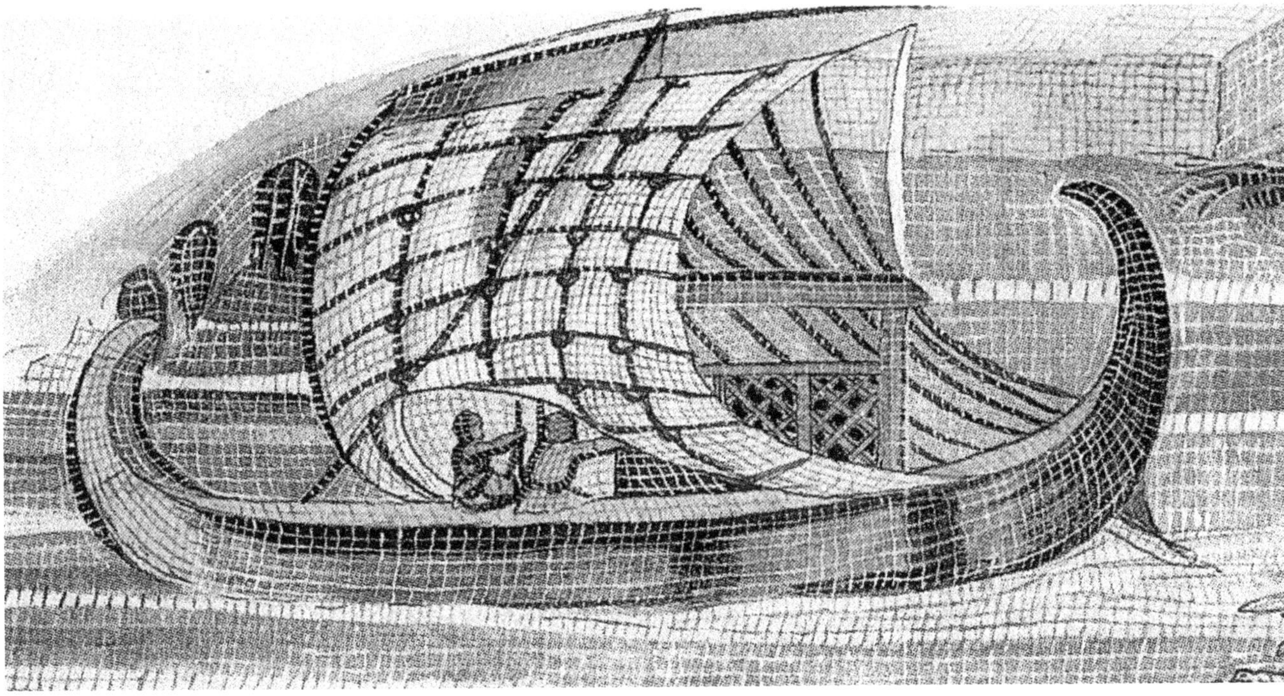

Fig. 3.6.7: dal Pozzo watercolor of the papyriform ship

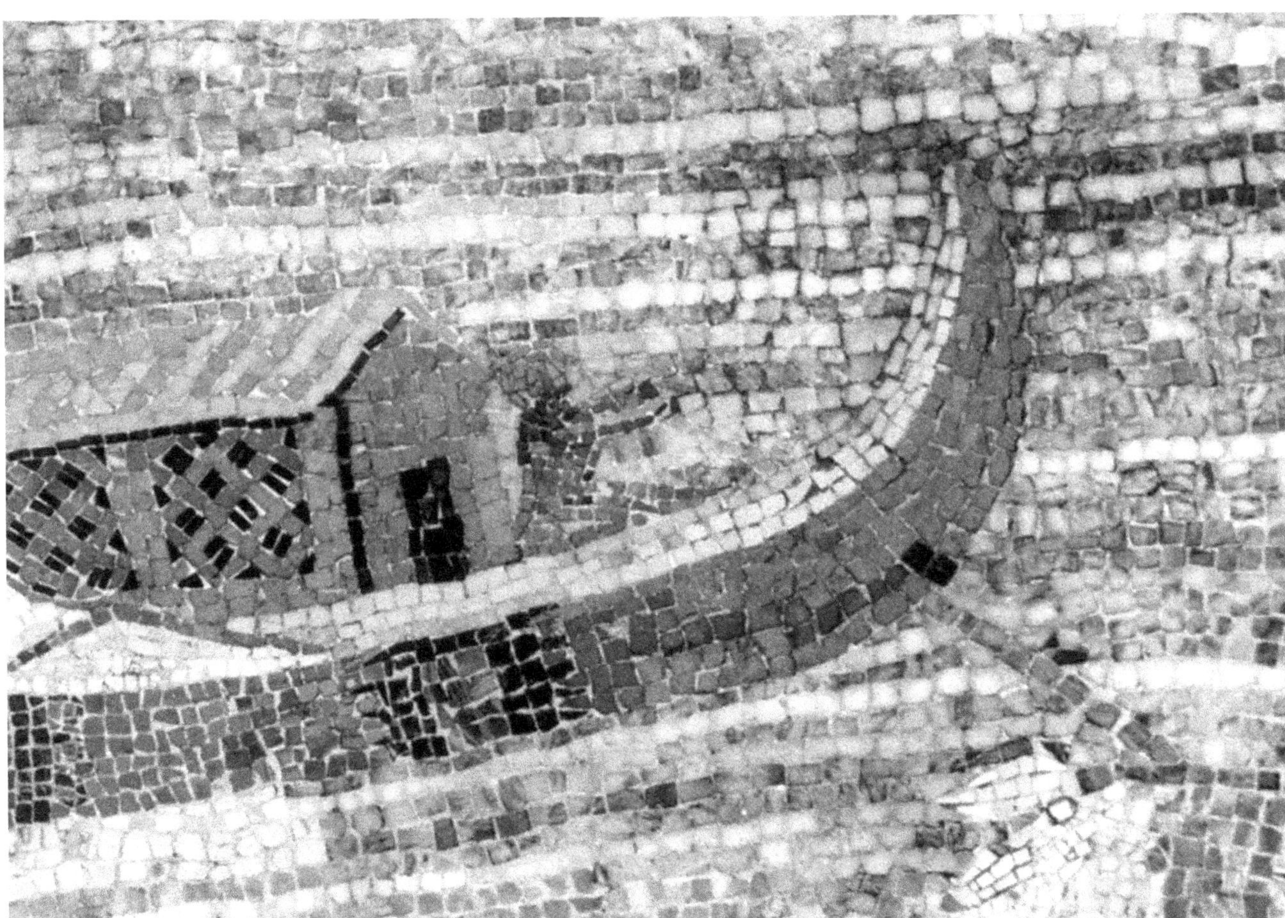

Fig. 3.6.8: Close-up of the helmsman working the steering oar

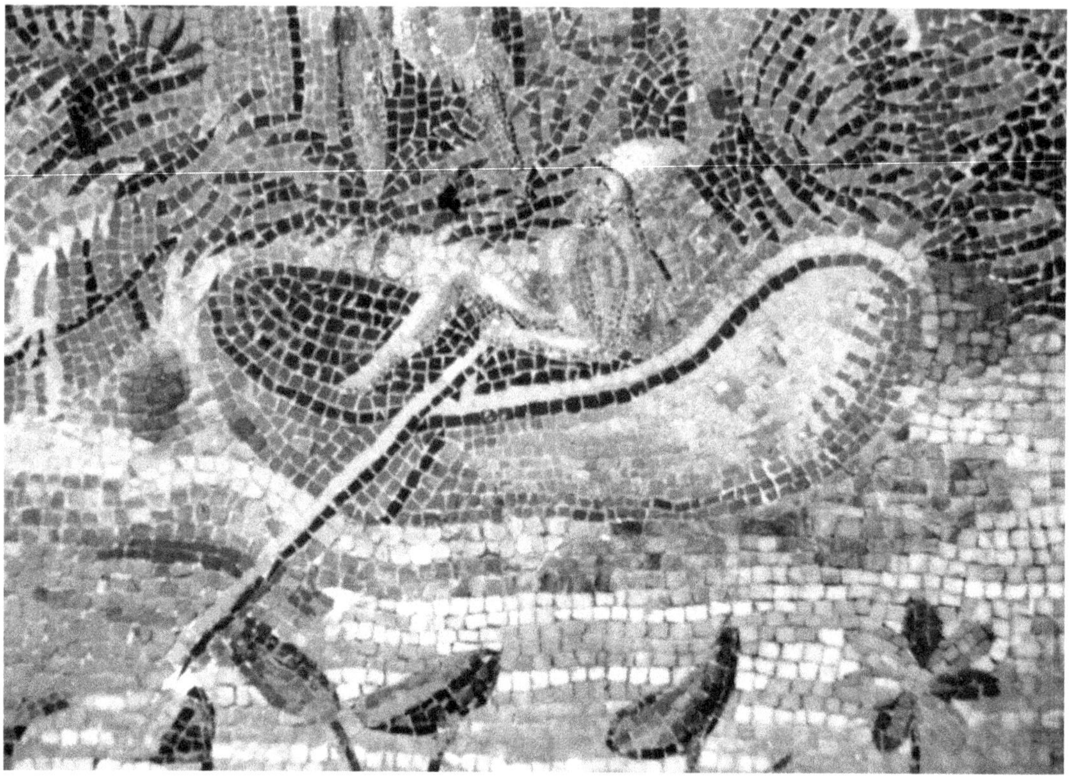

Fig. 3.6.9: The coracle boat

The boat probably represents a *coracle* made of a light frame of branches and laced together with withes or cords. Several small hides sewn together were stretched over the frame and when the boat was finished pitch or bitumen was used to coat the inner hull and make it watertight. The dark brown-black tesserae used for the inner coloring of the hull indicate the bitumen/pitch. The coracle was used for fishing and carrying the catch, or light cargo and also for personal transportation on the river. A larger version could carry up to 4 – 5 tons of burden.

The Warship: In the lower corner of the mosaic is depicted a complex maritime scene within an anchorage (fig. 3.6.3a). The warship is shown in an oblique perspective with the bow pointing to the lower edge of the panel, as viewed slightly from above. The hull is made with dark brown tesserae on its bottom and lighter brown on the prow. This dark hue on the bottom indicates the bitumen coating to make the hull watertight and also protect it against *teredo navalis* (ship worms). The concave stem ends with a projecting ram that is submerged in the water (fig. 3.6.10). The starboard outrigger oarbox is seen from its fore side while on the port one is depicted lengthwise the hull. There are twenty-seven oars projecting from the port top oarports. Only the first oar from the lower row is visible. The same number of oars may have been set in the lower row. The port oars arranged in *en-echelon* are set in two rows (fig. 3.6.10). The shafts are made with light and dark brown tesserae while the blades appear to be submerged in the water. The arrangement of the starboard oars is not very clear, but we may suggest as being similar to that on the port side (fig. 3.6.10). The vessel misses its steering gear.

The up-raised stempost is adorned with a back turned rounded tip. Between the stempost and both sides of the prow is depicted a trapezoid frame, outlined with one row of black stones. The remains of a large black dot visible in the middle of the port frame may indicate the *ophthlmos* that was a typical decoration on the stem of the Hellenistic warships (fig. 3.6.10). The altered pattern is the result of one of the restorations that the mosaic went through. The top port stern is outlined with brown tesserae and the field is made with dark grayish-green hues. The inner turned sternpost is adorned with four open branches *aphlaston*[25] crossed by two angled *stylis*;[26] both elements distinguish typical Hellenistic warships.

On the deck are seen seventeen warriors (*hoplites* or *epibatai*) equipped with elliptical shields and long lances or spears.[27] The shields are arranged along the port bulwark, as a protection for the warriors. The standing *prorates*[28] on the prow sounds a long trumpet, resting on his right arm and palm. In his left hand he holds two short spars. He is announcing the festive entrance of the ship into the harbor or anchorage (fig. 3.6.10). On the quarter, beneath the sternpost and the *aphlaston*, sits the helmsman or

[25] See Glossary.
[26] Ibid.
[27] The shields probably were rounded; due to the intended perspective they appear elliptical or indeed they were elliptical.
[28] See Glossary.

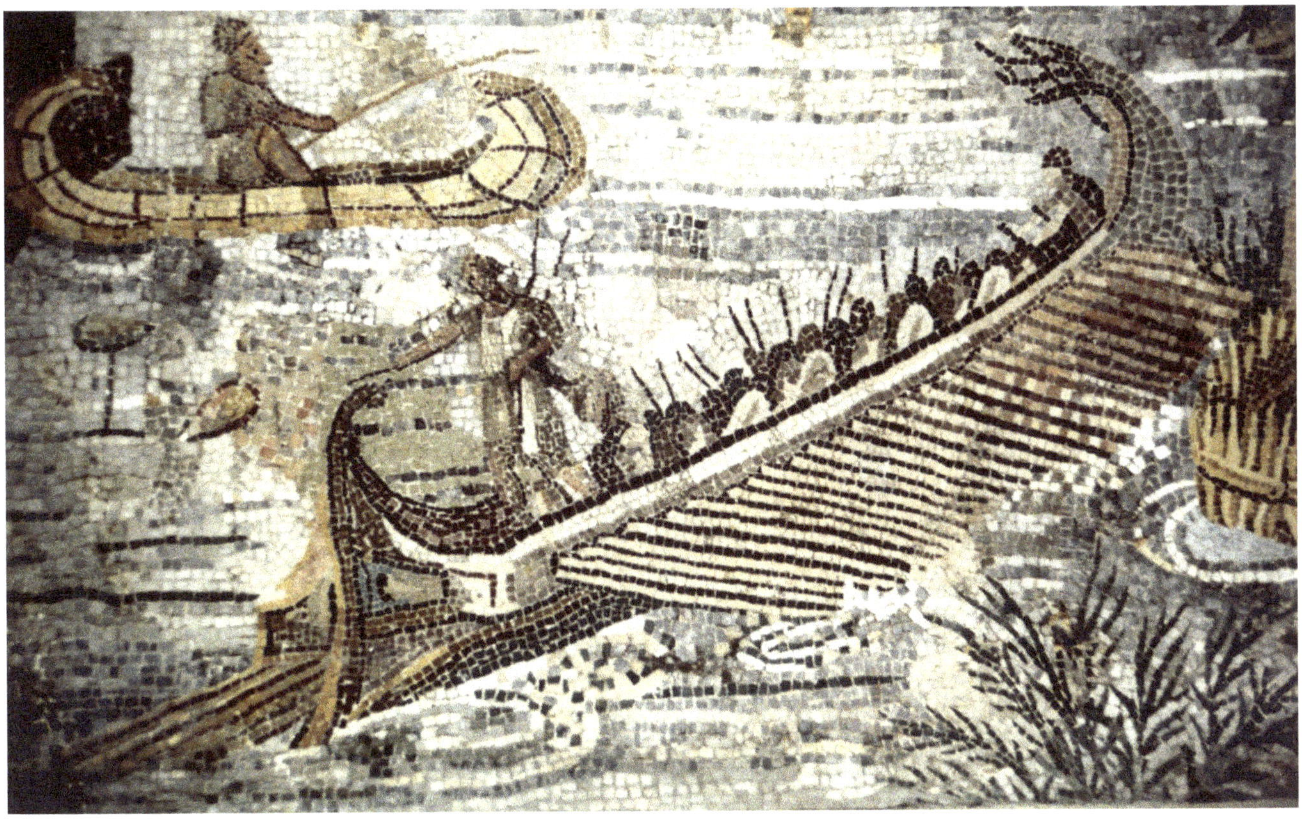

Fig. 3.6.10: The Warship

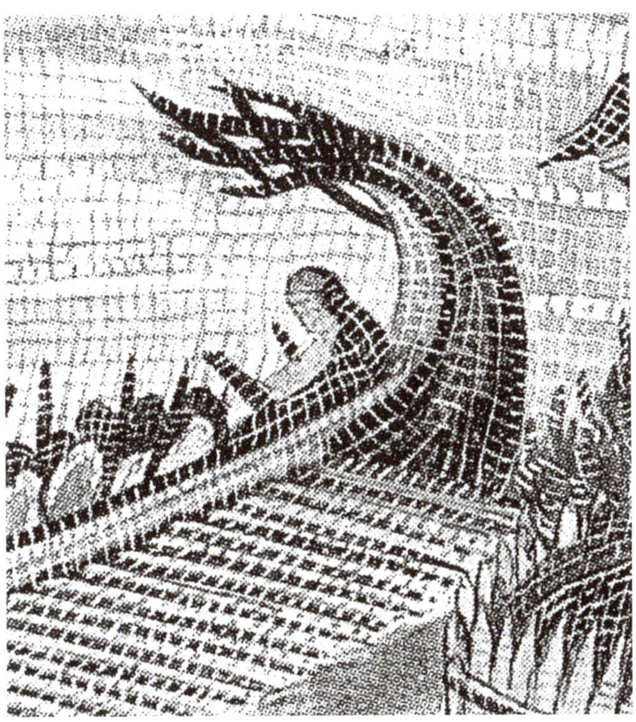

Fig. 3.6.11: Close-up of the helmsman in the dal Pozzowatercolor

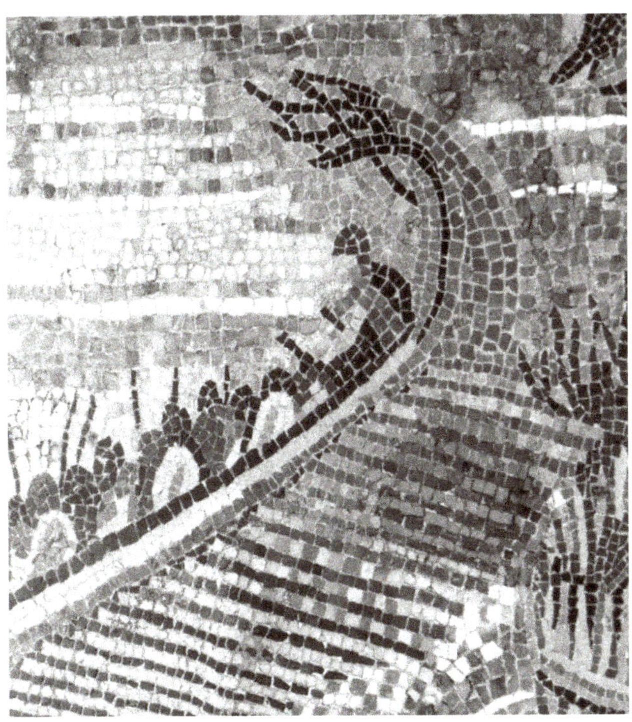

Fig. 3.6.12: Close-up of the keleutes/helmsman

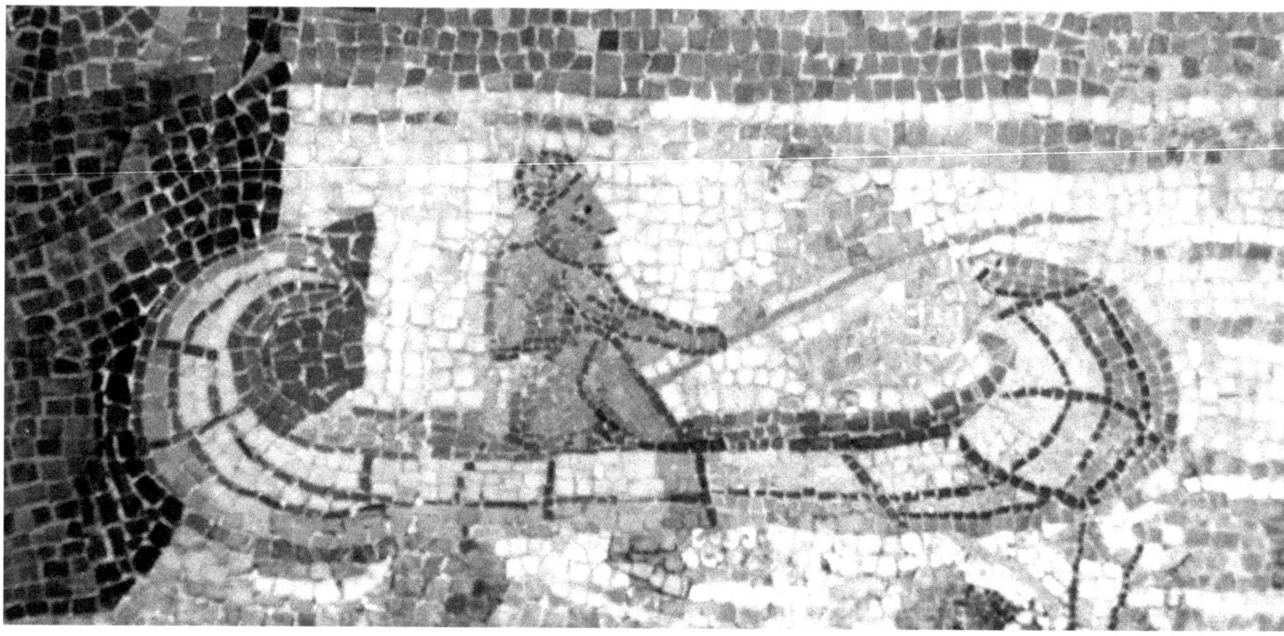

Fig. 3.6.13: Papyrus Boat 1

keleutes.[29] There is no indication of any steering gear. At some stage, this gear was probably misinterpreted and made into a rowing oar. Originally, the ship may have had at least one steering oar or rudder on the port quarter. In the dal Pozzo drawing the tips of two spars project above the hands of the helmsman. They probably are the looms of the rudders or may represent the tiller inserted into the head of each loom (fig. 3.6.11). The position in which the figure sits and the small sticks held in each hand as it appears in the mosaic today, may indicate that he is beating the drum for the rowing rhythm (fig. 3.6.12). This representation resulted from one of the restoration, whereas the restorer did not follow the dal Pozzo drawing. In Hellenistic warships *prorates* took command of the rowers and *keleutes* could also be the helmsman.[30] In the mosaic, the helmsman seems to beat the rhythm for the rower in maneuvering the vessel into the anchorage.

Papyrus Boat 1: This boat is placed above the warship, perpendicular to the quay (fig. 3.6.10). It has a long, narrow hull, flat bottom and both up-turned rounded ends finish with pointed tips (fig. 3.6.13). There is a slight difference between the ends of the boat. We may assume that the broader end is the stern while the thinner one is the bow pointed to the quay. Although, the anchor or a mooring line is missing the boat is anchored as deduced from its static position. The hull depicted with light brown, dark ochre and reddish-brown tesserae, indicates the papyrus or reed bundles bonded together by withes or cords. The vertical lines along the hull depicted with dark reddish-brown stones represent the cords (this scheme is seen in all the papyrus boats depicted in the mosaic). The boat has a shallow hold, as deduced from the dark brown coloring of the inner left side hull and the raised knees of the seated fisherman (fig. 3.6.13). The boat misses its propulsion gear (oars or sail). The fisherman facing the stern is fishing with a rod held over the stern.

Papyrus Boat 2: The boat is depicted in the lower-middle part of the mosaic beneath the arched pavilion (figs. 3.6.3a). Its hull is similar to that of Boat 1 (fig. 3.6.13), but slightly longer (fig. 3.6.14). The broad, rounded bow is turned back over the fore deck and finished with a rounded tip. The stern is slightly smaller and also has a reverse curvature, finished with a pointed tip. The papyrus bundles are tight with cords or withes as indicated by the vertical black strips. The flat deck made with black tesserae is also deduced from the feet of the standing figure. The black color represents the bitumen coating was used to make the boat watertight (usually the coating was made on the outer hull). On the fore deck are depicted papyrus bulbs. The man punts the boat from the port side in a standing position by using a long pole held with both hands, while his body leans forward, expressing the energy used in punting (fig. 3.6.14). The weight of his body is sustained by the bent left leg.

The life of the people who lived in the Nile Delta was not easy. The men used almost everything that the Delta provided for food, in a similar way as the papyrus bulbs transported in the Palestrina boat; this scene seems to follow a passage described by Strabo:

> *"They indeed live a hard life, go almost naked and are nomads. But some use grass as food as also tender things, lotus and reed-roots".*[31]

Herodotus gives detailed description about the use of lotus as part of the Egyptian's diet:

[29] Ibid.
[30] Casson, 1971, p. 301.
[31] Strabo 17.2.1-2.

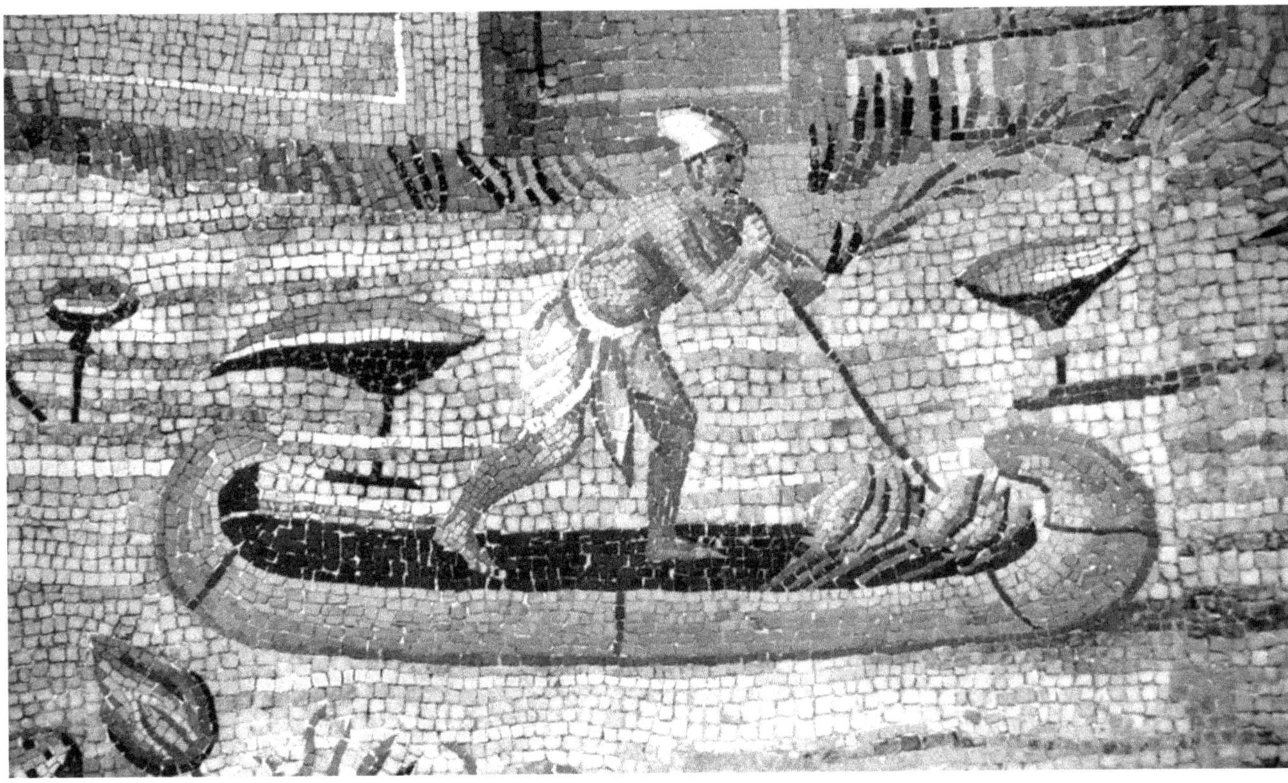

Fig. 3.6.14: Papyrus Boat 2

"They make their food less costly. When the river is in flood and overflows the plains, many lilies, which the Egyptians call lotus, grow in the water. They plant these and dry them in the sun, then they crush the pulp-like center of the plant and bake loaves from it. The root of this lotus is edible and of a sweetish taste. Other lilies also grow in the river, which are like roses ...; this produces many edible seeds as big as an olive stone, which are eaten both fresh and dried. They use also byblus, which grows annually; it is plucked from the marshes, the top of it is cut off and the lower part of a cubit length, eaten or sold. Those who wish to use the byblus at its very best, bake it in a red hot oven before eating".[32]

Papyrus Boat 3: This boat is placed almost in the center of the maritime scene in the mosaic, as sailing from right to left (fig. 3.6.3a). The hull is somewhat different from the boats described above; it has a high, rounded and tapered stem finished with an inner turned small volute (fig. 3.6.15). The rounded and very broad stern ends with an inner turned pointed sternpost. The stern of this boat is similar to that of Boat 1 (fig. 3.6.13). The horizontal platform is depicted with dark brown tesserae which indicates bitumen coating to make the boat watertight. An upright figure punts the boat using a long pole pushed forward from the port stem. The man leans forward as if putting all his energy into the punting, and also suggested by his upraised right leg. The boat is used for his transportation in the Delta.

[32] Herodotus II.92.

Papyrus Boat 4: The boat is set in an oblique position above the arched pavilion, with its bow pointing upwards (fig. 3.6.16). The hull is identical to that of Boat 1 (fig. 3.6.13). Both up-curved ends are finished with pointed tips. The black tesserae indicate the bitumen coating of the inner hull. The flat deck may be deduced from the seated naked figure who paddles the boat from the starboard side. The short loom visible between his right knee and the left palm indicates the paddle-oar (fig. 3.6.16). The boat is used for the man's transportation in the Delta.

Thalamegos (Cabin–Carrier): This is the largest and most luxurious vessel depicted in the Palestrina Nile mosaic (figs. 3.6.3a; 3.6.17). It sails from right to left with the prow pointed to the left. The broad hull and its rigging are different from the sailing ships represented above (figs. 3.6.4, 3.6.5). The hull built with long wooden strakes is indicated by dark reddish-brown, ochre, brown, grayish-green and black tesserae (fig. 3.6.17). The prow curves upwards at an angle (c.45 degree). The stempost is surmounted by a fine horse or seahorse figurehead, and probably casted in gold or bronze. The broad rounded stern has a high reverse arching over the quarter. The sternpost is adorned with a vertical fishtail/fin figurehead. A narrow yellowish banner coiled around its upper part of the tail gives the ship a festive appearance (fig. 3.6.17).

Beneath the port gunwale, there is a wide strake depicted with dark brown cubes. This may indicate the upper wale with its aft-end extending in a wing-like shape and probably supported by two or four short through spars, thus housing the rudder (fig. 3.6.17). A fancy cabin with a barrel roof

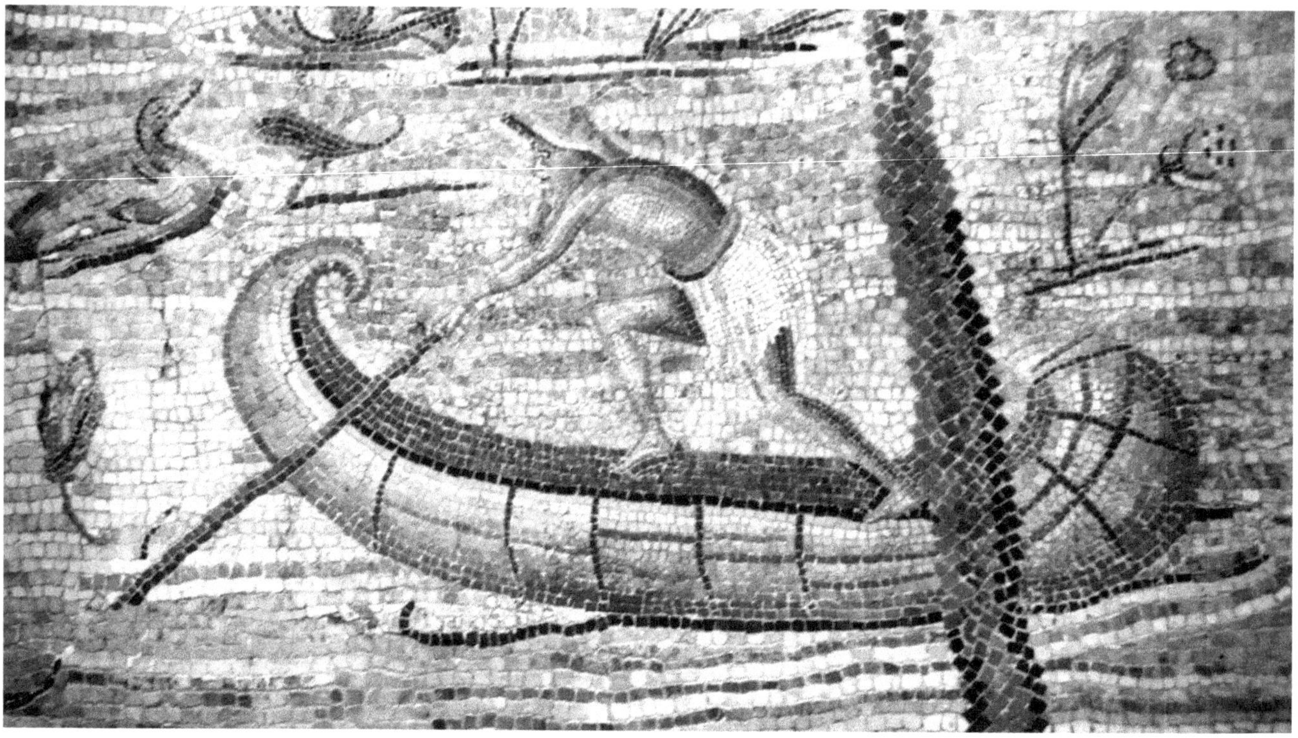

Fig. 3.6.15: Papyrus Boat 3

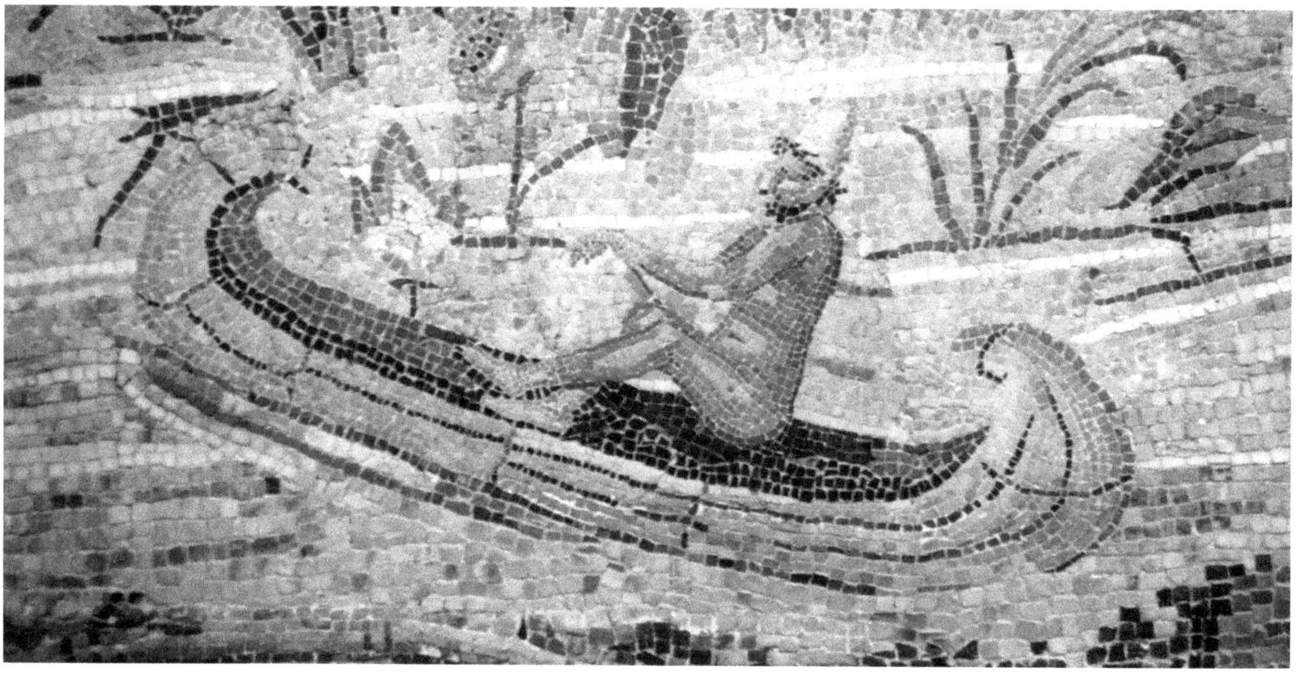

Fig. 3.6.16: Papyrus Boat 4

of terracotta tiles is placed on the mid-deck. Both arched pediments are decorated with *acroteria*, probably in the shape of *acanthus* leaves (a very common decorative element in classical architecture and art). On the port side of the cabin there are two Ionic columns that frame the half-height lattice screens. The fore side of the cabin seems to be made of bricks wall with the door is set in the center.

Three long spars with a forward slanting represent the rowing oars mounted on the port side. The shafts are fastened to the fore gunwale by rope or leather *grommets* with three hanging ends and bronze half ring rowlocks (fig. 3.6.13). The blades not visible are submerged in the water. Two seated men on the fore deck, facing the cabin work two of the oars. A third figure emerging from the cabin seems

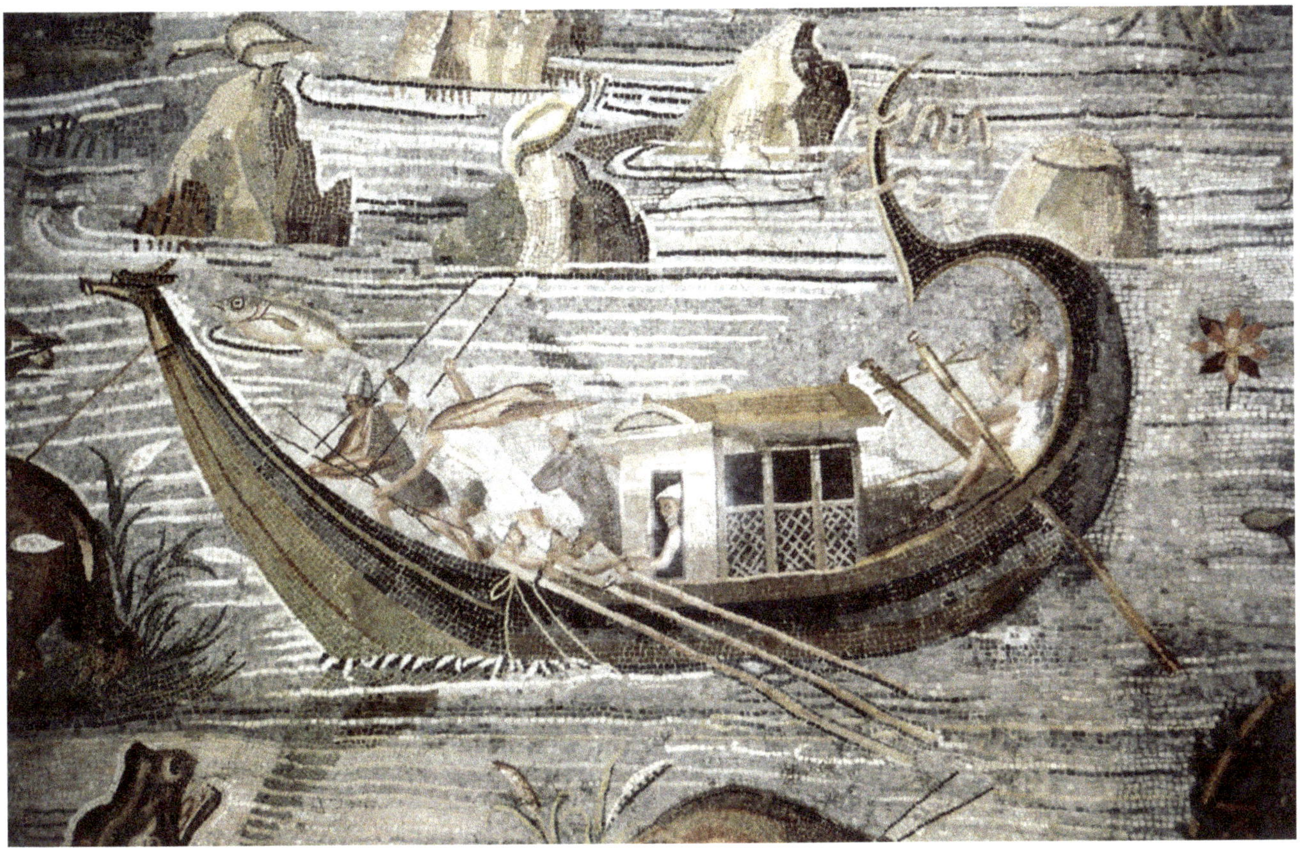

Fig. 3.6.17: The Thalamegos (cabin-carrier)

to be working the oar closer to the cabin, as indicated by his left hand holding the loom of the oar. We may assume that the same arrangement of row-oars may was true on the starboard side. Therefore each oar worked by one rower, indicates that the crew comprised six rowers and one helmsman. The rudders, one mounted on either quarter are worked by the helmsman who seats on a raised quarter bench beneath the arched sternpost. He maneuvers each rudder by a tiller inserted perpendicularly below the head of the looms (fig. 3.6.17). The shaft on the port rudder projects from behind the aft-wing of the top wale. The tips of both looms are decorated with some kind of a figurehead. Three other figures on the fore gangway are engaged in hunting hippopotami from both sides of the prow. Two of the men are in the act of throwing their spears at the hippopotami; one is on the right-hand shore while only the head of the second beast projects above the water (fig. 3.6.3a).

We learn about hippopotami hunting in the Nile Delta from a description by Diodorus, which is very similar to the scene represented in the mosaic:

> *"The animal called 'horse' (ἵππος) is not less than five cubits high and is four footed and clove-footed like an ox; ... and its body as a whole is unlike that of the elephant, and its skin is the toughest of almost any beast's. Being a river and land animal, it spends the day in the streams, exercising in deep water, while at night it forages about the countryside on the grain and hay, so that, if the animal were prolific and reproducing each year, it would entirely destroy the farms of Egypt. But it is caught by the united work of many men who strike it with iron spears; for whenever it appears they converge their boats upon it, and gathering about it wound it (the hippo) repeatedly with a kind of chisel fitted with iron barbs (harpoon), and then fastening the end of a rope of tow ... which had became imbedded in the animal, they let it go until it dies from loss of blood".*[33]

Discussion

The vessels depicted in the Palestrina mosaic were not used as decorative motifs only to embellish the surface rather they reflect evidence of actual vessels that plied the Nile, the Delta as well as the eastern Mediterranean coast in the late 2nd century BCE. Some of the vessels' characteristics permit us to associate them with types mentioned in ancient literature or compare them with ship depictions in other arts. The Palestrina vessels can be classified into four categories. Each vessel will be discussed below with comparative material:

1. Four papyrus boats and one small boat/*coracle*.
2. Two wooden sailing ships, each having a different hull shape.

[33] Diodorus I.35.9-10.

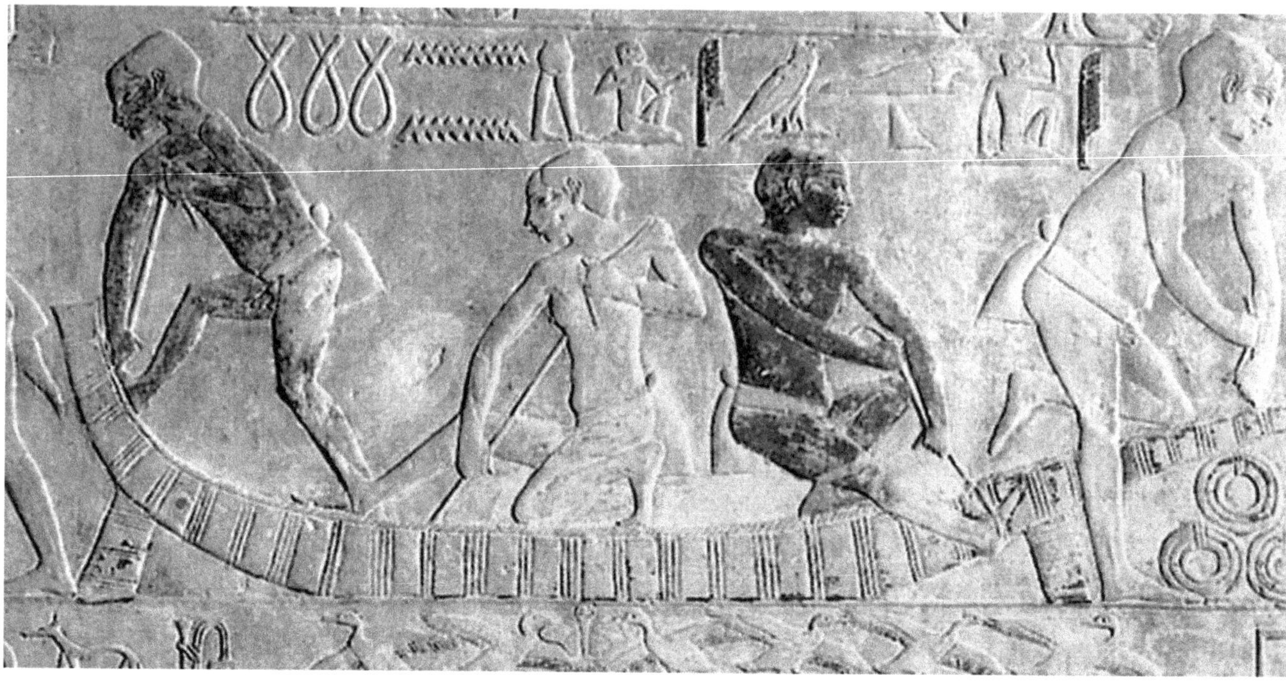

Fig. 3.6.18: Papyrus boat building from the Tomb of Ptahhotep at Saqqara

3. One Cabin–Carrier/*thalamegos*.
4. One warship.

Papyrus Boats show distinctly their building material and their hull shape. Reed bundle rafts or boats are among the oldest water vessels that were used in Egypt, Mesopotamia and Greece.[34] In Egypt, the construction material used for such boats was papyrus (*Cyperus papyrus*),[35] which grew in abundance in the marshes of the Delta as it did in Mesopotamia. Strabo mentions that in the 1st century BCE, "*byblus* (papyrus) grew in large quantities in the lower part of the Delta".[36] The only tool needed to build papyrus or reed-bundle vessels, was a simple blade (flint or metal) to cut the plants. The other required technique was the rope making, the ability to bind the bundles together and tie them with the ropes or cords in a similar way as depicted in a relief from the Tomb of Ptahotep at Saqqara (fig. 3.6.18; 5th Dynasty; 2350 BCE). The horizontal lines indicate the bundles and the vertical lines suggesting their binding by cords distinguish the papyrus or reed-bundle vessels in paintings, reliefs and models. The propulsion of the papyrus/reed boats in the dynastic Egypt comprised a long punts pole or long paddle oars with blades shaped as lance-arrows. Papyrus boats in the Pharaonic Egypt also could be rigged with a square sail set on bipod (A frame) mast[37] or a single mast pole. Such boats were extensively used in hunting and fishing activities carried out in the marshes of the Delta, as well as for funerary ceremonies, in ancient Egypt. They also were used in open sea transporting troops to Levant in the 8th century BCE, as referred in a passage in Isaiah (18:1, 2).[38] In reliefs from Mesopotamia, dated to the reign of Ashurbanipal (669 – 621 BCE), reed-rafts were used in sea battles that took place in the Assyrian marshes (fig. 3.6.19).[39] The collection of asphalt in the Dead Sea, in the 4th century BCE, was made from reed boats that also were used as patrol vessels:

> *"... when the asphalt has been ejected, the people who live about the sea on both sides are hostile to each other, making the collection without boats in a peculiar fashion. They make ready large bundles of reeds and cast them into the sea. On these not more than three men take their place, two of whom row with oars which are lashed on, but one carries a bow and repels any who sail against them from the other shore or who venture to interfere with them..."*[40]

The asphalt and the salt resources from the Dead Sea led to a war in 312 BCE between Antigonus, King of the Seleucids and Ptolemy I, King of Egypt that resulted in the supremacy of Egypt over the area. This war was waged from reed rafts:

> *"... for examining the Lake and apparently having found a source of revenue for his kingdom, he (Antigonus) placed Hyeronymus, the writer of the history, in charge and instructed him to prepare boats, collect the asphalt and bring it to a certain*

34 Probably the earliest evidence for the use of reed boats comes from Greece. Obsidian dated to c.9000 BCE was found in the Franchthi cave, in the Bay of Koilada, Argolis (Peloponnese). This obsidian originated from the Island of Melos. It is assumed that some time about 9,000 – 10,000 years ago, people who lived in the southeastern Peloponnese, traveled to the Cyclades in reed-bundle raft; Tzalas, 1995, p. 441.

35 McGrail, 2001, p. 21.

36 Strabo17.I.15.

37 Landstrom, figs. 31, 39, 132.

38 McGrail, p. 21.

39 There are several Assyrian reliefs depicting of military combats carried out from reed-bundle rafts; deGrave, 1981, figs. 51, 52, 60, 61, 62, 63, 64, 66.

40 Diodorus XIX.98-99.

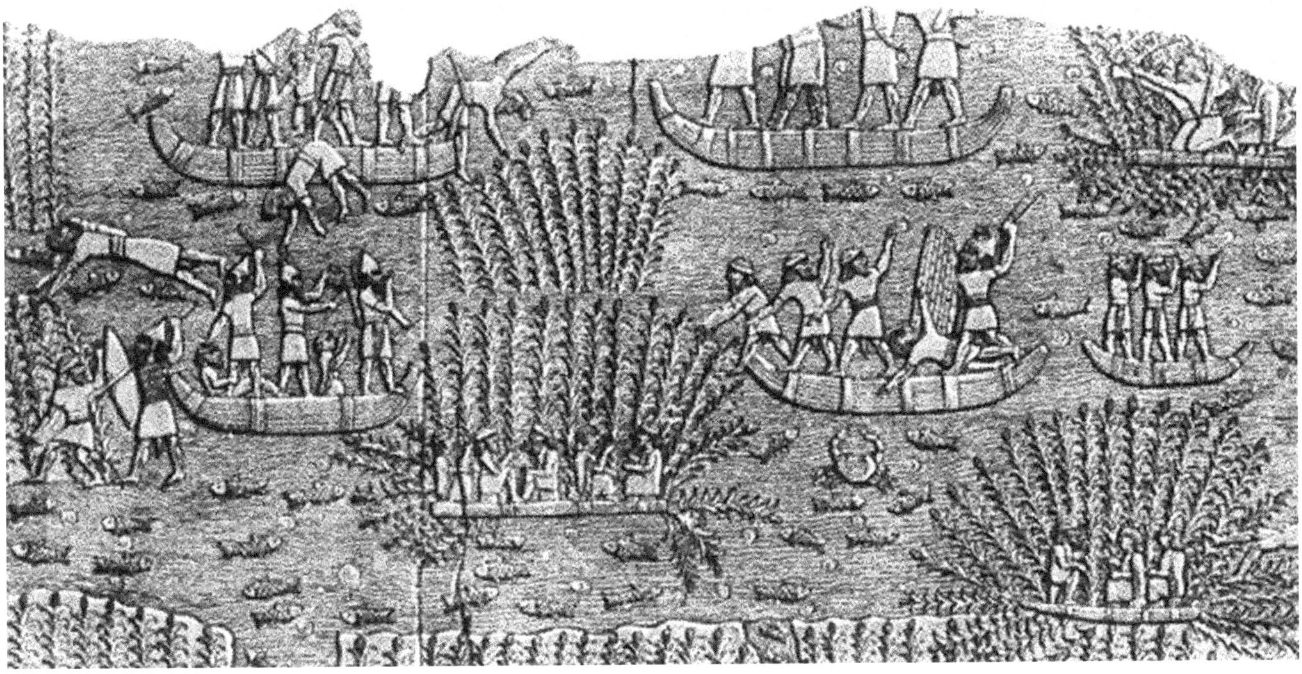

Fig. 3.6.19: Assyrian military combat carried out from reed-bundle rafts

> *place. But the result was not in accord with the expectations of Antigonus. The Arabs (Nabateans) collecting to a number of six thousand and sailing up in their rafts of reeds against those on the boats killed almost all of them with their arrows. As a result, Antigonus gave up his sources of revenue because of the defeat he had suffered".*[41]

Both passages from Diodorus show the bitumen resources and its importance for the Ptolemaic Egypt. The coloring of inner hull and deck of the Palestrina papyrus boats indicate the bitumen coating that its provenience probably came from the Dead Sea.

Many paintings, reliefs, seals and models from Egypt and Mesopotamia depict reed/papyrus vessels with their propulsions. They were either punt with a long pole from the prow or the stern, or they were paddled. One of the oldest depiction of a papyrus boat that is punt comes from the Tomb of Ptahotep (5th Dynasty), from Saqqara.[42] Additional example dated to the New Kingdom comes from the Tomb of Thutmosis III at Puirura (T– 39; Dynasty 18).[43]

Papyrus reed/boats encompassing decorative motifs were still depicted in tombs of the classical periods. A relief at Kom-Ombo, dated to the 2nd century BCE, showing Ptolemy VIII (Eueregetes II),[44] and the four boats in the Palestrina Nile mosaic (figs. 3.6.13 – 3.6.16), dated to 110 – 100 BCE, augment that similar vessels still sailed in the Nile and the Delta in the later Hellenistic period. They also indicate that these simple boats had not changed their hull construction and means of propulsion throughout the centuries, and as late as the 2nd century BCE. Papyrus/reed boats are still used today as ferries for people and goods in the Upper Nile, Lakes Tana and Chad,[45] and as far as the Eastern Islands, Mexico and Lake Titicaca (between Peru and Bolivia).[46]

Coracle: The elliptical boat depicted in the Nile mosaic is the only vessel known as *coracle* (fig. 3.6.9). Such a boat is a rare representation in any arts and especially in mosaics. Building such boat does not require a shipshed nor expensive timber rather only young flexible branches of willow or other such water-trees, animal hides (ox or sheep), and some ropes or cords. They were coated with pitch/bitumen (inside or outside) to make the hull watertight. The uniform hues of pink and browns shades in the Palestrina boat may indicate the animal hides stretched over the woven frame of branches. The black coloring inside the hull indicates the bitumen coating.

Herodotus gave a detailed reference of skin boats construction in Mesopotamia in the 5th century BCE, and what happened to them when they completed their journey. The frames of a skin boats were dismantled and sold, while the hides were packed on the back of donkeys and carried home overland:

> *"Their boats which ply on the river and go to Babylon are all made of skins and rounded. They make these in Armenia, higher up the stream than Assyria. First they cut frames of willows, then they*

41 Ibid., XIX.100.1-2.
42 Schultz and Seidel, 1998, fig. 87, p. 90; Tiradritti and De Luca, 1993, pp. 52 - 53.
43 Wilkinson and Hill, 1983, fig. 30.4.11.
44 Hornell 1946, p. 50; McGrail, 2001, pp. 52-52.

45 McGrail, p. 2.
46 Heyerdahl, 1971, figs. 1, 5, 6.

stretch hides over these for a covering, making as it were a hold; they neither broaden the stern nor the prow, but the boat is round like a shield. They then fill it with reeds and send it floating down the river with cargo. Two men standing upright steer the boat, each with a paddle, one drawing it to him, the other thrusting it from him. These boats are of all sizes, some small, some very great. There is a live ass in each boat, or more than one in the larger. So when they have floated down to Babylon and disposed of their cargo, they sell the framework of the boat and all the reeds; the hides are set on the backs of the asses, which are driven back to Armenia, for it is not by any means possible to go up stream by water, by reason of swiftness of the current; it is for this reason that they make their boats of hide and not wood". [47]

Herodotus referred to boats known as *quffa* that are still used today in Iraq and in the Arabian Peninsula; they are woven and coated with bitumen to make them watertight. The earliest depictions of *quffa* are found on Assyrian reliefs dated to the reign of Ashurnasipal (883 – 859 BCE), and also in the Palace of Senacherib, c.700 BCE.[48] Strabo mentioned woven boats with reeds.[49] These types had elongated hulls and were not rounded as the *quffa*. In the marshes of southern Iraq and on the Euphrates at Hir, boat-shaped rushes/reeds consolidated with an internal framework and coated with bitumen from the outside to make them watertight were still to be found in the 19th century.[50] The average size of a modern *quffa* is 4m in diameter, with a depth of 2.1m; larger sizes are 4.9 – 5.5m in diameter and could carry up to 5 tons of grain, and the smallest 2 tons or 30 sheep.[51] DeGrave mentions *quffa* of 5.5m in diameter that can carry loads of 16 tons.[52] A very fine example of a rounded *quffa*, dated to the 17th or the 19th century, can be seen in the Maritime Museum in Paris. This vessel was coated with a thick layer of bitumen (1 – 2 cm) on the outside, to make it watertight (fig. 3.6.20). Despite the lack of historical records of skin boats used in the Nile and the Delta, we may assume that such vessels were in common use as they did not require expensive wood for their building. Apparently the Mesopotamian and the Egyptian tradition of papyrus/reed boat building were very similar due to their way of life defined by the rivers and the abundant growing of these plants in the marshes. The similarity in boat building may also be associated with the trades of merchants (spice and aromatics) from Arabia to Ethiopia and Egypt. *Coracles* or *quffa* type boats could carry such cargo to the Nile Delta or sail through the connecting channels between the Nile and the Red Sea. The *coracle*/hide-boat depicted in the Nile mosaic (fig. 3.6.9) may be considered as a rare representation of such vessel in any arts and especially in mosaics that still was built and

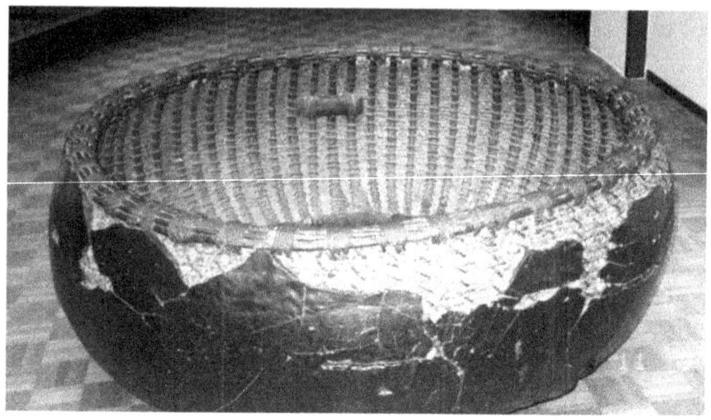

Fig. 3.6.20: Quffa with bitumen outer coating

sailing in the Nile Delta in the later 2nd century BCE. Hide boats/*coracle* are still built in different parts of the globe. They are built in various shapes: circular (India), elliptical (Arabia, Britain, Ireland), rectangular (Tibet, South America), and generally "boat shape" (Ireland, Arctic).[53]

The hulls shapes of the wooden planked vessels depicted in the Nile mosaic are varied. These vessels bring evidence of preserved boat-building traditions of river and seagoing ships that sailed in the Nile and the Delta. This tradition may be traced from Egypt's earliest history until the late Hellenistic – early Roman periods.

Sailing Ship 1: The wooden vessel having an elongated slim hull, flat bottom and a raised angled prow with a transom end (fig. 3.6.4), indicates that it followed a preserved tradition in Egypt since the 5th – 6th Dynasty (Old Kingdom).[54] Although, the hull of the Palestrina ship resembles similar ships of ancient dynastic periods in Egypt, it is a Hellenistic vessel. The raised inner curving stern adorned by a stylized fishtail/fin sternpost is a Greek influence, which became a characteristic element of Hellenistic sailing ships in the Mediterranean. Vessels with transom prow are typical river crafts that could take the ground (the beach), on the banks of the Nile when were not sailing (thus being anchored). The Palestrina Sailing Ship 1 is a typical river vessel because its hull was not meant to be put under stress of heavy sea waves and currents while sailing in open water (sea), but more for sailing like a barge carrying cargo and passengers on the Nile and the Delta (fig. 3.6.4). Therefore this ship represents a river *navis oneraria* that may be associated with a more specific type known in the ancient literature as *phasolos* or *kerkouros*.[55]

Papyriform Sailing Ship: Apparently when the early Egyptians built wooden vessels, they copied the hull shape from the papyrus boats and this resulted in a new type named

47 Herodotus I.194.
48 Casson, fig. 4; McGrail, fig. 3.15, p. 66.
49 Strabo 16.I.15.
50 McGrail, p. 67.
51 Ibid., p. 66.
52 DeGrave, 1981, p. 86.

53 McGrail, 1998, p. 173; Table 10.1, p. 174 lists the shapes of the coracle and their sizes: circular 1.52 - 4.27m (dia.); elliptical 1.22 x 0.91 - 1.83x1.37m; boat-shape 2.44 x 1.07 – 7.6 x 1.37m.
54 A fine example of a wooden planked boat with two transom ends and a raised bow (c.2400 - 2300 BCE), is depicted in the Tomb of Kaem'onkh at Giza; Landstrom, fig. 104, p. 40.
55 See Glossary.

papyriform.[56] The oldest excavated planked vessel in Egypt comes from a pit near the pyramid of Cheops (Khufu), and dated to the 4th Dynasty. This ship may be considered the as the first wooden *papyriform* example; its hull was not painted rather it shows the natural wooden planks. The sternpost has an angled inboard turn while the stempost is vertical; both posts have the characteristic ends of the papyrus umbel (fig. 5.1).[57] In later depictions of *papyriform* boats the hull scheme of colors was green amidships and gold or yellow for the ends, thus indicating the papyrus umbels tight to form the boat-shape. The propulsion of such vessels comprised paddling oars and steering oars. They also could be set with a square sail mounted to A-shape or a single pole mast. The *papyriform* ship in the Nile mosaic with its hull painted scheme of green, ochre and black (fig. 3.6.5), follows the same tradition of similar types known mostly from the Middle Kingdom in Egypt. The Palestrina ship does not have the characteristic stylized papyrus umbel ends of the Egyptian dynastic ships. The ship's raised stempost has a slight inboard turn finished with a rounded tip, whereas it follows the fashion of the Hellenistic sailing vessels. The vertical sternpost is also finished with a rounded tip. The rectangular cabin with barrel-like terracotta tile roof, *acanthus* leaves decorations at the tips of the pendants, and the lattice-screen framed by Ionic columns, resemble a Hellenistic seaside villas or inner garden kiosks.[58] In later periods this cabin became a functional unit for the crew and storage of cargoes.

The Palestrina *papyriform* vessel as depicted in the mosaic today seems to be a bit different from the dal Pozzo watercolor drawing. The sail in the mosaic shows very crude restoration of the middle brails on the fore bunt, as well as the arched foot of the sail (fig. 3.6.5). This representation indicates that the restorers did not use the dal Pozzo drawing as a relevant reference (fig. 3.6.7). The drawing was reproduced from the original piece of the mosaic that was transported to Rome for restoration in 1626 – 1630. The brails running through the brail-rings/loops overstep the yard and their lower ends extend to the starboard gunwale. Apparently when this section was restored either in 1853 – 1855 or in 1952 – 1953, the brails were erased and the helmsman was placed on the quarter deck as maneuvering the long shaft of the steering-oar (fig. 3.6.5). Also another figure was added in the mosaic; he is seated on the prow and works the forestay. The ship depicted in the drawing shows only two figures in the ship and both are working the double-lines halyard, probably lowering the sail as the ship sails into the harbor (fig. 3.6.7). The Palestrina ship represents a small to medium sized merchantman *(navis oneraria)* carrying light cargo and passengers. It also may be associated with *phasolos* or *kerkouros* type sailing on the Nile and the Delta, as well as along the Egyptian Mediterranean coast.

[56] Landstrom, p. 24.
[57] Friedman, 2007, pp. 131-132.
[58] Such structures are depicted in Roman seascape scenes, in wall paintings and mosaics from Pompeii, Sicily and North Africa from the 1st – 2nd centuries CE onwards.

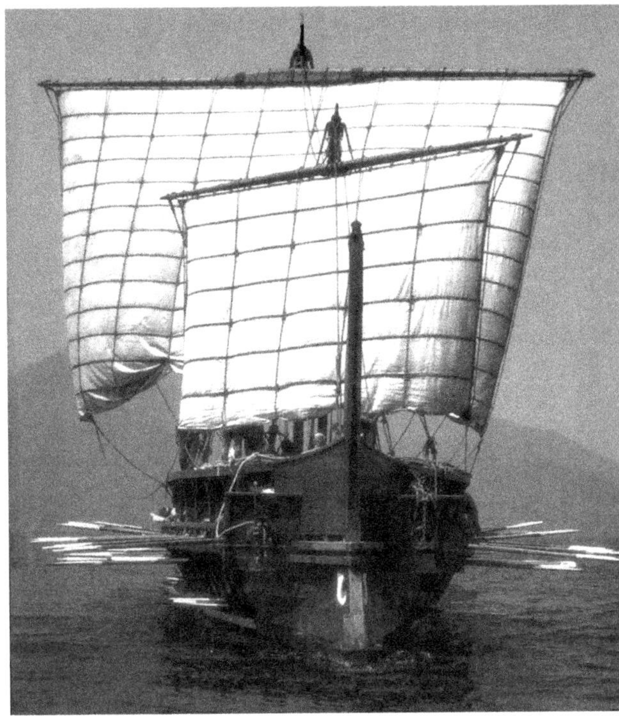

Fig. 3.6.21: Sail rigging of the Olympias

The sailing rig of the Sailing Ship 1 (fig. 3.6.4) and the *papyriform* (fig. 3.6.5) is very similar. The large open square sail is set on a mast, as the yard is supported by lifts running from either side of the masthead to the yardarms. The sail is adjusted to the wind conditions by the brails running through brail-rings/loops on the fore face of the bunt. The rings as they appear on the mosaic are accurate and realistic and can be compared to the sail rigging of the reconstructed trireme *Olympias* (fig. 3.6.21).

Thalamegos: This vessel depicted in the Nile mosaic is the largest and the most luxurious (fig. 3.6.17). It is engaged in the in hippopotami hunting during the flooded Delta. The very luxurious and expensive yacht known as *thalamegos* (literally means cabin-carrier), and resembling a great floating palace is described by Athenaeus. Such ship was built by Ptolemy Philopator (240 – 204 BCE) under the supervision of Archimedes:

> *"Philopator also constructed a great river boat, the so called "cabin-carrier" (thalamegos), has a length of 300 feet (99m) and abeam at the broadest part of 45 feet (14.85m). The height, including the pavilion when it was raised was little short of 60 feet (19.8m).*[59] *Its shape was neither like that of war galleys nor like that of merchantmen, but had been altered somewhat in draught to suit its use on the river. For below the waterline it was flat and broad, but its bulk raises high in the air; and the top parts of its sides, especially near the bow, extended in considerable overhanging. It had double bows and*

[59] Athenaeus V.204. The measurements were calculated from feet to meter considering that 1 foot=0.33m.

sterns, projecting upward to a high point, because the waves in the river often rise very high. The hold amidships was constructed with saloons for dining parties, with berths, and with all the other conveniences of living. Round the ship, on three sides ran the promenades.[60] Its mast had a height of 105 feet (34.65m) with sail of fine linen reinforced by a purple topsail".[61]

The description of the Ptolemaic *thalamegos* indicates a very large catamaran "*with double bows and sterns*" that the grandiose cabin was built on the connecting platform of the catamaran. The vessel was propelled by a very large sail set on a massive mast. As seagoing ships, triangular topsails also topped it.[62] Strabo informs us that at Schedia, on the Canopic Canal there was a marina for *thalamegoi*:

"Schedia is four shoeni distant from Alexandria; it is a settlement of city, and contains the station of the cabin-boats (thalamegoi) on which praefects sail to Upper Aegypt".[63]

A papyrus dated 257 BCE (*P.Ryl.* 558) mentions that a *thalamogos* used to transport officials on the Nile and had a crew of eight.[64] During the late Hellenistic period, when the greatness of Egypt was diminished, and especially after the Battle of Actium (31 BCE), the Egyptian shipbuilding industry probably shrank. As a result of these changes, grandiose and luxurious *thalamegos* vessels were reduced in size but still provided luxurious yachts for the Romanized nobles in Egypt.

The ship depicted in the Nile mosaic may be associated with a smaller size of *thalamegos*, a very fine ship, that its stempost was adorned with an expensive gold or bronze horse or seahorse figurehead (fig. 3.6.17). This adornment indicates a Phoenician tradition. Several reliefs from Mesopotamia, dated to the 9th – 7th centuries BCE,[65] depict *hippos* vessels with horse figurehead on the stem or stern post. They transported timbers or other cargoes and sailed in open sea, as well as on rivers.[66]

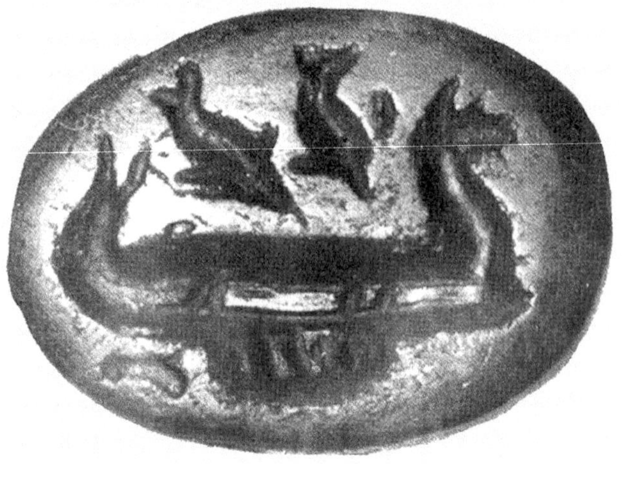

Fig. 3.6.22: Hippos Ship on carnelian seal

The sternpost of the Palestrina ship adorned with a stylized fishtail/fin figurehead indicates a Greek influence (fig. 3.6.17). The earliest fishtail decoration of the Greek ships sternpost dates from the mid-8th or early-7th century BCE; it appears as an arched post ornamented with small fins to the side.[67] A small version of a fishtail figurehead appears on both posts of three Cypriot clay ship models (8th – 7th century BCE).[68] A stylized fishtail decoration that seems to be a transition to a goose/swan head is found on the sternpost of a bronze shaped oil lamp dated to the 4th century BCE.[69] The tip of the vertical sternpost is decorated with a stylized form of a dolphin's tail.[70] A carnelian intaglio, probably of Phoenician origin, dated to the Hellenistic era (4th – 3rd century BCE), depicts a vessel with a broad hull. The stempost is adorned with a large horse figurehead, while the sternpost is surmounted by a fishtail (fig. 3.6.22).[71] This is the only comparable example yet for the Palestrina *thalamegos* (fig. 3.6.17). From 540 BCE onwards, the sternpost decoration of Greek ships appears as a goose head.[72]

The steering gear of the Sailing Ship 1 (fig. 3.6.4) and that of the *thalamegos* (fig. 3.6.17) are quite similar. The rudders of both vessels worked by long tillers inserted perpendicularly beneath the head of the looms reflect an Egyptian influence of the dynastic periods. This working method of the rudders by tillers apparently is an Egyptian invention. The head of the looms of the Palestrina *thalamegos* are decorated with some kind of figurehead. The steering oars of the *papyriform* vessels from the Middle and the New Kingdoms were worked by long tillers and the heads of the looms were decorated with figureheads of a king, falcon or ducks.[73] The rudders decorations of the ships depicted in the Nile mosaic

[60] Ibid., V.205.
[61] Ibid., V.206.
[62] A relevant illustration for such sails is depicted in the great merchantman in the Torlonia relief; Casson, fig. 144.
[63] Strabo 17.I.16.
[64] Casson, n. 66, p. 341.
[65] Stone blocks transported in hippos ships sailing on the river are found on reliefs from the reign of Shelmaneser III (858 – 824 BCE); deGrave, figs. 79, 80. Hippos ships sailing in the open sea and transporting wood logs are depicted in reliefs from Khorsabad, in the reign of Saragon II (722 – 705 BCE); deGrave, fig. 85.
[66] A relief from Karatepe, Turkey (dated to the reign of Tiglath-Pileser III, 745 – 727 BCE), evidences that also Phoenician warships had their sternpost decorated with horse figurehead; Casson, fig. 79; deGrave, 1981, fig. 82. The sternpost of a warship depicted in a relief in the Palace at Kuyundgik, Mesopotamia (669 – 629 BCE) is also adorned by a horse figurehead; Basch, 1987, fig. 672. The hippos ships are also connected with religious and mythological depictions of Lamashtu, an evil spirit who brought fever and sickness, especially to pregnant women and children. There are several plaques of Lamashtu with hippos boats. These plaques are dated to the reign of Ashurbanipal (669 – 629 BCE); deGrave, figs. 68, 69, 70, 71, 72.

[67] Basch, 1987, figs. 352, 353, 357, 257 A – B, 394.
[68] Ibid., figs. 551, 552, 553.
[69] Ibid., fig. 476.
[70] Ibid., fig. 524.
[71] The seal measures 12 x 8 mm; now it is found in the National Maritime Museum in Haifa, Israel.
[72] Basch, figs. 425, 428, 430, 432, 440, 442, 443, 452, 454, 460, 472.
[73] Landstrom, figs. 275, 288, 322, 324, 352, 354, 358, 361, 365, 369, 371, 393, 383.

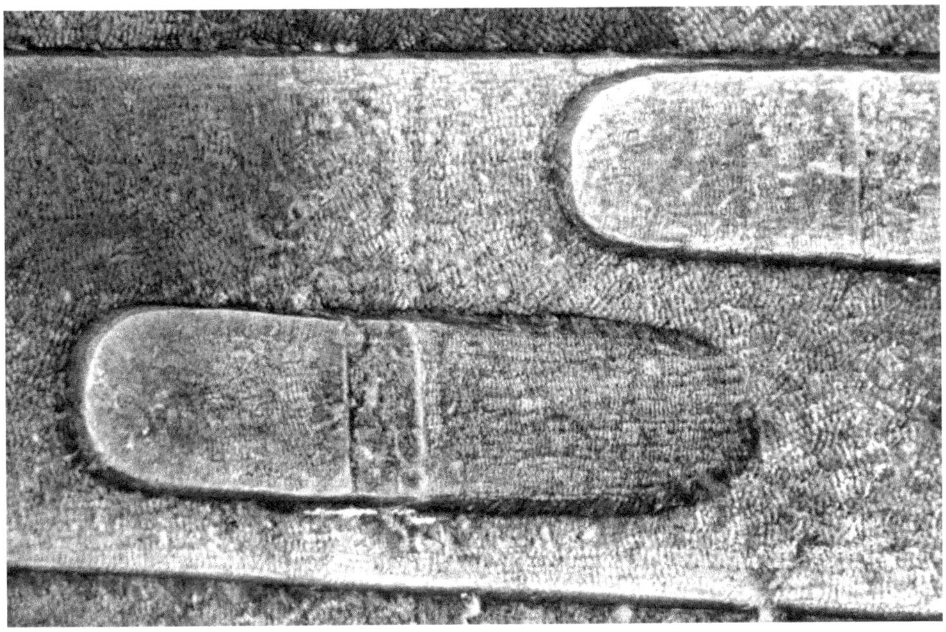

Fig. 3.6.23: Close-up of the en echelon oarports arrangement on the Samothrace monument

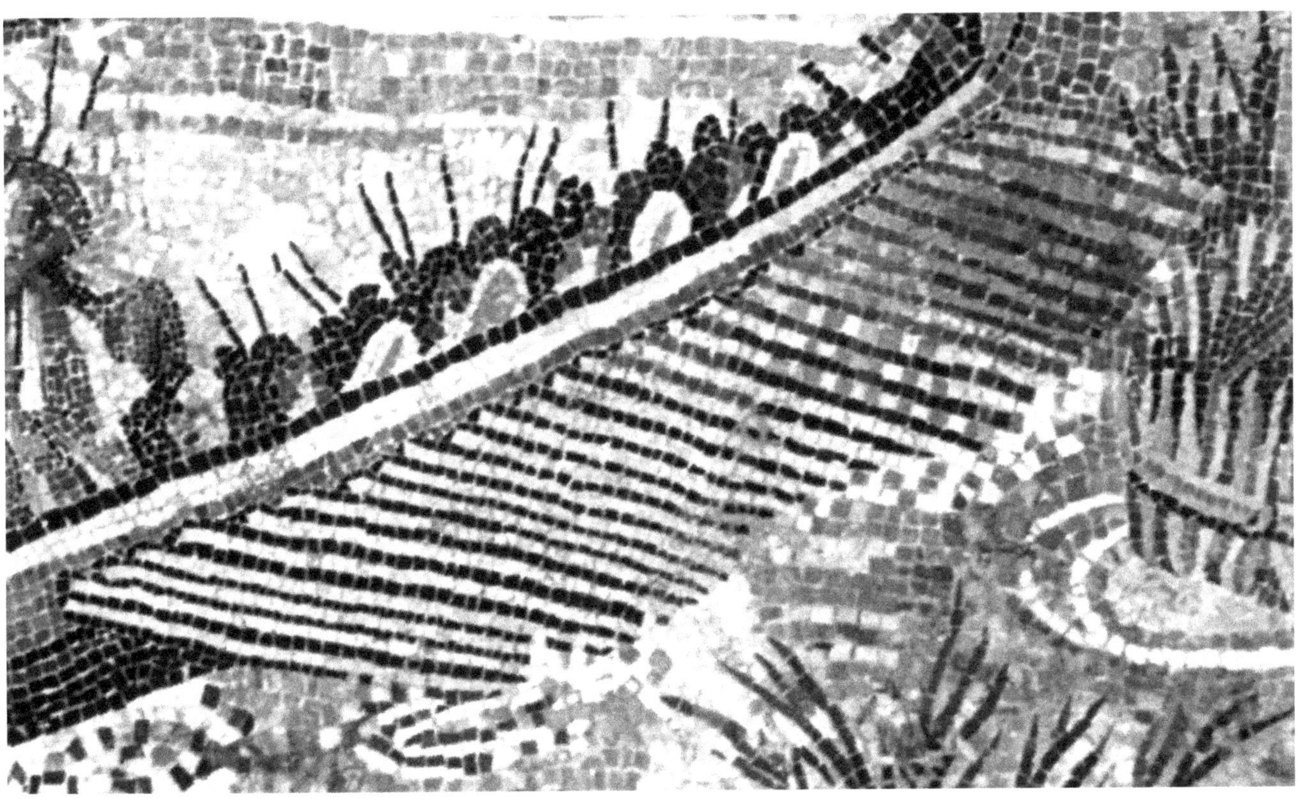

Fig. 3.6.24: Oars arrangement of the Nile Warship

emphasize the continuity of Egyptian tradition throughout the later Hellenistic period.

The Warship: This vessel is shown with all the distinct elements of a typical Hellenistic warship; it rather has a festive appearance than being engaged in a sea combat (fig. 3.6.10). The standing *prorates* in the bow and sounding a long trumpet announces the festive entrance of the ship into the harbor. The projecting heads of the warriors on the deck who are equipped with elliptical shields and long lances seems to indicate that they are ready to disembark and take part immediately in the flood festival. The *keleutes* seems to beat the rhythm for the rowers to sail the ship or steer it into the harbor.

Two rows of angled shafts projecting through the outrigger oarbox represent the port oars. The *en echelon* arrangement of the port oars in the Palestrina warship is

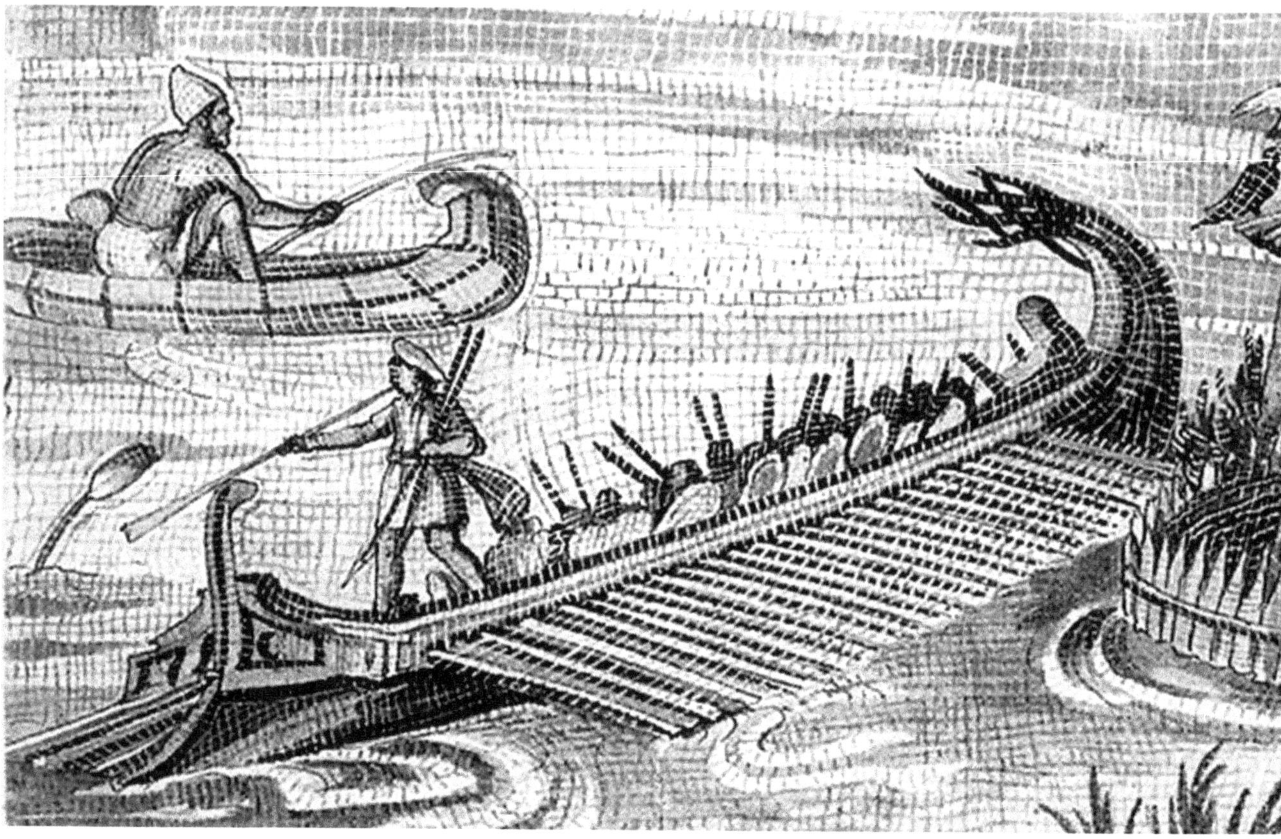

Fig. 3.6.25: dal Pozzo watercolor drawing of the Nile Warship

clearly augmented by the oarports depicted on the later face of the Samothrace monument (200 – 160 BCE), now in the Louvre Museum, Paris (fig. 3.6.23). The *tholepins* are shown in each oarport on the monument. They are aft the middle point in each oarport, thus indicating that the oar was worked forward the tholepin. The oars were worked against the strop and not against the pin.[74] We may assume that the oars in the Palestrina ship could be worked in a similar fashion as shown by the Samothrace arrangement.

In a trireme there were 170 rowers arranged in three files [62 *thranite* (31x2), 54 *zygite* (17x2) and 54 *thalamite* (17x2)]. Besides the rowers, there were about 30 men of whom 15 were warriors (*hoplites* or *epibatai*) and the remainder comprised the captain (*kybernetes*), *prorates,*[75] the helmsman who also served as *keleutes,*[76] the ship carpenter (*naupegos*),[77] and auxiliary personnel responsible for various tasks on the ship. Based on the view of the oars arrangement on the port side of the Palestrina warship we may suggest that only the *thrantes* (top rowers) and the lower ones the *zygites* (thwarts) are depicted. Thus, we may assume that the Palestrina warship comprised 108 rowers on board [54 *thranite* (17x2) and 54 *zygite* 17x2)], each working one rowing oar; the crew would comprise a total of 130 personnel (108 rowers, 17 warriors, one *prorates*, one helmsman /*keleutes* and at least three auxiliary personnel).

[74] Morrison and Coats, 1996, p. 220.
[75] The prorates was in charge of the rowing personnel and responsible for their training and morale; Casson, p. 304.
[76] The keleutes beat the rhythm for the rowers.
[77] Casson, p. 304.

The port oars in the Palestrina warship, with the aft oars much longer that the fore ones (fig. 3.6.24), brought Morrison to assume that the warship is a rare representation of a *hemiolia* (one and half) or *trihemiolia* (two and half), attested in ancient literary.[78] Morrison's assumption was based only on iconographic representations without reference to the history of the mosaic. In the dal Pozzo watercolor drawing made after the original mosaic, all the port oars have the same length at water level (fig. 3.6.25).[79] The depiction of the aft-oars being longer than the fore oars is a result of one of the mosaic's restoration phases. The dal Pozzo drawing emphasizes that the Nile warship is a typical Hellenistic bireme or *dikrotos* with two files of oars arranged in *en echelon*, with the distinct *aphlaston* and *stylis* on the sternpost ornament, as well as the *oculus* decoration on either side of the stem. The coloring of the lower hull with dark brown and black tesserae, just above the water line, may indicate the pitch/bitumen coating used to make the hull watertight and protect it from *teredo navalis* (ship worms) or other damage.[80]

[78] Morrison, 1980, pp. 121 - 22; Morrison and Coates, 1996, pp. 73 – 4; Casson, 1971, n. 113, p. 128; n. 115,
p. 129.
[79] A detailed description of the ship and the oar arrangement; in forthcoming article.
[80] There are many ancient literary references, and several wrecks that were found with lead sheathing; Nemi Barges; Casson, n.40, p. 20; Kyrenia ship, Grand Conglue, Antichithera, Madhia, Titan; Casson, App. I, pp. 214 – 5; the Punic warship from Marsala; Frost, 1981, pp. 65 – 75; McGrail, 2001, p. 153.

3.7 Ostia, Piazzale Delle Corporazioni - Italy

Historical Notes

The ancient settlement of Ostia originally was founded to exploit the salt from the salt marshes (*salariae*) around the mouth of the Tiber, probably extracted from the Middle and Late Bronze Age (1400 – 1000 BCE). Small villages near the salt-pans may have existed also in the Early Iron Age (1000 – 700 BCE). Historical sources from the Republican Period to the Middle Ages indicate that Ancus Marcius founded Ostia as the first Roman colony in 620 BCE.[1] The name Ostia derives from the Latin word *ostium* (river mouth) with reference to the Tiber.[2] Dionysius of Halicarnassus (1st century BCE), in his work "*Roman Antiquities*", pointed out that the river received its name from Tiberinus, who reigned for a period of eight years. At a battle that was fought nearby, the king was slain and "*being carried away by the stream, gave his name to the river, which previously was called the Albula*".[3] Dionysius also mentioned that the Tiber descending from the Apennine Mountains and flowing close to Rome discharges its waters "*upon the harborless and exposed shores made by the Tyrrhenian Sea*".[4] He wrote that Ancus Marcius built the first port at the river's mouth because seagoing ships of great burden could not sail the Tiber, which was navigable quite up to its source for river boats that sailed as far as Rome.[5] The location of the site also followed economic and political factors, reflecting the need to control the lower valley of the Tiber both strategically and commercially:

> *" ... oared ships however large and merchantmen up to three thousand bushels (literally "three thousand amphorae") burden enter at the mouth of the river and are rowed and towed up to Rome, while those of a larger size ride at anchor off the mouth, where they are unloaded and loaded again by river boats. Upon the elbow of land that lies between the river and the sea the king built a city and surrounded it with a wall, naming it from its situation Ostia, or as we should call it, thyra or "portal"; and by this means he made Rome not only an inland city but also a seaport, and gave it a taste of the good things from beyond the sea".*[6]

The earliest archaeological remains at Ostia date from the beginning of the 4th century BCE.[7] In 369 BCE, the Romans conquered the Etruscan Veii and immediately consolidated their domain in the lower valley of the Tiber by building a fortification near the river's mouth.[8] In 267 BCE, Ostia became the seat of the *quaestorea classici* (officials taking care of the fleet), the *quaestor Ostiensis*. Ostia became the main naval base during the 2nd Punic War (217 – 201 BCE), when the Romans carried out the campaign against Hannibal. In the second half of the 2nd century BCE, Rome obtained supremacy in the Mediterranean and Ostia gained a larger commercial role, which changed it gradually into a commercial harbor.[9] At the beginning of the Imperial era, the maritime trade had increased and the modest river post could no longer provide an anchorage for the larger ships. The lack of a natural bay near the river's mouth made it very difficult to guide ships into the Tiber, especially in poor visibility.[10] This complicated process of unloading cargoes from large ships (anchored in open sea, at the month of the Tiber) and reloading them into smaller vessels for their transportation to Ostia and Rome was not very profitable. Therefore Ostia needed a commercial oriented harbor.

Emperor Claudius completed the project of the seaport in 42 CE that was planned by Caesar or Augustus.[11] The artificial Claudian harbor comprised a huge basin that was protected by two arched moles and a lighthouse at its entrance. The completion of the construction was celebrated in 64 CE, during the reign of Nero. During a strong storm 200 ships perished within the Claudian harbor basin in 62 CE, probably caused by a tsunami (in the same year Pompeii was struck by an earthquake). This harbor failed to provide a safe haven for moored ships due to the extensive dimensions of the quays, which led Emperor Trajan to build a second harbor, between 106 and 113 CE, that included improvements of the Claudian harbor.[12] The hexagonal basin was located further inland and was therefore more protected. This harbor linked the Claudian port via an artificial channel. Originally the Trajan's harbor was built as a port facility, but the complex gradually turned into a city. From the second half of the 3rd century CE onwards this harbor began to replace Ostia as the focal point of the sea trade for goods heading to Rome.[13] The building of the Trajan's harbor was meant to facilitate the centralization and administration of the corn import under imperial control.[14] Trajan's work at Ostia increased the harbor area and the security of the inner basin which made it possible to bring larger merchant ships carrying grain from Alexandria. The harbor district was controlled by an Imperial official, the *procurator Portus Ostiensis*, also called *procurator Portus Utriusque* (of both harbors), after the construction of the second port. During the 2nd century, Ostia became very prosperous and its importance increased by the building of administrative and commercial structures such as warehouses for the storage of goods (*horrea*), and offices for shippers and traders for centralizing and controlling the commercial activities. Although, Ostia harbor was not as large as Alexandria or Piraeus, it became the focal link in the great sea-route from east to west.[15] In the second half

1 Pellegrino, 2000, p.7; Ancus Marcinus ruled during the 7th century BCE. This reference comes from ancient authors, e.g.: Ennius, Livius, Cicero, and Dionysius of Halicanassus.
2 Pellegrino, p. 7.
3 Dionysius of Halicarnassus, Book I.71.2.
4 Ibid., Book III.44.1.
5 Ibid.
6 Ibid., Book .III.44.3-4;
7 Pellegrino, p. 7.
8 Ibid.

9 Ibid., p. 8.
10 Ibid.
11 Ibid., p. 11.
12 Ibid.
13 Ibid.
14 Meiggs, 1973, p. 55.
15 Ibid., p. 61.

of the 2nd century, during the reign of Septimius Severus, Ostia's economy and trade flourished. Piazzale delle Corporazioni or the Square of the Guilds is assumed to have been founded during this period. Ostia turned into the principal harbor of the Roman Empire and the main center of imported goods (wheat, exotic animals, ivory, silk, etc.) from the provinces.

The decline of Ostia during the second quarter of the 3rd century resulted from the political and economical crises that hit the entire empire. Portus became the important harbor to which all commercial activities had been transferred. These crises led to the abandonment of Piazzalle delle Corporazioni and Ostia gradually became a residential center. In the course of time, the Tiber continued to carry great quantities of sediment that gradually silted up the Claudian harbor, leaving only traces of the two moles.[16] Trajan's harbor, being cut off both from its links to the Tiber and from the sea, lost its usefulness and became a lake.[17] In the following centuries Ostia became a quarry for stones and marble which were used to build the Cathedral of Pisa (11th century) and that of Orvieto (14th century).[18] The heavy floods of 1557 that inundated the site brought to the final destruction of Ostia, cutting it off from communication with the sea.[19]

The Excavations

The first archaeological excavation at Ostia began in the last quarter of the 18th century, in a marshy area that was infested with malaria. These excavations were meant to find artworks to be sold to antique dealers or for private collections. The Scottish artist Gavin Hamilton conducted the first recorded investigations in 1774 – 1775 at Porta Marina in the remains of the *thermae maritimae*.[20] Robert Fung, an English artist, who was also the British consul, began excavations at Ostia in 1794 near the site selected by Hamilton. The work continued until 1800 and numerous works of art were discovered during this period. While the majority of the artifacts were sent to the Pontifical collections, a few were also sent to England.[21] Pope Pius VII decreed that excavation by private persons should no longer be permitted and official work began in the winter of 1801 under the direction of Giuseppe Perini.[22] Later excavations were carried out in 1924 – 1925 (to the SW of the modern village) in the area of the necropolis, along the line of the road leading from Via Ostiensis to Via Severina.[23] Piero Ercole began the first scientific excavations in 1866, being assisted by his nephew Carlo Ludovico Visconti, in the area outside of what was then believed to be Porta Romana; they discovered numerous tombs. Between 1864 and 1870, more tombs were discovered along the coastal road of Via Severina and some remains of the *thermae* near Porta Marina. Reports were published in *Annali dell'Instituto* and the objects collected in these investigations were stored in two rooms in the Lateran, with the idea of founding a local museum.[24] The Italian government continued sporadic works in 1871 – 1878 and 1889 by clearing Via della Fontana. In 1907, Prof. Vaglieri published monthly reports in *Notizie degli Scavi*. The only comprehensive work on Ostia was written by Dr. L. Paschetto in *Ostia Colonia Romana* that did not include the discoveries after the summer of 1911.[25] Dante Vaglieri began his work in 1908 and continued it until his death in 1913. His mission was to reveal the entirety of the ancient Ostia and to establish its topography and history.[26] Guido Calza continued Vaglieri's work in the same scientific spirit from 1914 till 1946. Calza not only carried out the excavations but also had time to publish his discoveries in reports and articles.[27]

Piazzale delle Corporazioni – Regio II, Insula VII (II.VII.4)

The historical evolution of Piazzale delle Corporazioni or the Square of the Shippers may be traced through six phases which will be summarized below (fig. 3.7.1):[28]

Phase I

The Piazzale delle Corporazioni (Square of the Guilds or Corporations) was built together with the theatre, in the early Augustan period (8 – 14 CE). The double colonnade portico is contemporary with the original construction of the theater, in the reign of Augustus; it measured 107 x 78 meters, and the outside wall was built of *opus reticulatum*.[29] Originally this place was designed as a *porticus post scenaum* where people could go for a stroll.[30] There was a monumental entrance in the north part, towards the Tiber.

Phase II

During the reign of Claudius the level of the area behind the theater was raised and a colonnade of one row of brick columns set on travertine bases were built. Subsequently, this place became the focal point of the commercial activities conducted between Ostia and the Mediterranean provinces. Fifty rooms may have been separated by wooden partitions, but it is also quite possible that there were no rooms, and this area may have been used as a *porticus post scaenam* (a standard *porticus* behind a theatre). Mosaics were placed on the floors of the *porticus*. Only in four places it was possible to examine them (in rooms 52, 53, 57, 58).[31]

16 Testaguzza, 1964, p. 175.
17 Ibid., p. 176.
18 Ashby, 1912, p. 160.
19 Ibid.
20 Ibid., p. 161.
21 Ibid., p. 162.
22 Ibid.; galley slaves from Civitavecchia were employed in the work.
23 Ashby, p. 162.

24 Ibid., p. 164.
25 Ibid., p. 165.
26 Rowel, 1948, p. 36.
27 Ibid.
28 The majority of the mosaics are not set at the original level of the Piazzale delle Corporazioni; beneath them there is an earlier pavement that could not be examined during the excavations.
29 Meiggs, p. 283.
30 Pellegrino, p. 22.
31 Ibid.

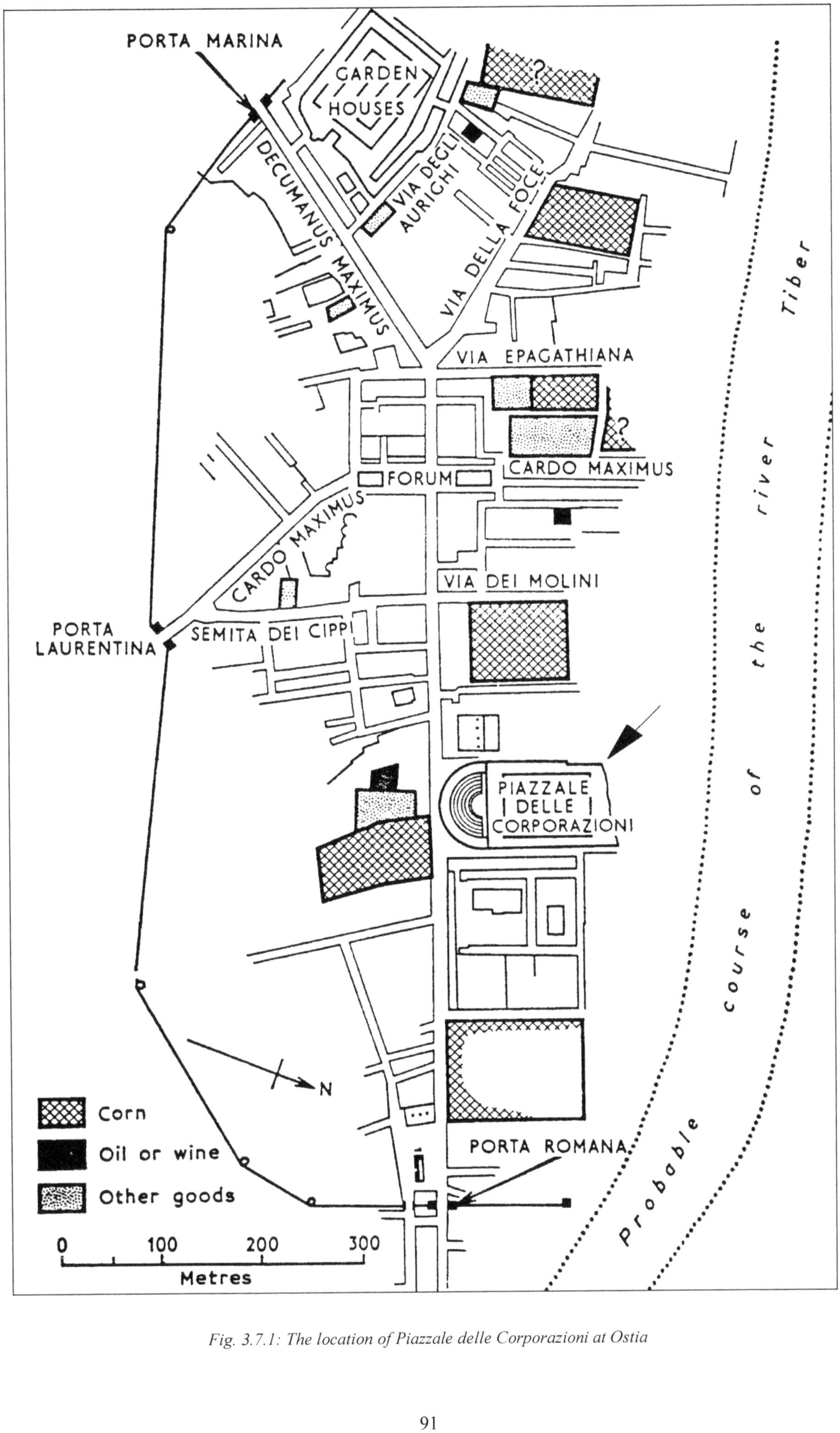

Fig. 3.7.1: The location of Piazzale delle Corporazioni at Ostia

Phase III

In Domitian's period a temple was erected in the centre of the square, facing the theatre (II,VII,5). The area around the temple may have been a garden, adorned with statues.

Phase IV

The excavations carried out in the area of Piazzale delle Corporazioni revealed evidence of overseas shipping before the end of the 2nd century CE. A marble pediment with the inscription "*naviculari Africani*" was found on the east side of the colonnade.[32] A study of the lettering indicated that it was not later than the Hadrianic period.[33] In the period of Hadrian the level of the square and the *porticus* was raised by 40 centimeters (the dating results from excavations in the 1980's). The entrance at the north end was closed, and the *porticus* was doubled. As in the period of Claudius, rooms may have been created by wooden partitions, but this remains uncertain.

Phase V

Most mosaics at the higher level were installed later than the period of Hadrian, presumably in the reign of Commodus (190 – 200 CE) or later.[34] In this period the *porticus* had 60 rooms or more (fig. 3.7.2). The theatre was rebuilt by Commodus, and it was inaugurated by Septimius Severus and his son Caracalla, in 196 AD.

Phase VI

The mosaic floors in the offices were restored and altered in antiquity. In the Severan period dividing masonry walls (*opus vittatum*) were set on top of the mosaics, probably replacing the wooden partitions.

The mosaic decorations made in black and white technique clearly indicate the occupation and hometown of the shops owners that were linked with overseas trade, especially wheat. The earlier mosaics (probably the original phase) are different; they depict mythological themes such as Minerva and the Stag, or the Nereids.

The mosaic floors of the offices depict ships, African animals, lighthouses, Latin inscriptions referring to the commercial activities, the guilds (*collegia* or *corpora*), shippers (*navicularii*) and traders (*negotiantes*), and the grain traders (*modii* or corn measures). These rooms represented the commercial offices or meeting points for all the traders that came to Ostia from the provinces. Calza attested that the traders and shippers who were involved with supplies to Rome were concentrated at Piazzale delle Corporazioni by the imperial authority under Augustus. Apparently this official control of the overseas trade remained unchanged during the 3rd century CE[35], and grouping them at Ostia, reflects the earliest direct imperial control on the trade and the commerce in the Roman Empire.[36] Traders who were located at Ostia came from North Africa, Gaul, Sardinia and the Adriatic, or Alexandria. Three of the inscriptions relate to the direct connection between the shippers and the traders "*navicularii et negotiantes*". It is assumed that the *negotiantes* dealt with the ordering of goods to be bought in their home district and then shipping them to Ostia.[37] Most of the offices were of foreign origin and were not intended for the Ostian shippers and traders. Some local groups of workers served the offices and had their shops in the Piazzale (*restiones*, *caudicarii*, *stupatores*, *pelliones*).

Ship Depictions in the Offices

Piazzale delle Corporazioni comprises 61 rooms/offices arranged on the three sided colonnades. The rooms are numbered from right to left, starting with the SE corner (fig. 3.7.2). The mosaics did not survive in all the offices. Only those offices that their mosaic floors are decorated with ships will be discussed bellow. The ships depicted on these floors are found either as an individual pattern, in pairs, in pairs with rounded or stepped structures between them, or showing harbor activities. Some of the inscriptions following the ships are set in one or two lines, which indicate the origin of the ships or shippers. The ships will be described as following the plan of the colonnade. Only 16 offices depicting ships have survived. There were probably many more rooms with ship depictions; now some of them lack the mosaic floor and asphalt has been used to fill in the floor. All the ships depicted in the offices lack any mooring or anchoring system. From their static position we may deduce that they are anchored and not sailing. Few human figures found with some of the ships are engaged in varied harbor activities.

Station 3

On top of the panel, there is a long *tabula ansata* with a two line inscribed inscription: "NAVICULARIORUM LIGNITORUM", referring to timber shippers who were trading and transporting wooden logs to Ostia and Rome. Beneath the inscription, within a black frame are depicted two ships one on either side of a central cylindrical structure with flames emanating from it (fig. 3.7.3). Their prows are turned towards the tower, as also facing each another. The structure set on a raised platform probably was placed at the harbor's entrance which may symbolize a lighthouse. The hulls of both ships are very similar, with rounded ends and flat bottoms.

Ship 1 (left-hand): The ship has an up-curved stern surmounted by a vertical sternpost. The prow points to the right. The angled stem is surmounted by a block-like stempost outlined with one strip of black tesserae. The

32 Meiggs, p. 285.
33 Ibid.
34 Ibid., p. 289; Becatti, (SO IV) dates most mosaics to the period 190-200 CE. He mentioned that the Hadrianic changes of the square were dated in the period of Commodus.
35 Meiggs, p. 283.
36 Ibid.
37 Ibid., p. 287.

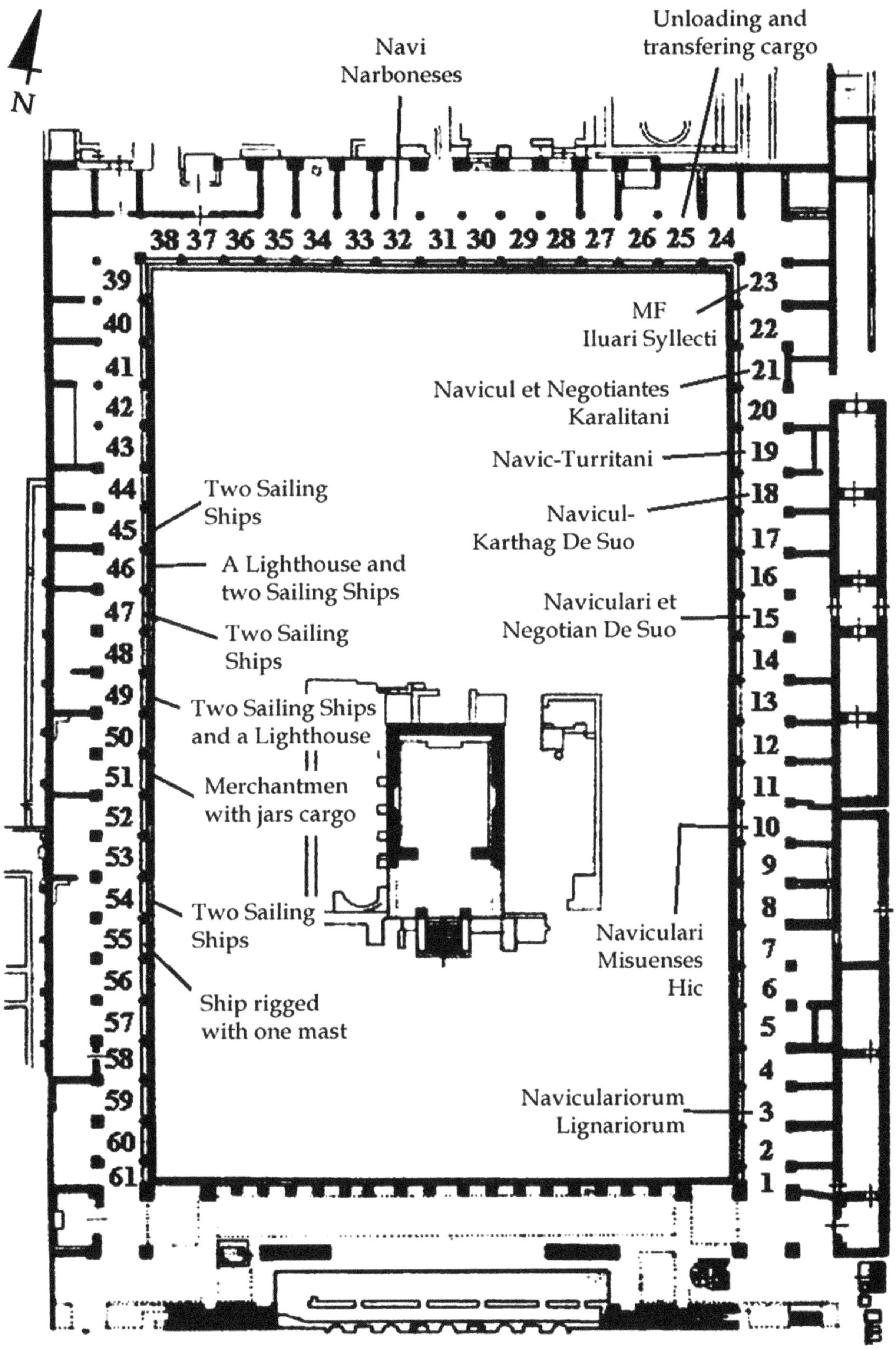

Fig. 3.7.2: Plan of the shippers' offices at Piazzale delle Corporazioni

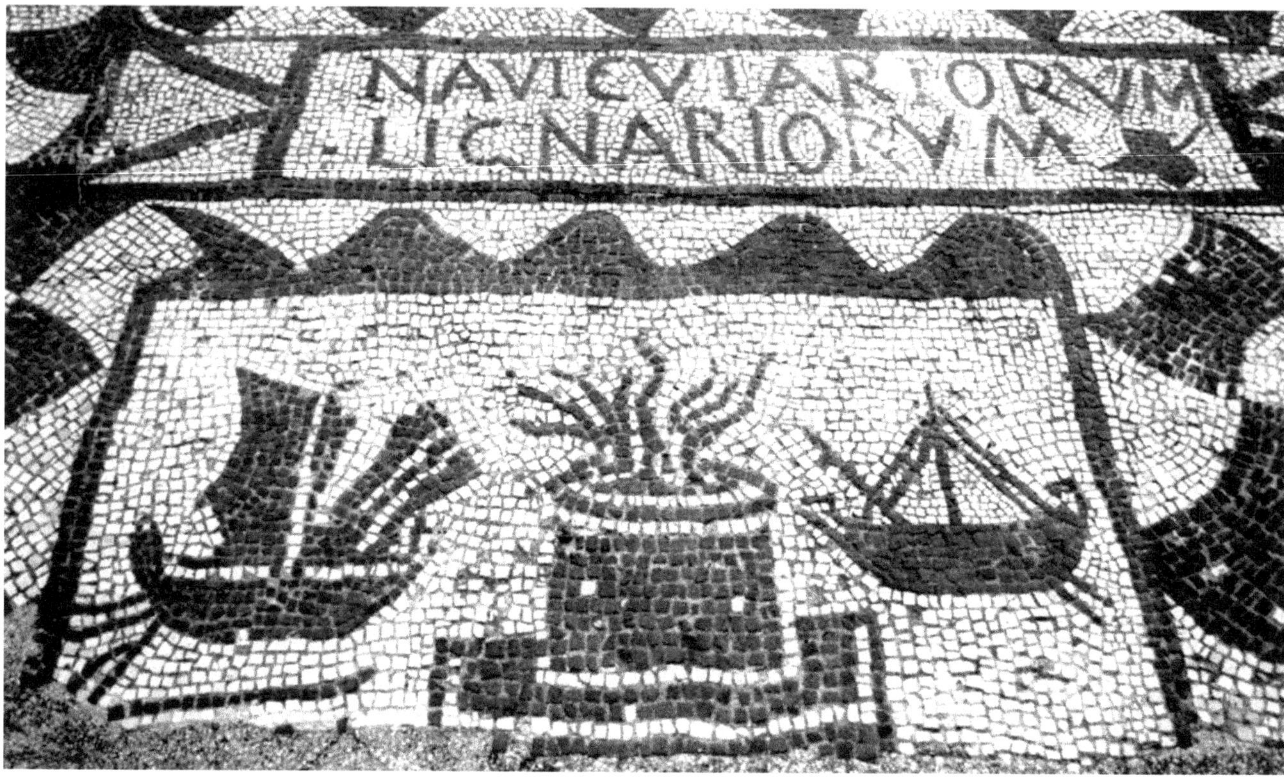

Fig. 3.7.3: Station 3 – Naviculariorum Lignariorum

planks are not indicated and the hull is depicted only with black tesserae (fig. 3.7.4). The rigging comprises two masts and quadrilateral sails, and a pair of steering oars mounted on the quarters. A vertical white strip indicates the main mast stepped slightly forward amidships. The yard and the head of the sail are depicted as one unit, slightly angled with the port yardarm raised. The tip of the starboard yardarm and the upper corner of the sail are damaged. Near this part, it is possible to distinguish some restorations made in the mosaic. The elongated square sail made with black tesserae is seen from its lee face, deduced from the white mast (fig. 3.7.4). The concave arch of the port leech emphasizes the billowing sail, thus indicating that the wind blows from astern or the starboard quarter. The port sheets and part of the sail suffered some damage. This section was replaced by white tesserae, much bigger than the original ones in the background, and no attempt was made to reconstruct the shape of this part. The foot of the sail is not very clear, probably is being hidden by the rectangular cabin placed on the mid-deck; the entire section is a black mass. The artemon mast and sail, angle forward above the stem. Two angled white strips depicted along the black artemon sail probably indicate the halyards or they outline the mast seen from the lee side (fig. 3.7.4). The yard and the head of the sail are indistinct. The artemon sail is similar to the main sail but slightly smaller. The tip of the port yardarm and the corner of the sail are damaged. The white tesserae work between the upper sails indicated some restoration of this part.

The steering oars differ. The port oar, indicated by a tapered blade, with concave upper edge, is crossed longitudinally by the lower shaft rendered with one row of white tesserae. It is almost horizontal, projecting behind the stern. A thin black shaft and an elliptical blade indicate the starboard oar (fig. 3.7.4). The lower part of the shaft made with one row of white tesserae cross the blade longitudinally. The oar has a rearward extension, projecting from beneath the stern. Apparently this ship replaced an earlier pattern in the mosaic. Therefore the mosaicist may have used the pattern of the port oar while he misinterpreted the shape of the starboard oar.

Ship 2 (right-hand): The hull of this ship is quite similar to that of Ship 1. The bow points to the left. The upraised curved stern is surmounted by a slightly inward turned sternpost (fig. 3.7.5). Both stem-and-stern posts are outlined with one strip of black stones and the field is made with white tesserae. The hull is rendered with black tesserae without any indication of the strakes. The rigging comprises two masts, four lines and a pair of steering oars. The main tapered mast is stepped amidships and secured by two forestays and two backstays extending to either gunwale (fig. 3.7.5). The artemon mast has a forward rake above the stem. No standing rigging is shown, nor the sail. Thin black spars with backward slanting projecting beneath the hull represent both steering oars, one set on either quarter. The blades apparently are submerged into the water. No shore or sea level is indicated.

The rounded structure with the flames atop and placed between the ships may indicate a lighthouse or a beacon

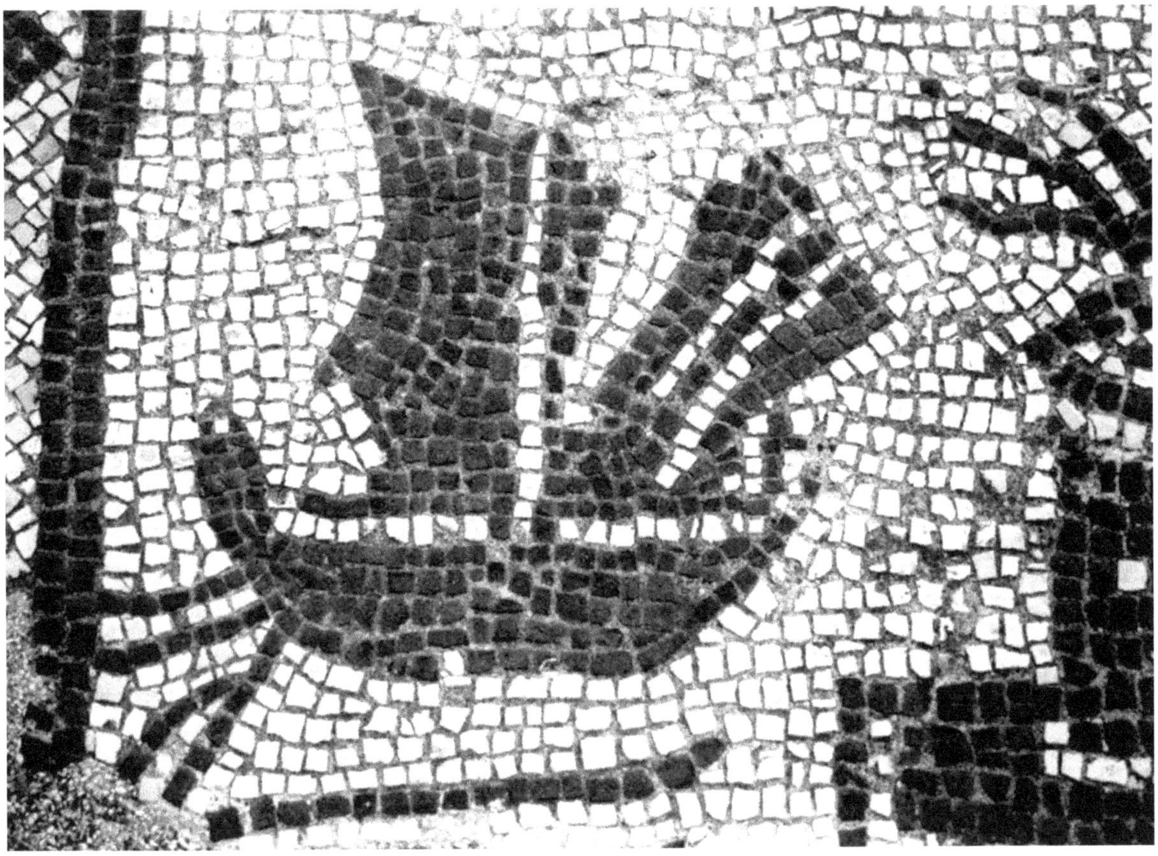

Fig. 3.7.4: Ship 1 – Station 3

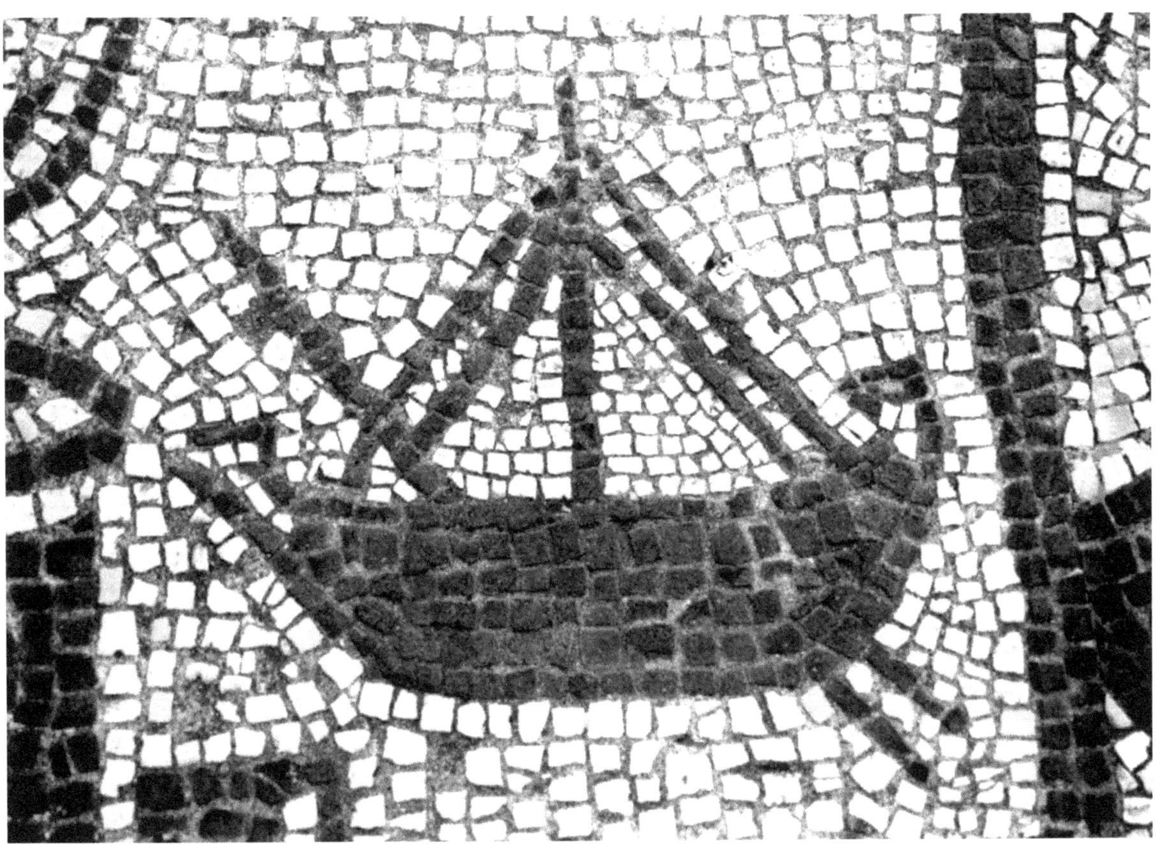

Fig. 3.7.5: Ship 2 – Station 3

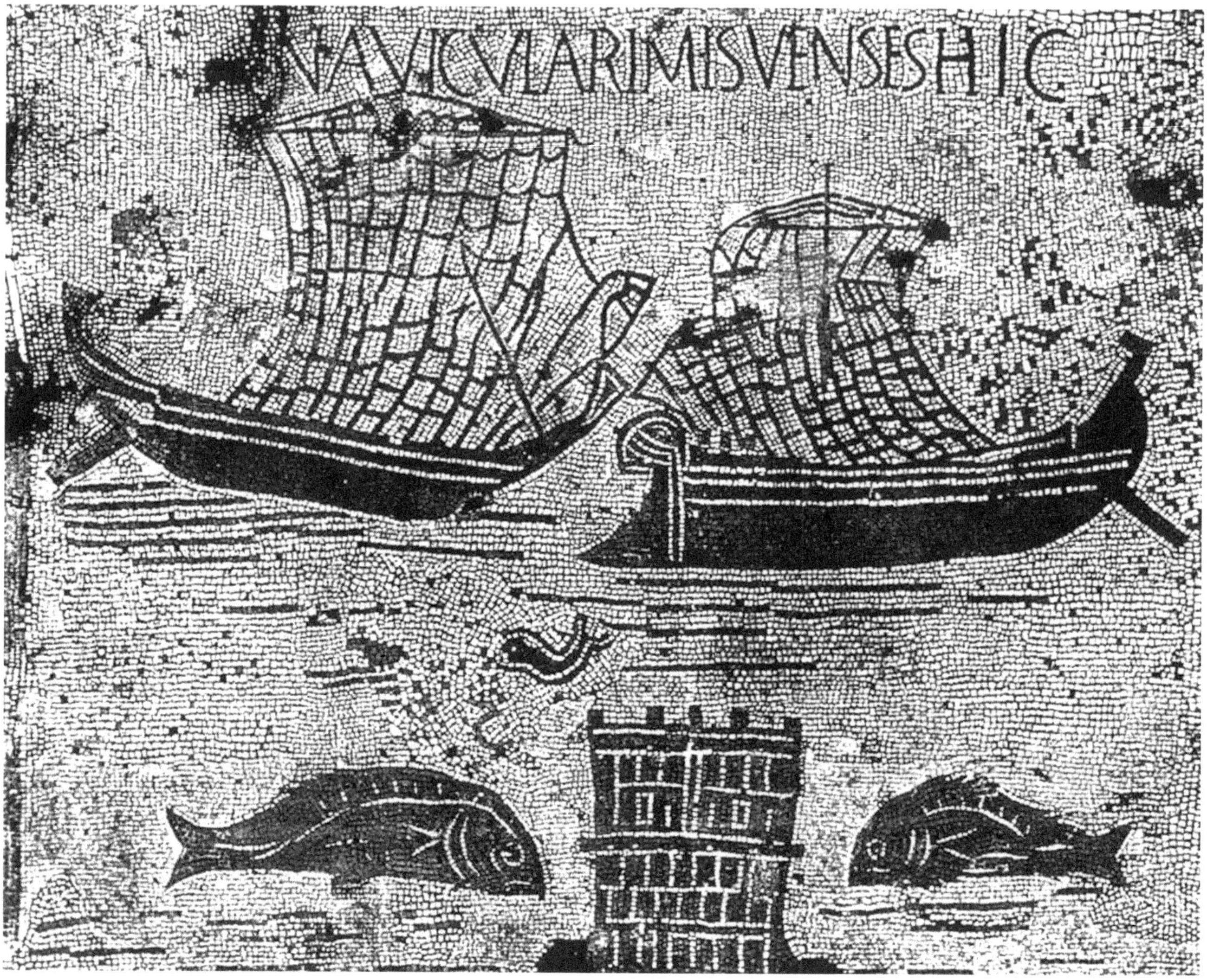

Fig. 3.7.6: Station 10 – Naviculari Misuenses Hic

set at the entrance to the harbor to guide vessels towards the harbor (fig. 3.7.3). The mosaicist has intentionally given larger dimensions to the structure in relation to the ships, thus adding some perspective to the scene. Probably the ships are seen inside the harbor basin, as the observer is positioned facing the large structure at the entrance of the harbor.

Station 10

The mosaic depicts a long inscription, two sailing ships, a rounded crenellated structure and one fish on either side of the structure (fig. 3.7.6). The inscription "NAVICULARI MISUENSES HIC" is not inserted in a *tabula ansata* but forms a long strip above the ships. It indicates that the office belonged to the shippers from Misua, in North Africa. The ships with different hulls face each other.

Ship 1 (left-hand): The bow points to the right. The ship is depicted in an angle with the prow pointed downward (fig. 3.7.7). It has a long slim hull with both rounded ends. The raised stempost outlined with one strip of white tesserae has a straight cut tip. It is higher than the sternpost. Two parallel horizontal white strips outline the top wale. There are no indications for other strakes. The up-curved stern is surmounted by a short sternpost with a shallow inward turn.

The ship is rigged with two sails and one steering oar. The main fully open square sail is seen from its fore face. On the bunt are depicted eight brails and eight reinforcing-bands (fig. 3.7.8). All the lines are rendered with one strip of black tesserae. A thin horizontal black strip indicating the yard is supported by one lift stretching from the masthead to either tip of the yardarms. Several small black arches atop the yard may indicate the fairleads or the top ends of the brails running over the yard and down to on the lee face of the sail. The mast is not visible and is only represented by the tip of the masthead projecting above the lifts. The angle line running from the middle of the yard to the inner side of the stempost is the forestay that secured the mast in place (fig. 3.7.8). The port leech and the middle part of the sails have undergone some repairs that altered the shape of three brails, and about three reinforcing-bands. The short black angled line stretching between the upper port leech and the tip of the artemon's port yardarm indicates the port brace. It was misplaced, probably during the restoration of the sails.

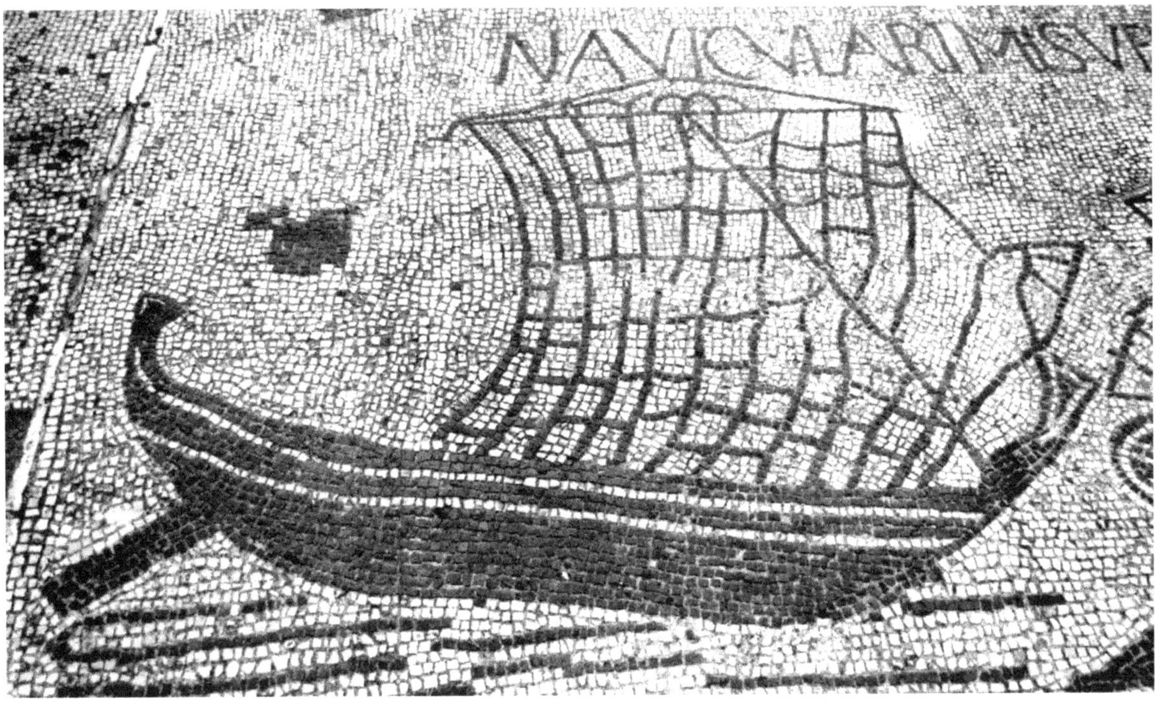

Fig. 3.7.7: Ship 1 – Station 10

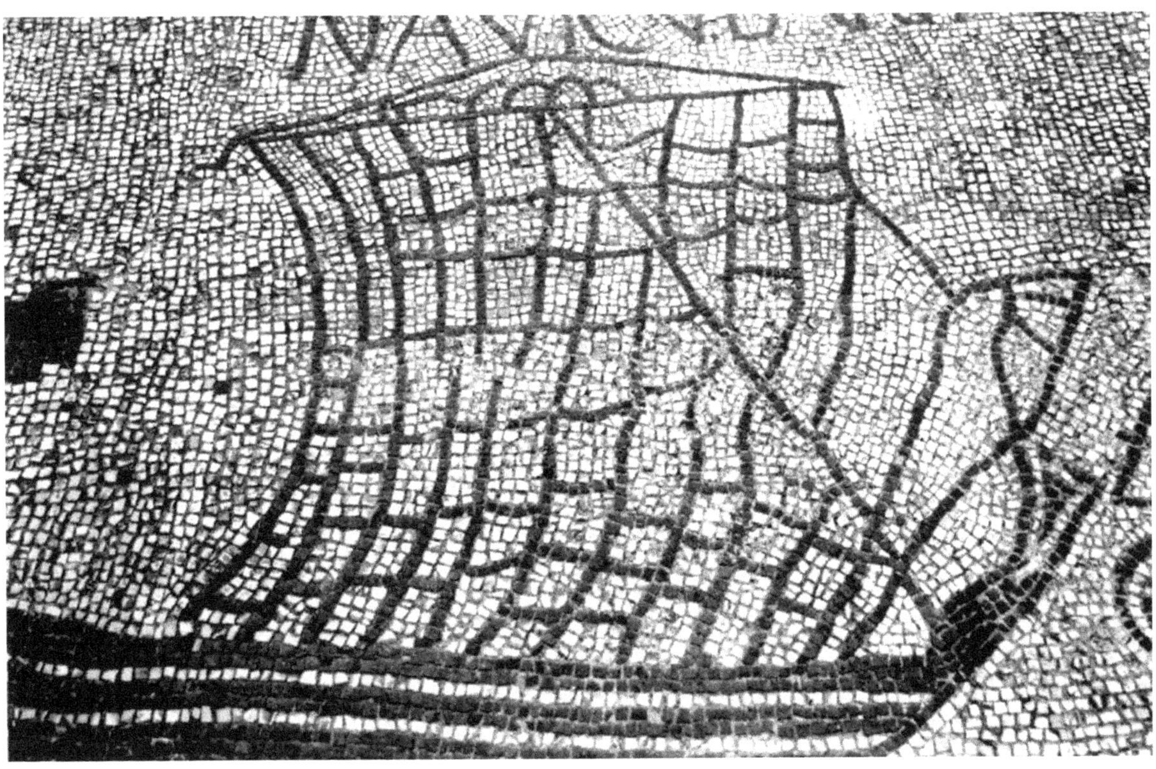

Fig. 3.7.8: Close-up of the sails

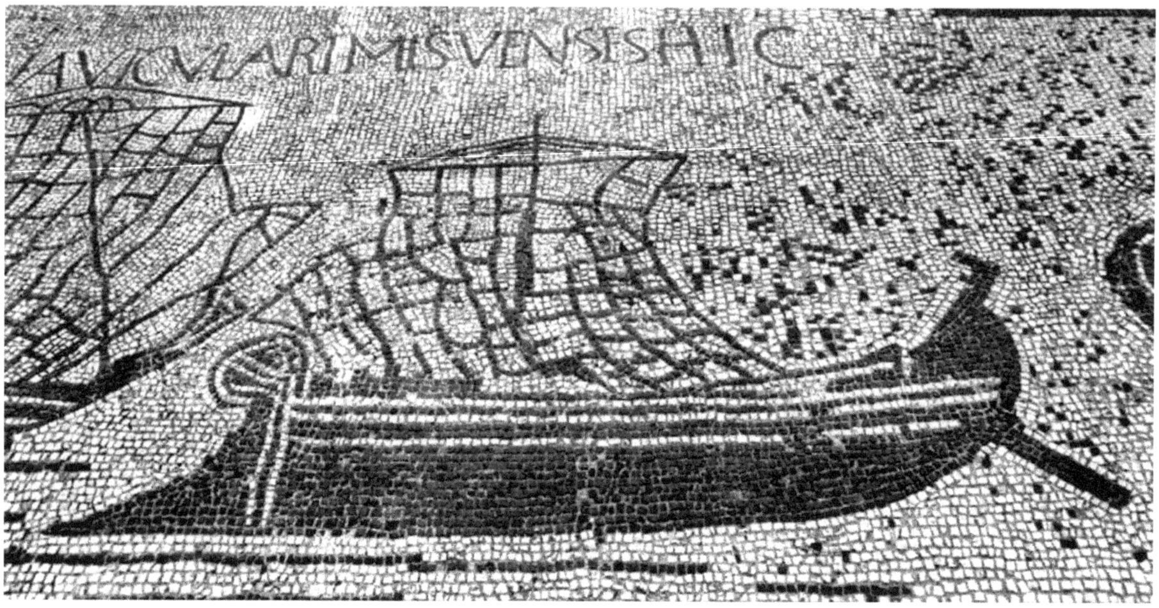

Fig. 3.7.9: Ship 2 – Station 10

The artemon sail is quite distorted and not distinct. The mast and the sail seem to make one large tapered pattern. The sail may be deduced from the black lines depicted on the right side (fig. 3.7.8). The distorted shape of the artemon rigging probably resulted from one of the restorations made on the mosaic and the mosaicist's lack of knowledge to understand such rigging.

A black elongated rectangular blade stretching to the rear beneath the stern represents the steering oar. We may assume that the ship was rigged with two steering-oars or rudders. No figure is associated with the ship.

Ship 2 (right-hand): The hull of this ship is different from that of Ship 1. It is broader, with a very rounded upraised stern and a concave stem finished with a projecting pointed cutwater (fig. 3.7.9). The bow points to the left. A very rounded head (probably an inverted volute), projecting above the gunwale, indicates the stempost. Between the port stem and the hull there are two white vertical lines, which also follow the rounded head of the stempost. They may distinguish the prow. Two bitts used to secure the sail's rigging project above the short fore-bulwark. The top aft tapered strake set above the port quarter indicates the quarter bulwark or the aft-wing housing the shaft of the port rudder (fig. 3.7.9). The top wale, just beneath the gunwale, is outlined with one strip of white tesserae and stretches alongside the port stern. The white strip beneath the wale may indicate the lower edge of the upper strake or the seam between the strakes. Apart from this strip, no other seams are indicated.

The ship is rigged with two sails seen from their lee sides. The sails and the masts are not very distinct due to the heavy restorations, observed on the brails and the reinforcing-bands of the main bunt and the unclear representation of the artemon rigging (fig. 3.7.10). Apparently the restorers did not understand the function of these elements and they tried to follow the original depiction, with unsatisfactory results. This ship was probably made directly at the site and it was inserted into an earlier mosaic that is no longer distinguishable. The main mast indicated by a fragmented tapered black spar is stepped slightly forward amidships. The square sail is fully open, billowing over the starboard side. On its lee face are depicted six brails and eight reinforcing-bands (fig. 3.7.10). The yard, equals the length of the sail's head, is secured by three lifts stretching from either side of the masthead to the yardarms. The standing (fore-and-back stays and/or shrouds) and running (braces, halyard) rigs are missing or were altered during the restoration. The artemon rig is quite disordered. It may be deduced from the forward angle of the sail and the raked yard (fig. 3.7.10). The thicker undulating black strip depicted on the lee face of the bunt suggests the artemon mast. Both lines to the left of the mast may represent the artemon brails. Two fragmentary horizontal strips suggest the reinforcing-bands.

The steering gear comprises an elongated black rectangular blade with a rearward extension beneath the stern (fig. 3.7.9). It is not possible to distinguish on which quarter the rudder is mounted, or the mosaicist did not differentiate each rudder. The ship may have had two rudders, one placed on either quarter. Probably the shaft of the port rudder was housed by the aft-wing extension of the port top strake.

Station 15

A long inscription (not inserted in a *tabula ansata*) is depicted above a sailing ship: "NAVICUR ET NEGOTIAN DE SUO" is a distinct reference to shippers and traders office (fig. 3.7.11). Probably the *negotiantes* ordered the

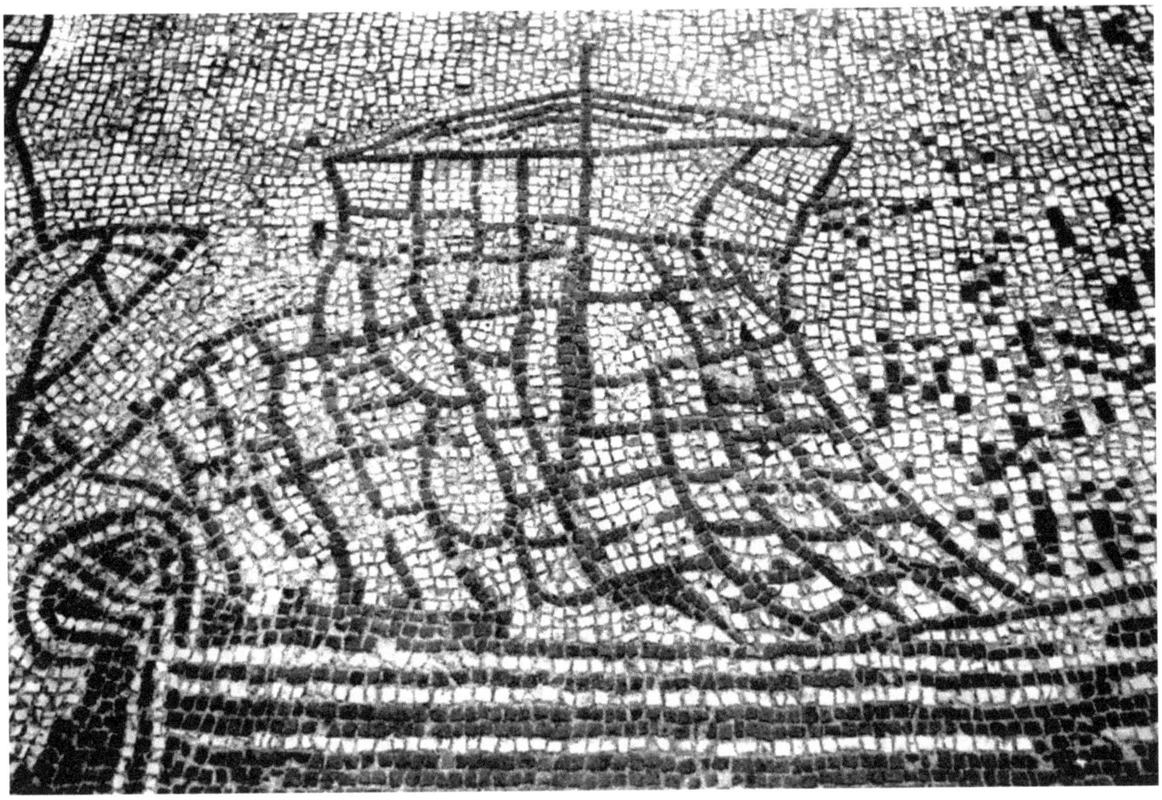

Fig. 3.7.10: Close-up of the sails

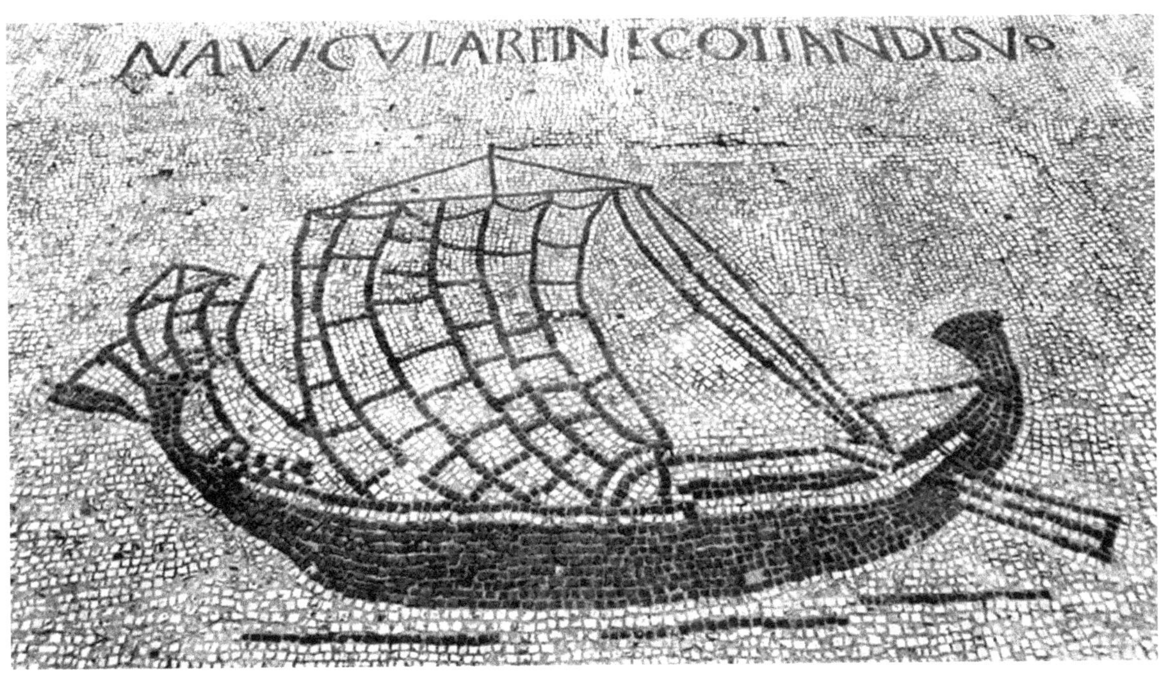

Fig. 3.7.11: Station 15 – Navicular et Negotian De Suo

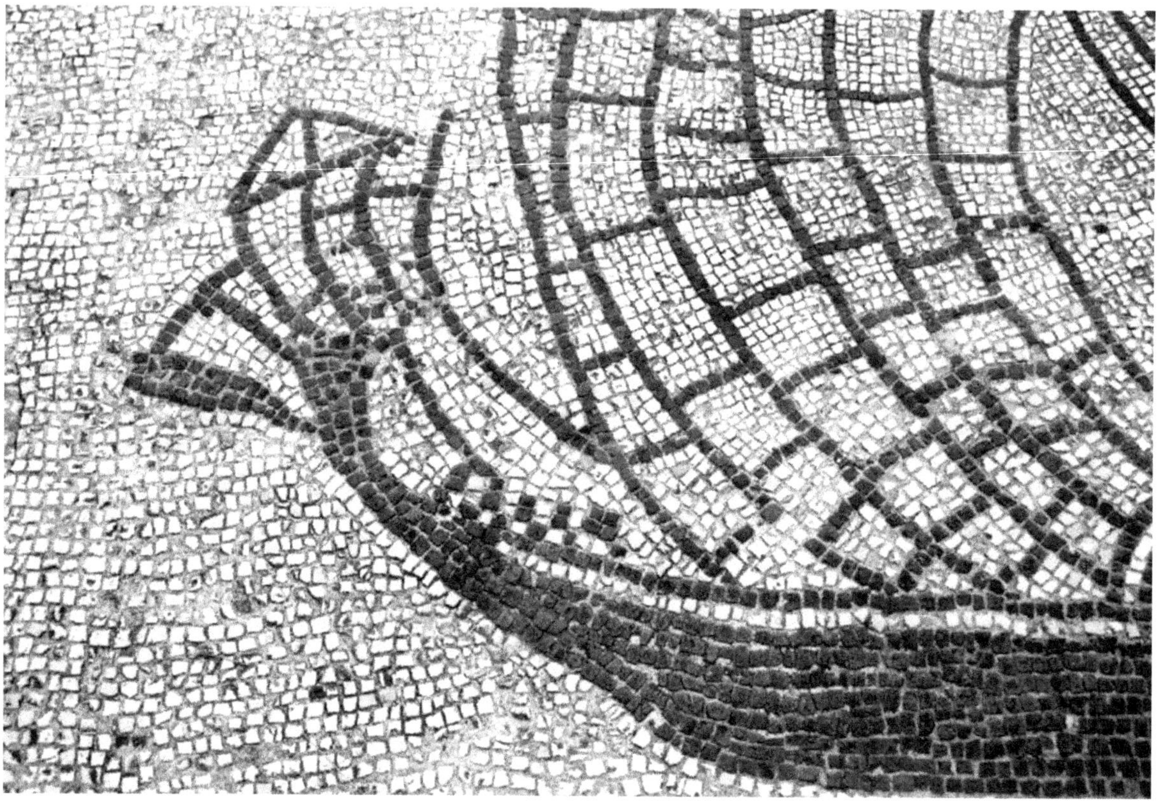

Fig. 3.7.12: The artemon rigging

goods in their home district to be merchandized in Italy and then engaged the *navicularii* (shippers) to ship the products to Ostia and Rome.

The single ship depicted in the mosaic has an elongated and tapered hull, slightly narrower at the stern, is broader at the stem (fig. 3.7.11). The bow points to the left. The stem has an outward angle while the rounded stern rises vertically. The stem-and-stern posts have a slight inward turns, both adorned with block shaped posts with straight cut tops. The gunwale is indicated by one strip of white tesserae. Slightly aft amidships, a wide strake is attached above the port aft gunwale, forming the raised aft bulwark. This strake is made with two rows of white and black tesserae. Five small spars projecting above the port fore-gunwale indicate the bitts. These spars depicted like a chess board, probably resulted from the mosaicist's lack of understanding the function of such elements and produced them like some decoration. Two angled spars projecting forward from the tip and beneath the outer stempost may indicate the bowsprits used for the sails rigging. The lower spar tapers towards the stempost. Probably there were three bowsprits; the artemon sail hides the right-hand one. They are connected by a through spar at their fore-tips (fig. 3.7.12). The black horizontal strip (made with one row of tesserae), probably is the flat roof of the cabin placed on the stern and used by the crew, as well by the helmsman who worked the rudders from there (fig. 3.7.11).

The rigging comprises two sails, three lines and one steering oar or rudder. The sails are seen from their fore face. The fully open main square sail billows forward. The head of the main sail equals the length of the yard. On the fore bunt are depicted five brails and nine reinforcing-bands (fig. 3.7.11). The yard is secured by two lifts stretching from either tip of the masthead to the tips of the yardarms. This white triangles also may indicate the top sails set above the yard. No reinforcing-bands or brails are depicted on these bunts. The masthead projects above the yard. A short band or flag attached to its tip was used to gauge the wind direction or probably as some festive decoration. The mast is hidden behind the sail. It was probably stepped amidships as it may be deduced from the position of the masthead. Three lines stretch between the tip of the port yardarm and the port quarter gunwale: the right-hand line (the extreme one) is presumably the port brace and the other two may represent the brails passed over the yard to work the sail. The port sheet appears to be attached to the fore tip of the quarter bulwark.

The artemon rigging comprises a fully open small square sail seen from its fore face and the mast with a forward rake above the stem (fig. 3.7.12). The mast is indicated by a thin black strip parallel to the stempost; its lower end is hidden by the bitts. The yard and the sail's head are equal in length. Two triangles set above the yard may indicate the top sails or the lifts to secure the yard. Two vertical lines depicted on the middle bunt may represent the central brails and the three fragmented horizontal lines indicate the reinforcing-bands. We may assume that this sail also is seen from its lee side. The starboard leech of the artemon sail has an upper extension that its function is

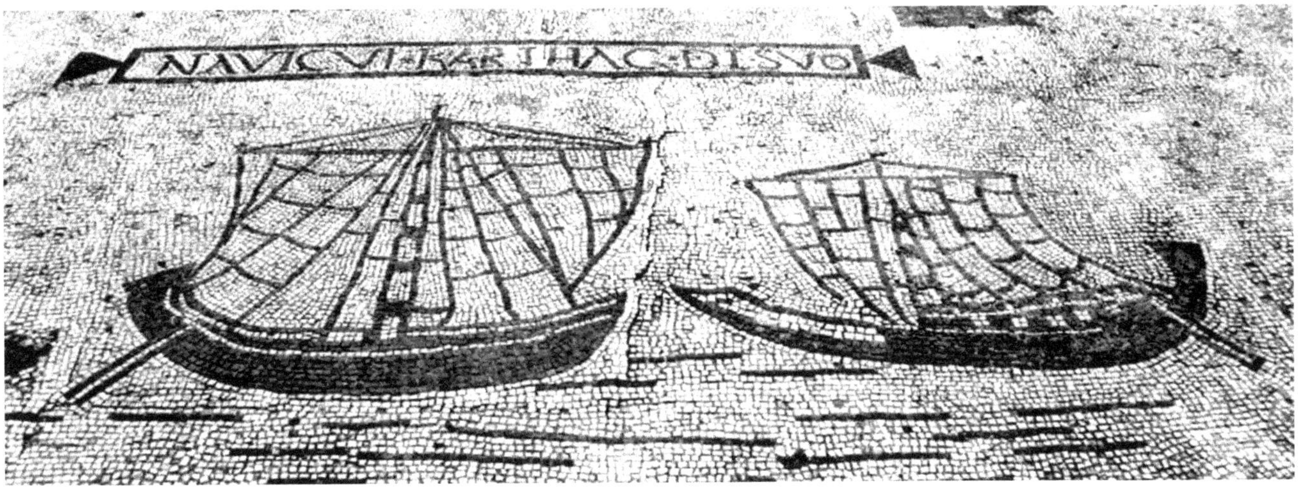

Fig. 3.7.13: Station 18 – Navicul-Karthag-De Suo

not understood. It appears that the lower line attached to the clew may represent the starboard sheet. Its lower end appears to go behind the lower starboard leech of the main sail. Indications of repairs to the mosaic are observed on both sails, especially distinguished on the upper and lower part of the main sail, on the brails and the reinforcing-bands (fig. 3.7.11). The artemon sail also suffered from the same repairs as may be deduced from its brails and the reef-bands (fig. 3.7.12). This ship probably replaced and older vessel that was erased. The mosaic work of the white tesserae, between the chine of the ship and the top line beneath the bottom of the ship indicating the wave, follows the remains of the projecting forefoot of a previous vessel (fig. 3.7.11). The white tesserae used in this section are slightly bigger than the original ones used in the background and the mosaic is quite crude.

The steering gear comprises only the port rudder mounted on the quarter with a rearward extension. The shaft projects from behind the aft-wing extension of the top port plank. The loom of the rudder (outlined with a black strip) projects above the quarter gunwale and its tip reaches the aft end of the quarter bulwark (fig. 3.7.11). The elongated rectangular blade is outlined with one strip of black tesserae, while the lower part of the shaft bisects the blade longitudinally, thus indicating both wings. The same steering arrangement may have been true on the starboard quarter. The white tesserae work beneath the blade and stern, apparently filled in the blade of the previous ship (fig. 3.7.11). Two fragmented horizontal strips made with black tesserae symbolize the sea waves, thus suggesting that the ship is anchored in open waters.

Station 18

This office belonged to the shippers from Carthage, as designated by the long inscription set into a *tabula ansata* "NAVICUL KARTHAG DE SUO". The inscription is found above two sailing ships facing each other (fig. 3.7.13). Both vessels are shown with their fully open square sails seen from their lee faces. The sea is symbolized by several short black strips, as representing the sea waves, thus suggesting that the vessels are anchored in open waters. Although no anchor or mooring line is shown, the vessels are anchored, as deduced from their static position.

Ship 1 (left-hand): The hull has a banana shape with a rounded upraised stern and almost a vertical stem finished with a pointed tip (figs. 3.7.14, 3.7.14 a). The prow pointed to the left appears to be broader than the stern. The stern is higher than the stem. The upper and lower edges of the top wale are outlined with one strip of white tesserae. No other strakes or seams are shown on the black hull.

The rigging comprises a fully open square sail, a mast and one steering oar. The tapered pole, stepped amidships is depicted with alternate black and white bands that probably indicating a composite mast, girdled with wooldings at fixed intervals.[38] The yard has almost the same length as the hull and is parallel to the deck. It is secured by two lifts stretching from either side of the masthead to the tips of the yardarms. The fore-and-back stays stretching from either side of the masthead are secured to the fore and quarter gunwales; the port and starboard shrouds secure the mast laterally (fig. 3.7.14a). The sail is very broad and billows over the port side, thus suggesting that the wind blows from the starboard quarter. On the lee bunt are depicted seven brails and four to five reinforcing-bands. Indications to repairs in the mosaic are observed on the sail, especially emphasized in the representation of the standing-and-running rig (fig. 3.7.14a).

The steering gear comprises only the starboard rudder with a rearward extension mounted to the quarter. The lower shaft (indicated by the white strip) bisects the black blade longitudinally, while its upper part appears to be hidden by the aft-wing extension of the quarter wale (figs. 3.7.14, 3.7.14a). The same steering arrangement was true on the port quarter.

[38] Casson, p. 231, n. 31.

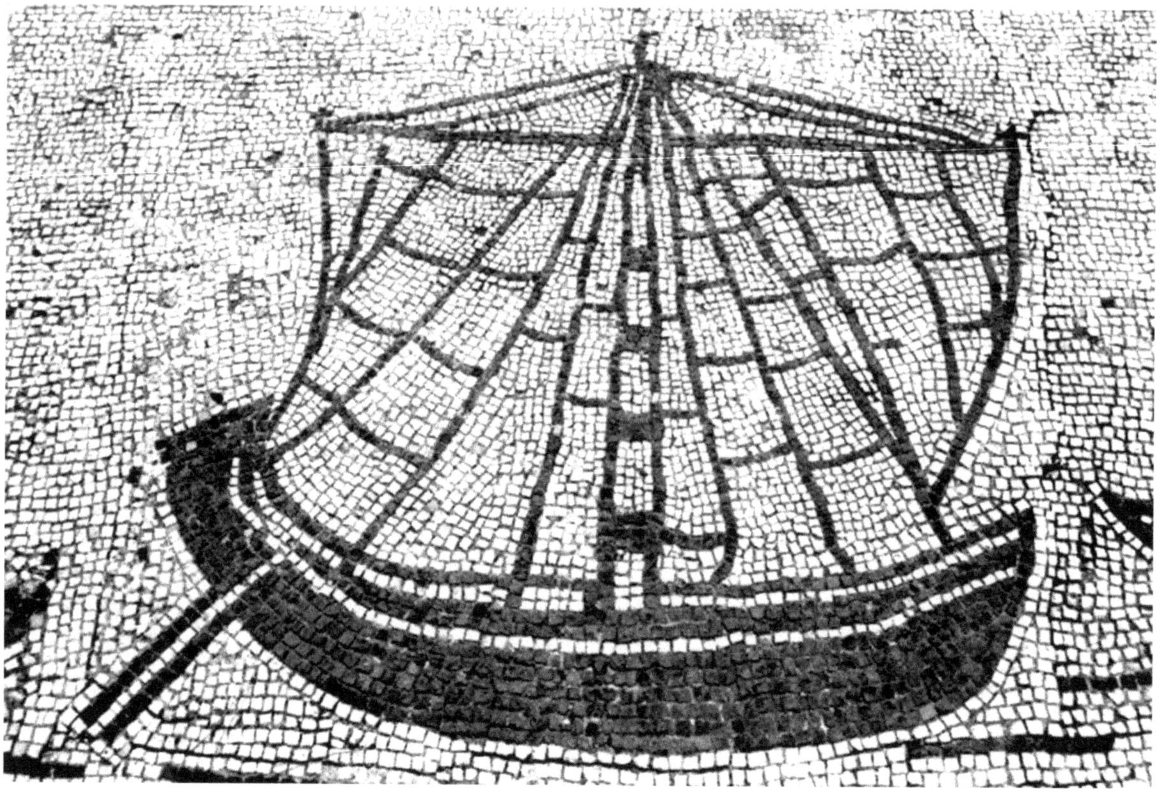

Fig. 3.7.14: Ship 1 – Station 18

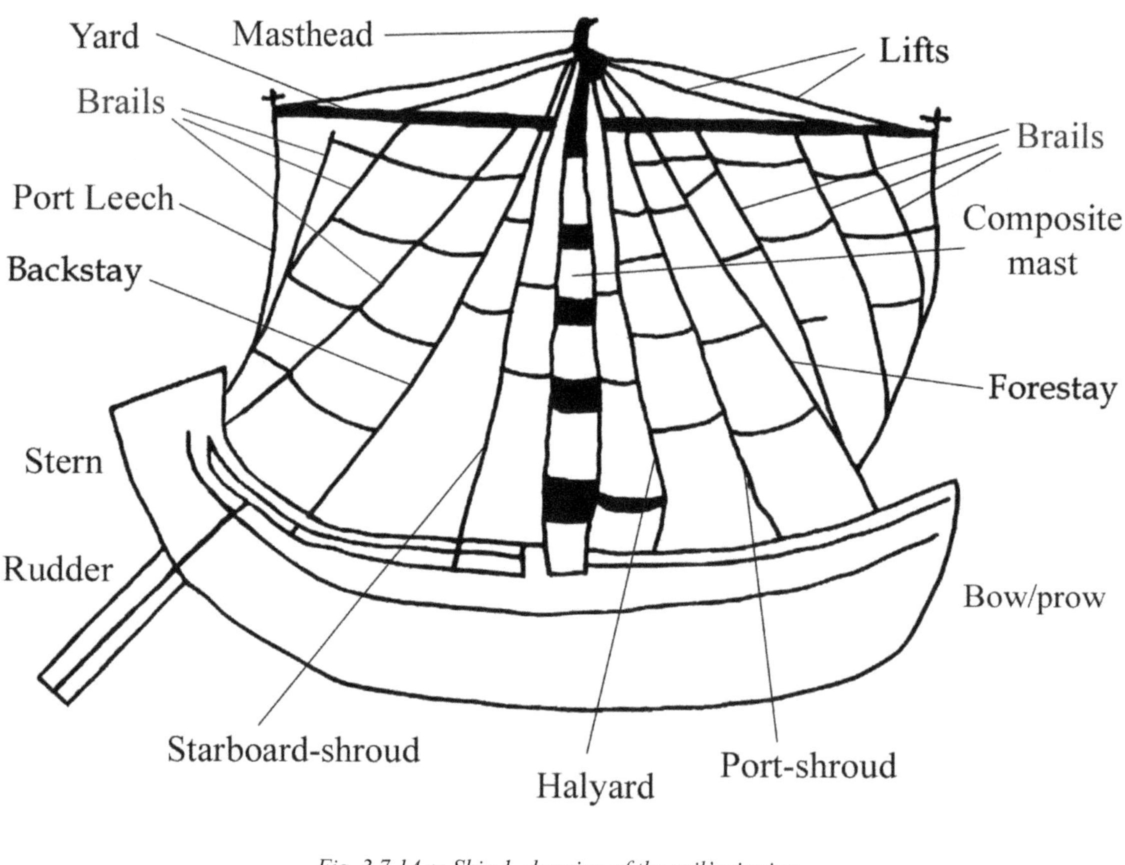

Fig. 3.7.14 a: Ship 1, drawing of the sail's rigging

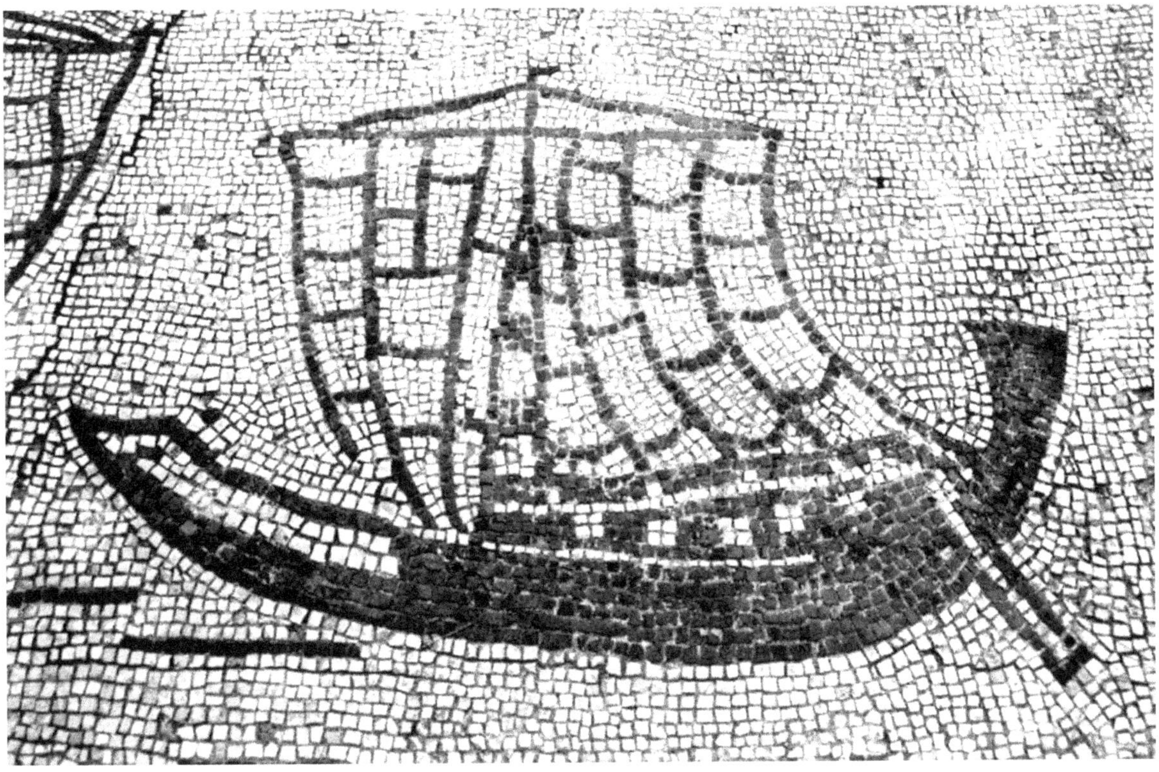

Fig. 3.7.15: Ship 2 – Station 18

Ship 2 (right-hand): The hull of this ship has a more pronounced banana shape, with a tapered stem and a broad up-raised stern adorned with a vertical block-shaped sternpost (fig. 3.7.15). The bow points to the left. On the fore upper side of the stem is depicted a white rectangular strake. It probably was inserted later, as suggested by the bigger size of the white tesserae that differ from the original ones in the background and the sail (fig. 3.7.15). A cabin with a pitched roof of terracotta tiles (indicated by the small oblique white lines and the outline the lower edge of the roof) is set between amidships and the quarter. On the port side of the cabin are depicted five small square windows made with white stones each.

The rigging comprises a tapered composite mast (indicated by alternated white and black bands) projecting above the fore side of the cabin's roof, a fully open square sail seen from its lee face, and one steering-oar (fig. 3.7.15). The sail billows over the starboard side, thus indicating that the wind blows from the starboard quarter or astern and the ship sails into the wind. The standing (fore-and-back stays and the shrouds) and running (halyard, braces, sheets) rigs are missing. On the lee bunt are depicted six brails and four reinforcing-bands. The irregular setting of the lines on the lee bunt and the mast indicate the restoration carried out on the sail and the lack of restorer's understanding of these rigging which he rendered it more like some decorations. The yard has the same length as the sail's head. It is secured by one lift stretching from either side of the masthead to the tips of the yardarms. A small flag is attached to the tip of the masthead, to gauge the wind direction.

The steering gear comprises only the port oar with a rearward inclination that is mounted on the quarter. The loom appears to be hidden by the back wall of the cabin (fig. 3.7.15). The shaft made with one strip of white tesserae bisects the rectangular blade longitudinally. The same steering arrangement may have been true on the starboard quarter. The helmsman probably worked the rudders from behind the cabin.

Station 19

The inscription set into a *tabula ansata* "NAVIC-TURRITANI" indicates that the office probably belonged to shippers from Turris Libisoni (Porto Torres) in Sardinia.[39] Beneath the inscription is depicted a sailing ship with two sails (fig. 3.7.16). The elongated hull has two rounded ends; the forward extended stem and the upraised stern are surmounted by short block-like posts with a cut top. The top wale is outlined with one row of white tesserae. Three small bitts project above the port quarter gunwale. The ship is rigged with two masts and sails, ant two steering oars. The main mast is stepped amidships and visible between the foot of the sail and the gunwale; it is a broad pole rendered with black tesserae and has a slight backward rake (fig. 3.7.17). The yard, the head of the sail and the masthead are cut by the lower edge of the *tabula ansata*. The fully open main square sail is seen from its fore face and appears to be set parallel with the gunwale. On the bunt are depicted six brails and three reinforcing-bands. On either side of the mast are depicted two black lines. Both lines to the left side

[39] Houston, 1980, n. 70, p. 170.

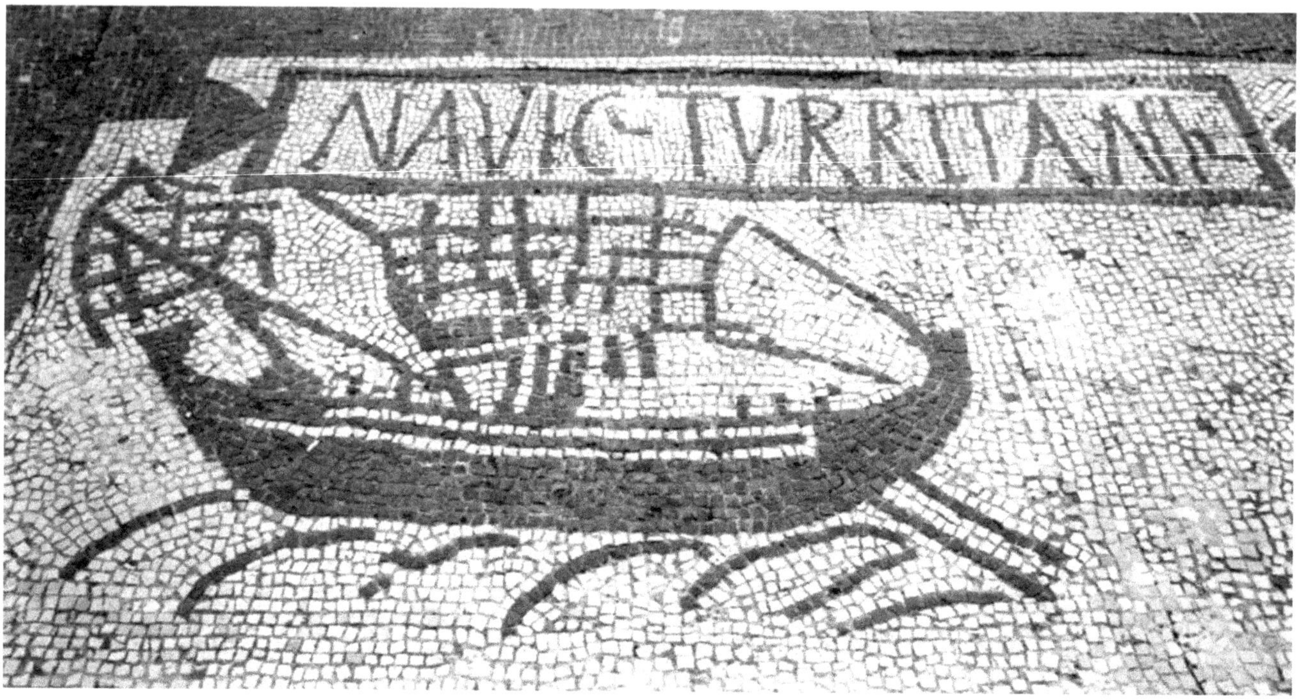

Fig. 3.7.16: Station 19 – Navic-Turritani

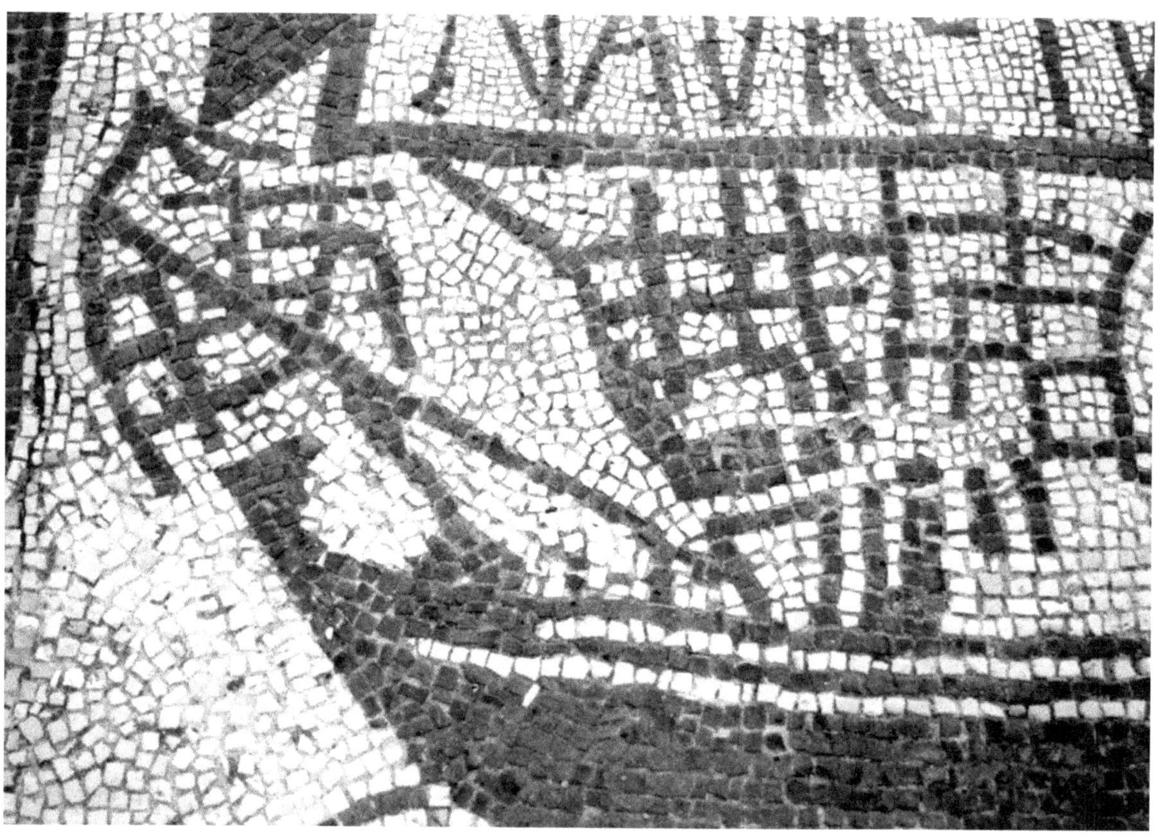

Fig. 3.7.17: The artemon rigging

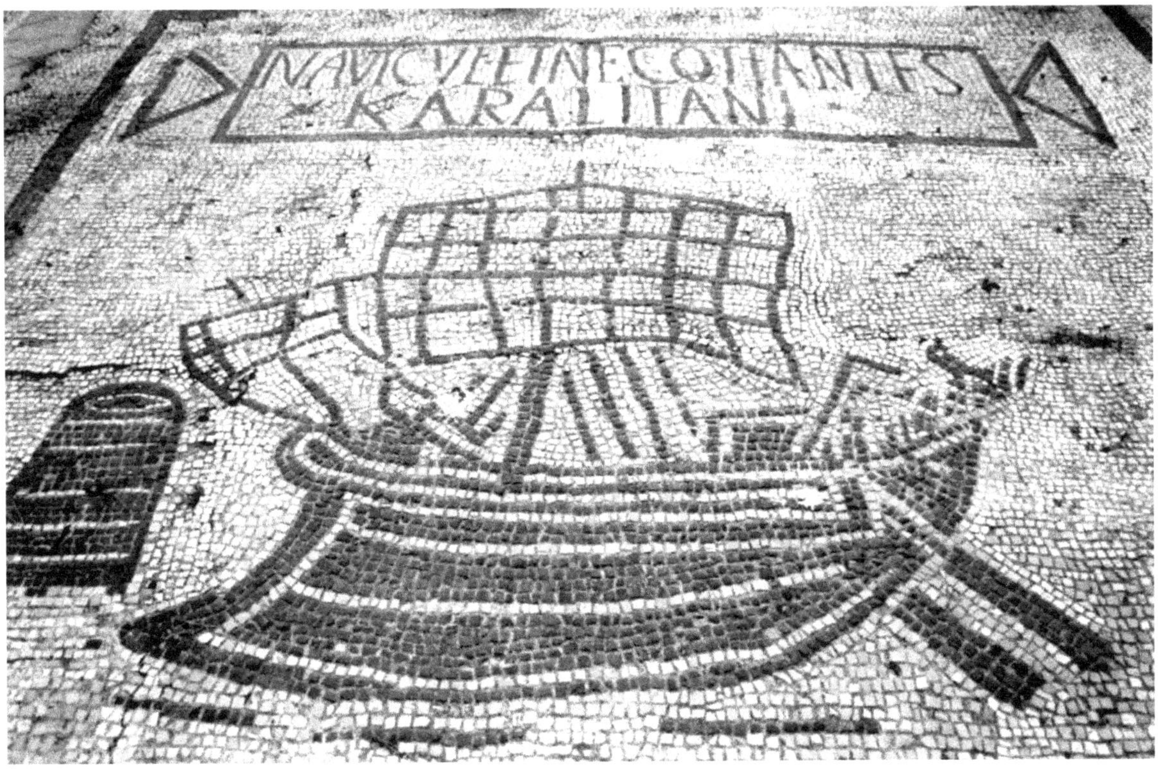

Fig. 3.7.18: Station 21 – Navicul et Negotiantes – Karalitani

of the mast may indicate a double-lines halyard. The short ends of two lines projecting from beneath the foot of the sail and on the right side of the mast probably represent two brails passed over the yard and worked from the lee side of the sail. The extreme line stretching between the upper port leech and the tip of the sternpost represents the port brace, while the port sheet stretches between the clew and the inner curvature of the sternpost (fig. 3.7.16). The angled line stretching between the lower left side of *tabula ansata* and the starboard top leech is probably the starboard brace, or the mosaicist extended the leech upward. The starboard sheet stretches vertically from the clew to the gunwale, near the left side of the mast.

The artemon mast has a forward rake and parallel to the stempost. The lower part of the mast seems to be stepped into a triangular small projection above the fore gunwale (fig. 3.7.17). The top fragmented horizontal line beneath the left corner of *tabula ansata* may indicate the yard and the head of the sail, whereas the lower line indicates the foot of the sail. The yard is supported by one lift stretching from either side of the masthead to tips of the yardarms (fig. 3.7.17). The masthead projecting above the yard has a forward rake. This depiction is misinterpreted or probably the mosaicist intended to give some perspective to the sail. The small square sail is fully open and appears to be seen from its lee face. On the bunt are depicted two brails and two reinforcing-bands. The short lines projecting from each clew indicate the port and starboard sheets. The line beneath the foot of the sail and closer to the port sheet probably shows the lower end of one of the brails. The angled line stretching between the tip of the port yardarm and the starboard sheet of the main sail may represent the artemon's port brace. This line, most likely, was meant to represent the backstay for securing the artemon mast. The mosaic shows traces of some repairs, clearly visible on the rigging lines of the main sail and the artemon sail (fig. 3.7.16). Short arching strips beneath the bottom of the ship indicate the wavy sea. The static position of the ship suggests that it is anchored, though it does not show a mooring or an anchor line. The lines depicted on both bunts show a crude restoration, whereas the mosaicist not understanding their function rendered them more like a decorative pattern. Probably this ship replaced an earlier vessel when the shop may have changed owners. This restoration in the mosaic also is noticed in the varied size of the white tesserae used in the background, as well as the mosaic work.

The steering gear comprises two oars with a rearward extension, projecting beneath the stern. They are indicated by thin shafts made with one row of black tesserae. The blades probably are submerged in the water (fig. 3.7.16).

Station 21

A two-line inscription inserted into a *tabula ansata* indicates that the office belonged to shippers and traders from Cagliari in Sardinia, "NAVI ET NEGOTIANTES KARALITANI".

The inscription is set above a sailing ship with its bow pointing to the left (fig. 3.7.18). This is the second inscription that refers to *navicularii et negotiantes* (shippers and traders). On either side of the ship is a cylindrical object

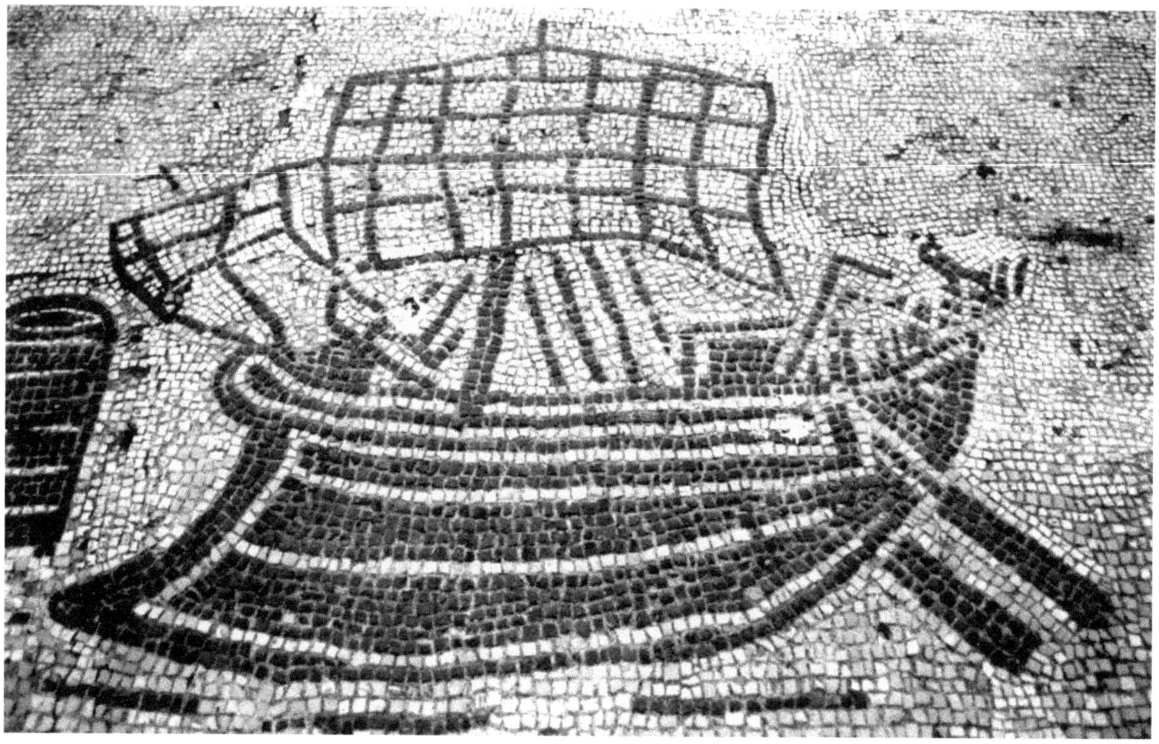

Fig. 3.7.19: Station 21 – The Ship

with a rounded rim set on three small legs. They represent the *modii* used for grain measure; we assume that this office belonged to grain traders that shipped it from Sardinia to Ostia and Rome.

The tapered hull, broader at its bow is slimmer at the stern (fig. 3.7.19). The concave stem is finished by a pointed cutwater, and it is turned to the left. One white strip of tesserae designates the prow, the hull and the keel of the ship. The narrow top strake outlined with one white strip and projecting over the stem finished with a rounded tip may indicate the bulwark. The top wale is outlined with one row of white cubes. The lower white strip probably indicates the lower wale just above the waterline or the seam of the joint strakes. On the port quarter is placed a rectangular cabin that its upper part and the left-hand wall are outlined with one strip of white tesserae. The aft extended poop is distinguished by a lattice fence. Above the aft-end of the poop fence projects the head of a goose or a swan looking backward.

The ship is rigged with two masts and sails, and a pair of rudders mounted on the quarters. The main mast made with two rows of black tesserae is stepped forward amidships with a slight rearward rake. It is visible between the foot of the sail and the port gunwale. The square sail is seen from its fore face billowing forward, thus indicating that the wind blows from the starboard quarter (fig. 3.7.19). On the bunt are depicted six brails and three reinforcing-bands, each made with one row of black tesserae. The yard and the head of the sail are depicted as one unit and both have the same width. A triangular topsail is set above the yard. Four brails are shown on the bunt. The tip of the masthead projects above the tip of the topsail. On the right side of the mast are depicted four lines that stretch between the foot of the sail and the gunwale. The line closer to the right side of the mast probably represents the halyard. The right-hand line (extreme right) stretching to the bottom fore-side of the cabin indicates the backstay. Both lines in the middle may represent the brails passed over the yard and worked from the lee side of the sail, or they indicate double-lines port shrouds. To the left side of the mast, three lines stretch to the fore gunwale (fig. 3.7.19). The top fragmented line (extreme left) may represent the forestay (its location is not correct; it should have been stretching forward the bunt to the stempost). Both lines may indicate the starboard shrouds. The port sheet goes behind the aft-side of the quarter cabin's roof. The starboard sheet, hanging free, seems to stretch towards the lower part of the main mast. No braces are shown.

The artemon mast has a forward rake above the stem. Its lower part is hidden by the short fore-bulwark plank. The open small square sail is seen from its lee face, emphasized by the overturned port leech and side of the sail (fig. 3.7.19). This depiction appears to be anachronistic, whereas the main sail is seen from its fore face while the artemon sail is shown from its lee face. This representation probably was meant to give some perspective to the ship and some kinetic appearance to the sails or the mosaicist did not understand the proper representation of this sail. One brail and one reinforcing-band are depicted on the artemon bunt. Three reinforcing-bands are depicted on the small overturned port-edge of the sail. The yard and the sail's head is one unit. The tip of the masthead projects above the yard. The port sheet stretches between the clew and the rounded tip of the forward extended top wale. The starboard sheet appears to the tight to the lower end of the main sail's forestay (fig. 3.7.19). The starboard brace stretches between the tip of the yardarm and the middle section of the main sail's starboard leech.

The steering gear comprises two rudders with a rearward extension; one mounted on either quarter. The elongated rectangular blade projecting beneath the stern indicates the starboard rudder. Its lower edge has a straight cut. The shaft of the port rudder (outlined with one row of white tesserae) projects from behind the aft-wing of the top strake. The elongated blade with straight shoulders and a rounded edge is a bit thinner than starboard one. This depiction is not accurate because both rudders were identical. The different widths of the blades resulted from the mosaicist's misinterpretation. A shorter spar inserted perpendicularly into the head of the loom indicates the tiller worked by the helmsman (fig. 3.7.19). The same arrangement was probably true for the starboard rudder. Segments of black horizontal strips indicate the sea waves.

Station 23

Two-lines inscription found between two sailing ships and two dolphins beneath the ships feeding on the same octopus are not inscribed into a *tabula ansata* (fig. 3.7.20). The inscription "MF IUI ΛRI SYLLECTI", is a misinterpretation; the complete inscription may have been "(NAVIC)ULARI SYLLECTI(NI)".[40] The letters MF in the first line are misinterpreted; originally they may have referred to NE (*negotiantes* or *negotiatores* ET).[41] Although, the inscription in this mosaic is fragmented, it is possible to understand that the shippers were from Syllectum (Byzacium province) in North Africa. It appears that in a later restoration N was transformed in an M and E into F; the initials of NE were misinterpreted and made with the letters as they appear today (fig. 3.7.20). If the correct inscription referred to *navicularii et negotiantii*, then this is the third office indicating its purpose. Besides the inscription, we may trace different stages of restorations in the mosaic, especially significant in the left-hand dolphin, the inscription, and the area beneath the octopus (fig. 3.7.21). The ships facing each other have different hulls and both are depicted with their fully open square sails.

Ship 1 (left-hand): The prow points to the right, thus revealing its starboard hull. The broad concave stem finishes with a projecting pointed cutwater (fig. 3.7.21). The stempost is adorned with an outer turned volute, emphasized by a strip of white tesserae that continues along the stem. On the lower starboard stem is depicted a white figure that may represent an *oculus* or a small dolphin (fig. 3.7.21).

The rounded stern is upraised almost vertically. The top wide wale outlined with one strip of white tesserae extends alongside the starboard hull and the quarter. The end of the sternpost is damaged and therefore its shape is not distinct. Two white strips depicted beneath the top wale indicate the strakes or the seams. We may assume that the middle strip probably is the lower wale just above the waterline, while the second strip may have designated the keel or the guardboard. A plank attached above the gunwale forms the bulwark. It appears as being made of two pieces, which is the opposite. The white spacing, closer to the quarter resulted from one of the restorations. Three small projections depicted on the fore part of the bulwark are the bitts.

The rigging is a bit different than that of the previous vessels described above. It comprises three masts and sails, and two rudders, one mounted on either quarter. The black sails with their white rigging lines are seen from their fore face (fig. 3.7.21). The main mast made with three rows of black tesserae, visible between the foot of the sail and the gunwale, is stepped vertically forward amidships. The main square sail is fully open and billows forward, thus suggesting that the wind blows from astern or the port quarter. On the bunt are depicted five white brails and six white reinforcing-bands. The yard equal to the width of the sail's head and is indicated by one white strip. The starboard sheet is attached to the fore edge of the aft bulwark. No standing (fore-and-aft stays, and/or shrouds) or running (braces or halyard) riggings are shown. Black triangular topsails are set on top of the yard. Two white lifts are depicted on the fore face of each sail. A small flag surmounting the tip masthead was used as an indicator for the wind direction or as a decoration. The artemon (fore) sail has a forward rake following the position of the mast that is hidden by the sail. This sail is slightly smaller than the main sail. On the black bunt are depicted four brails and four reinforcing-bands; all the line are rendered with white tesserae. A horizontal white strip indicates the yard and the head of the sail (fig. 3.7.21). Atop the yard, is set a triangular black topsail; two white lifts are shown on the fore face of the bunt. The tip of the masthead projects above the tip of the topsail. The *mizzenmast* (rearmost) stepped on the quarter has a slight forward inclination. It is the smallest of the ship's rigging. The mast is visible between the foot of the sail and the aft bulwark (fig. 3.7.21). Three white brails and two reinforcing-bands are represented on the bunt. A white horizontal strip indicates the yard or the head of the sail. The tip of the masthead projects above the yard. The starboard and port sheets stretch vertically towards the top edge of the aft-bulwark (fig. 3.7.21).

The steering gear comprises two rudders placed on the quarters with a backward slanting. The elongated port blade projects beneath the keel. The shaft of the starboard rudder, outlined with one white strip, projects beneath the aft-wing of the top wale (fig. 3.7.21). The blades of both oars are identical; it appears that their bottom edge was rounded. The projecting figure above the stern may indicate the helmsman seated on the quarter behind the mizzen rig and working the rudders; he probably faces the sail (fig. 3.7.21). This decoration also may represent the head of a swan with

[40] Calza, 1914, p. 285.
[41] Ibid.

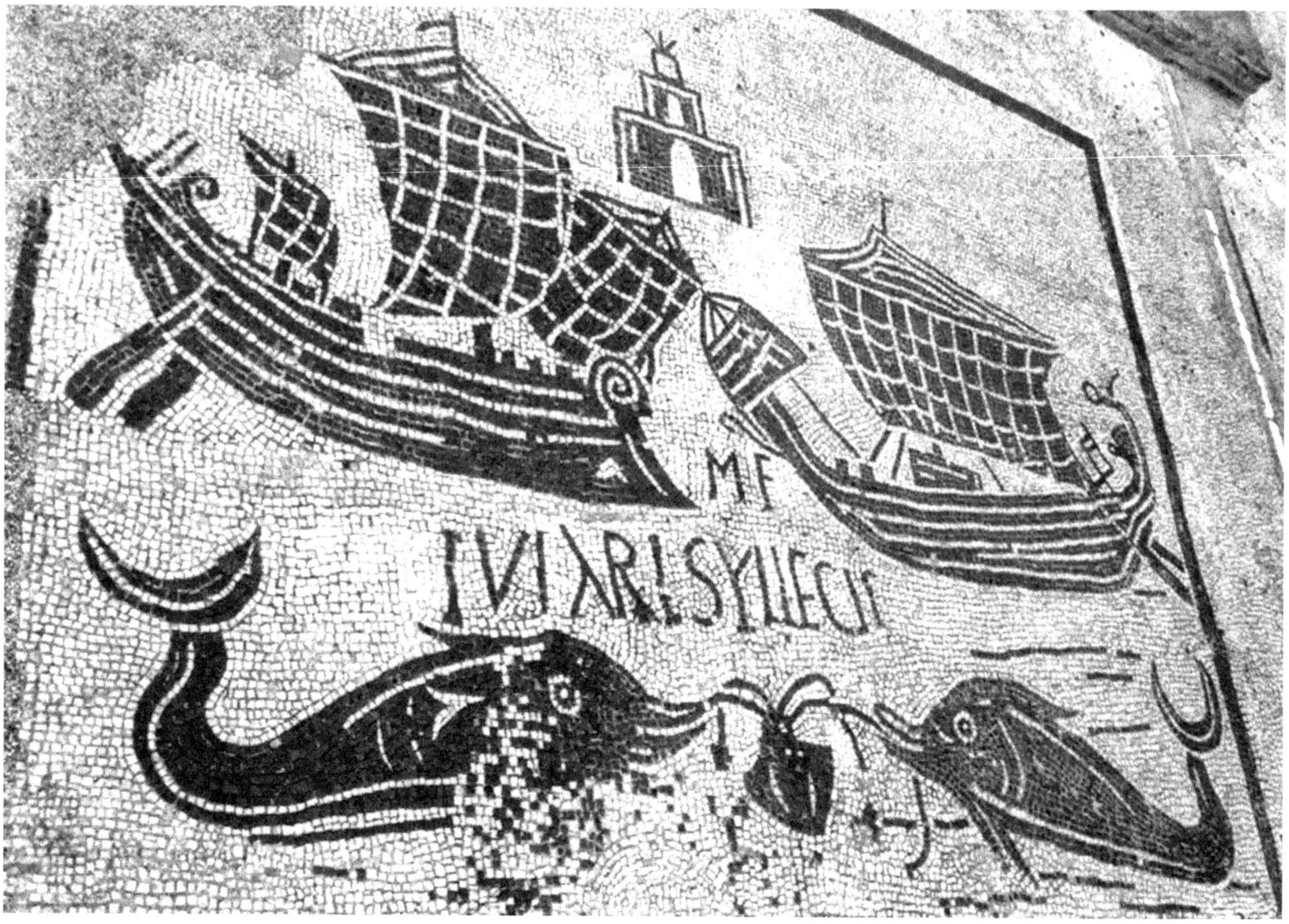

Fig.3.7.20: Station 23 – MF Iulari Syllecti

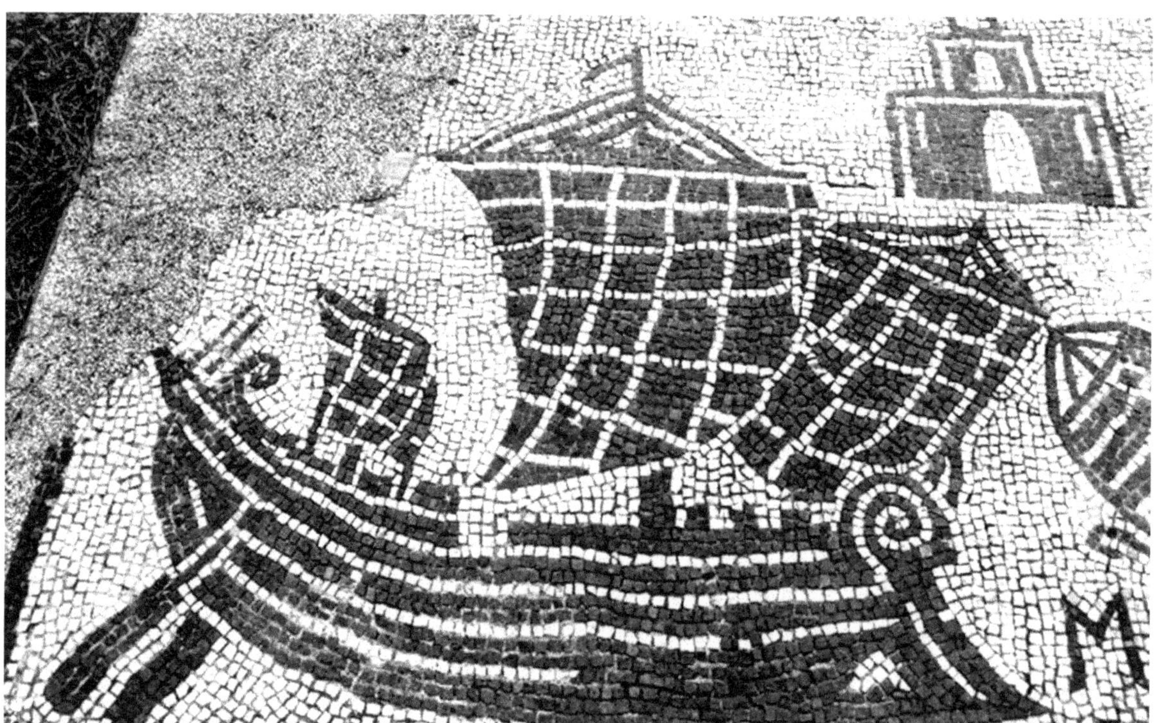

Fig. 3.7.21: Ship 1 – Station 23

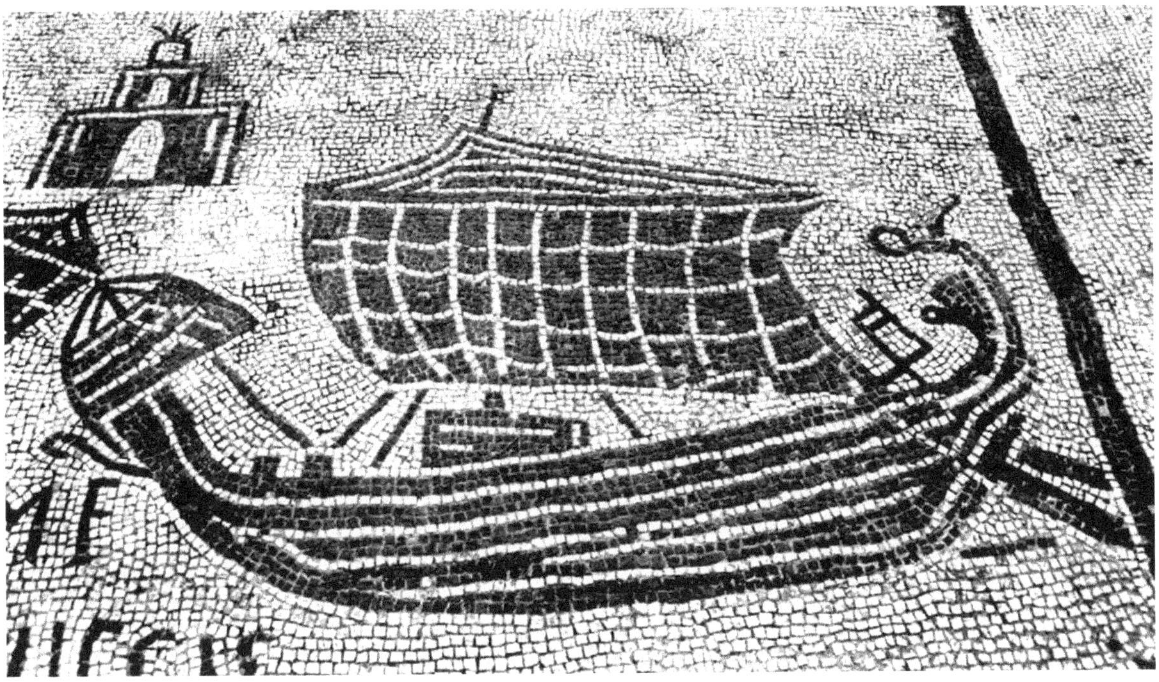

Fig. 3.7.22: Ship 2 – Station 23

his head turned backwards. This figure is a characteristic element of Roman merchantman. Two short vertical spars projecting behind the figure are not understood. They probably were part of an aft cabin in a previous vessel that was replaced the present one. They have no other meaning for the ship.

Ship 2 (right-hand): The prow points to the left. The tapered hull is broader at the stern (fig. 3.7.22). The upraised stem is surmounted by a broad stempost. The curved upraised stern is adorned by an inward-turned block-shaped sternpost. The rod with a hook at its upper end and attached to the outer edge of the stempost, almost parallel to it, may represent the bowsprit that was used for fastening the sail's tackle lines. A kind of a rod with a loop at its top and attached to the outer tip of the sternpost probably represents a decoration or it may suggest a stylized goose figurehead adornment (fig. 3.7.22). The upper edge of the top wale outlined with a white strip, seemingly also forms the gunwale. This strip also follows the shape of the stempost. The aft end of the wale has a vertical cut edge and apparently forms the aft-wing that houses the shaft of the port rudder. The other white strips indicate the seams between the strakes or they outline the lower wale just above the waterline. The bottom white line outlines the keel or it represents the guardboard. Two small square bitts project above the port-fore gunwale. A rectangular cabin is set forward amidships on the deck. Two white strips indicate seams of the cabin's wooden planks.

The ship is rigged with two masts and black sails, and two rudders, one mounted on either quarter (fig. 3.7.22). The black sails are viewed from their fore face. The main mast is stepped forward amidships projecting from above the fore roof of the cabin. The fully open, square main sail is very broad and billows forward, thus indicating that the wind blows from the starboard quarter or astern. On the bunt are depicted nine brails and four reinforcing-bands rendered with white tesserae. On the starboard side of the mast are shown two lines stretching beneath the foot of the sails and the fore gunwale. The extreme line, stretching between the starboard clew and the top of the second bitt, represents the starboard sheet. The line closer to the top fore-edge of the cabin probably is the forestay. It appears to stretch from behind the bunt; this is a misinterpretation of the mosaicist. In a proper depiction this line should have been shown forward the bunt, while its lower end should be tight to the stempost or the bowsprit. The line stretching behind the aft side of the cabin (visible between the foot of the sail and the gunwale) is probably the backstay (fig. 3.7.22). Apparently the port sheet is attached to the quarter gunwale just behind the loom of the port rudder. The yard appears to be rendered with two rows of black tesserae. A black triangular topsail is set above the yard. Two white strips depicted along each leach indicate the lifts supporting the yard and the sail. On the topsail are depicted three horizontal white strips probably representing the reinforcing-bands. This depiction suggests that either the mosaicist or the restorer did not fully understand the function of these lines and for that reason they rendered them in a kind of labyrinth. The tip of the masthead projects above the tip of the topsail.

Only the forward inclined sail indicates the artemon rigging, which is laid on the stempost (fig. 3.7.22). The mast is not distinct, or it is aligned with the stempost. The yard and the head of the sail are indicated by one strip of white tesserae. It is secured by two lifts stretching from either side of the masthead to the yardarms. On the small square sail, seen from its fore face, are depicted two reinforcing-bands and on the right side of the bunt are traces of two brails. It

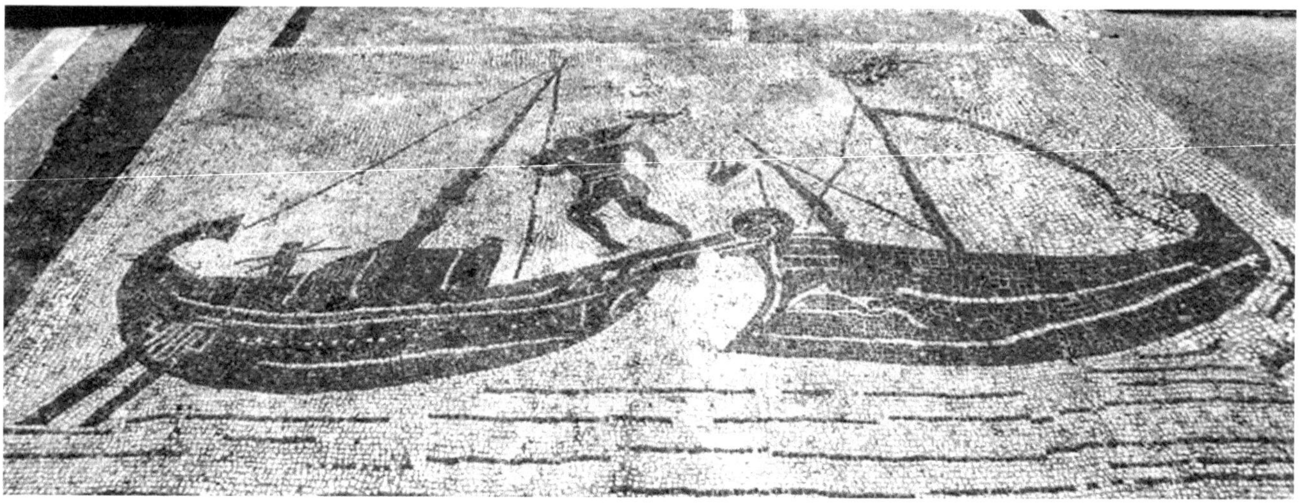

Fig. 3.7.23: Station 25

appears that this sail went through some restoration, as may be deduced from the rigging lines. The masthead projects above the yard. The line stretching between the foot of the artemon sail to the second bitt most probably indicates the artemon backstay (fig. 3.7.22).

The rudders are mounted on the quarters with a backward slanting. The upper and thinner elongated blade projecting beneath the stern represents the starboard rudder. The shaft of the port rudder projects from beneath the aft-wing of the top wale (fig. 3.7.22). The lower end of the blades is cut by the right-hand frame of the mosaic. On the port quarter, behind the main sail is depicted a kind of rectangular frame whose function is not entirely clear. We may suggest that the lower horizontal line projecting from the head of the port loom may indicate the tiller. The other pattern is not clear and therefore no suggestion is appropriate for the function of this element.

Station 25

No inscription is found in this mosaic. Two ships facing one another and the activity carried out indicate that the office probably belonged to the *caudicarii*, the shippers who operated the *navii caudicarii*, used as service vessels to unload cargoes from the big freighters anchored in the open sea and subsequently transport them on the Tiber to Ostia and to Rome. The *caudicarii* guild was one of the largest at Ostia, as attested by archaeological evidence from the 3rd - 4th centuries.[42] The stevedore in the mosaic who is carrying a jar on his right shoulder and going from one ship to the other indicates the amphorae cargo unloading from a large seagoing merchantman anchored at the month of the Tiber and then reloading it into a river transporter to be brought to Ostia or Rome (fig. 3.7.23). The mosaic shows traces of several repairs, especially distinct between the stern of the right-hand ship and its rigging. Black horizontal strips indicate the sea waves, thus emphasizing the environment of the activity.

[42] Meiggs, p. 293.

Ship 1 (left-hand)**:** The bow points to the right. The rounded hull is shaped like a spoon with the prow slimmer than the stern (fig. 3.7.24). The curved stem is surmounted by a horizontal forward extended stempost. The upraised rounded stern is adorned by an inward-turned sternpost over the quarter. The end of the post resembles a short papyrus umbel of Egyptian *papyriform* ships type. The top wale is outlined with white strips. The white dots alongside the middle of the port hull may suggest the nails drawn through the planks into the inside frames strengthening the hull. The lower white strip may indicate the lower wale just above the water or the waterline. Two square openings depicted on the starboard quarter, probably are the quarter oarports whereas the shafts of the steering oars go through (fig. 3.7.24). A cabin with an arched roof is found between the mid-deck and the quarter. Behind the cabin, is set a capstan with four horizontal bars used to turn it round. The rigging comprises a tapered mast stepped amidships with a slight forward inclination (fig. 3.7.24). It projects from above the mid-roof of the cabin. On either side of the mast are depicted several cleats to enable the seamen to climb it when necessary to work the line rigging and the sail (missing in the mosaic).[43] The mast is secured in place by a backstay that stretches from the masthead to the tip of the sternpost and the forestay that stretches from the masthead to the deck forward of the cabin's entrance. The mosaicist may have depicted this line closer to the cabin because he did not want to interfere with the stevedore on the bow. The yard and the sail are missing. Both ships are anchored, though that both of them lack any anchor or a mooring device. A gangplank is set between the bows of both ships and a stevedore walks over the prow of the merchantman carrying an amphora on his right shoulder (fig. 3.7.24).

The steering gear comprises two steering oars placed on the starboard quarters. The mosaicist mistook this depiction or probably he wanted to show both oars as viewed from

[43] Ancient large merchant ships were equipped with a rope ladder abaft the mast to enable the crew to climb and work the yard, the sail or the tackle lines when necessary.

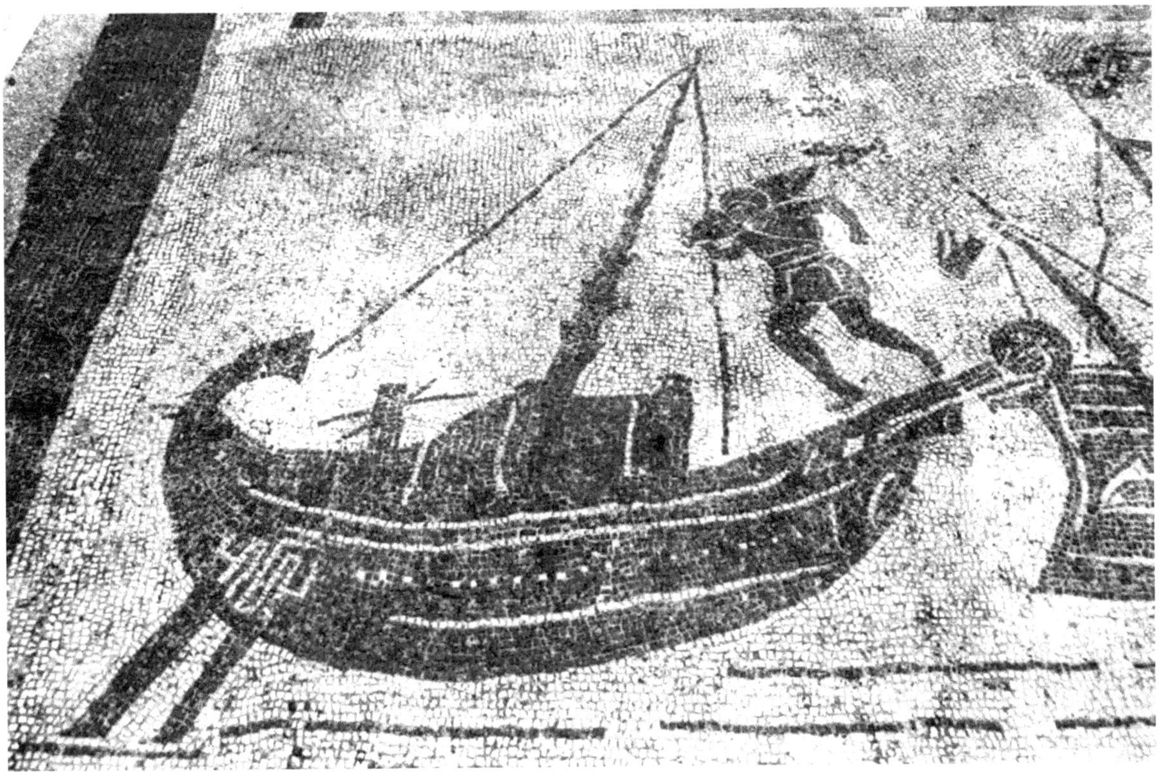

Fig. 3.7.24: Ship 1 – Station 25

the stern though to give a perspective view of the vessel (fig. 3.7.24). The shafts projecting through square oarports outlined with a strip of white tesserae. The port elongated blade is slightly wider than the starboard one. The lower ends of the blades are submerged in the water as suggested by their horizontal lower ends.

Ship 2 (right-hand)**:** The tapered hull is wide on its fore part and slimmer at the stern (fig. 3.7.25). The prow points to the left. The concave stem is finished with a short projecting pointed cutwater. An outer turned volute completes the stempost over the stem. Beneath the volute is a short projecting beam supporting the gangplank (fig. 3.7.25). Four short white strips depicted on the upper fore-stem, probably represent the bulwark above the gunwale and the lower edge of the top plank. The stem and the rounded head stempost are outlined with a white strip that follows the shape of the bow. A dolphin is depicted in the middle of the port stem with its head pointing forward (fig. 3.7.25). The highly upraised curved stern is adorned with a short vertical sternpost. Two white strips depicted along the port side may symbolize the top and lower wales.

The ship is rigged with two masts and four lines. The main tapered mast is stepped forward amidships with a slight forward slanting (fig. 3.7.25). A flag indicated by the small, black, wavy rectangle seems to be attached at the tip of the masthead. The mast is secured in place by undulating forestay, stretching from the masthead to the base of the volute. The arched thicker line stretching from the masthead to the sternpost indicates the backstay. Both lines went through repairs, especially noticeable on the lower part of the backstay and the white background between the lines and the mast (fig. 3.7.26). The yard and the sail are missing. The tapered artemon mast has a forward rake above the bow. The lower part suffered some damage and it was replaced with white tesserae. It is secured by the forestay stretching towards the top of the volute, and the backstay that stretches towards the lower left side of the main mast. To the left side of the artemon mast is depicted by a segmented perpendicular strip (two rows of black tesserae) that may represent the lowered yard with the furled sail beneath it (fig. 3.7.25). Two fragmented arched lines depicted above the left corner of the yard may represent the starboard lifts. This section shows clear indications of the repairs made on the mosaic which distorted significantly the artemon rigging and the standing rig of the main mast.

A trapezoidal pattern, outlined with black tesserae, is found beneath the stern and the lower right edge of the mosaic. Its function is not clear, especially since repairs are visible in this section. We may assume that it represents a mooring device on the quay. The steering gear was also altered by the repair of the mosaic and appears to have been inserted into the mooring device and replaced by white tesserae (fig. 3.7.25).

Station 32

One-line inscription, flanked on either side by an ivy leaf, refers to the shipper's office from Narbo (France), "NAVI NARBONENSES". The inscription is set above the frame that surrounds the maritime scene. The sailing ship faces a tall stepped tower-like structure (fig. 3.7.26). The bow

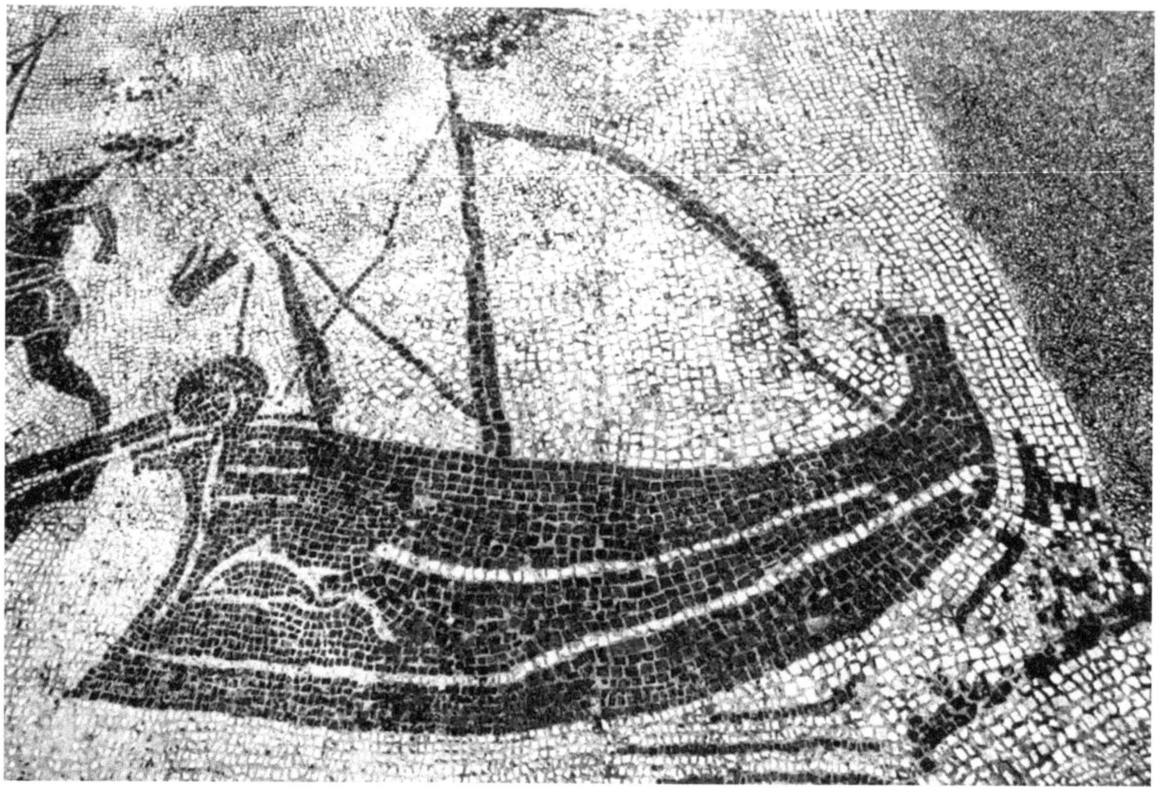

Fig. 3.7.25: Ship 2 – Station 25

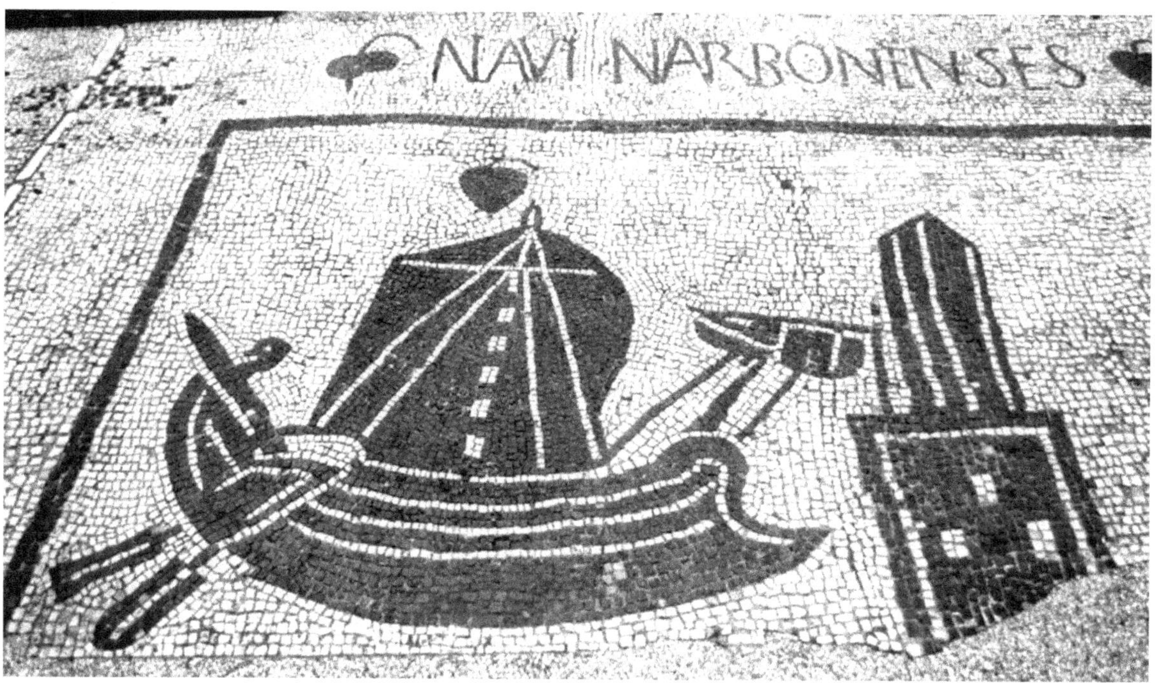

Fig. 3.7.26: Station 32 – Navi Narbonenses

points to the right, thus revealing the starboard side of the ship. The rounded hull has a concave stem finished with a pointed projecting cutwater. The short stempost has a rounded head resembling a stylized volute. The prow, stern and the top wale are outlined with one strip of white tesserae. The lower white strip may indicate the seam of the adjacent planks or the lower wale just above the waterline. A white arching strip set between the quarter gunwale and the aft plank may indicate the aft-wing housing the shaft of the starboard rudder. Above the stern is set a plank with

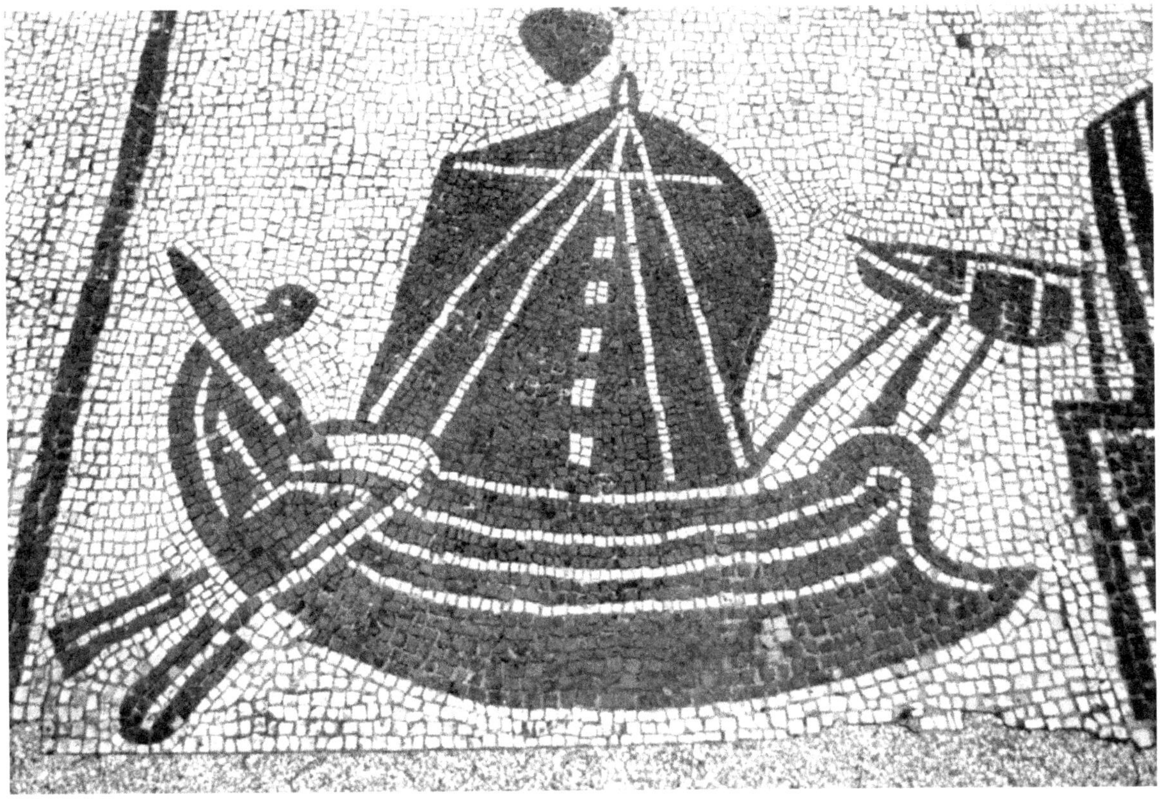

Fig. 3.7.27: Station 32 – The Ship

a rearward extension forming the fenced poop. The head of a duck or goose turned backwards, projects above the fence (fig. 3.7.27).

The ship is rigged with a fully open black square sail, two masts and a pair of rudders mounted on the quarters. The main sail and the mast are seen from the lee side, thus the sail billows over the port side. The alternated black and white bands of the spar indicate a composite mast.[44] The masthead is adorned with a small elliptical object, outlined with one row of black tesserae. This device may indicate a top lantern used for night navigation or it is the trademark of the ship. The yard and the head of the sail are made with one strip of white tesserae. Four lines depicted with white tesserae form the standing rig of the mast. The extreme left-hand line stretching from the masthead to the end of the starboard aft-wing indicates the backstay, while the extreme right-hand line is the forestay (fig. 3.7.27). The lines closer to the mast that their upper ends are attached beneath the yard at the joint point with the mast, may represent the port and starboard shrouds. The running rigs of the sail (halyard, braces, brails, sheets) are not depicted. On top of the yard is set a triangular black topsail. The forward inclined, tapered artemon mast is stepped on the fore deck. It seems to be used as a crane for loading or unloading cargoes to or from the structure facing the ship (fig. 3.7.26).

The steering gear comprises two rudders, one mounted on each quarter, have a backward slanting. The shaft of the starboard rudder (outlined with a white strip) seems to project from beneath the white aft-wing extension of the top wale (fig. 3.7.27). The lower end of the shaft, depicted with white tesserae, bisects the black blade longitudinally. The blade has a horizontal cut shoulder and a rounded end. The elongated rectangular blade projecting beneath the stern and a small section of the shaft indicates the port oar. The longitudinal edges of this blade have concave cuts (fig. 3.7.27). The shoulder and the end have a straight cut. The lower end of the shaft, made with a white strop, bisects the blade longitudinally. The depiction of two different blades is a misinterpretation of the mosaicist. The starboard blade, the lower part of the shaft, and the bottom of the ship show some repairs on the mosaic. Apparently this blade was remodeled. We may assume that originally both blades had the same shape and size and similar to the port blade.

The structure facing the ship is a two-storey stepped building. The base of the structure is missing. The rectangular first floor is outlined with one strip of white tesserae. Three narrow rectangular windows, depicted with white tesserae, are set on different levels, suggesting that it was reached inside to the top floor from a spiral stairway. The top floor, a tower-like structure, is probably hexagonal in shape judging by the angled top edge and the three vertical white lines that may indicate the corners. On top of the first floor (on the left side, facing the ship) is depicted a device that may represent a crane; it is indicated by the vertical pole and an extended horizontal arm above the ship's prow (fig. 3.7.27). Beneath the arm is an elongated hanging bundle (made with black tesserae,

44 See Glossary.

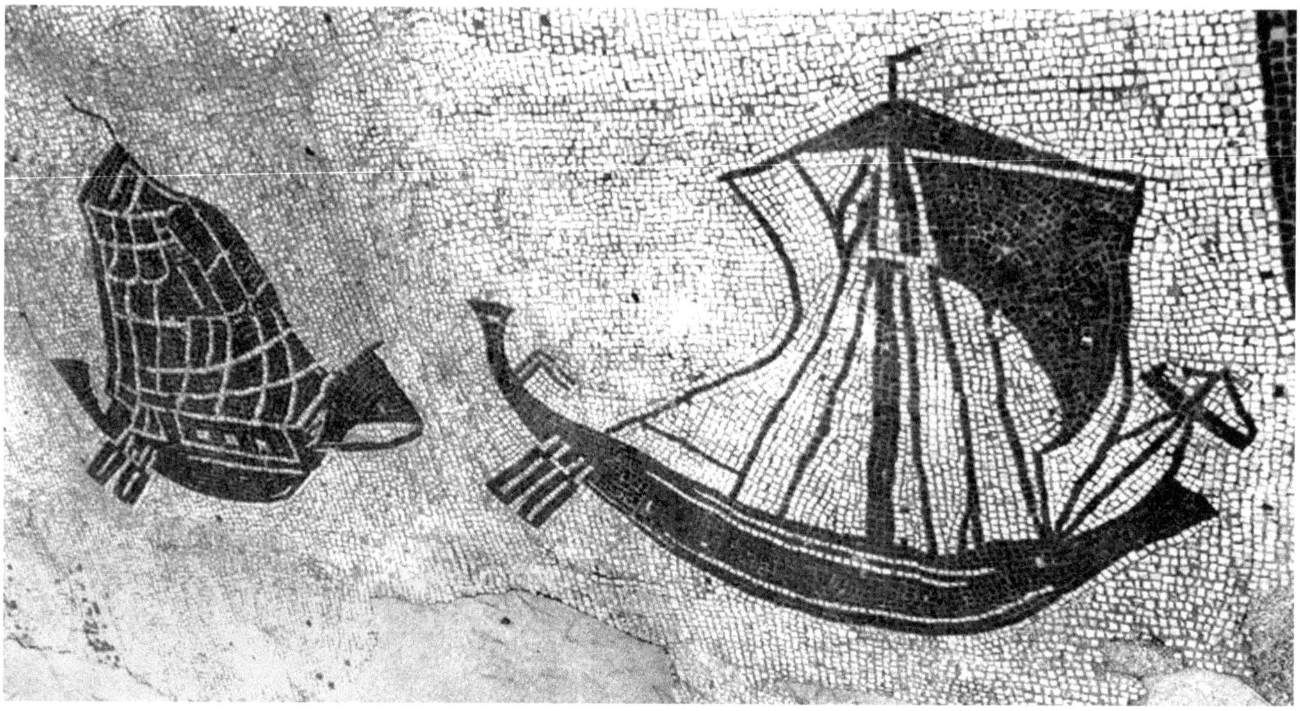

Fig. 3.7.28: Station 45

while two vertical and a partial horizontal white strip seem to be the ropes tightening the bundle), probably is in the process of loading or unloading the cargo from the ship or the storehouse. Apparently, the artemon mast was also used as a crane, and the lines hanging on either side are the ropes to maneuver the crane. Though that anchoring or mooring lines is missing we may assume that the ship is anchored near the structure to be the warehouse, as deduced from its static position.

Station 45

No inscription is found in this office but only two sailing ships of different sizes following one another with their prows turned to the right, thus showing their starboard side (fig. 3.7.28).

Ship 1 (left-hand)**:** The short hull is very rounded, almost symmetrical (fig. 3.7.29). The upraised stern is adorned by a vertical block-shaped sternpost. The shape of the stempost is not very clear. Beneath the starboard gunwale there is a short, broad, rectangular plank outlined with strips of white tesserae. This element may indicate an outrigger oarbox. Three small white squares depicted within the black strake may represent the represent oarports. The white strip beneath the plank may indicate a lower seam between planks or the wale just above the waterline. A short arching white strip (not continued) depicted on the lower fore-hull probably was meant to indicate the keel.

The rigging comprises two sails and a pair of steering oars mounted on the quarters. The main sail made with black tesserae is seen from its fore face (fig. 3.7.29). The rigging lines are depicted with white tesserae. Eight brails and five reinforcing-bands are shown on the fore bunt. The irregular arrangement of the brails and the reinforcing-bands indicate that the sail suffered some repairs; the mosaicist or the restorer not understanding the functions of these lines, made them more like a decoration on the sail (fig. 3.7.29). This restoration is particularly noticeable on the upper side of the sail. The brails, halyard, braces and the sheets of the main sail are missing. An arched white strip, that also may represent the head of the mainsail, indicates the yard. The shortened yard may have resulted from the restoration or the mosaicist who wished to represent the sail in a perspective view. A triangular black topsail is set above the yard. The leeches of the sail or the lifts shown on the bunt are made with white strips. At the tip of the topsails, probably the masthead, is attached a short black band or a small flag blown by the wind. The angled white strip of tesserae distinguishes the furled port foot of the main sail from the artemon sail. The artemon rigging is not very clear. The forward angled black trapeze above the stempost probably represents the artemon sail supported by the angled mast projecting above the bow. The sail hides the mast as well as the stempost. The thick arched line between the tip of the starboard yardarm and the root of the stempost, most probably represents the artemon's starboard brace (fig. 3.7.29).

Both steering oars are placed on the starboard quarter. This is a misinterpreted depiction. Probably the artist intended to show the rear of the ship in perspective, and for that reason he placed both oars on the same side. The shafts are outlined with white strips. They seem to project from beneath the aft-wing of the top strake with backward slanting (fig. 3.7.29). Both blades are almost symmetrical

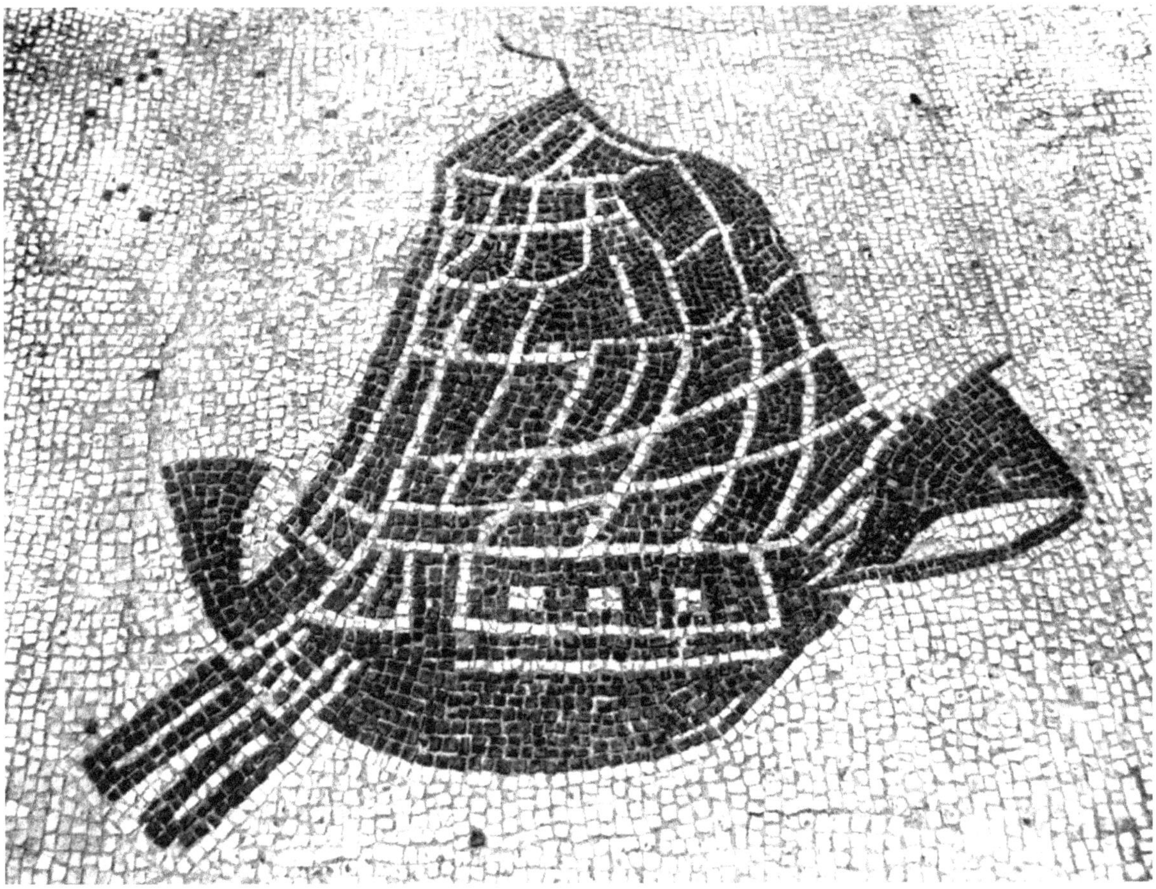

Fig. 3.7.29: Ship 1 – Station 45

and of the same size. The lower ends of the shafts (made with a white strip) bisect them longitudinally.

Ship 2 (right-hand)**:** The hull of the ship is a large banana-shape with the stem slightly broader and higher than the stern (fig. 3.7.30). The rounded stem is surmounted by a short block-shaped stempost. The rounded stern is surmounted by an almost vertical sternpost finished with a papyrus umbel-like or a fishtail tip adornment. The top narrow wale is partly outlined by two white strips. The arching white strip following the rounded bottom seems to outline the keel.

The rigging comprises two masts and sails, with a pair of steering oars or rudders mounted on the starboard quarter. This depiction is misplaced, probably due to the intention of the mosaicist to show the ship from its stern with each oar mounted on the quarters (starboard and port), as giving some perspective to the ship. The tapered main mast is stepped forward amidships (fig. 3.7.30). The yard has a slight downward curvature. A triangular black topsail is set above the yard. The tip of the masthead projects above the joint ends of the topsail and a small flag is attached to its end. The main sail is shown only partially on the right side of the mast. It appears to be in the process of furling; the port side is completely furled beneath the yard, while the remaining open starboard side is still billowing forward, indicating that the wind blows from the port quarter (fig. 3.7.30). The starboard sheet stretches vertically to the fore-gunwale. The arched line extending between the tip of the port yardarm and the quarter gunwale is the port brace. Between the mast and the port brace are depicted two lines. The line closer to the brace may indicate the backstay, while the one closer to the left side of the mast is the port shroud. The lines between the sail and the right-side of the mast are associated with the running and standing rig. The line closer to the mast probably represents the halyard. The line closer to the sail and the starboard sheet is the forestay, while the line in the middle with its bisected lower end indicates the starboard shroud (its lower end may have passed through a deadeye and the adjacent ends were attached to the port gunwale (fig. 3.7.30). The upper part of the mast, beneath the yard and the lines, shows evidence of repairs made to the mosaic. The whole section is moved slightly to the right (fig. 3.7.30). The trapezoid element formed by the horizontal black line, stretching between the port brace and the backstay probably represents the roof of the quarter-cabin. The lower ends of the brace and the backstay apparently form the lateral sides of the cabin. There is no other explanation for this element.

The forward angled artemon mast and parallel to the stempost is represented by one black strip (fig. 3.7.30). A perpendicular strip made with two rows of black tesserae attached to the masthead is the yard and probably the furled sail beneath it. The yard is secured by one lift stretching from either side of the masthead to the tips of the yard. The standing rig of the artemon mast is indicated by the

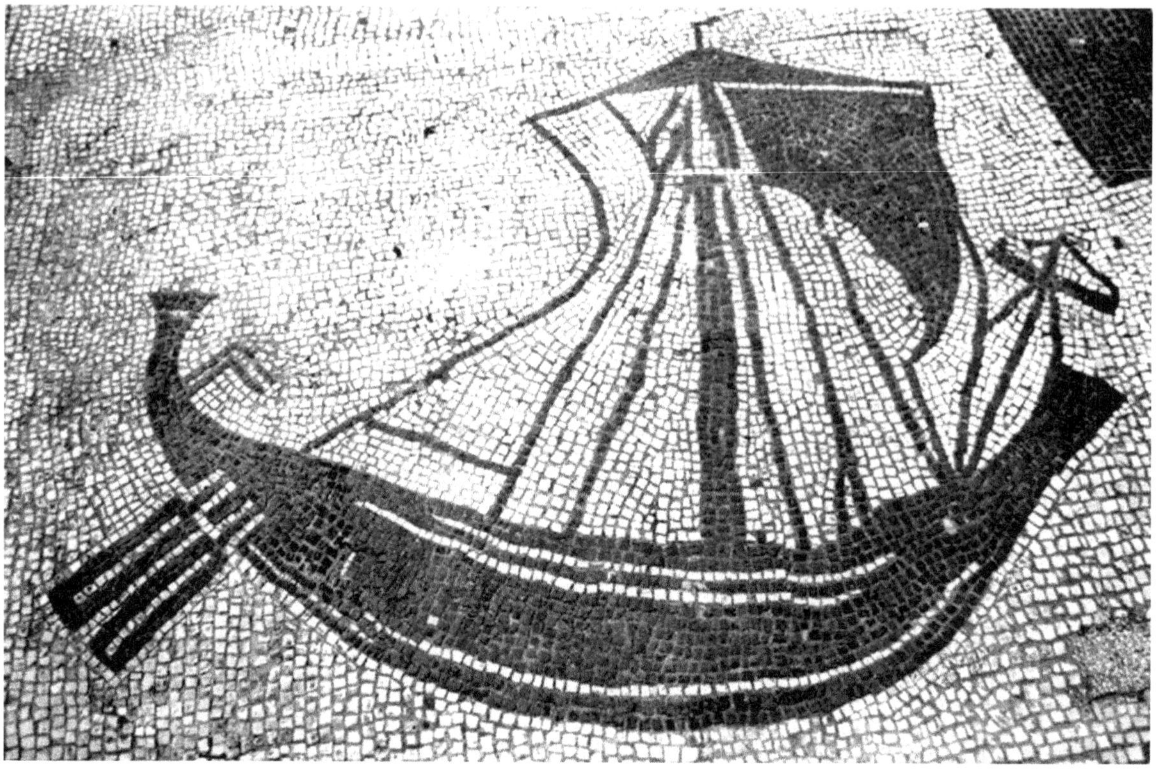

Fig. 3.7.30: Ship 2 – Station 45

short angled black lines on either side of the mast; the top line may indicate the backstay and the lower one above the stem is the forestay.

Both steering oars are placed on the starboard quarter (fig. 3.7.30). This misplacement resulted from the mosaicist's intention to give some perspective to the stern. The shafts were meant to be mounted on either quarter. The shafts are partly outlined with white tesserae and appear to project from the lower hull. Both elongated rectangular blades are similar in shape and size. The lower ends of the shafts made with white strip bisect them longitudinally. Two parallel short spars project from the inner arched sternpost (fig. 3.7.30). Smaller spars are attached perpendicularly to their tips. This element probably represents the bitts set on the quarter gunwale, or probably it formed the quarter-cabin before repairs were made to the mosaic. The repairs are most distinct on the white background, especially emphasized around Ship 1, its sail and stern, as well as on the left side of the port brace. Both ships probably replaced an earlier design that decorated this floor.

Station 46

A very thick black frame, delineating the room's perimeter, surrounds the maritime scene. On the white background are depicted a five storey stepped building, and two sailing ships facing each other are depicted underneath the building; a large dolphin with the tail shaped like an *acanthus* leaf is shown beneath the ships and his head points to the left (fig. 3.7.31). The structure built with large stone square blocks, indicated by the white vertical and horizontal stripes, probably is a lighthouse. On top of the structure are depicted the flames of the beacon. Both ships are illustrated with fully open square sails and different hull shapes. The horizontal black stripes beneath the ships and the dolphin indicate the sea waves.

Ship 1 (left-hand): The rounded hull has a concave stem finished with a projecting pointed cutwater is pointed to the right (fig. 3.7.32). The slightly raised rounded stempost appears to be a stylized volute. Three strips made with white tesserae indicate the gunwale, the middle wale just above the waterline and the outlining of the keel. A large *oculus* outlined with white tesserae is depicted on the starboard stem, between the gunwale and the keel; the pupil within the iris is indicated by a horizontal short white strip. The broad and rounded stern is adorned with a vertical sternpost shaped like a swan's head turned backward. A lattice fence is placed on the extended starboard poop. Two bitts are depicted on the starboard fore-gunwale, while a third one is set on the aft gunwale before the fore end of the quarter fence.

The rigging comprises the main sail and mast, the artemon rigging and two rudders mounted on the quarters. The main mast (only its lower section seen above the gunwale) is slightly stepped off amidships. Its width is made with two rows of black tesserae while the height is made with six rows. The square main sail depicted with black tesserae is seen from its fore face (fig. 3.7.32). On the bunt are rendered six white brails and five reinforcing-bands. The yard and the head of the sail are depicted as one unit. A white triangular topsail is set above the yard, deduced and

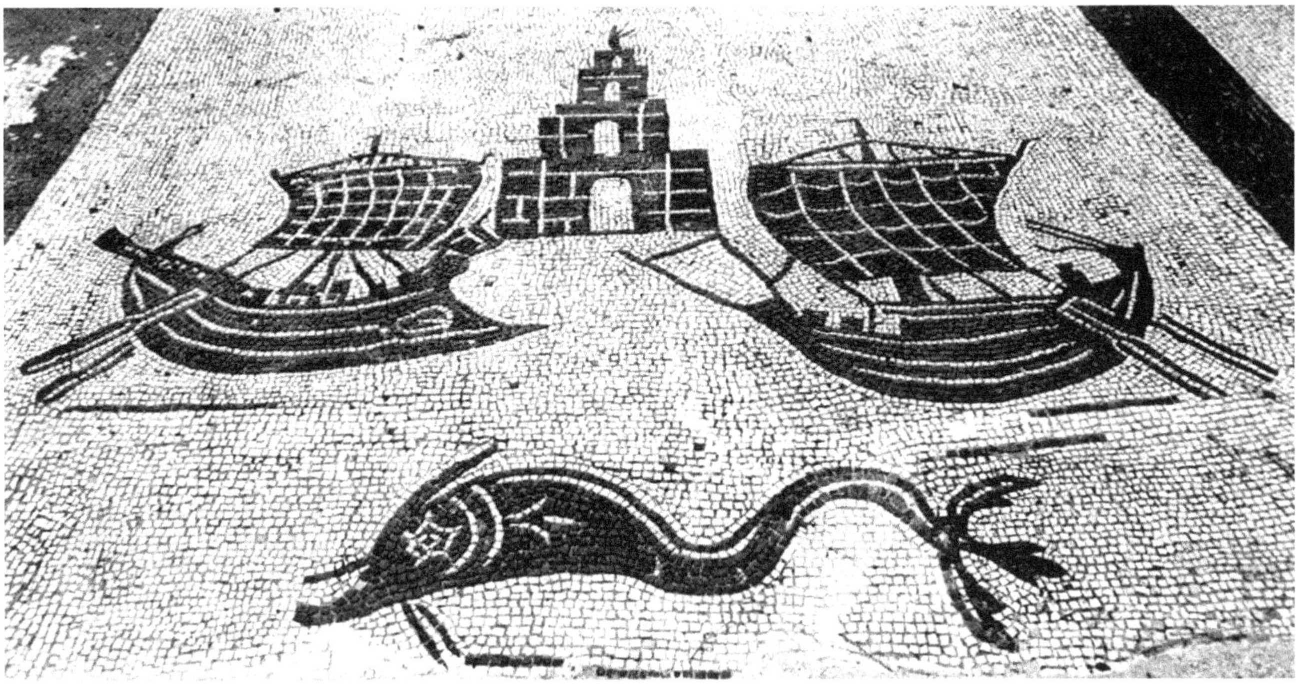

Fig. 3.7.31: Station 46

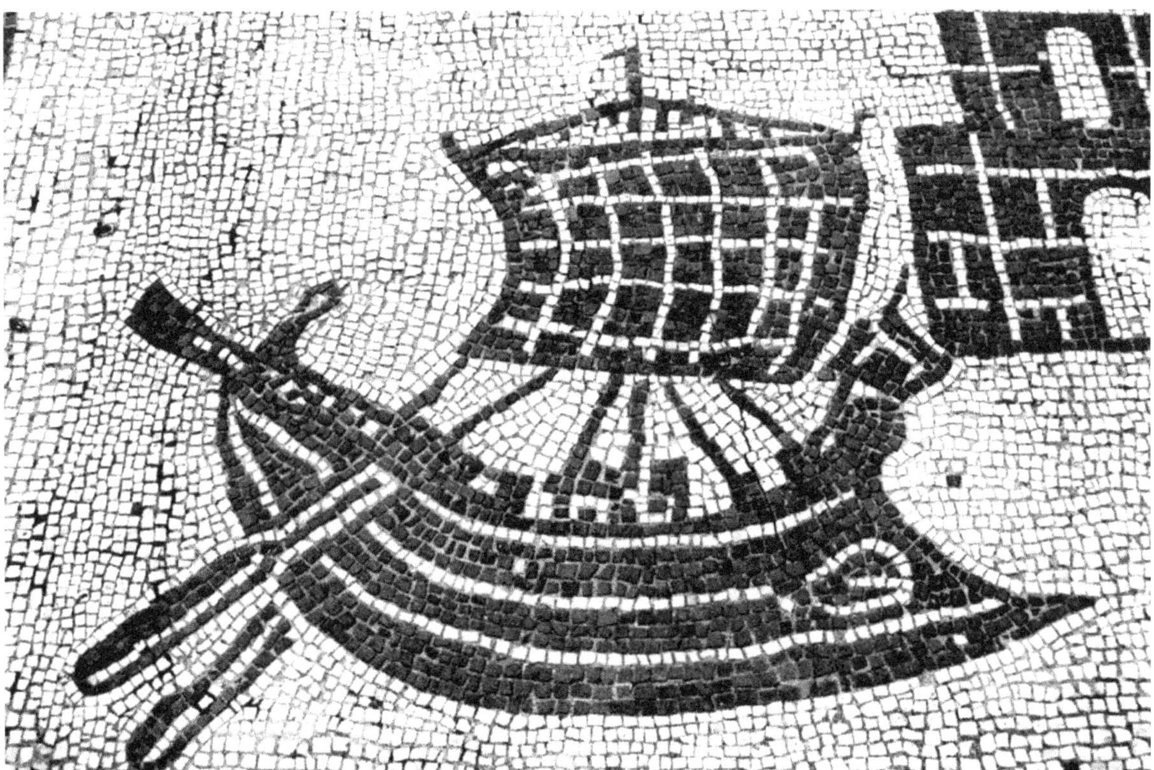

Fig. 3.7.32: Ship 1 – Station 46

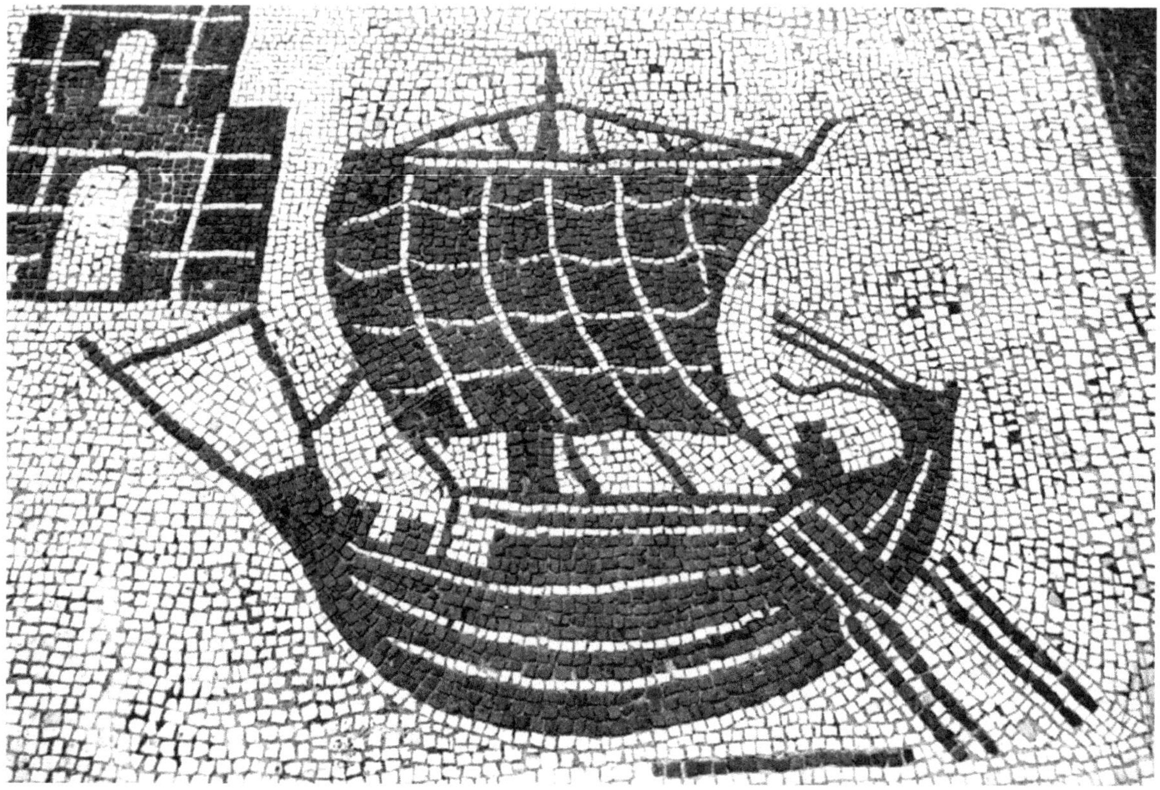

Fig. 3.7.33: Ship 2 – Station 46

seven short black brails are depicted on the bunt. A small flag is attached to the tip of the masthead, as well as to either tip of the yard (fig. 3.7.32). Six lines associated with the standing and running rig, are visible beneath the foot of the sail. The line stretching between the starboard clew and the top edge of the aft-fence is the starboard sheet. The line with a backward stretch and closer to the sheet may indicate the backstay. Both lines in the middle probably represent the port and starboard shrouds stretching laterally amidships and their lower ends are attached to either gunwale. The line closer to the right-side of the mast may indicate the halyard that its lower end is tight around the second bitt. The right-hand line (extreme) is the forestay and its lower end is tight around the first bitt on the fore-gunwale (fig. 3.7.32). The lower ends of the brails and the braces are missing. A thin black stripe adjacent to the inner side of the stempost indicates the forward inclined artemon mast. The artemon sail is represented by a small black square probably viewed from its fore face. On the bunt are depicted two white brails and one reinforcing-band. The short black strip projecting above the tip of the starboard yardarm may indicate the masthead; probably the mosaicist intended to give a perspective view to the sail and therefore he misplaced the masthead (fig. 3.7.32).

The steering gear comprises two steering oars placed on the quarters; both have a backward slanting. Only the elongated blade projecting beneath the stern indicates the port rudder. The lower end of the shaft, made with one white stripe, bisects the blade longitudinally. The starboard oar has a long shaft outlined with one strip of white tesserae projects beneath the quarter fence (fig. 3.7.32). Its lower end made with a white strip bisects the black elliptical blade longitudinally. The shoulder of the starboard blade is rounded, whereas the ends of both blades are rounded.

Ship 2 (right-hand): The hull is short and very rounded, resembling a spoon-shape; its prow points to the left (fig. 3.7.33). The gunwale and two seams or wales are made with white strips. A broad black strip set above the gunwale, between amidships and the stern may indicate a bulwark or the cabin. On its upper part two white strips probably outline the roof of the cabin or it indicates the top edge of the bulwark. Two short bitts are set on the fore gunwale. The upraised stern and stem are finished with block shaped stem-and-stern posts. Two black spars projecting from each corner of the stempost probably represent the bowsprits. A cross spar is attached between their tips thus to strengthen and support them (fig. 3.7.33). Three black strips are attached to the inner corner and at the top of the sternpost, two are parallel and the lower one is undulated. They may represent a stylized three open branched *aphlaston* or some decorative banners. The black object with a small head projecting above the port quarter gunwale probably is the capstan minus its spokes.

The rigging comprises a tapered mast stepped slightly forward of amidships, and projecting above the fore end of the cabin (fig. 3.7.33). The fully open, square sail, depicted with black tesserae, is seen from its fore face and bellowing forward, thus indicating that the wind blows from starboard quarter or astern. On the bunt are depicted six brails and

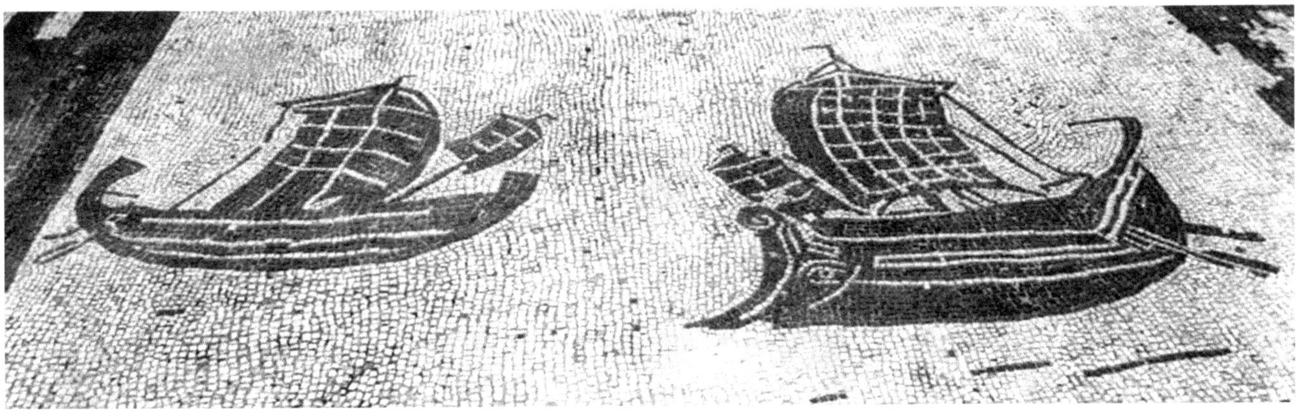

Fig. 3.7.34: Station 47

four reinforcing-bands rendered with white tesserae. The yard and the head of the sail are depicted with a white horizontal strip. It is secured by two lifts stretching from either side of the masthead to either tip of the yardarms. This triangular pattern also may suggest a white topsail set above the yard, whereas the brails are made with black tesserae. At small flag is attached to the tip of the masthead, to indicate the wind direction. A small ball-like element above the tip of the lifts indicates the *parrel*.[45] The standing and running rig are also shown. The short line visible between the starboard leech of the sail and the top bowsprit may indicate the forestay that its lower end is tight around the top bowsprit. The line stretching between the starboard clew and the fore corner of the cabin's roof is the starboard sheet; its lower end seems to be tight around the second bitt. The port sheet stretches to the quarter gunwale, just to the base of the capstan. The line closer to the right side of mast may represent the halyard. The line in the middle, between the halyard and the port sheet, probably represents the backstay (fig. 3.7.33).

The reversed U-shape pattern depicted on the port quarter and outline with white strips indicates the quarter oarport. Two rudders with a backward slanting are mounted on the quarters. The lower edge of the port shaft appears to form the lower side of the oarport. Both slim elongated blades are bisected by the lower end of the shafts. The port blade has a rounded shoulder. The starboard blade probably had also a rounded shoulder, but it shows some repair which were made on the mosaic in this section (fig. 3.7.33). The ends of the both blades are submerged in the water.

Station 47

In the middle of the white background are depicted two sailing ships facing each other (fig. 3.7.34). They have different hulls and their sails are depicted with black tesserae, while the rigging lines are made with white strips.

Ship 1 (left-hand): The elongated, almost symmetrical hull has rounded ends (fig. 3.7.35). The prow points to the right. The upraised stem is surmounted by a broad short block-shaped stempost. The raised sternpost has a slight inner turn with a straight cut tip apparently resemble the end of papyrus umbel or a stylized fishtail. Three horizontal white strips depicted on the starboard side define the gunwale, the wale just above the waterline, whereas the third one outlines the keel.

The rigging comprises two sails, one mast and two steering oars (fig. 3.7.35). Both sails rendered with black tesserae are seen from their fore face. The main square sail billowing forward, suggests that the wind blows from the starboard quarter or astern. Apparently the sail is depicted from its lee side, deduced from the arched port leech and the bellied starboard edge of the sail. On the inner bunt are depicted two white brails and three reinforcing-bands. The top white horizontal strip indicates the head of the sail and the yard (fig. 3.7.35). The mast is not shown. A black triangular topsail is set atop the yard and the tip of the masthead projects above. A small flag is attached to the tip of the masthead. The leeches of the topsail or the lifts are made with white tesserae. The white line in the middle may represent the central brail or the masthead. The tip of the port yardarm extends over the sail's leech. The line extending from the tip of the port yardarm to the quarter gunwale and parallel to the leech represents the port brace (fig. 3.7.35). The back-and-fore stays, the shrouds, the halyard, the sheets and the starboard brace are missing. The white space between the port leech and the brace, apparently replaced the port side of the main sail, or the mosaicist intended to give a perspective view to the bunt. The white tesserae setting are different from those used in the sail and the background. The forward inclined artemon mast, made with one black strip, is stepped forward amidships. The small fully open square sail is probably seen from its lee face, deduced from the concave port leech and the convex starboard leech. The white vertical strip may indicate the upper part of the mast or the middle brace on the lee face of the bunt. Both white horizontal segments may indicate the reinforcing-bands. The yard and the sail's head are indicated by one white strip. A small triangular black topsail is set above the yard. A small flag is attached to the tip of the masthead. The standing or running rigs are missing.

[45] See Glossary.

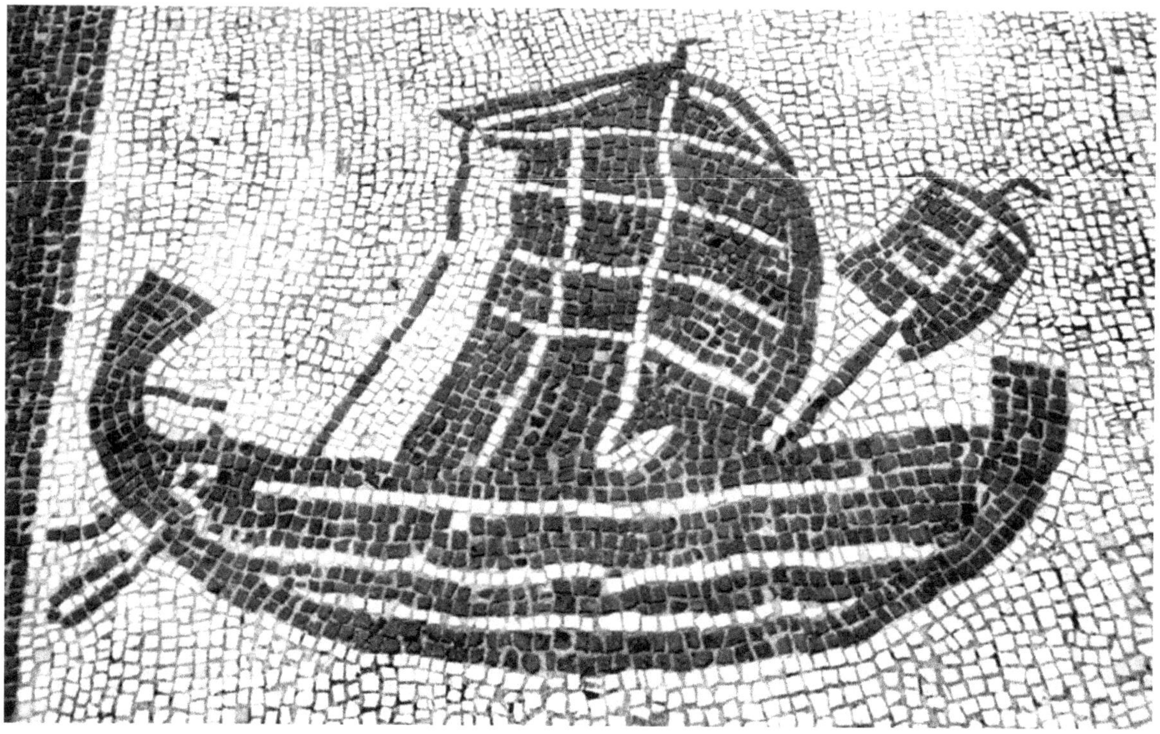

Fig. 3.7.35: Ship 1 – Station 47

Two oars are mounted on the quarters with a backward slanting (fig. 3.7.35). A small arched black strip projecting behind the stern indicates the port oar. The small white square depicted on the upper starboard quarter indicates the oarport with the starboard shaft going through and visible between the blade and the stern (fig. 3.7.35). The lower end of the shaft, indicated by a white strip, bisects the small elliptical black blade longitudinally. The different shapes of the oars probably resulted from the misunderstanding of the mosaicist when he rendered them, or the port oar was not completed when the mosaic underwent through some repairs which are distinct in the white background beneath the stern and around the sternpost. The horizontal short black line projecting from the inner curved sternpost probably indicates the flat roof of the quarter cabin; its right-hand wall is missing, but being suggested by the short projection above the quarter gunwale.

Ship 2 (right-hand): The broad hull, wider at its stern has a concave stem finished by a projecting pointed cutwater (fig. 3.7.36). The prow points to the left. A small outer turned volute finishes the stempost. The stem and the volute are outlined with a white strip. The upraised, rounded stern is topped by a thin strake or wale with a backward extension. The wide bulwark on the port side is outlined with a white strip. The top wale is also outlined with white tesserae. It forms an angle on the port quarter, probably indicating an attempt of a perspective view of the stern and the aft-wing housing the port rudder (fig. 3.7.36). The horizontal white line beneath the wale may represent the lower wale above the waterline. An *oculus* with eyebrow and a rounded pupil is depicted on the port stem beneath the wale. A similar *oculus* was mounted on the starboard stem. The horizontal black strip above the port quarter may indicate the flat roof of the quarter cabin.

The rigging comprises two masts and sails, the standing and running line rigging and a pair of rudders mounted on the quarters. The main mast visible between the foot of the sail and the gunwale is stepped amidships and has a slight forward rake (fig. 3.7.36). The fully open square main sail billowing forward is seen from its lee face deduced from the arched port leech and the backward turned edge of the sail. This depiction was meant to show the sail in perspective and emphasis that the wind blows from astern or the port quarter. The checker pattern formed by the brails and the reinforcing-bands was meant to be seen on the fore face of the sail. The lack of nautical knowledge of the mosaicist not understanding properly this rigging he produced it on the lee side of the sail. On the black bunt are depicted five white vertical strips that the left-hand one indicates the port leech while the other four are the brails. Nine horizontal white strips represent the reinforcing-bands. A white strip indicates the yard and the head of the sail. A triangular black topsail is set above the yard (fig. 3.7.36). The lifts or leeches of the topsail are depicted with white strips. The double vertical white lines shown in the middle of the topsail may represent two brails or they outline the masthead. The tip of the masthead projects above the tip of the topsail and a small flag is attached to its end. The short black lines stretching from beneath the flag to the top leech of the topsails appear to be the lifts supporting the yard. The line stretching from beneath the tip of the starboard yardarm to the quarter gunwale is the starboard brace. The line stretching from the middle of the starboard leech to the cabin's roof probably is the backstay (fig. 3.7.36). The

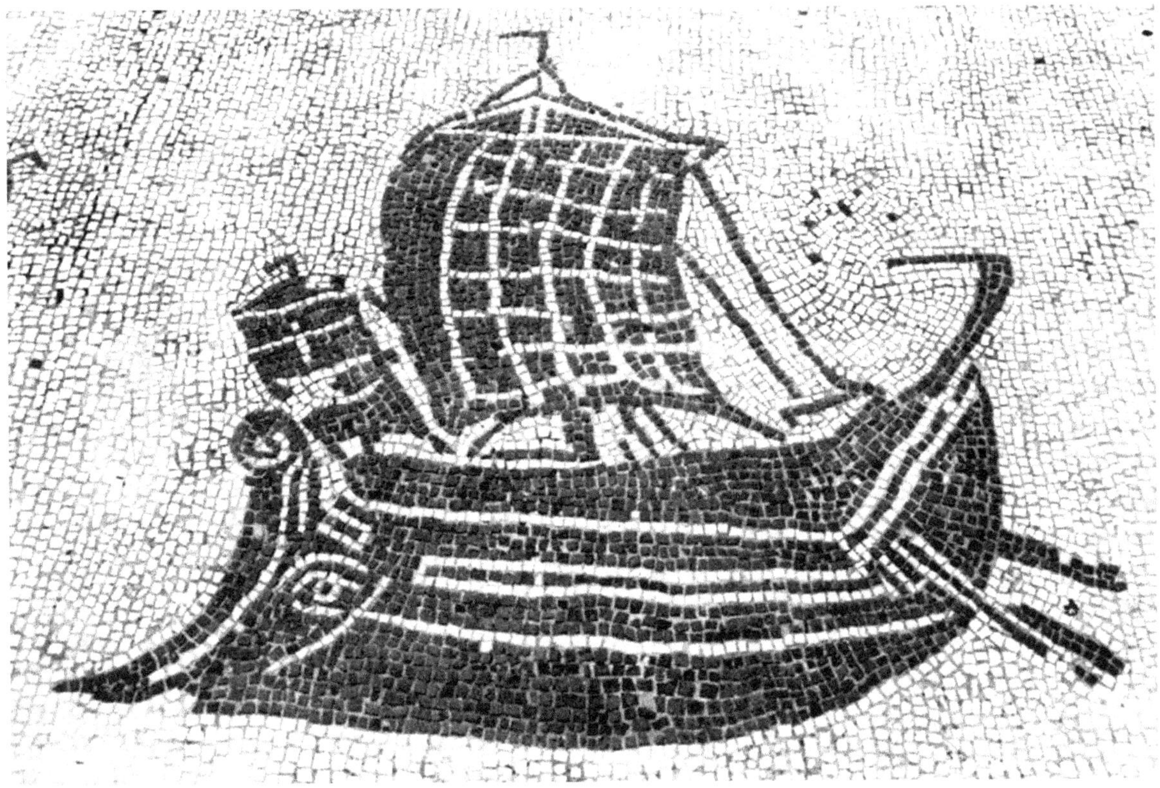

Fig. 3.7.36: Ship 2 – Station 47

lower end of the starboard sheet is tight to the lower fore-edge of the cabin. The lower ends of three lines beneath the foot of the sail are depicted between the right side of the mast and the starboard sheet. The line closer to the right side of the mast probably is the halyard, while the other tow line may indicate the brails passed over the yard and worked from the lee side of the sail, or they may represent the shrouds. The short arched line close to the port clew may indicate the port sheet; it is misplaced due to repairs on the mosaic (fig. 3.7.36). The forestay is not shown.

The artemon mast is stepped forward and has a forward rake above the stem. The fully open black elongated sail is seen from its fore face. A white horizontal strip represents the yard or the head of the sail. The topsail is set above the yard. The masthead projects above the tip of the topsail and a small flag it attached to its tip. On the bunt are depicted one central brail and one reinforcing-band. On the lower part of the sail is depicted a white pattern, probably resembling a sea turtle with his hind legs stretched backward, or else the artist has made two of the reinforcing-bands and the brail into a single unit (fig. 3.7.36). The line stretching between the tip of the port yardarm and the middle section of the main sail's port edge is probably the artemon's port brace. The port sheet is attached to the fore gunwale near the port sheet of the main sail. The starboard leech is attached to the fore edge of the top plank behind the stempost.

The depiction of the main sail seen from its lee side and the artemon sail from the fore face probably was made intentionally by the mosaicist to give a perspective view to the ship. This representation also may suggest that each sail could be adjusted independently to the wind direction to set the ship on the desired sailing course. The rendering of the ship is not fine but suggests the attempt of the mosaicist to give a complex view of the sailing ship.

One rudder with a backward slanting is depicted on either quarter. They have different shapes. The elongated, black blade projecting behind the stern indicates the starboard oar. The shaft of the port steering oar outlined with white tesserae projects through a square oarport beneath the aft-wing of the top quarter wale (fig. 3.7.36). The elongated blade of this rudder has angled shoulders. The blades of the rudders, the bottom of the forefoot, the sails, and especially the white background around the lower part of the ship show evidence of repairs on the mosaic. Probably this vessel replaced an older ship or other pattern, when the floor was remodeled when probably the shop change ownership.

Station 49

The mosaic comprises two scenes: *1.* on the top is depicts a Nereid riding on a dolphin; *2.* in the lower scene is depicted a four storey stepped lighthouse placed between two sailing ships and a dolphin is found beneath the left-hand vessel (fig. 3.7.37). The sea waves are indicated by several short horizontal black strips. The maritime scene is a later addition, as deduced from the theme and the mosaic work, especially distinct in the tesserae setting in the white background.

Ship 1 (left-hand): The short and broad hull has a sickle shape ending with a shallow concave stem and a short

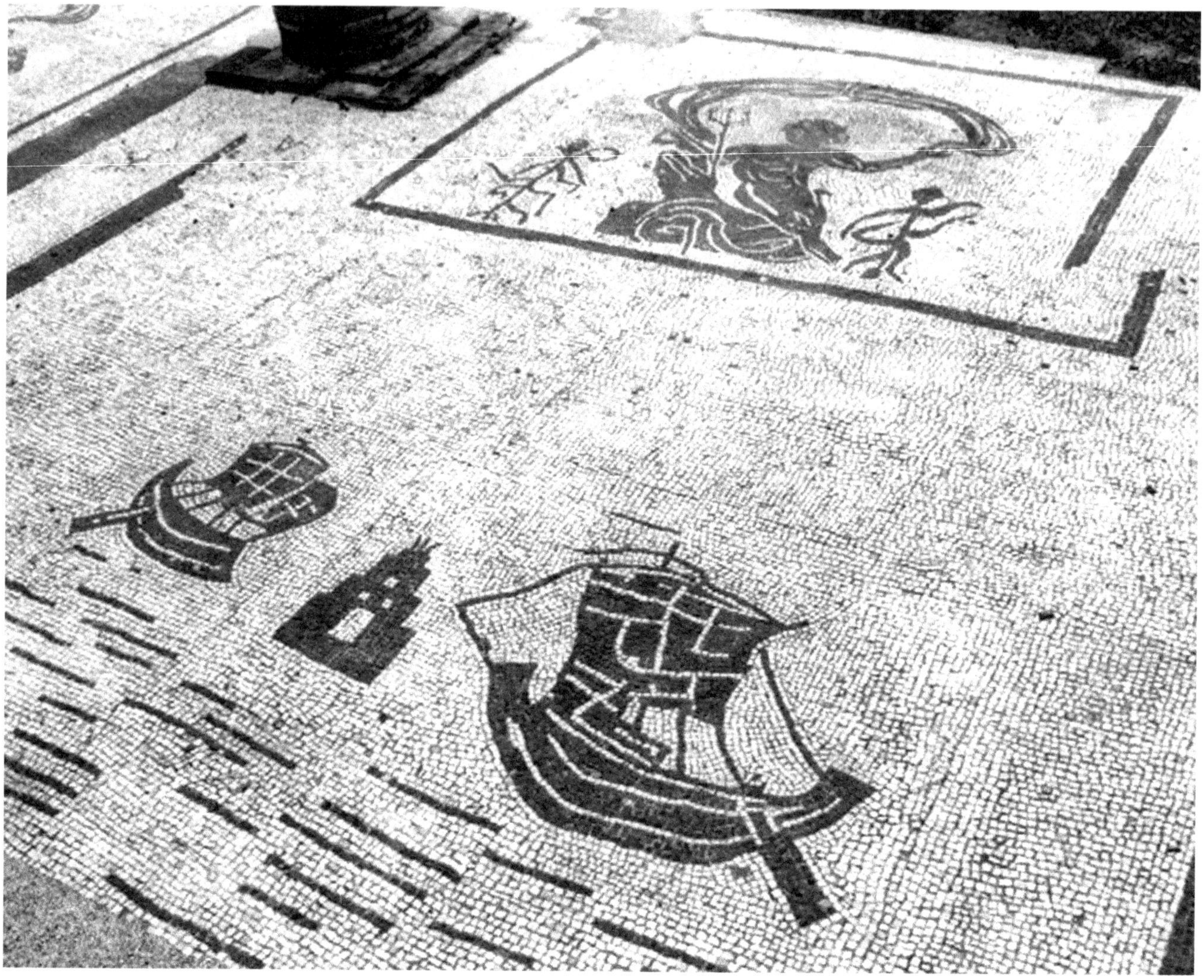

Fig. 3.7.37: Station 49

pointed cutwater (fig. 3.7.38). The prow points to the right. The stempost ends at the gunwale level, while the upraised rounded stern is adorned by a short vertical sternpost. Two curved white strips following the shape of the hull represent the upper wale while the lower line outlines the keel.

The rigging comprises two sails and six lines associated with the running rig, and one rudder. The main mast made with a vertical line of black tesserae is stepped forward amidships. The open black square sail seen from its fore face is deduced from the concave starboard leech and the convex port leech. Apparently the mosaicist wanted to suggest that the wind blows from the port quarter. On the bunt are depicted two brails and two reinforcing-bands. A white arched strip indicates the yard and the head of the sail. A very narrow triangular topsail is set above the yard (fig. 3.7.38). A small flag is set at the tip of the topsail. Three lines depicted to the left side of the mast and seen beneath the foot of the sail represent part of the standing and running rig. The starboard sheet stretches between the clew and the stern gunwale. The angled line closer to the sheet may indicate the backstay. The line between the mast and the backstay probably indicates the port sheet stretching to the port gunwale, thus emphasizing that the wind blows from the port quarter or stern, or it may be the halyard stretched backwards. The black strip set to the right side of the main mast and parallel to it probably is the artemon mast (fig. 3.7.38). The forward angled open artemon sail also indicates that the wind blows from the port quarter or astern. On the bunt are depicted one brail and two reinforcing-bands. The topsail and the masthead are depicted as a black triangular mass. The port sheet is attached to the tip of the stempost.

The steering gear comprises only the starboard rudder indicated by an elongated rectangular blade projecting beneath the quarter wale (fig. 3.7.38). The shaft made with one white strip bisects the blade longitudinally. A similar arrangement may have been true on the port quarter.

Ship 2 (right-hand): The hull of this vessel has a banana shape, being wider and longer than that of Ship 1. The prow points to the left. Both rounded ends have a slight outer extension. The stem appears to be higher than the stern. Block-shaped stem-and-stern posts surmount the ends. A forward projecting spar attached to the left-hand tip of

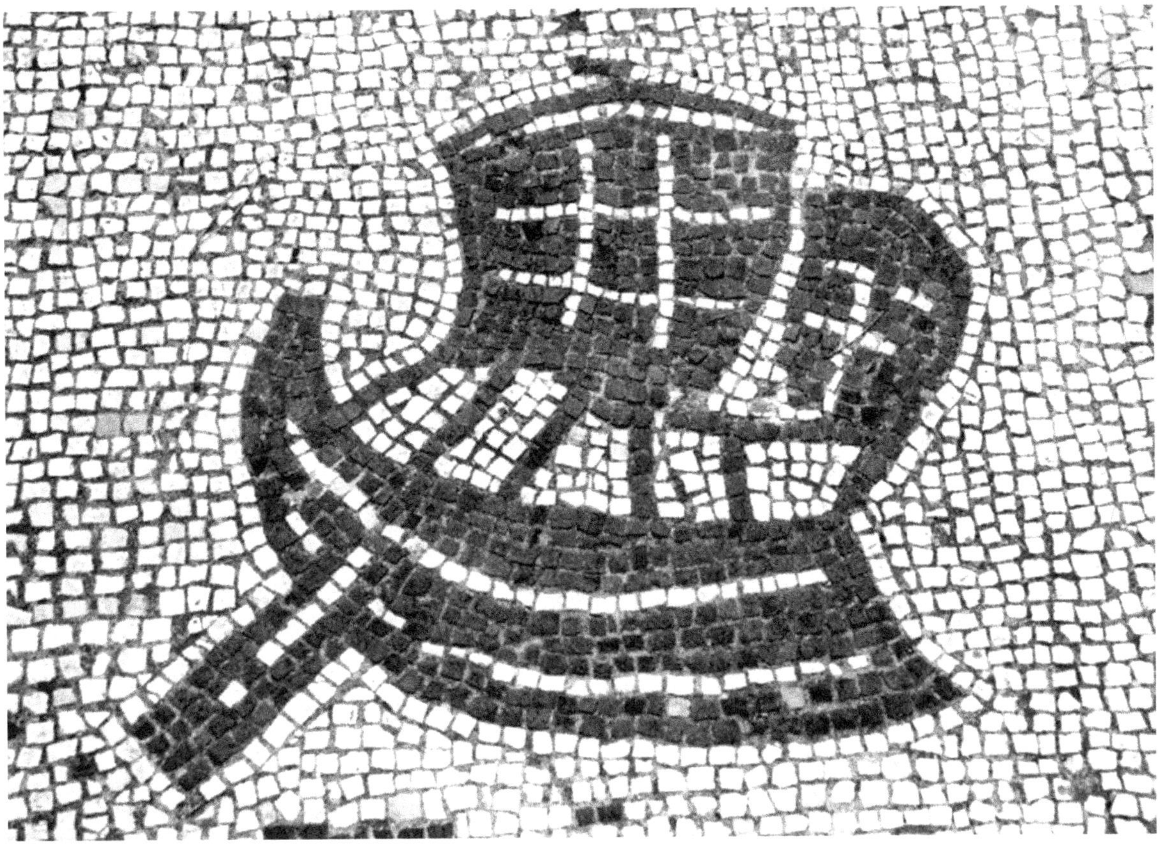

Fig. 3.7.38: Ship 1 – Station 49

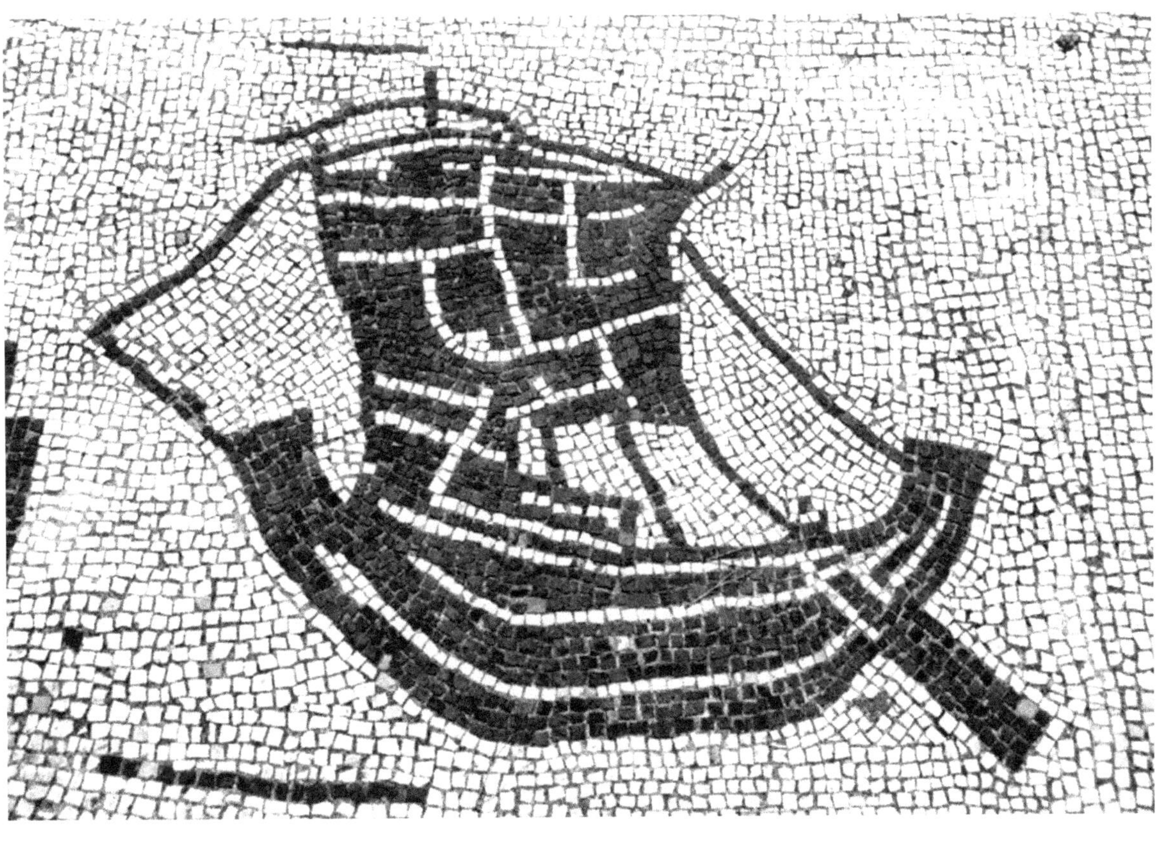

Fig. 3.7.39: Ship 2 – Station 49

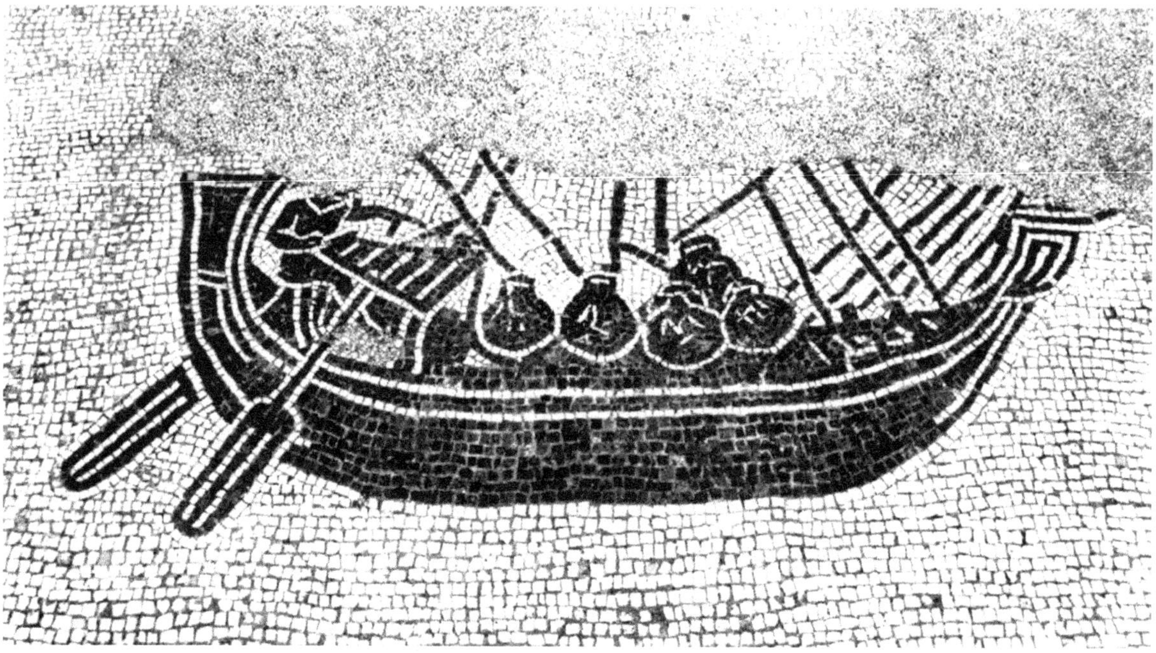

Fig. 3.7.40: Station 51 – Merchantman

the stempost represents the bowsprit (fig. 3.7.39). A short, broad plank set above the fore and mid-gunwale forms the bulwark. Three white strips depicted on the port hull indicate the gunwale, the top wale, while the lower strip outlines the keel, the stem and the stern.

The ship is rigged only with one black fully open black square sail, some rigging lines and one rudder. The mast (not distinct) seems to be stepped forward amidships with a slight forward rake, deduced from the angled sail. On the bunt seen from its lee side (assumed from the concave starboard leech, thus the wind blows from the port quarter or astern) are depicted three brails and four reinforcing-bands made with white tesserae (fig. 3.7.39). The sail's head and the yard indicated by a white strip appear to be secured by two lifts stretching from the masthead to each tip of the yard. These lines may also represent the leeches of the triangular topsail set above the yard. The masthead seems to be offset, projecting above the tip of the topsail. The black arched line above the topsail is not part of the rigging. It probably belonged to the topsail of a previous vessel depicted in the mosaic and replaced by the ship modified to the shape as seen today. A short band or flag is attached to the tip of the port yardarm. The irregular rendering of the brails and the reinforcing-bands on the main bunt attest that the mosaic went under some remodeling (fig. 3.7.39). Apparently the mosaicist or the restorer did not understand the function of these lines and depicted them more as a decorative pattern on the bunt. The top half-bulwark and the port side of the sail's foot also bear witness to the restoration or remodeling of the mosaic. The line stretching from the tip of the port yardarm to the tip of the bowsprit is the port brace. The black line stretching from beneath the upper corner of the starboard leech to the inner tip of the sternpost probably represents the backstay or the starboard brace; apparently it was misplaced. The starboard sheet seems to be tied around a short bitt on the quarter gunwale. The line almost parallel to the starboard sheet and stretching to the port gunwale probably indicates the backstay. The short line visible between the middle foot of the sail and the top edge of the bulwark may indicate the halyard or probably the mast.

The single steering oar mounted on the port quarter has a thick shaft (outlined with white tesserae) appears to project through a square oarport beneath the quarter top wale (fig. 3.7.39). The elongated rectangular blade is made with black tesserae. A similar oar may have been mounted on the starboard quarter.

Station 51

The mosaic is partly damaged. It depicts a merchantman loaded with a cargo of bag-shape amphorae and a man registering the cargo on the deck (fig. 3.7.40). The prow of the vessel points to the right. The hull has survived in its entirety but the ship's sailing rig is damaged, as well as the man's head. The surviving lower ends of the lines and the mast provide partly information of the sailing rig. The broad, elongated hull has two rounded ends, both outlined with a white strip of tesserae. The stem has a slight forward angle being surmounted by a vertical block-shaped stempost, decorated with double white reversed triangular frames. The upraised stern is surmounted by a vertical sternpost, outlined by a white frame; it is slightly higher than the stempost and resemble a papyrus umbel or a stylized fishtail. Two short horizontal parallel white strips indicate the root of the block-shaped sternpost (fig. 3.7.40). The narrow strip outlined with one row of white tesserae, indicates the top wale alongside the starboard hull. The fore-deck is partially covered, and three short thwarts outlined with white tesserae are set on the stem and seen

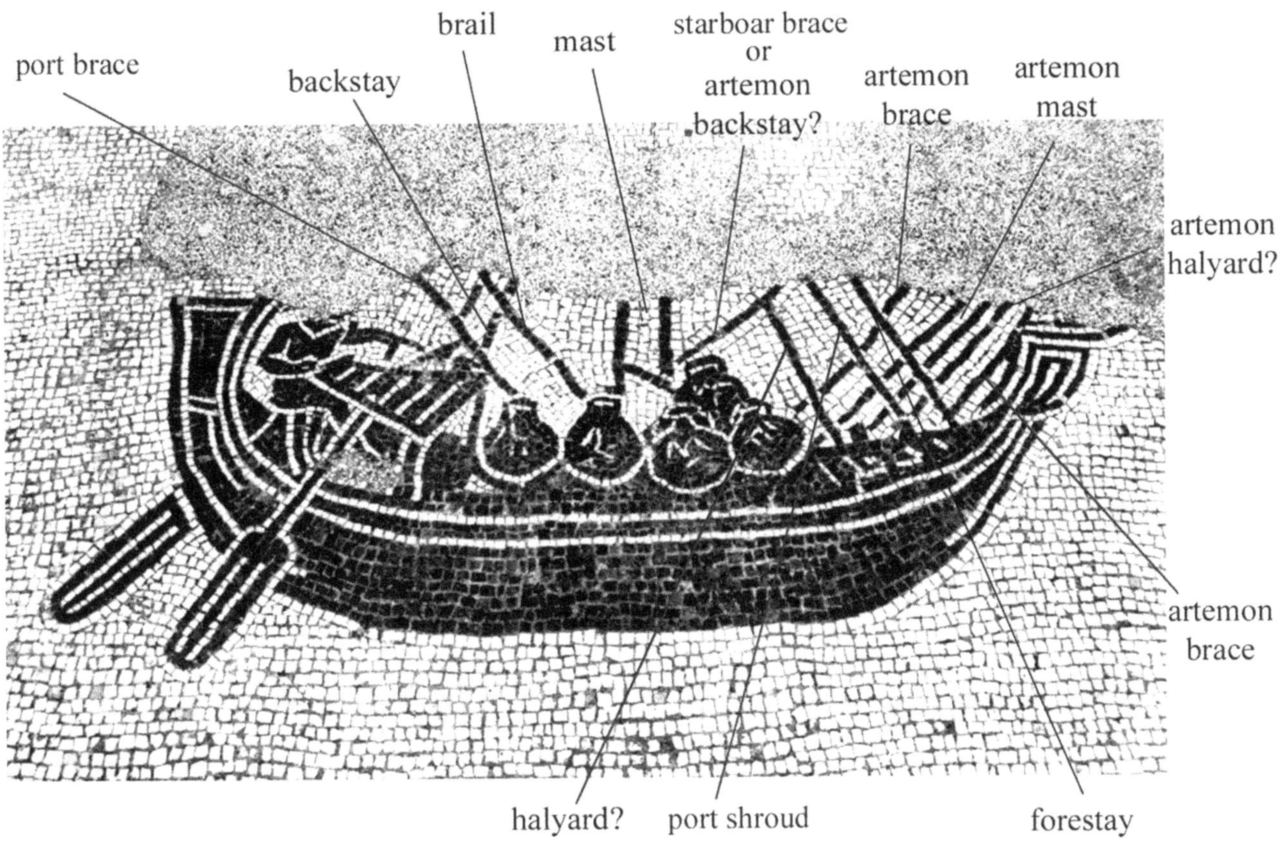

Fig. 3.7.41: Reconstructed standing-and-running rigging

from above indicate that the mosaicist intended to show the ship from bird-eye view to give some perspective to the vessel. Two short bitts are depicted on the port fore-gunwale. The cargo of the ship is symbolically represented by seven bag-shape jars placed on the top deck. This cargo was probably loaded in the hold as well as on the deck. A man sits on a chair placed on the quarter and holds a rectangular board with its fore-edge supported on two legs, whereas the aft-edge is laid on the man's lap (fig. 3.7.40). His left arm stretches forward, while in his right hand he holds a *stylus* (writing instrument). The man is counting the cargo of the jars with his left hand and then he records the amount of cargo on the wax surface set within the wooden board. On the tablet are several lines, thus suggesting the written record (fig. 3.7.40).

Although, the upper part of the rigging is damaged and missing we may reconstruct it from the remaining visible parts in the ship (fig. 3.7.41). The tapered main mast, outlined with black strips, is stepped just forward off amidships. Probably the sail was furled beneath the yard while the lower ends of the tackle lines stretch to the deck. On the right side of the mast are three angled lines that stretch to the fore port gunwale. The line closer to the fore-jars probably is the halyard, the middle lines may represent the port shroud that its lower end is tight around the first bitt, and the right-hand line (extreme) is the forestay (fig. 3.7.41). Two short parallel lines stretching towards the fore-port side of the registration board may represent the lower ends of the double-lines backstay (left-hand extreme); this assumption should be treated with great care. The lines intersecting the backstay and touching the rim of the jar near the *tabula* probably represent the port brace (the extreme line. The second parallel line that its lower end is attached to the rim of the jar near the left-hand side of the mast is one of the brails. Both lines are the running rigging of the furled sail (fig. 3.7.41). Some of the lines with a forward rake though parallel to the stempost may be associated with the artemon rigging. The tapered mast is outline with two black strips. The lines on either side of the mast may indicate the port and starboard braces used to maneuver the yard, while the one closer to the right side of the mast is the artemon halyard. The angled right-hand line stretching towards the right-side of the main mast may represent the artemon backstay or the main starboard brace (fig. 3.7.41).

The steering gear comprises one oar mounted on the quarter. The starboard oar has a long shaft outlined with white tesserae and an elliptical blade. The tip of the loom appears to be attached to the starboard side of the *tabula*. The port oar is represented only an elongated blade with rounded lower edge that projects behind the stern. The starboard blade has rounded shoulders and lower edge. Both blades are bisected longitudinally by the lower shafts outlined with two white strips (fig. 3.7.41). We may assume that each blade was made of two wings that were inserted within a groove cut alongside the lower edges of the shafts and then

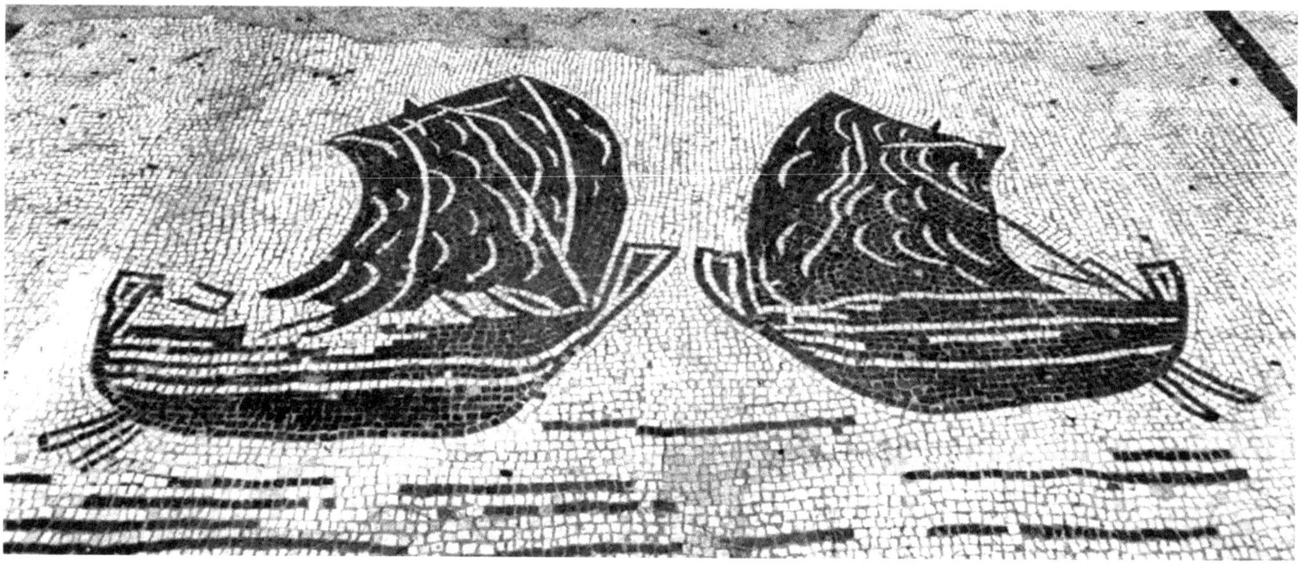

Fig. 3.7.42: Station 54

they were secured and reinforced by wooden treenails or bronze nails. From the static position of the ship we may deduce that it is anchored, although that such a device is missing.

Station 54

On the lower quarter of the mosaic are depicted two sailing ships facing each other (fig. 3.7.42). Both hulls and rigging are very similar. No inscription is found with the ships.

Ship 1 (left-hand): The squared broad hull has a forward angled stem (fig. 3.7.43). The bow points to the right. The high stempost has a block-shape and its edge is outlined with a white strip that forms a reversed triangle. The almost vertical stern is surmounted by a block-shaped sternpost also outlined with a strip of white tesserae similar to the stempost. The gunwale is indicated by a partial white strip on the starboard quarter (fig. 3.7.43). The white shorter strip placed above the quarter gunwale may indicate the planks of the quarter bulwark. The fore bulwark is represented by a wide black plank extending towards amidships. Apparently the bulwark suffered some repairs and at amidships it was replaced by white tesserae, thus indicating a broken plank. The top wale is outlined with two white strips that its upper edge follows the shape of the stem. The hull is made with black tesserae. No other seams or wooden planks are indicated.

The rigging comprises only the main sail, some rigging lines, and two rudders. The fully open square black sail is seen from its fore face as billowing forward (fig. 3.7.43). On the bunt are depicted three distinct white longitudinal lines. The line stretching from the middle of the yard to the inner side of the stempost's root represents the forestay. The arching lines on either side of the forestay following the billowing bunt may represent the brails or misplaced braces. The free hanging line overlapping the forestay may represent the halyard, which is also misplaced. The short ripple white lines depicted length-wise on the fore bunt were probably meant to emphasis the billowing sail between the brails and the reinforcing-bands (fig. 3.7.43).[46] The white strip beneath the yard or the head of the sail may present the top reinforcing-band. The angled yard and the length-wise white ripples on the face of the bunt indicate that the wind probably is blowing from the port quarter or astern. The sail hides the mast. The projecting tip of the masthead above the yard indicates that the mast is stepped forward with a backward rake. The middle of sail's foot appears to be furled. The short black line extending from beneath the sail's foot towards the top edge of the fore-bulwark is not clear and we cannot speculate as to its function. The starboard sheet and corner of the sail show marks of repair on the mosaic. Therefore, the mosaicist who did not understand the sailing rig rendered this section as being free hanging and blown by the wind. The rectangular element projecting from the inner side of the sternpost also is misinterpreted and probably it belonged to a design replaced by this ship (fig. 3.7.43). Traces of repairs can be distinguished at the foot of the sail, the middle bulwark, the sternpost, and in the background on the left-side of the ship.

The steering gear is suggested by two elongated trapezoidal blades with a backward slanting beneath the stern. The narrow blades made with one strip of back tesserae are bisected longitudinally by the lower shafts rendered with white tesserae (fig. 3.7.43). Although the shafts are not shown we may deduced that they projected from beneath the quarter top wale.

Ship 2 (right-hand): The hull is somewhat shorter and a bit broader than that of Ship 1; the prow pointed to the left is slightly tapered and higher than the stern (fig. 3.7.44). The angled stem is surmounted by a vertical block-shaped stempost. The upraised rounded stern is topped by a

[46] A similar sail appears in a merchantman depicted in a fresco from a villa near Sirmione, Italy (end of the 1st century BCE); Casson, fig. 138.

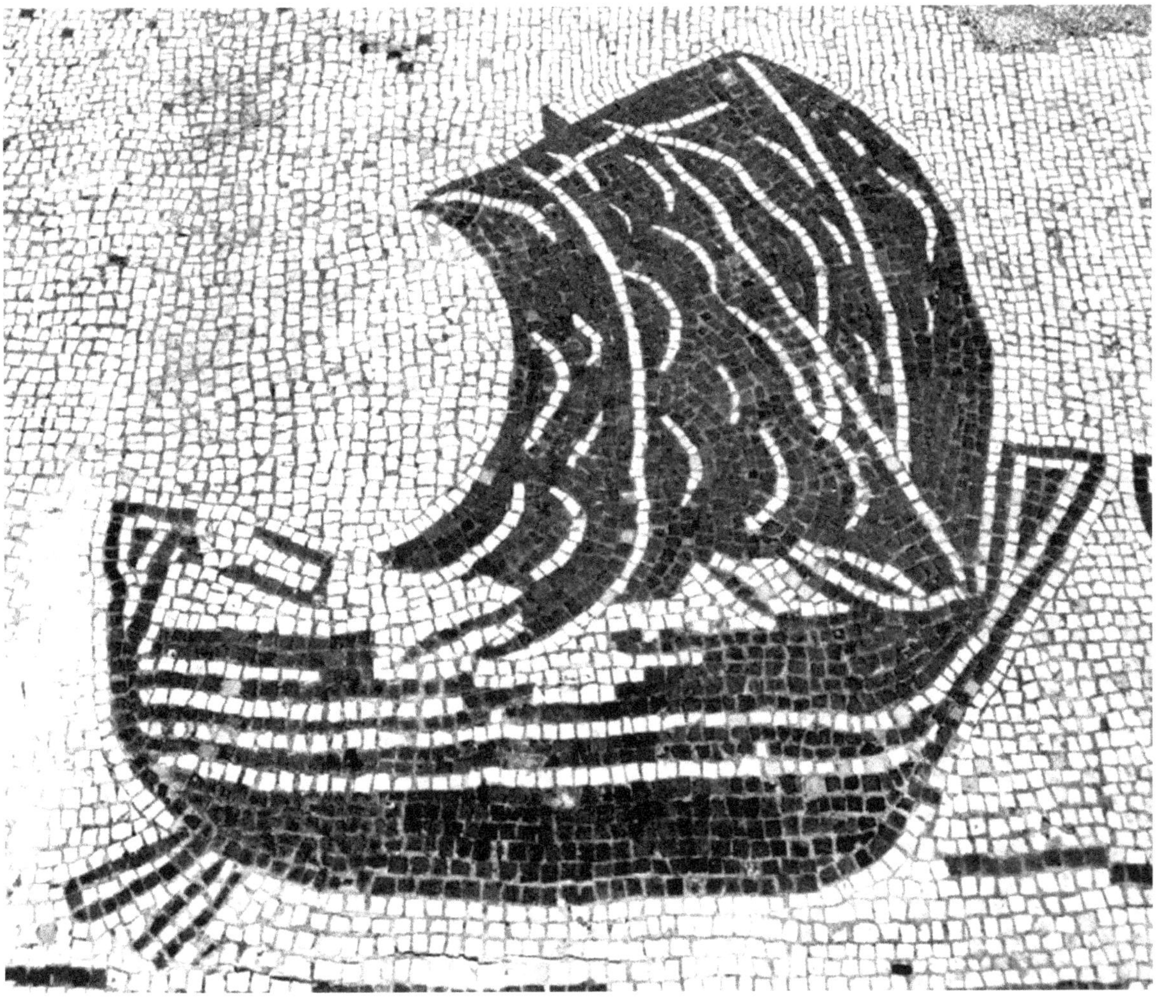

Fig. 3.7.43: Ship 1 – Station 54

vertical sternpost. Both stem-and-stern posts are outlined with a white strip which resembles a reversed triangle. The gunwale is outlined with one white strip. The continuous upper white strip and the fragmentary one beneath outline the top wale. The lower white strip may indicate either the lower wale just above the water line or possibly the waterline itself.

The fully open square black sail billowing forward is seen from its lee side. The angled and shortened yard indicates the action of the wind on the sail that probably is blowing from the port quarter or astern (fig. 3.7.44). The mast is hidden by the sail but the position of the masthead tip above the yard may indicate that it was stepped forward amidships with a backward rake. The angled line (made with white and black tesserae) stretching from beneath the yard to the port quarter gunwale represents the backstay. The segmented horizontal white strip beneath the yard probably represents the top reinforcing-band on the lee bunt. The rippled white longitudinal lines probably represent the billowing sail between the brails and the reinforcing-bands depicted on the fore face of the bunt.[47] The function of the angled line stretching from the lower starboard leech to the lower end of the backstay is not clear; we may suggest that it is the starboard brace which was misplaced by the mosaicist or the restorer (fig. 3.7.44). The function of the narrow white rectangle outlined with a black strip and placed above the lower end of the backstay is not clear.

Two narrow rectangular blades extending to the rear and projecting beneath the stern represent the steering gear; one oar was mounted on either quarter. The lower end of the shafts depicted with white tesserae bisects the blades longitudinally.

The depiction of both ships attest for repairs on the mosaic or the ships may have replaced an earlier design when the earlier office changed its function. Some of the unclear elements of the ships probably are reminiscences of earlier previous designs, as well as the mosiacist's lack of nautical knowledge to understand the ships' elements.

Station 55

This is the last office with a ship depiction in mosaic. The prow points to the right. The broad hull with two rounded ends is slightly tapered at the stern. Both the stern and the surrounding frame of the mosaic are damaged and the

47 Ibid.

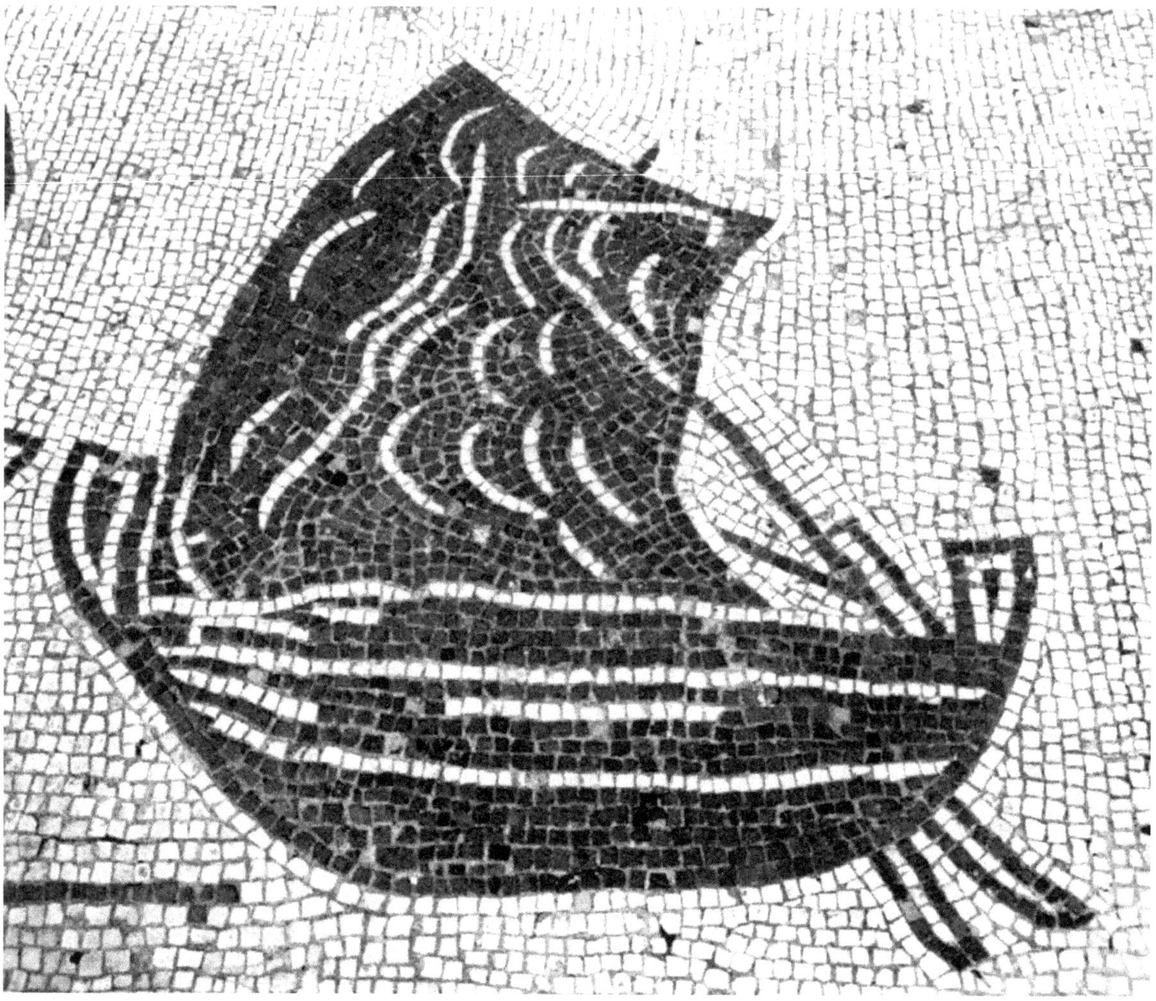

Fig. 3.7.44: Ship 2 – Station 54

restoration carried out in this section has not been done accurately – the black stones are not set in their original place, rather they have been mixed with the white ones (fig. 3.7.45). The sternpost is not distinct, due to the unclear restoration of this part. The prow has a slight forward angle. The vertical stempost is block-shaped. The white crenellated pattern on the upper part of the hull may suggest the upper edge of the lead sheathing or pitch coating, or some decoration on the hull (fig. 3.7.45). The horizontal white strip depicted on the lower hull may represent a lower wale or perhaps the waterline.

The vessel is rigged with a tapered mast stepped forward off amidships with a slight backward slanting (fig. 3.7.45). The sail and the yard are missing. The lines on both sides of the mast are associated with the standing rig. To the left side of the mast are depicted three angled lines that are intersected by five almost horizontal lines. On the right side of the mast are depicted three parallel angled lines intersected by three others. The extreme line (top), to the left side of the mast, most probably represents the backstay, and the one closer to the mast may be the port shroud or the halyard (fig. 3.7.45). The function of the middle line (between the backstay and the shroud) is not exactly clear because its upper end is attached to the second upper horizontal line. The extreme top line on the right side of the mast stretching to the fore gunwale may indicate the forestay. The middle one probably is the starboard shroud, while the function of the lower line is not clear. All the intersecting lines with the standing rig create a kind of cobweb, whereas the function of the intersecting ones is not distinct. We may assume that they represent a series of ratlines that the crew climbed to work the sail, yard and mast when needed.

Two fragmented parallel angled white strips depicted on the quarter may represent the shaft of the starboard steering oar. The blade is missing. Originally the steering rig comprised two steering oars or rudders.

Discussion

The ships depicted in the mosaic floors in the offices at Piazzale delle Corporazioni represent seagoing merchant vessels of various sizes and smaller crafts engaged in harbor activities. Although that no mooring or anchoring devices are shown with the vessels we may assume that all of them are anchored and not sailing. Only two mosaics with ship depictions show human figures engaged in harbor activities and not maneuvering the sailing vessels. The inscriptions associated with some of the offices indicate the home place

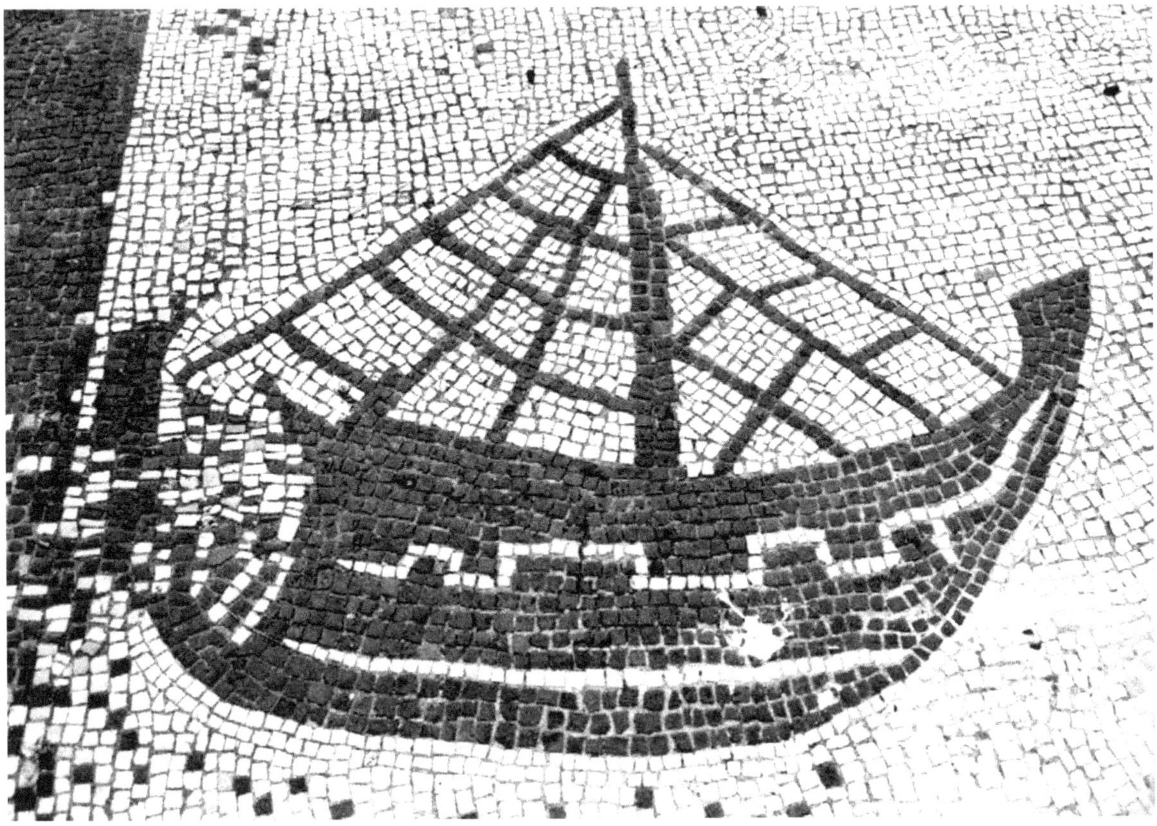

Fig. 3.7.45: Station 55

of the shippers and traders, or they suggest the kind of imported merchants to Ostia. Several ships are also depicted with a stepped lighthouse. The vessels appear either as single or in pairs facing each other. One of the ships is shown with its cargo of bag-shaped jars (fig. 3.7.40); others reveal harbor activities such as unloading cargo from a large merchantman to a smaller one (fig. 3.7.23). This operation may have taken place in the open sea in the vicinity of the Tiber's mouth where large merchantmen anchored and then the cargoes were unloaded into shuttle vessels to be transported upriver to Ostia and Rome. The scene is very similar to the description of Dionysius of Halicarnassus:

> *"Accordingly, oared ships however large and merchantmen up to three thousand bushels (three thousand amphorae) burden enter at the mouth of the river and are rowed and towed up to Rome, while those of a larger size ride at anchor off the mouth, where they are unloaded and loaded again by river boats".*[48]

Strabo has a similar description of activities carried out at the mouth of the Tiber.[49] There are ships engaged in log trades as indicated by the inscription in Station 3 (fig. 3.7.3). Names of North African shippers and traders, mostly engaged in grain trade are indicated in the offices from Misua (fig. 3.7.6), Karthago (fig. 3.7.13) and Syllectini (fig. 3.7.20). Two offices represent the traders from Sardinia, Turris (fig. 3.7.16), and Karalitani/Cagliari (fig. 3.7.18). The offices with inscriptions indicating shippers from North African and Sardinian provinces are found in the eastern side of the Piazzale. The office of Narbo, Gaul, is found in the northern side (fig. 3.7.2). The shippers' offices on the western side of the Piazzale representing different provinces are not found with inscriptions, as those on the eastern side (fig. 3.7.2). Only a fragmentary inscription of Alexandria (Station 40) is found in the NW corner of the Piazzale; most of the mosaic is damaged and replaced by asphalt.[50] We assume that this office was mainly engaged in the grain trade, although no ship or *modii* have preserved. The ships trading with North Africa also were engaged in transporting exotic wild animals, such as elephants, tigers, rhinos, deer, bears, etc., for the games held at the Colosseum and Circus Maximus in Rome. None of the ships depicted at Ostia show such a cargo as it may be seen in the mosaic of the Great Hunt at Piazza Armerina in Sicily (figs. 3.8.20 – middle section, 3.8.21, 3.8.25), and in the Dermech mosaic from Carthage (fig. 3.8.28). The elephant depicted in Station 14 followed by the inscription "STAT. SABRATENSIS", suggests that either elephants or ivory were traded from Sabrata (Libya). An elephant, a deer and a boar appear on the mosaic floor in Station 28 but no ship is depicted.

The shape of the ships' hulls can classify the Ostia vessels in the categories bellow:

[48] Dionysius of Halicarnassus, III.44.3.

[49] Strabo V.3.5.

[50] Personal observation of the present author while carrying out the research of the ships at Ostia in 2002.

A. Hull ended by a rounded and angled stem

In this category the hull has a rounded or a forward angled stem. The stempost is either surmounted by a block-shape or it ends at the gunwale level:

1. Rounded stem: figs. 3.7.14, 3.7.16, 3.7.22, 3.7.24, 3.7.29, 3.7.30, 3.7.33, 3.7.35, 3.7.40, 3.7.45.
2. Angle d stem: figs. 3.7.4, 3.7.5, 3.7.7, 3.7.11, 3.7.15, 3.7.39, 3.7.43, 3.7.44.

The usual rigging of any merchant ship comprises one main mast and a square sail. Few of the Ostia ships bear such rig: figs. 3.7.14, 3.7.15, 3.7.33, 3.7.39, 3.7.43, 3.7.44. The artemon rig (mast and sail), was an additional equipment that helped to sail the ship in an optimal manner. Especially it aided when a ship was sailing in or out of the harbor, where the main sail would be cumbersome. The ship from Narbon shows that the artemon mast could also be used as a crane on certain occasions (fig. 3.7.26).

B. Long hull with concave stem ending with a projecting pointed cutwater

The stempost of most of these ships is finished with an outer turned volute (figs. 3.7.9, 3.7.21, 3.7.25, 3.7.36). The stempost of three ships have a rounded tip (figs. 3.7.19, 3.7.27, 3.7.32). All the vessels in this category have high rounded sterns. A fence with a backward extension forming an extended poop appears on five ships (figs. 3.7.9, 3.7.19, 3.7.27, 3.7.32, 3.7.36). Four of the ships are depicted with a goose/swan head projecting above the aft fence (figs. 3.7.19, 3.7.21, 3.7.27, 3.7.32). Few of the ships are decorated with a dolphin or an *oculus* on the stem. These features indicate typical seagoing vessels where the decoration is used as an *apotropae*, a "weapon" against the "evil eye" to dispel the terrors of the sea and for a successful journey,[51] two vessels are decorated with a dolphin (figs. 3.7.21, 3.7.25) and two with an *oculus* (figs. 3.7.32, 3.7.36).

The majority of Ostia vessels are rigged with two masts and sails: figs. 3.7.4, 3.7.7, 3.7.9, 3.7.11, 3.7.16, 3.7.19, 3.7.22, 3.7.27, 3.7.29, 3.7.30, 3.7.32, 3.7.35, 3.7.36, 3.7.38, 3.7.40. Some vessels are rigged only with one or two masts and their standing rigging: figs. 3.7.5, 3.7.24, 3.7.25, 3.7.45. Representations of ships with three masts and sails are very rare in any arts and especially in mosaics. Such a vessel is depicted in the office of shippers from Syllectum (fig. 3.7.21), which indicates a super-galley and mainly known from ancient literature as *myriagogos* or *myriophoros* (capacity of 10,000 amphorae).[52] Athenaeus gives a detailed description of a ship named *Syracusia* that was rigged with three masts, had three decks, and carried about 1,700 – 1,900 tons of grain (its total burden was 3,650 tons; mixed cargo, crew, passengers, military).[53] Hieron II, tyrant of Syracuse (301 – 215 BCE) ordered the construction of the ship under the supervision of Archimedes, *c.*240 BCE. In his description Athenaeus indicates that the masts were also used as defensive equipments, or as shown by the Narbo ship, they had multi-purpose uses, being converted as cranes when needed:

> *"Now the ship was built to hold twenty banks of rowers, with three gangways.[54] The ship carried three masts, from each of which two-stone hurling cranes were suspended".[55]*

The ship was sent to Alexandria with a mixed cargo, but its principle load was the grain, which was offered as a present to Ptolemy II during a severe famine period in Egypt. When the ship arrived at Alexandria its name was changed to *Alexandris*:

> *"The ship was named Syracusia; but when Hieron sent her forth, he changed the name to Alexandris. On board were loaded ninety thousand bushels of grain, ten thousand jars of Sicilian salt-fish, six hundred tons of wool and other freights amounting to six hundred tons... He determined to send it as a present to king Ptolemy of Alexandria; for there was in fact a scarcity of grain throughout of Egypt".[56]*

Classes of Ostia Ships

The ships depicted in the mosaics at Piazzale delle Corporazioni can be divided into two classes:

1. Seagoing Ships: These vessels are merchantmen with broad hulls and a concave stem finishing in a projecting pointed cutwater, or both rounded ends. They are also identified by the decoration of an *oculus* or dolphin on the stem. Their rigging comprises one to two masts and sails, and rarely three. They also carried a triangular topsail set above the yard: figs. 3.7.7, 3.7.9, 3.7.11, 3.7.14, 3.7.19, 3.7.21, 3.7.27, 3.7.29, 3.7.32, 3.7.36, 3.7.38, 3.7.40. Seneca wrote about such vessels with reference to Alexandrian vessels:

> *"The Alexandrian boats, no matter how great the crowd of vessels, are seen by the very trim of their sails. For they alone may keep spread their topsails, which all ships use when out at sea, because nothing sends a ship along as its upper canvas".[57]*

The mast in large merchantmen was generally composite, girdled with wooldings at fixed intervals.[58] Some of the Ostia ships are depicted with composite masts rendered with intercalated white and black bands (figs. 3.7.14, 3.7.15, 3.7.27). The sails of the Ostia ships were reefed by a series of brails, mostly set on the fore bunt, as we can see in many depictions on reliefs from the 3rd century

[51] Landstrom, p. 35.
[52] Casson, p. 396.
[53] Turfa and Steinmayer, 1999, p. 106, Table 1.
[54] Athenaeus, V.207.
[55] Ibid., V.208.
[56] Ibid., V.209.
[57] Seneca, Epist., 77.1-2
[58] Casson, p. 231, n. 31.

CE, and wall paintings from the 1st century BCE to the 3rd century CE.[59] All the ships are steered by a pair of rudders or steering oars mounted on the quarters. The aft-wing extension of the top strake housed the shafts of the rudders. The position of the steering oars or rudders do no show their mounting system, but we may deduce that in general the system was the aft-mounting. This system consists of two through beams, with the lower beam set slightly aft of the upper one. The shaft rested on both offset through beams and was then secured by lashing (fig. 3.7.46).[60] Usually, a tiller inserted perpendicularly into the head of the loom worked the oars. The only example of the Ostia ships rigged with rudders worked by a tiller appears in the office of the shippers from Karalitani/Cagliari, Sicily (fig. 3.7.19) and probably Ship 2 – Station 23 (fig. 3.7.22). A conclusive aft-mount steering system is found in the left-hand ship in the Copenhagen sarcophagus (fig. 3.7.47). One of the characteristics of Roman seagoing merchant ships is the goose, swan or duck figurehead set on the quarter behind the aft-fence or as an adornment of the sternpost (figs. 3.7.19, 3.7.21?, 3.7.27. 3.7.32).

2. River and Harbor Vessels: The hulls of these vessels are broad and both ends are either rounded or the stem has a slight forward angle. They were used to unload cargoes from the large merchantmen anchored in the open sea and then transport them on the Tiber to Ostia and Rome. Such ships were towed when the winds and currents were not favorable for sailing upriver. Apparently when favorable winds blew inland, then sailors abroad the transporters spread the sail to aid the toiling of oxen or straining oarsmen to tow these vessels from the Tiber banks.[61] The mast of a towboat was tapered and had a series of cleats on the pole to prevent the slipping of the towline passed around it, or permitted mariners to climb the mast when needed to work the rigging lines (fig. 3.7.24). There are three reliefs, depicting ships with a similar mast, which were found at Rome, Isola Sacra and Salerno (3rd – 4th century CE).[62] In the Salerno relief, the moored vessel is shown in the process of unloading. During this operation the mast with the cleats was lowered on the deck when it was no longer in use (fig. 3.7.48). In seagoing merchantmen the mast was climbed via a rope ladder abaft. The capstan (usually placed on the fore deck) could control the towline. In Ship 1 from Station 25, the capstan is placed on the stern (fig. 3.7.24), as well as in Ship 1, Station 46 (fig. 3.7.33). Besides the sailing rig, river and harbor vessels were also rigged with several rowing oars when wind conditions were unfavorable or for entering/leaving a harbor. One of the Ostia ships is depicted with distinct oarports on the starboard side (fig. 3.7.29). The same arrangement was true on the port side.

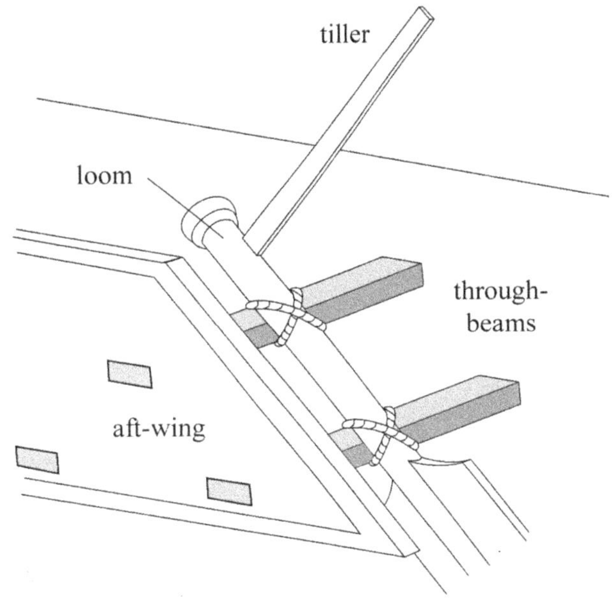

Fig. 3.7.46: Reconstructed aft-mount system of a quarter rudder

Types of Vessels and Harbor Activities

The ships from Piazzale delle Corporazioni are *navis oneraria*/merchantmen of different sizes representing seagoing ships and river/harbor crafts. During the first half of the 2nd century there was a building boom in Rome and other Italian cities, and therefore the Romans had to look for other sources of wood supply, especially after the heavy depletion of timber on the Apennine Mountains in both Punic Wars.[63] The need to import such merchandise is attested by the inscription in Station 3, "NAVICULARIORUM LIGNARIORUM" and both associated sailing ships (fig. 3.7.3). This office attests that such agency was necessary at Ostia, the Roman imperial harbor. Trading of exotic animals from North Africa (Stations 14, 28) was made together with the grain trade from Egypt, North Africa and Sardinia. The transport of grain was done in ships that were probably similar to the vessel called *Isis Giminiana*, depicted in a fresco found outside Porta Lauretina at Ostia, and dated to the 3rd century (fig. 3.7.49). The ship has a broad hull very similar to many of the vessels represented in the mosaics at Piazzale delle Corporazioni. A long tiller inserted into the head of the looms provided an easier maneuverability of the *Isis* Ship rudders. The average cargo capacity of a merchantman was 60 – 70 tons.[64] The bigger ships had a load capacity of 100 – 140 tons[65] (15,000 – 22,400 *modii*) and the largest 150 –500 tons.

The harbor activities were varied and complex. The grain, which was the main imported cargo to Ostia and Rome was controlled by the *praefectus annona* and confined to three areas:

[59] Casson, figs. 143, 144, 147, 149, 156.
[60] Mott, 1997, p. 28.
[61] Eubanks, 1930, p. 690. In a relief from Avignon, 3rd century CE, is depicted a boat being towed from ashore by two men. The towline is fastened to the top of the towing mast; Casson, 1994, fig. 97, p. 132.
[62] Casson, 1965, Pls. III.1, IV.2, V.1; ibid. 1994, fig. 94, p. 127.

[63] Meiggs, p. 190.
[64] Rickman, 1980, p. 267.
[65] 1 ton of wheat (on average) = 150 modii; Rickman, p. 275.

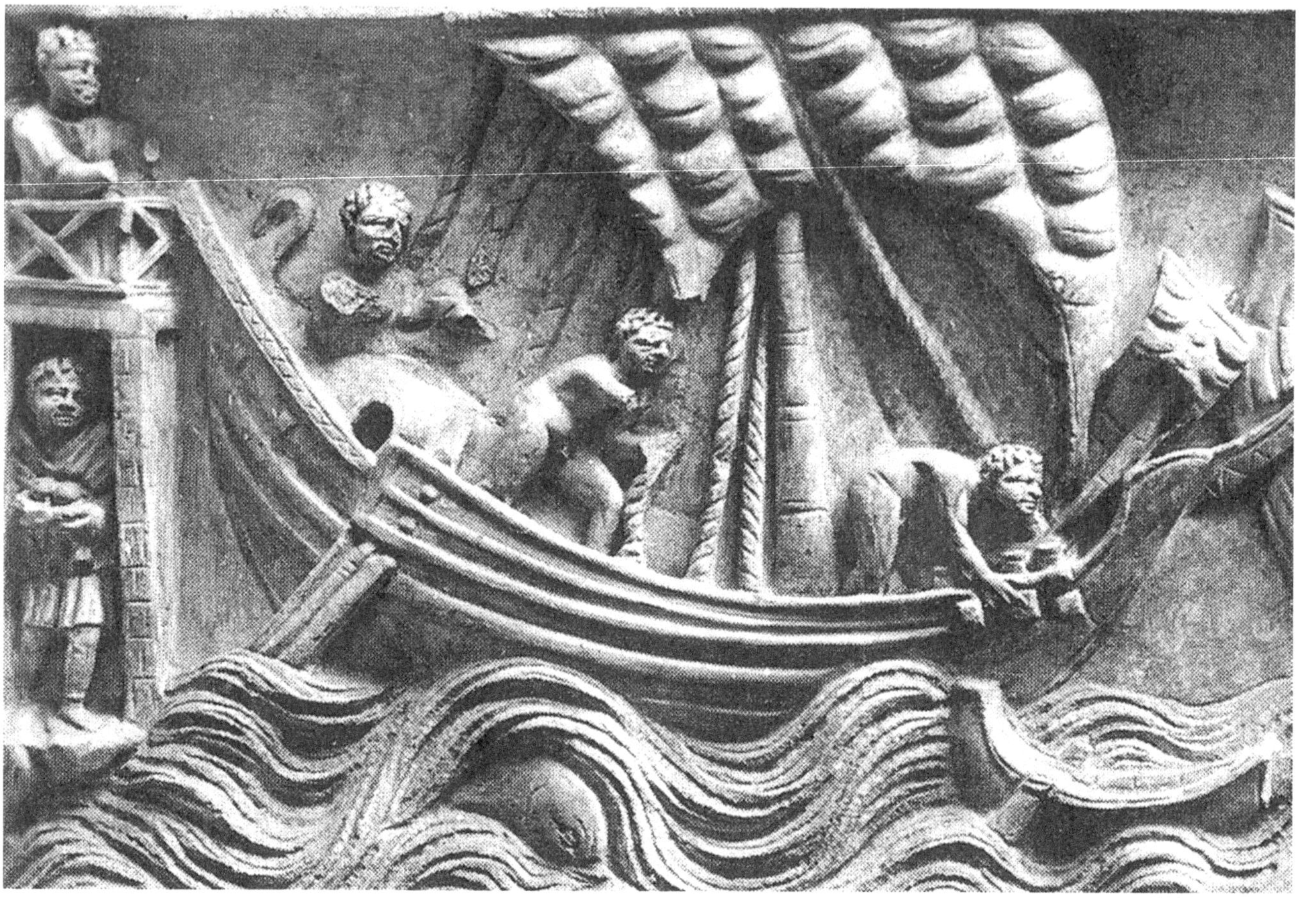

Fig. 3.7.47: Detail of an aft-mount rudder; left-hand ship, Copenhagen sarcophagus

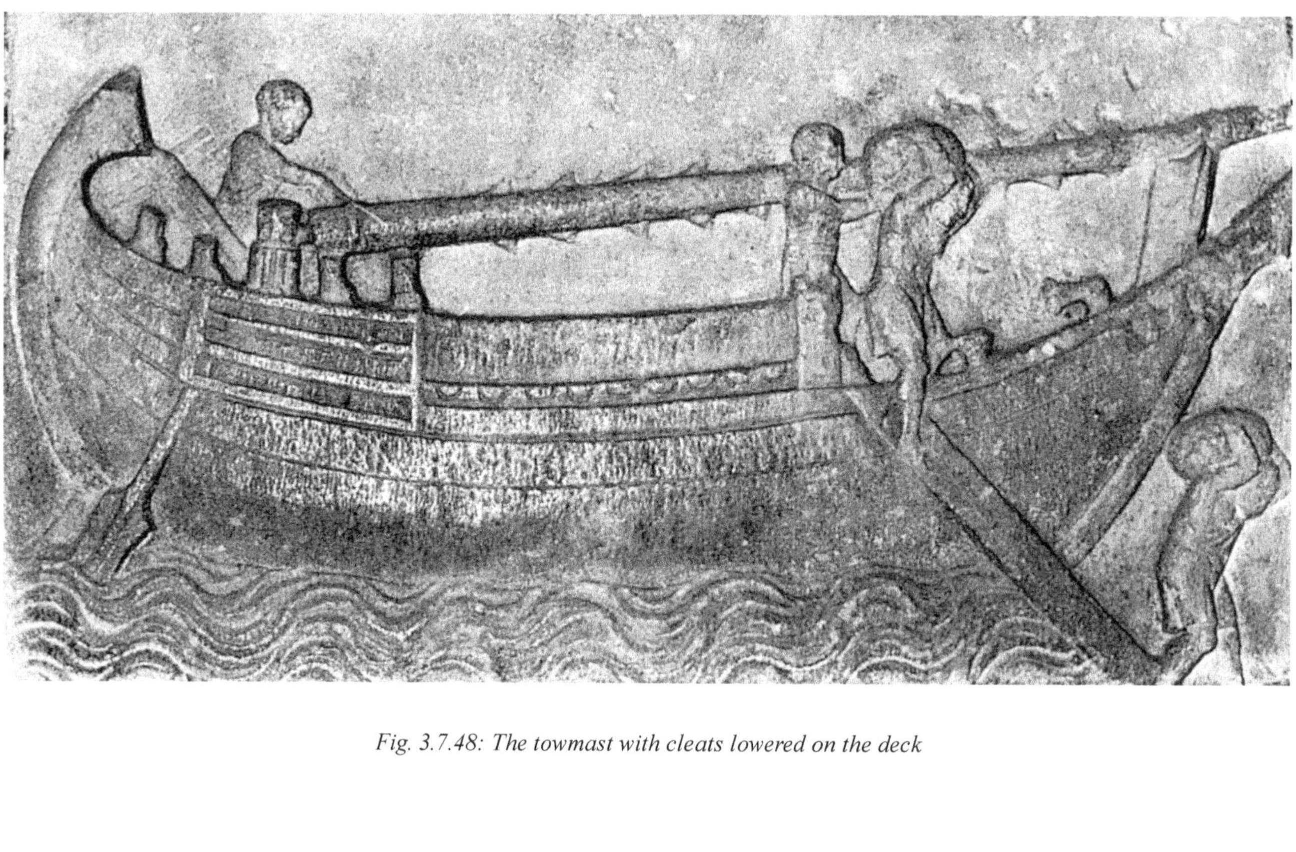

Fig. 3.7.48: The towmast with cleats lowered on the deck

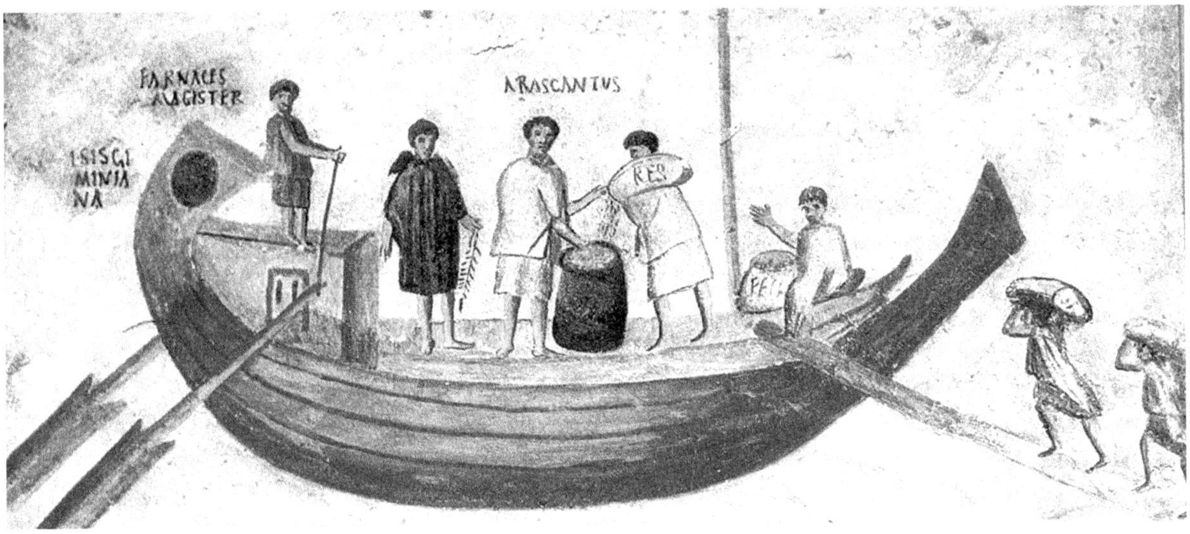

Fig. 3.7.49: The grain ship "Isis Giminiana"

1. The payment of individual *navicularii* (shippers or traders) employed to transport grain for the state;
2. The purchase of additional supplies as required by the state for the *frumentationes* from independent merchants;
3. Monitoring the total amount of grain brought in by independent traders and how far it corresponded with the needs of the capital.[66]

The *tabularius* was a book keeper supervising *rationes.*[67] *Praefectus annona* had to keep lists of all regular imports of grain and they had to be as accurate and updated as possible.[68] The same registration system may have been true for other imported merchandise brought to Ostia and Rome. The control and registration of amphorae cargo before being unloaded or shipped away as depicted in the mosaic in Station 51 augments the importance of such operation (fig. 3.7.40). The sited figure on the quarterdeck writes the account of the jars on a wax board. These tablets were then taken to the customhouse and kept for the records.

The unloading of the cargo from merchant ships to river vessels was done by the *sacarii* (stevedores or porters). The mosaic in Station 25 is a typical representation of such activity (fig. 3.7.23). Additional examples come from the Torlonia relief[69] or the fresco depicting the *Isis Giminiana* merchantman (fig. 3.7.49). The ship depicted in the office of the shippers from Narbon, representing the Gaul province,[70] shows that the artemon mast could be used as a crane when cargoes were loaded or unloaded (fig. 3.7.26). The merchantman depicted in the Torlonia relief (3rd century CE)[71] clearly shows how the artemon was maneuvered as a crane (figs. 5.2, 5.2a).

The ships facing each other, especially associated with a lighthouse may be considered as indicating the meeting between the vessels of *lenuncularii tabulatorii auxiliarii* (checking or recording clerks) and the seagoing ships anchored at the mouth of the Tiber (figs. 3.7.6, 3.7.13, 3.7.20, 3.7.28, 3.7.31, 3.7.34, 3.7.37, 3.7.42). The *lenuncularii tabulatorii* (taken in tow) were also in charge of towing operations[72] as we also can see in modern harbors where large ships are towed by tugboats to their moorings near the quay. There were also *lenuncularii pleromarii* who operated vessels which transported cargoes from incoming freighters to the river vessels and then transported them to Ostia and Rome. The vessel depicted in the office of the shippers from Narbon may indicate that small (50 – 80t) and medium size ships (100 - 150t) could sail up the Tiber and be unloaded in the port of Ostia near a warehouse. The device depicted on the top of the rectangular structure facing the ship, indicates that cranes were used for loading or unloading (fig. 3.7.26).

The Ostia seagoing ships may be associated with more specific types: *keles* (cargo dispatches), *phasoloi* (passenger transport) or *kerkouroi* (particular cargo that required rapid transport). The *keles* was built mostly for speed. It was a single-banked vessel with few oars and rigged with one or two square sails, carrying modest loads of cargo.[73] Such vessels were also used as naval auxiliaries and favored by pirates.[74] The *kerkouros* (or *cercurus* in Latin) also carried few oars on board besides its sail rigging. The average size of vessels had a load capacity of 130 – 150 tons[75] and the largest types were 275 –500 tons.[76]

66 Rickman, pp. 270 – 271.
67 Ibid., p. 271.
68 Ibid.
69 Casson, fig. 144.
70 The Gaul traders were not engaged in the grain trade but probably in wine and Spanish oil. Sextus Fadius Secundus Muse is known as a rich Gallic shipper, from shards found at Monte Testaccio and an inscription on the base of a statue in his honor at Narbo. The shards indicate that they were from amphorae filled with Spanish oil and shipped to Rome; Meiggs, p. 289.
71 Casson, fig. 174.
72 Ibid., 1965, p. 35.
73 Casson, p. 160 and n. 17.
74 Ibid., p. 161.
75 Ibid., p. 183.
76 Ibid., p. 164.

3.8 Piazza Armerina – Sicily

Location and the Discovery of the Mosaics

The Roman Villa del Casale was decreed a UNESCO World Heritage Monument since 1997, at the 21st meeting of the World Heritage Committee in Naples. Italy, 1 – 6 December.[1] The villa was constructed on the remains of an older villa in the first quarter of the fourth century, probably as the centre of a huge *latifundium* covering the entire surrounding area. Its owner was possibly a member of senatorial class or may have been of the imperial family itself. The Roman villa of Piazza Armerina is situated about 550m above sea level, in the Casale district in central Sicily, and 5½ km SW of the modern town of Piazza Armerina. The villa was built near the Gela River, following the topography of the land at the foot of the Magone Ridge. It occupies an area of about 4000 square meters and the mosaic pavements extend for over 3500 square meters. The plan of the villa can be divided into four areas (fig. 3.8.1):

1. The *thermae* or the Baths.
2. The guest rooms.
3. The rooms of the villa's owner and his family.
4. A large *triclinium* (dining room), used for guests and festivities.

The main group of rooms is located to the west, towards the Nociana stream, which flows 200m from the villa. To the north side of the villa is the Magone Ridge (777m above sea level) and to the east it is surrounded by Mt. Saldano (727m, above sea level), only to the south the site is less hilly.[2] Paolo Chiaranda, a native historian of Piazza Armerina, already wrote about the remains of the large estate in the 17th century:

> *"...in the area of Casal dei Saraceni, numerous structures belonging to a very old building appear above the ground".*[3]

The site is mentioned again by local antiquarians in the middle of the 18th century. The earliest investigations were made in 1761, when an ancient temple was discovered. It was adorned with ancient mosaics and paved with colored marbles. Pieces of mosaics and some columns found early in the 19th century led to some excavations that were carried out later in that century Sabatino del Muto carried out diggings in 1812 in the area of the Basilica where he found numerous gold and silver objects that in earlier periods were passed around the noble families of Piazza Armerina.[4] Realizing the importance of the building, the city of Piazza Armerina decided that scientific campaigns should be carried out. The work was entrusted to the engineer Pappalardo in 1881. During the following campaigns, the mosaic in the three-apsidal building was uncovered (fig. 3.8.1/*46*). It was only in 1929 that systematic work on this villa began under the direction of Paolo Orsi, who was a pioneer in the archaeology of Sicily.[5] Giuseppe Cultrera, Suprintendente of antiquities of Syracuse, continued the excavation in 1935, 1938 and 1940 - 1941, when the oval court (fig. 3.8.1/*46*) some of the flanking rooms and the south apse of the Great Hunt Hall were exposed (fig. 3.8.1/*36*c). During the Second World War the work stopped. The rest of the villa as we know it today was unearthed by G. V. Gentili during five campaigns carried out between1950 and 1960.[6]

The Date of the Villa

The dating of the villa complex given by Gentili was mostly based on the stylistic criteria of the mosaic floors rather than archaeological evidence. In 1970, Andrea Carandini dug some trenches to reveal more information about the chronology of the building and also to study other aspects of the villa left unpublished by Gentili. This study still continues.

The villa complex spreading over an area of 150 x 100m (1 ½ hectares) was paved with beautiful polychrome mosaics, most of them being figurative, whereas frescoes and marble veneers decorated the walls. The entrance to the villa complex was from the south through an elaborate archway, which apparently was the first example used in a Roman villa (fig. 3.8.1).[7] The excavations of the 1950's and the cut trenches in 1970's revealed that there was an earlier villa, probably erected either at the end of the 1st or during the 2nd century CE.[8] The villa is a single-story building, and the *peristyle* is the center whereas almost all the main public and private quarters were built around. By studying a great number of parallels from North African and Italian mosaics, Carandini suggested that the villa dates between 320 and 370 CE.[9] Several Corinthian capitals from Piazza Armerina are closely related to capitals from the Palace of Diocletian at Spalato and the Baths of Diocletian at Rome. Heinz Kähler suggested that they were produced in the same eastern shop which supplied imperial architectural items during the period of 300 – 310 CE.[10] Coins found beneath the mosaics were dated mostly to the second half of the 3rd centuries and like the pottery they were pre-Constantine.[11] Above the mosaics were found coins of Maxentius (306 – 312 CE) and the second Flavius (312 – 363 CE). According to Gentili, a coin of Maximian (the end of the 3rd century CE) found in the mortar that cemented the

---

1 http://whc.unesco.org/en/news/180
2 Wilson, 1983, p. 14.
3 Ciurca and Bologna, 1997?, p. 5.
4 Ibid.

5 Wilson, p. 14.
6 Ibid., pp. 14-15.
7 Polzer, 1973, p.141. The study of the villa with its elaborate mosaics and some numismatic evidences brought Polzer to suggest that the villa has a tetrarchic origin that began under Maximian when he was Augustus (286 – 305 CE). Diocletian made him partner in 286 CE, after Maximian's conquest of Baguades, and thus he ravaged the Gaul; Polzer, p. 143. The copper coins issued by the Carthage mint in 296 CE, around the time of Maximian's activity in North Africa, comprised four shops identified by the four letters: P, S, T, Q (*prima, secunda, tertia, quarta*), assigned by Diocletian, Maximian, Constantius and Gelarius; Polzer, p. 45.
8 Wilson, p. 34; the archway was dated to 100 – 200 CE.
9 Ibid.
10 Polzer, p. 141.
11 Wilson, p. 36.

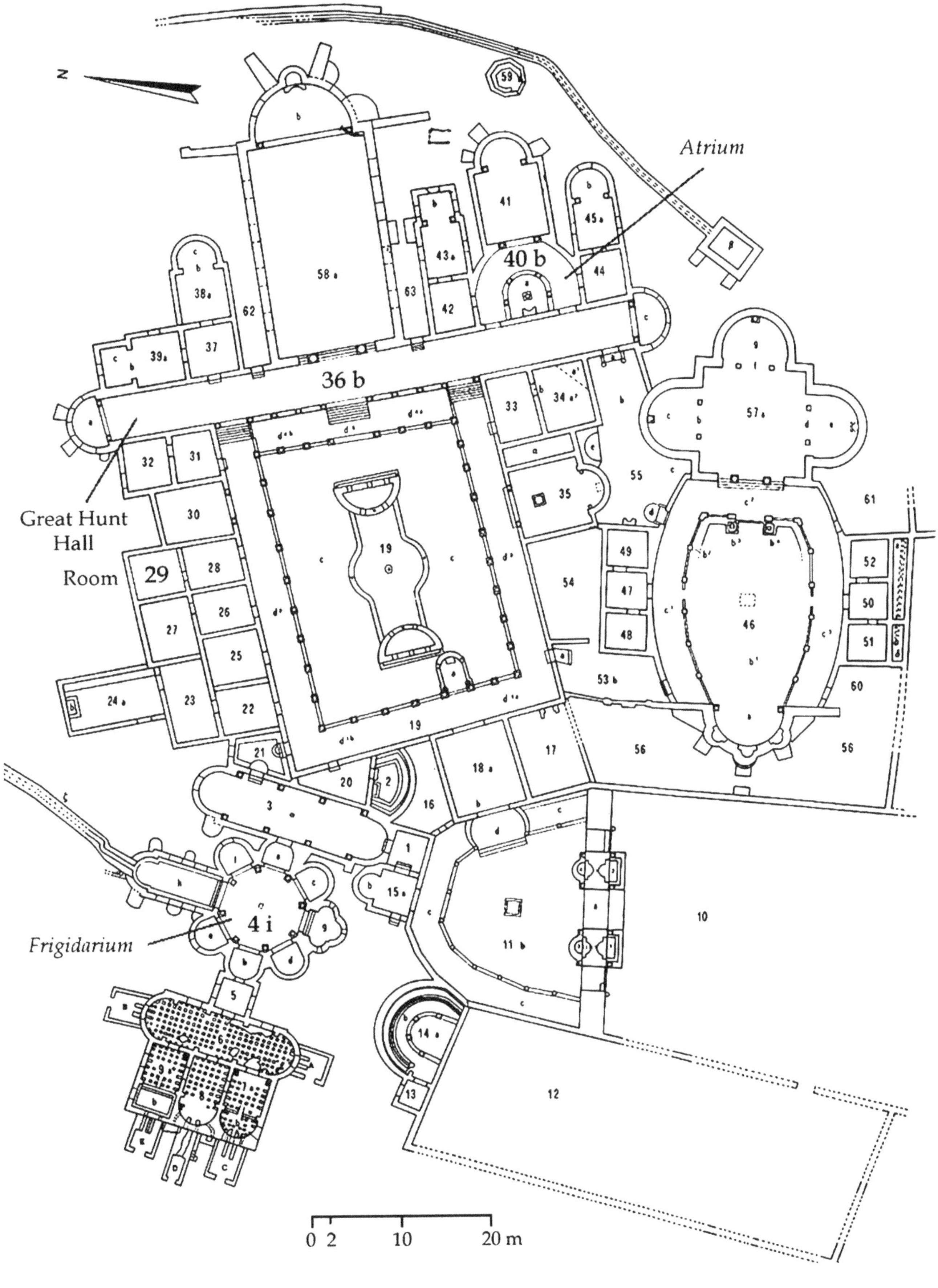

Fig. 3.8.1: Plan of the Villa Complex

marble slab of the threshold in the SE corner of the exedra in the *frigidarium* was considered to give the building date of the villa. A single stratified coin cannot be considered as a final evidence of dating the villa.[12] Excavations made in other places, namely in North Africa, indicated that the *terra sigillata* pottery did not appear before 310 CE, and the stamped ware not before 320 – 325 CE.[13] The pottery finds at Piazza Armerina (including *terra sigillata*), the date suggested by Carandini and the tetrarchic origin of the villa suggested by Polzer, indicate that the chronology of the villa, as it is preserved, should be placed in the first quarter of the 4th century CE.[14]

Since its discovery, the villa went through several minor or major repairs, mostly evidenced in the mosaic work. Extensive repairs were made in the center of the Great Hunt Hall, probably at the end of the 4th century.[15] In Room 38, known as the "*mosaic of the bikini girls*" (fig. 3.8.1/*34*), the pavement was laid on top of an earlier geometric mosaic dating from the mid-4th century.[16] The most significant restoration was the addition of the reinforcement pier behind the apse in the northern corner of the Great Hunt Hull and the arches of the aqueduct. All these repairs followed the aftermath of the earthquake in 365 CE, when part of Sicily,[17] Kenchreai,[18] in Greece, Cyprus, and other parts of the Eastern Mediterranean were also affected. Gentili assumed that besides the continuity of the site, the mosaics did not suffer weathering or iconoclastic interfere because they had been covered by silt and debris from at least the end of the 4th century CE.[19]

Large quantities of Byzantine, Arab and Norman pottery were found at the site, as well as coins of the Byzantine emperors Heraclius (610 – 614 CE) and Constantine II (654 – 659 CE), along with coins of Roger II (1130 – 1154).[20] During different periods of occupation, damage was caused mostly by digging-pits, for example in Rooms 10, 21 and 24 (fig. 3.8.1). A pottery kiln dated to the Norman period was found in Room 16. In Room 19, the mosaic floor was completely destroyed and replaced by a slab floor, probably before the Norman period.[21] It is assumed that activities at Piazza Armerina villa had been maintained for about 150 years, until the early 6th century CE.[22] The life of the villa may have ended in 800 CE, possibly by a fire and also by silt deposited after heavy flooding of the Gela River.

[12] Ibid.
[13] Ibid., pp. 36 –37.
[14] Wilson, p. 37. The identity of the villa's owner remains a subject of debate. However three individuals are mentioned: Proculus Populonius, governor of Sicily from 314 to 337; Caeionus Rufus Volupsianus, also called Lampadius, an influential wealth man, and Sabucinus Pinianus, probably of Roman birth; Polzer, p. 141.
[15] Wilson, p. 41.
[16] Ibid.
[17] Ibid.
[18] See the site of Kenchreai, pp. 81 - 82 above.
[19] Wilson, p. 42.
[20] Ibid.
[21] Ibid.
[22] Ibid.

The Mosaics and the Mosaicists

The detailed study of the mosaics at Piazza Armerina show strong parallels with those in the Roman provinces in North Africa. It appeared that the mosaic pavements in the villa were made using patterns of North African workshops, but actually African artists were brought to Piazza Armerina to carry out the work. Gentili suggested that most of the tesserae used in the mosaics are African stones and not Sicilian.[23]

It is assumed that in antiquity mosaic workshops used pattern books, which contained a large range of motifs and geometric designs that circulated all over the provinces through sea trade connections. Such patterns found at Piazza Armerina were altered and adapted to suit the taste of the villa's owner. More than one master designer was responsible for planning the overall composition of each floor made the complicated designs of the mosaics in the villa. Each designer or master craftsman would have supervised several workers who laid the tesserae in the floors. The mosaic work was probably carried out directly at the site and did not require the work of *emblema* made in a workshop. The floor in the Great Hunt Hall may be considered as the best conclusive example of the mosaic making directly on site at Piazza Armerina.

Mosaics with the Maritime Scenes

The mosaics in the villa can be classified by their subject depiction. Only the mosaics representing the maritime scenes that are found in different units of the complex and serving different functions will be described in this monograph. These scenes mainly comprise two subjects: 1. putti fishing from boats are found in the *frigidarium* (fig. 3.8.1/*4i*), in the baths complex (fig. 3.8.1/*4i*), Room 29, in the owner's private complex and in the semi-circular *atrium* (fig. 3.8.1/*40b*); 2. catching of wild beasts in Africa, Egypt and India and then transporting them in large merchant vessels to Rome for entertainment in the Colosseum and Circus Maximus (fig. 3.8.1/*36b*).

Frigidarium

The cold room with an octagonal plan (9 x 8.95m) was surrounded by several apsidal dressing rooms and two swimming pools (fig. 3.8.1/*4i*). Originally, the central area was covered by a dome decorated with glass mosaic tiles that was set on arches and supported by granite columns with Corinthian capitals. During the excavations carried out in 1950's many fragments of vitreous paste in various colors were found on the floor of the *frigidarium,* as well as in both pools.[24] The mosaic in the octagonal floor depicts a beautiful and complex maritime scene with Nereids, Tritons, putti fishing, dolphins, fish, Centaurs, sea lions and hippocampi. A similar bath layout is known from North Africa, especially in the *thermae* found NW of the theater

[23] Ibid., p. 44.
[24] Ciurca and Bologna, 16.

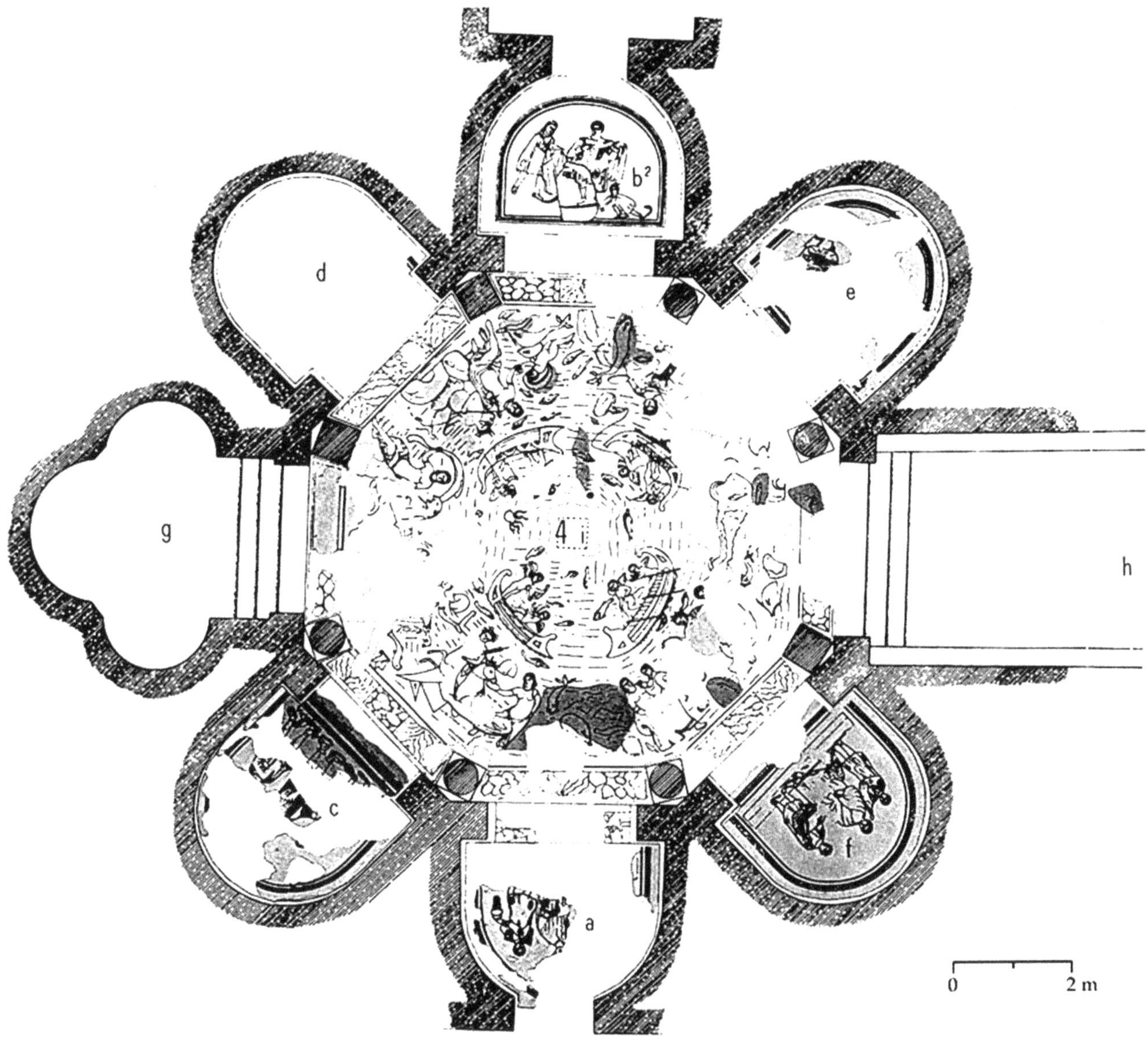

Fig. 3.8.2: Plan of the frigidarium *with the fishing mosaic floor*

at Bulla Regia.[25] In the center of the Armerina *frigidarium*'s floor are depicted four fishing boats arranged in a circle, one following the other (fig. 3.8.2). Two putti engaged in fishing with a net or rods occupy each boat. All the figures are naked, with blond or light brown curly hair and a V mark on their forehead, above the bridge of the nose. They are adorned with necklaces and bracelets on their arms and feet (figs. 3.8.3, 3.8.4, 3.8.5, 3.8.6). A pair of open wings is attached to their backs. Three boats have similar hulls with a concave prow finished by a projecting pointed cutwater. The rounded upraised stern is surmounted by a slightly inward-turned sternpost with a rounded tip (figs. 3.8.3, 3.8.6) or a fishtail tip (figs. 3.8.4). The almost vertical stempost of all the boats is adorned with a rounded tip or a stylized small volute. One boat has a short broad hull with both rounded ends (fig. 3.8.5). The stem and stern of all the boats is painted in red. The propulsion of all the boats consists only of rowing oars that also are used for steering.

Boat 1: The concave bow with a pointed forefoot is oriented to the right. The vertical stempost is adorned with an out-turned rounded head that probably is an abstract depiction of a volute (figs. 3.8.3). A white pattern that may represent an *oculus* is depicted within the trapezoidal frame on the starboard stem. A dolphin with its head pointing towards the waterline is inscribed within the triangular frame on the starboard stern (figs. 3.8.3). The top wide plank decorated with a wavy pattern depicted in perspective is emphasized by the light and dark green, red and white tesserae. The lower hull is decorated with two long strips of ochre and green tesserae, probably representing the strakes. The covered deck is at the gunwale level, but there is a rectangular opening slightly off amidships. The left leg of the putto working the oars goes through the opening and rests within the hold of the boat while the right leg rests on the deck (figs. 3.8.3). His back is turned towards the stem while he pools a pair of oars, indicated only by their looms and projecting above the amidships gunwale. The blades are probably submerged in the water. The second putto facing

[25] Carandini *et al.*, 1982, p. 343.

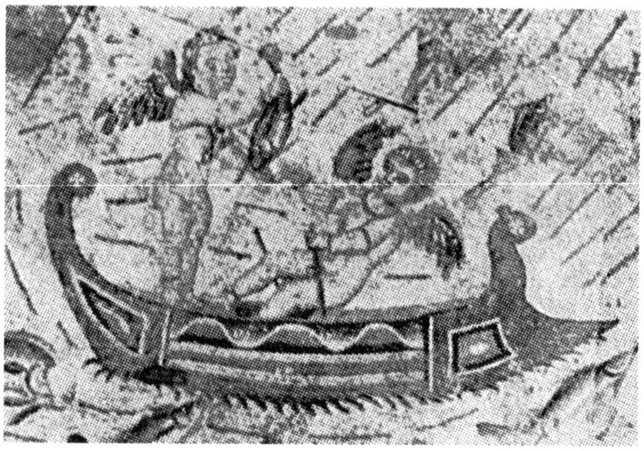

Fig. 3.8.3: Boat 1 – frigidarium

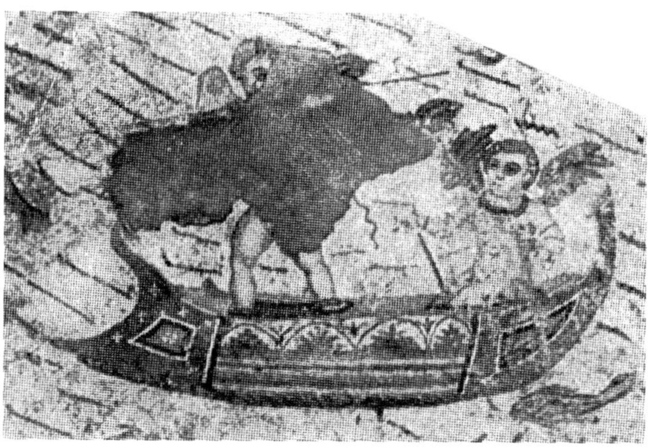

Fig. 3.8.4: Boat 2 - frigidarium

forward stands on the quarterdeck and holds a fish by its tail (fig. 3.8.3).

Boat 2: The bow points to the left. The concave stem is finished with a projecting pointed cutwater (figs. 3.8.4). The stempost arches out over the stem and is adorned with an inward turned rounded head. The top of the stempost as well as the standing putto of the fore-deck suffered some damages. The upraised rounded stern is surmounted by a vertical sternpost adorned with a fishtail tip. A small, white, star-like or flowery pattern is inscribed within the trapezoidal frame decoration on the port stem. This pattern may represent a stylized *oculus*. A green dolphin with his head pointed forward is depicted in a triangular frame on the stern. Four arches of *acanthus* leaves are depicted in the upper plank (figs. 3.8.4, 3.8.4a). The lower hull of this boat has a similar decoration to Boat 1. The deck is covered up to the gunwale level apart from two rectangular openings in the fore and aft decks. A putto stands on the fore deck with his legs going through the square opening on the deck rest his feet on the bottom of the boat. He is shown in a three-quarter view. It appears that he is in the act of throwing a harpoon or a long spear over the port stem. To keep his balance, the putto has his left knee slightly bent forward while the right leg stretches backwards. Most of the figure's body is damaged and only his right-side head, upper wings and legs have survived (fig. 3.8.4). The second putto seats on the edge quarter opening with his legs resting on the floor of the boat. He is depicted in three-quarter view facing the standing putto. The figure steers the boat by the oars mounted on either quarter. The port oar is mounted to the quarter gunwale by a small rowlock ring or a strop (fig. 3.8.4a). Only the loom visible above the quarter gunwale represents the starboard oar. The blades are missing or probably are submerged in the water.

Boat 3:[26] The bow points to the right. The hull of this boat is slightly different. It is shorter and both ends are rounded (fig. 3.8.5). The upraised stem is surmounted by a high curved stempost, adorned with an out-turned rounded head or a small volute. The rounded stern is surmounted by a vertical sternpost adorned by a fishtail-tip. The upper part of the hull is decorated with a pattern resembling the ends of five projecting through beams rendered in perspective (fig. 3.8.5). The lower part of the hull probably has the same decoration as Boats 1 and 2. It suffered some damage on its fore part. The deck platform seems to be at the gunwale level, with a rectangular opening in its aft part. Both legs of the standing putto on the quarter, facing the stern, pass through this opening, concluded from the revealed knees, while his legs rest on the boat's floor. He is depicted in three-quarter view, and his body is slightly inclined forward, thus showing the stress put on the fishing rod held in his right hand; the weight of the fish at the end of the line arches the rod (fig. 3.8.5). A second putto standing on the mid-deck, facing forward is also engaged in fishing by throwing a trident harpoon, indicated by his forward inclined body, whereas his weight is supported by the right bent leg and the backward stretched left leg. The boat lacks its oars.

Boat 4: The bow points to the left. The rounded elongated hull can be deduced from the curved bottom (fig. 3.8.6). The concave stem finishes with a projecting pointed cutwater. The short vertical stempost is adorned with an inward-turned rounded head or volute decorated with a reversed white swastika pattern. The broad upraised stern is surmounted by a vertical sternpost adorned with an inward turned rounded head, also decorated with a small white swastika. The upper plank is decorated with a series of small vertical dolphin heads pointing upward (fig. 3.8.6). The lower part of the hull is decorated with two strips of yellow tesserae. The bottom is made with brownish green stones, probably indicating the pitch coating to make the hull watertight. Rippled waves depicted beneath the bottom of the boat suggest that it sails on the waves. The standing putto on the covered deck that is at the same level as the gunwale is pulling the line of a fishing net full with fish. The strain of his action is emphasized by his backward lined body while his weight is supported by the forward stretched right leg and the backward bent left leg. Apparently in the

[26] This boat was studied from the black and white picture produced in Carandini *et al.*, Pl. LIX. The mosaic *in situ* had some patina on top it was in a poorly visible stage.

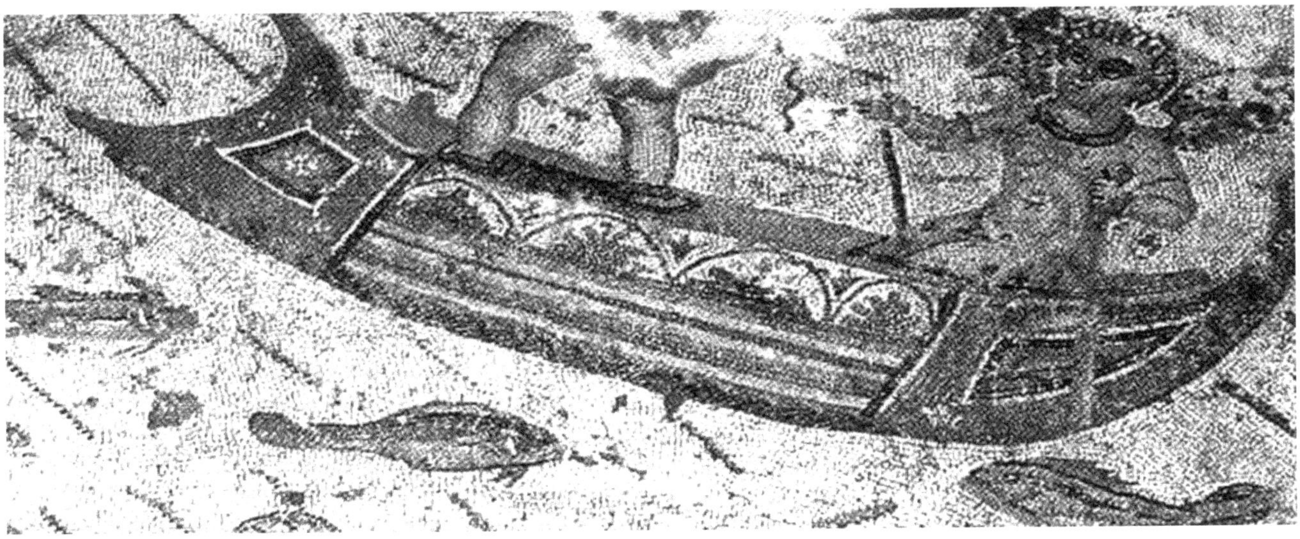

Fig. 3.8.4 a: Boat 2, detail of the helmsman – frigidarium

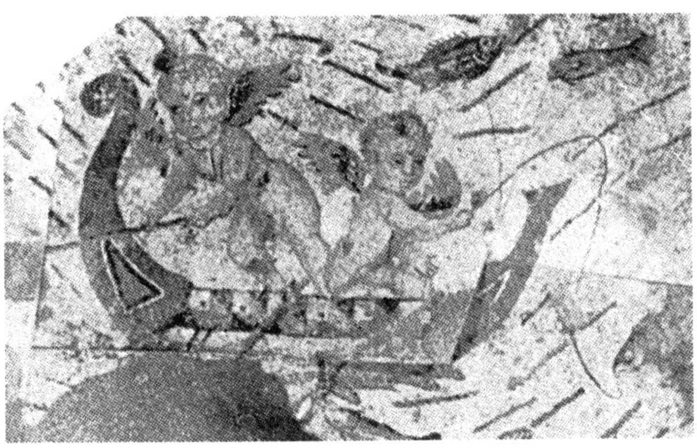

Fig. 3.8.5: Boat 3 – frigidarium

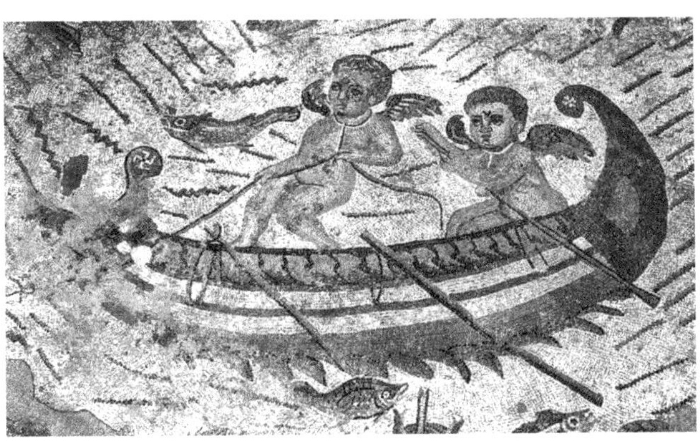

Fig. 3.8.6: Boat 4 - frigidarium

aft-deck is a square opening, suggested by the seated putto that his legs go through the opening and his feet rest on the bottom of the boat. He works the port steering oar by holding the loom with both his hands though trying to stabilize the boat, while it is unbalanced by the putto pulling the line of the fishing net (fig. 3.8.6). On the port gunwale are mounted three oars; two thicker rounded shafts indicate the rowing oars, while the aft thinner shaft with a small tapered blade distinguishes the steering oar. The oars have a backward slanting. The loom of the fore oar is mounted by a half-ring thong and lashed by a strop that its ends hang beneath. The other two oars apparently are not secured by any means, but beneath them are depicted two black strips symbolically indicating their mounting to the gunwale by a rope or leather strop.

Room 29

This is the only room in the private complex having a mosaic floor decorated with a maritime scene (fig. 3.8.7). The room (5.4 x 5.05m) is paved with a beautiful polychrome mosaic depicting fishing activity in a harbor faced by a colonnade portico. The fishing is carried out from four boats, each occupied by three putti. The scene is designed in three registers that form a whole picture. The upper part is occupied by a long waterfront structure with a colonnade, an elongated building with a pitched tile roof flanking each end of the portico, and another building behind the arching colonnade in the middle. Rich fauna with fish and dolphins populates the sea. The boats are arranged in two registers, the bows of the top boats point to the left, while those in the lower part point to the right. This arrangement was meant to give some perspective to the scene. All the boats have similar hulls and their decoration resembles the boats in the *frigidarium* (fig. 3.8.2). Boat 4 has suffered the most damage, only its bow and a small part of the fore hull have survived.

All putti have blond or light brown curly hair, a V mark

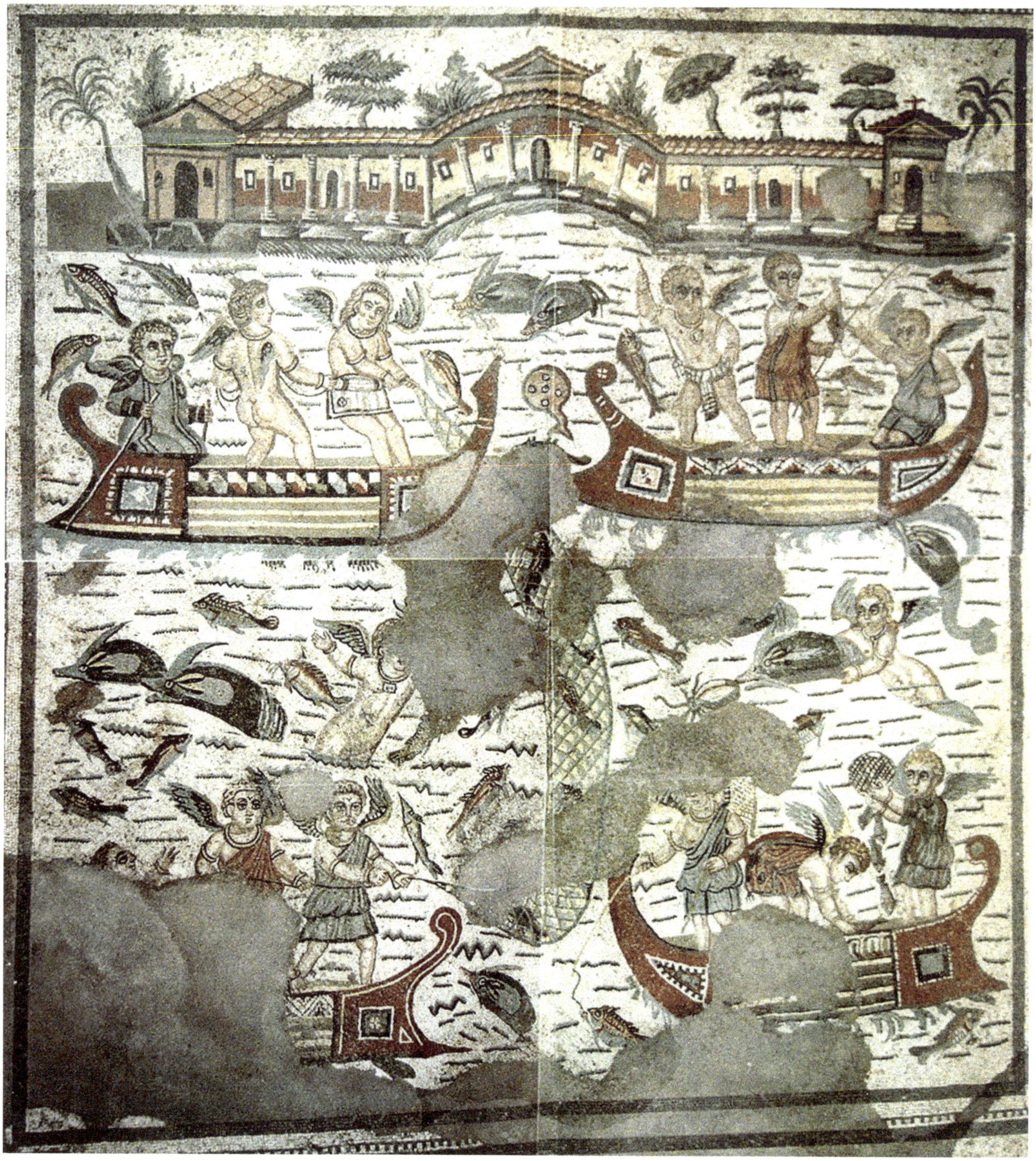

Fig. 3.8.7: Room 29

above the bridge of the nose and wings beneath the shoulders. They are adorned with rings around their arms and necks. Some are dressed in tunics, loin covering or are naked. The description of the boats follows the clockwise direction. The ripples beneath the bottoms suggest that the fishing is carried out while the boats are sailing, or they resulted from the strain of the activities that brought to the shaking of the boats.

Boat 1: The bow of the top left-hand boat points to the left. The concave stem is finished with a projecting pointed cutwater (fig. 3.8.8). The stempost has an outer angle and is adorned with an inward-turned rounded head or a stylized volute. On the starboard stem is depicted a rectangular frame with a white ivy leaf inscribed inside. The rounded upraised stern is surmounted by a vertical sternpost adorned with a fishtail tip. The stern is damaged and therefore the decoration is not clear. The upper plank is decorated with ten projections like ends of through beams depicted in perspective, as viewed from below, though symbolically supporting the deck; they are made with white, red, pink, hues of green and black tesserae. The lower part of the hull

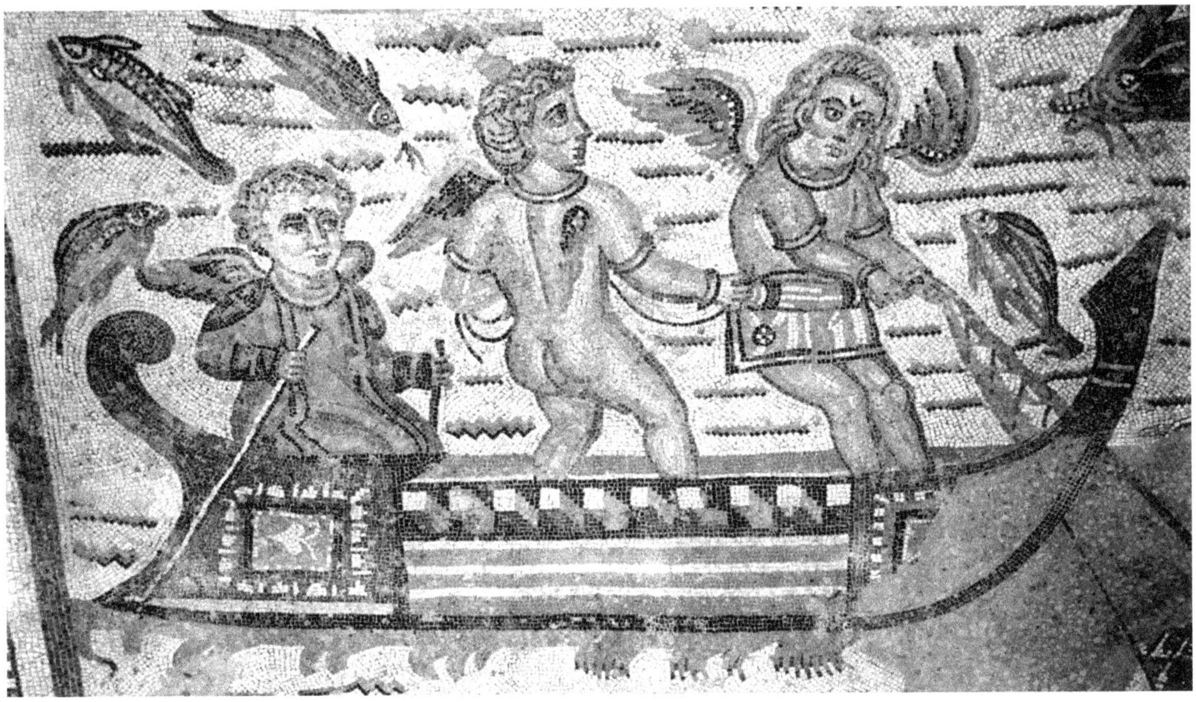

Fig. 3.8.8: Boat 1 – Room 29

is made with three light yellow strips alternating with three strips of ochre tesserae that probably indicate the lower strakes. The deck is just beneath the gunwale, deduced from the hidden feet of the putti. Three figures inhabit the boat. Two standing putti are engaged in pulling the fully loaded fishing net full with fish. The feet of the seated figure in the bow go through a square opening on the fore-deck and rest on the bottom of the boat. The rower is facing the standing putti while working the oars indicated by their thin shafts, and trying to stabilize the boat (fig. 3.8.8). The blades probably are submerged in the water.

Boat 2: This boat is very similar in shape and decoration with Boat 2 in the *frigidarium* (fig. 3.8.4). The upper plank of this boat has a similar *acanthus* leaf decoration depicted in a zigzag pattern but not arched as Boat 2 in the *frigidarium*. The deck is at the gunwale level, and a rectangular opening is found on the quarter (fig. 3.8.9). Three putti are engaged in fishing. The standing putto on the mid-deck has just pulled his rod out of the water and has caught a fish. He holds the fish by its tail in his right hand, while in his left hand he still holds the rod. The figure in the bow is in the process of throwing a trident harpoon at a fish on the port side of the boat. His body leans forward as emphasizing his action. The legs of the seated putto on the quarter penetrate the opening in the deck and rest on the bottom of the boat. He faces the standing putto on the mid-deck and apparently supervises the activities of both putti (fig. 3.8.9). The boat lacks its oars.

Boat 3: The bow points to the right. Part of the hull and its upper sternpost are damaged, as well as the head of the putto standing on the quarter (fig. 3.8.10). The upper plank has the same decoration as Boat 1(fig. 3.8.8) with stylized projecting ends of the through-beams, though supporting the deck. The lower strakes are made with white and dark green tesserae. Three putti standing on the deck just below the gunwale are engaged in fishing. The figure standing on the prow, facing the stern, empties the fish from a rounded basket trap (fig. 3.8.10). The figure on the mid-deck seems to be storing the fish in the hold of the boat. His legs are damaged. The putto on the quarter, also facing the stern is fishing with a rod in his right hand over the port stern. A fish is caught at the end of the line, beneath the stern. Apparently the fish is heavy and therefore he is uses the fishing net with a long handle held on his left shoulder (fig. 3.8.10). There are no oars in the boat or any anchor.

Boat 4: This is the most damaged boat in the maritime scene. The preserved bow points to the left. The upper plank seems to be decorated with a zigzag *acanthus* leaves made with white, red and dark green stones. The deck appears to be at the gunwale level, deduced from both standing putti on the prow and mid-deck. The figures on the bow and mid-deck are pulling the lower end of the fishing net. The legs of the standing putto on the mid-deck as well as almost the entire figure of the sited on the quarter are damaged (fig. 3.8.11). The seated putto on quarter and facing the standing figures is the helmsman who works the oars from the quarter. Only his head and the right hand survived the damage.

Boats 1 (fig. 3.8.8) and **4** (fig. 3.8.11) are engaged in fishing with a seine net (fig. 3.8.12).[27] The net is full with fish and the putti in Boat 1 pulls the rope attached to the upper end,

[27] A seine net is a long and narrow net used when fishing for surface fish. It hangs vertically in water, having floats at its upper edge and sinkers at its bottom; Kemp, 1988, p. 769.

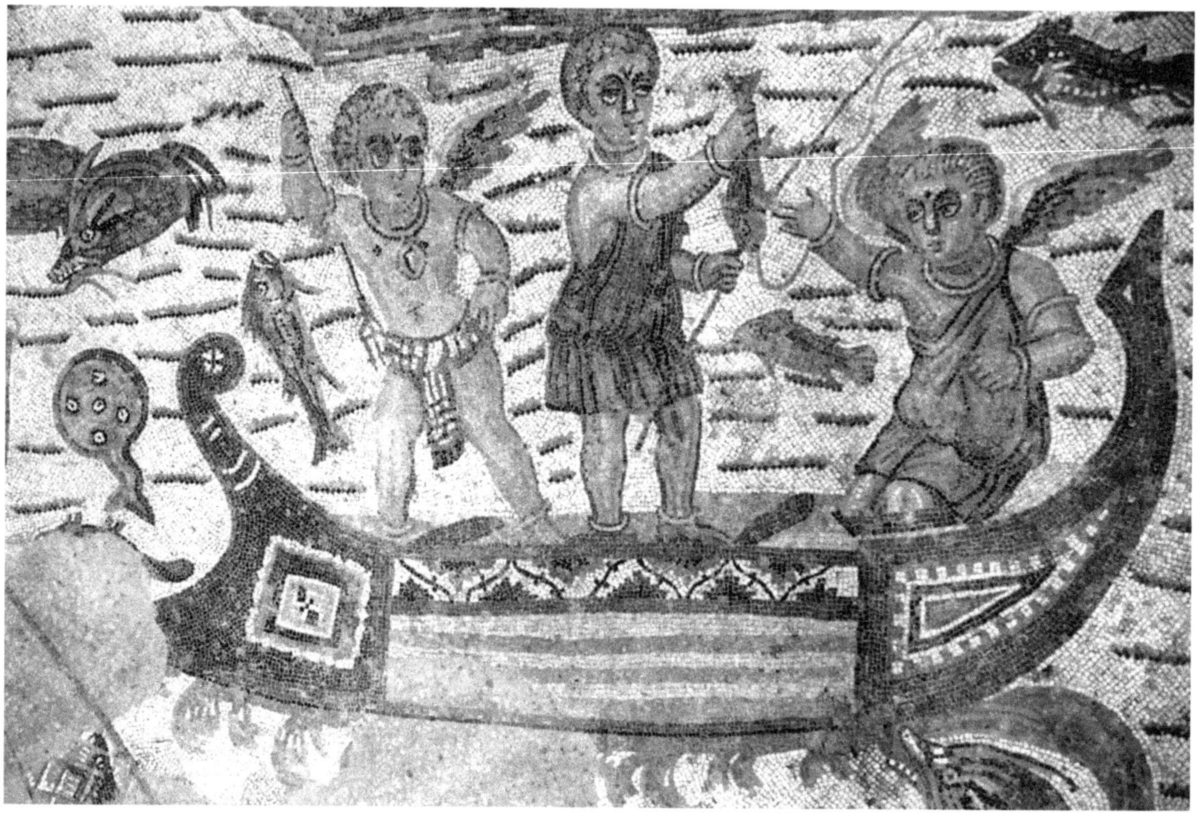

Fig. 3.8.9: Boat 2 – Room 29

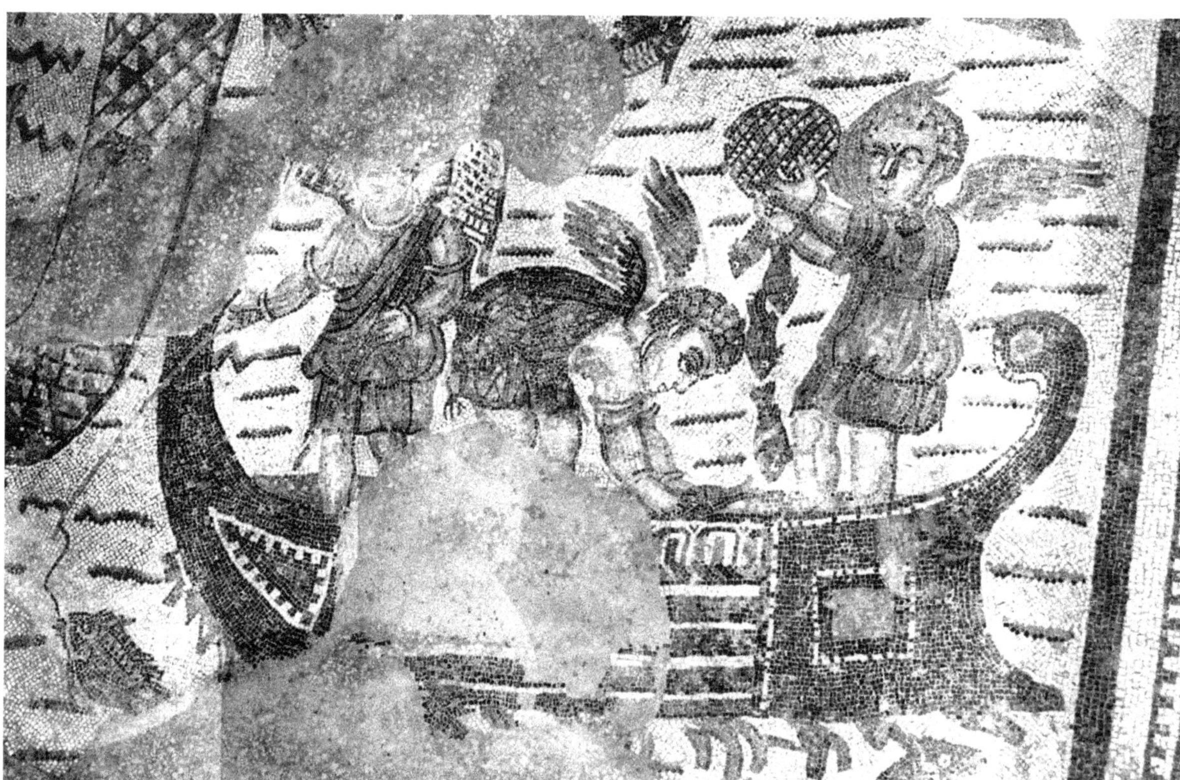

Fig. 3.8.10: Boat 3 – Room 29

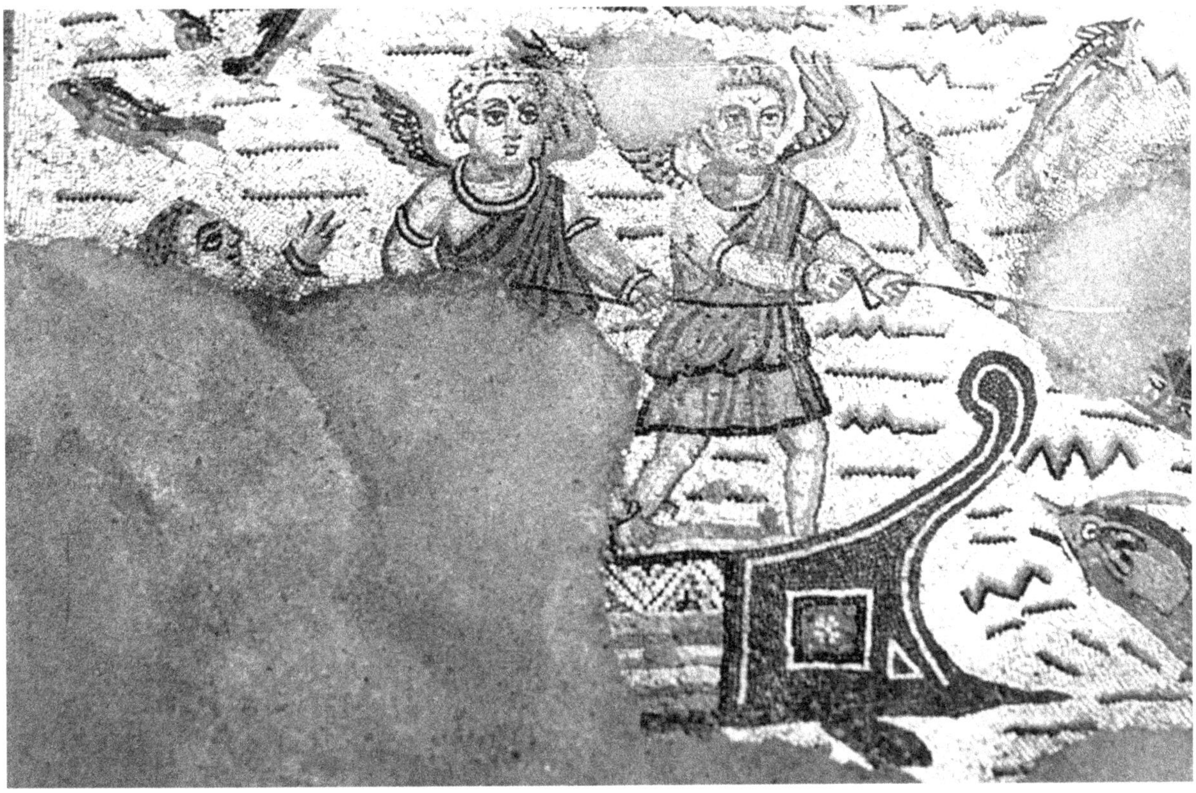

Fig. 3.8.11: Boat 4 – Room 29

while the putti in Boat 4 pull the lower end. The fishnet is the only connecting element in the fishing scenes and probably was meant to give some dynamics to the scene.

Semi-circle Atrium and the Boats

This part of the complex belongs to the private quarters (fig. 3.8.1/*40b*). The *atrium* comprises two sections:[28]

1. The open space (*impluvium*) is paved with slabs of local limestone. It served both as a collector of rainwater and as a small *nymphaeum*.[29] The *atrium* was planned as a *tetrastyle* in a semi-circular portico. Ionic capitals top the marble columns. Access to the Great Hunt Hall is by two doors (fig. 3.8.1/*36b*).
2. The portico is paved with a mosaic depicting fishing scenes in a harbor faced by arched colonnades that follow the shape of the shore, while the fishing is carried out by putti from boats (fig. 3.8.13). The water is populated with a rich fauna of fish, eels, catfish and dolphins. Besides the putti fishing from the boats there are other figures swimming and playing with ducks in the water.

This scene comprises two registers. The waterfront colonnade forms the upper part of the scene, with rectangular and rounded structures set between them. Within the maritime scene there are six small fishing boats,

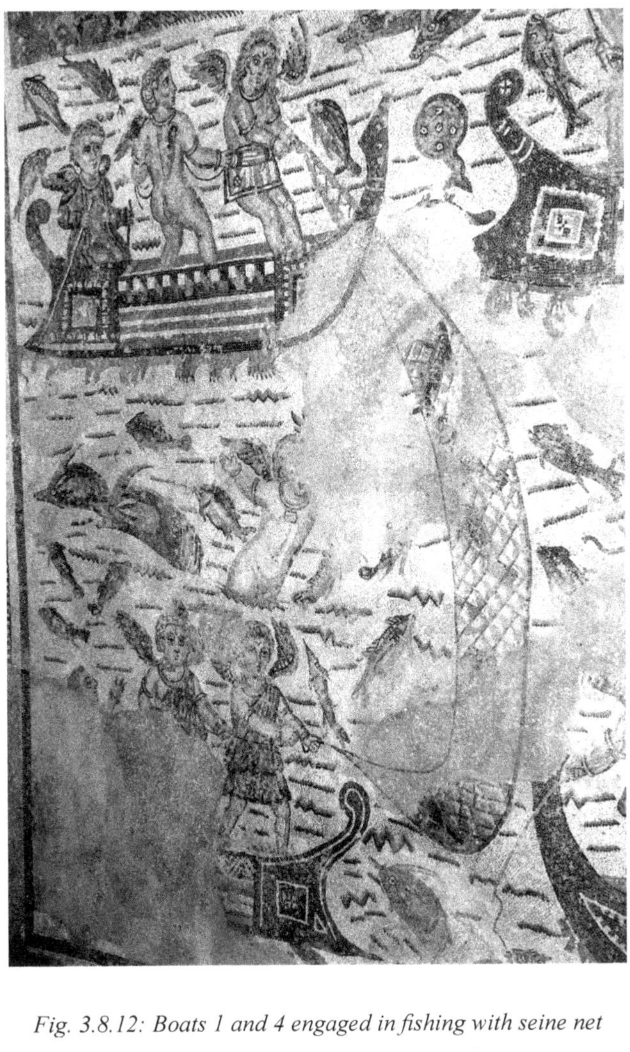

Fig. 3.8.12: Boats 1 and 4 engaged in fishing with seine net

[28] The dimensions of the atrium are: maximum width on the E-W axis is 7.9m; length on the N-S axis is 11.4m and the width of the mosaic band is 3m; Carandini, de Vos, 1982, p. 245.
[29] Ciurca and Bologna, p. 85.

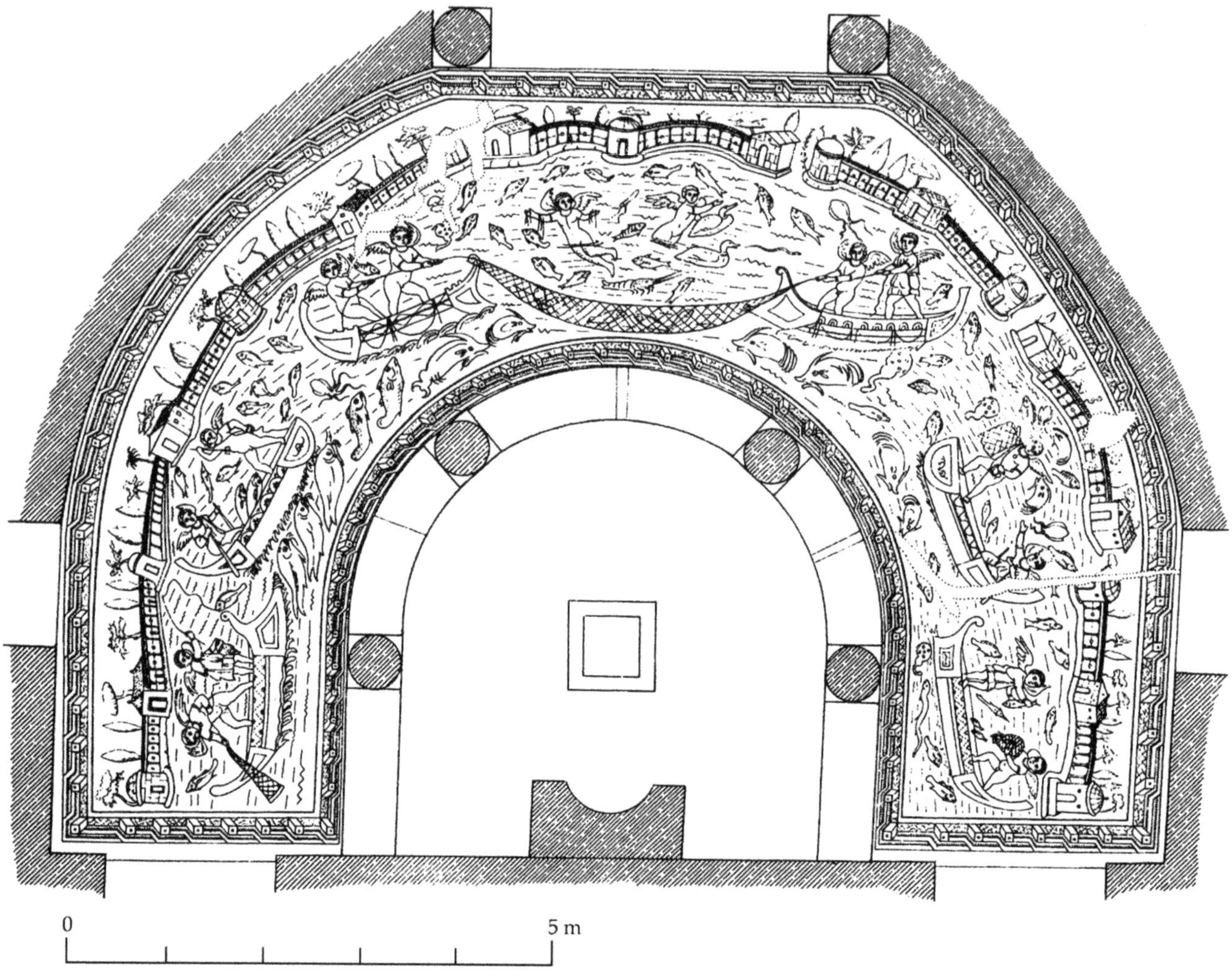

Fig. 3.8.13: Plan of the atrium *with the fishing mosaic floor*

each occupied by two putti engaged in different fishing activities. All the figures have blond curly hair and a V mark on the forehead, above the bridge of the nose. Several of them are naked and only five are dressed with a short tunic. They all have wings beneath their shoulders and are adorned with golden rings around their wrists, arms, necks and feet. Four boats are similar in shape and decoration to the boats depicted in the *frigidarium* (fig. 3.8.2) and Room 29 (fig. 3.8.7). Two of the boats have different stems that will be discussed bellow. The boats are described from left to right, in counter-clockwise order and following the shape of the *atrium*. The ripples beneath the bottom of the boats suggest that the fishing is carried out from the sailing vessels.

Boat 1: The prow points to the right. The concave stem is finished with a projecting pointed cutwater with its tip submerged beneath the rippling waves (fig. 3.8.14). The outward angled stempost is adorned with an inward-turned rounded head or volute. The center of this head is decorated with a star-like pattern. The rounded upraised stern is surmounted by a vertical sternpost adorned by a fishtail tip. The decoration of the upper plank with zigzag *acanthus* branches is similar to Boat 2 in Room 29 (fig. 3.8.9). The deck is just beneath the gunwale, as deduced from the hidden feet of both putti. The standing figure on the prow, facing the stem, fishes with a rod held in his right hand over the stempost. A fish is caught at the end of the line and is pulled to the prow. In his left hand the putto holds a conical basket, probably used to store the caught fish. The standing putto on the mid-deck, facing the quarter, pulls the upper line of a seine net from the starboard quarter (fig. 3.8.14). His body leans forward, thus putting the effort to pull the line of the net in consequent movement of his hands on the line. The body weight is sustained by the forward bent right knee and the backward stretched left leg. Rippling waves depicted beneath the bottom of the boat add some movement to the scene, thus illustrating that pulling the heavy net causes some instability to the boat.

Boat 2: The stem of this boat is different from all the vessels presented above. It has a transom prow, pointed to the right (fig. 3.8.15). The transom board is depicted with reddish-purple tesserae. In the middle is inscribed a reversed white semicircular frame, following the shape of the transom board. The head of Oceanus or other sea god is depicted

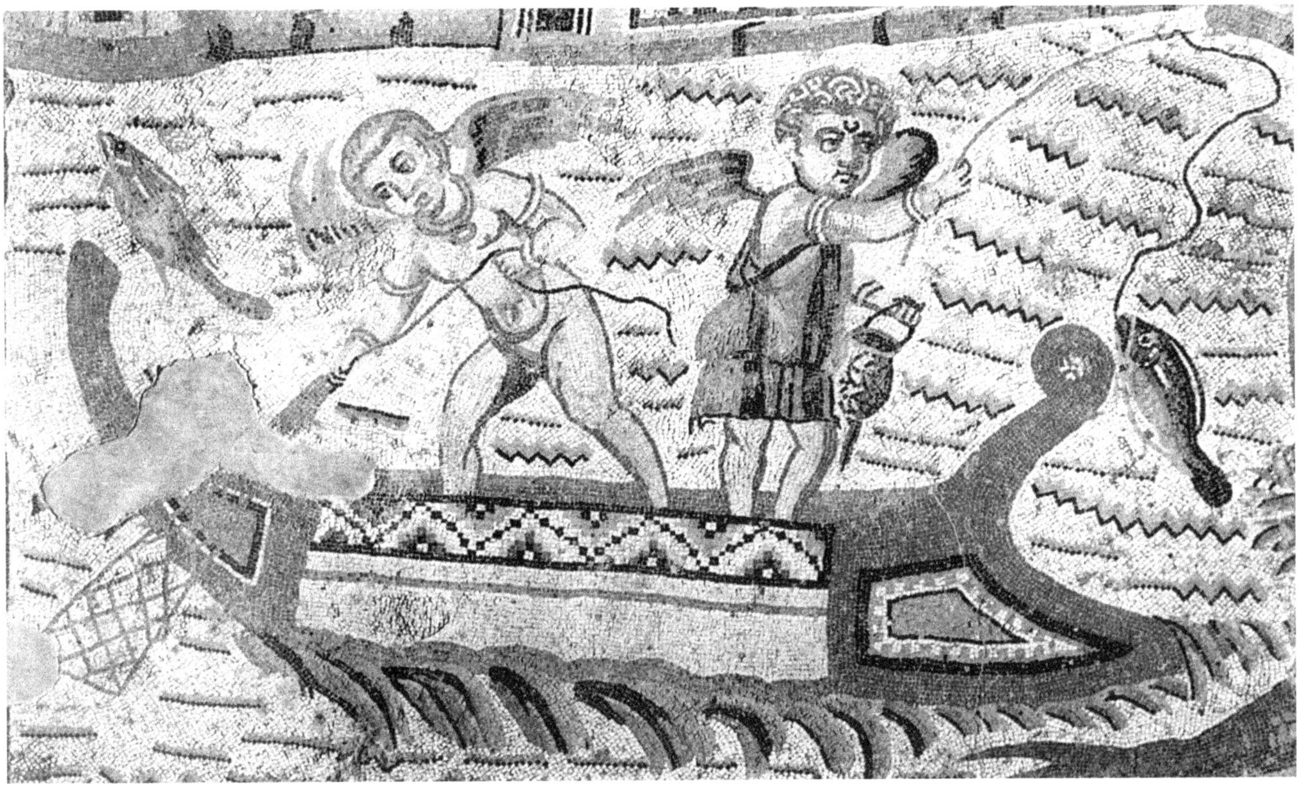

Fig. 3.8.14: Boat 1 - atrium

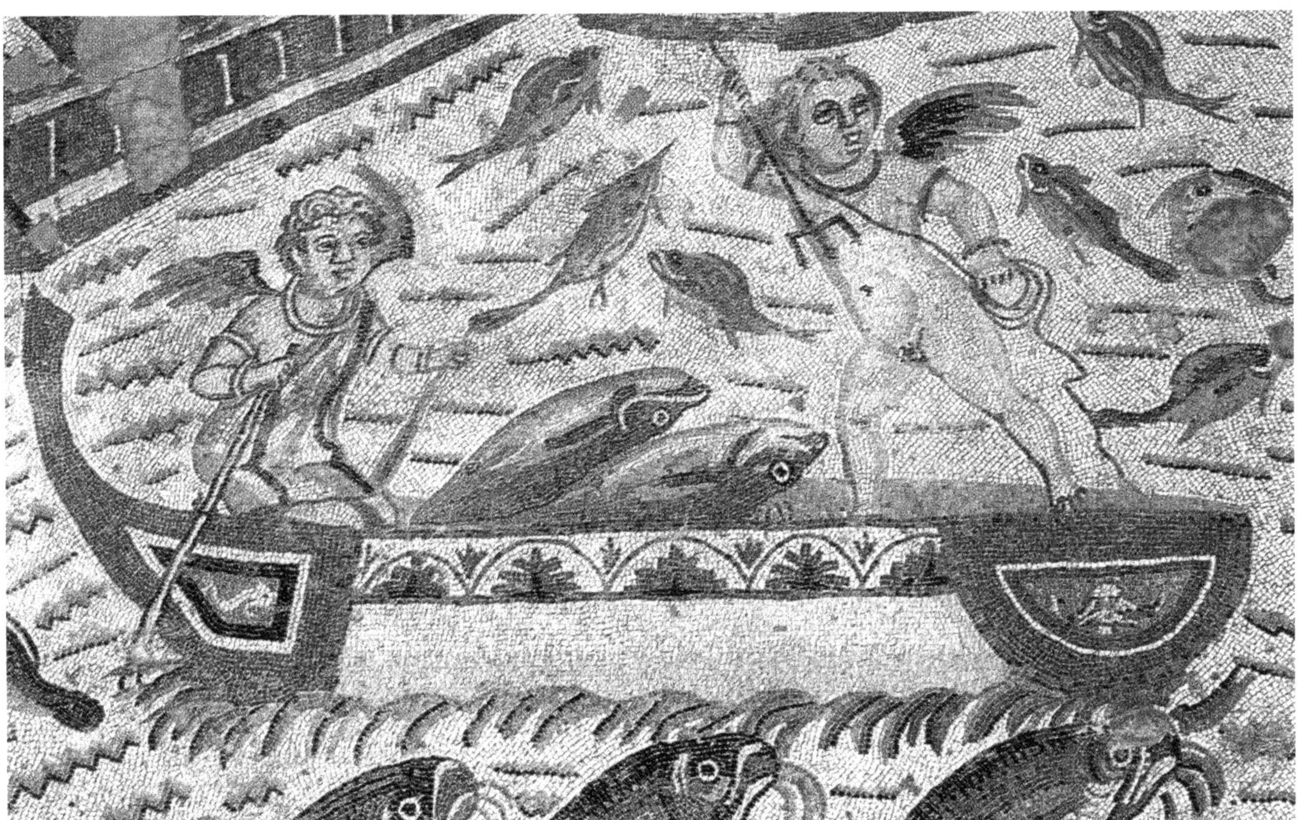

Fig. 3.8.15: Boat 2 - atrium

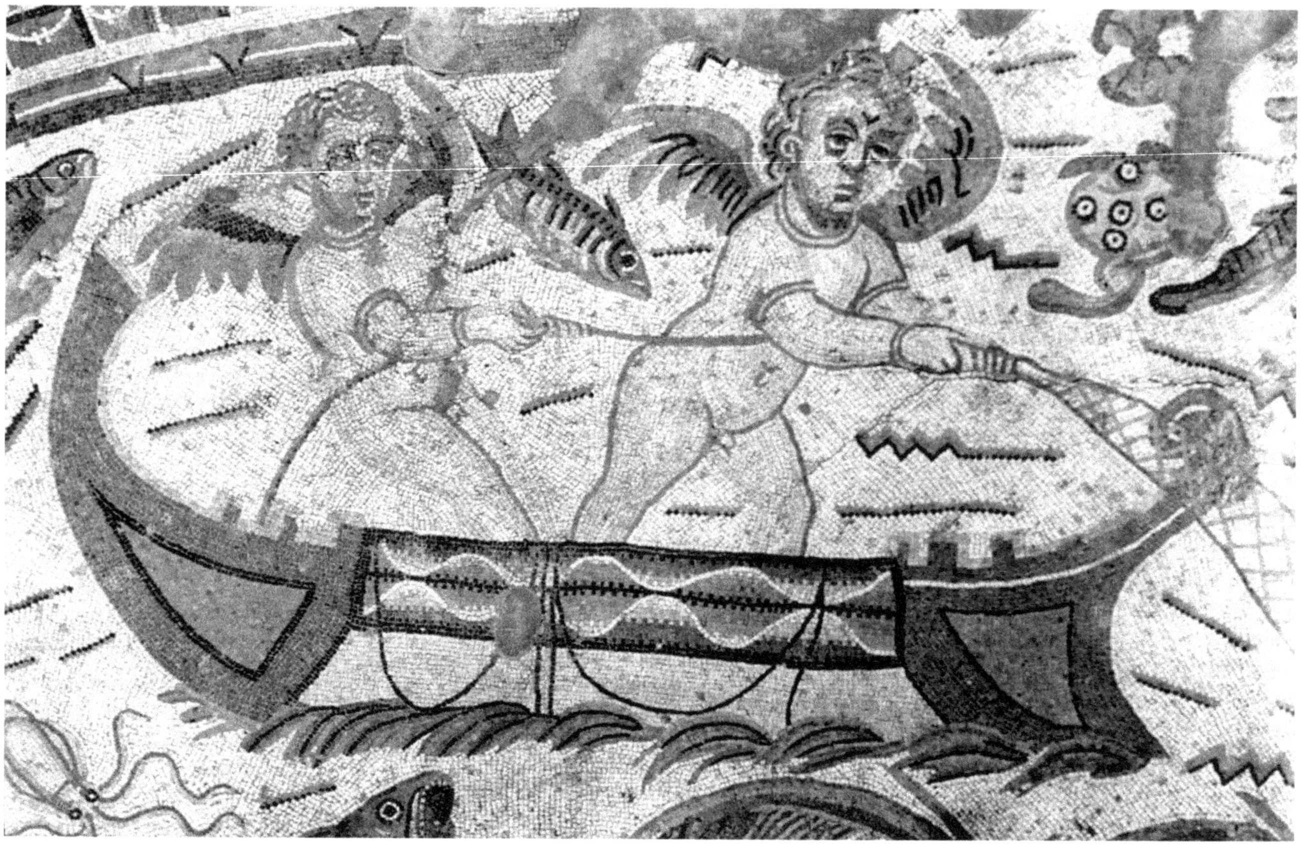

Fig. 3.8.16: Boat 3 – atrium

within the frame. The rounded upraised stern rendered with reddish-purple stones is surmounted by a vertical sternpost adorned by a fishtail tip. Within the trapezoidal frame on the stern is depicted a white dolphin with his head pointing forward. The starboard upper plank of this boat is decorated with a similar arched *acanthus* leaves as the plank of Boat 2 in the *frigidarium* (figs. 3.8.4, 3.8.4a). The deck seems to be at the gunwale level with a rectangular opening on the quarter. The feet of the seated helmsman are resting on the floor of the boat, deduced from the projecting knees above the gunwale. He works the looms of the oars mounted on the quarters by ring rowlock/thongs; the blades are submerged in the water. The helmsman steers the boat to stabilize it, while the putto on the prow, facing forward, is in the act of throwing a trident harpoon with his right hand and holding the looped securing line of the harpoon in his left hand (fig. 3.8.15). Two very large fish probably tuna turned on their backs lay on the deck. Three similar large fish are swimming beneath the boat.

Boat 3: The bow points to the right (fig. 3.8.16). The hull of this boat is similar to that of Boat 1 (fig. 3.8.14). The concave stem is finished with a short projecting pointed cutwater. An inward-turned small volute adorns the forward curved stempost. A white strip emphases the shape of the stempost and the volute adornment. On the stem and stern is depicted a green triangle frame decoration. The top wale is decorated with undulating waves in perspective, as seen in a mirror image. The horizontal axial black strip emphasizes the mirror effect (fig. 3.8.16). Two bitts project above the fore starboard gunwale and one above the port side. Two other bitts are depicted on the starboard and port quarter gunwale. To add some perspective to the bitts, the mosaicist gave them depth by a strip of pink tesserae (fig. 3.8.16). On the starboard side of the hull are depicted two large black loops with their ends forming vertical lines. They may represent stylized strops usually used to secure the oars to the gunwale. The boat lacks oars and therefore this pattern is more like a decoration. On the mid-deck, facing forward are two standing putti pulling the line of a seine net (fig. 3.8.16). The inclination of their bodies show the strain put in pulling the line of the net full with fish.

Boat 4: The bow points to the left. The concave stem is finished with a projecting pointed cutwater (fig. 3.8.17). A small outward-turned volute adorns the forward projecting stempost. A strip of white tesserae emphasizes the shape of the prow and the stempost. The rounded upraised stern is surmounted by a vertical sternpost finished by a fishtail tip. The port quarter and the sternpost are outlined with a white strip, similar to the prow. A white dolphin with its head pointing forward is depicted within a triangular frame on the port stern (fig. 3.8.17). The top plank is decorated with eight thin black arches. Within the arches are inscribed small-intercalated green or red squares. This pattern seems to indicate stylized oarports. Two yellow bands represent the lower strakes and the dark green strips between them may represent the seams. On the port side of the hull are depicted three large loops made with one row of black tesserae. At the joints of the loops are two vertical black

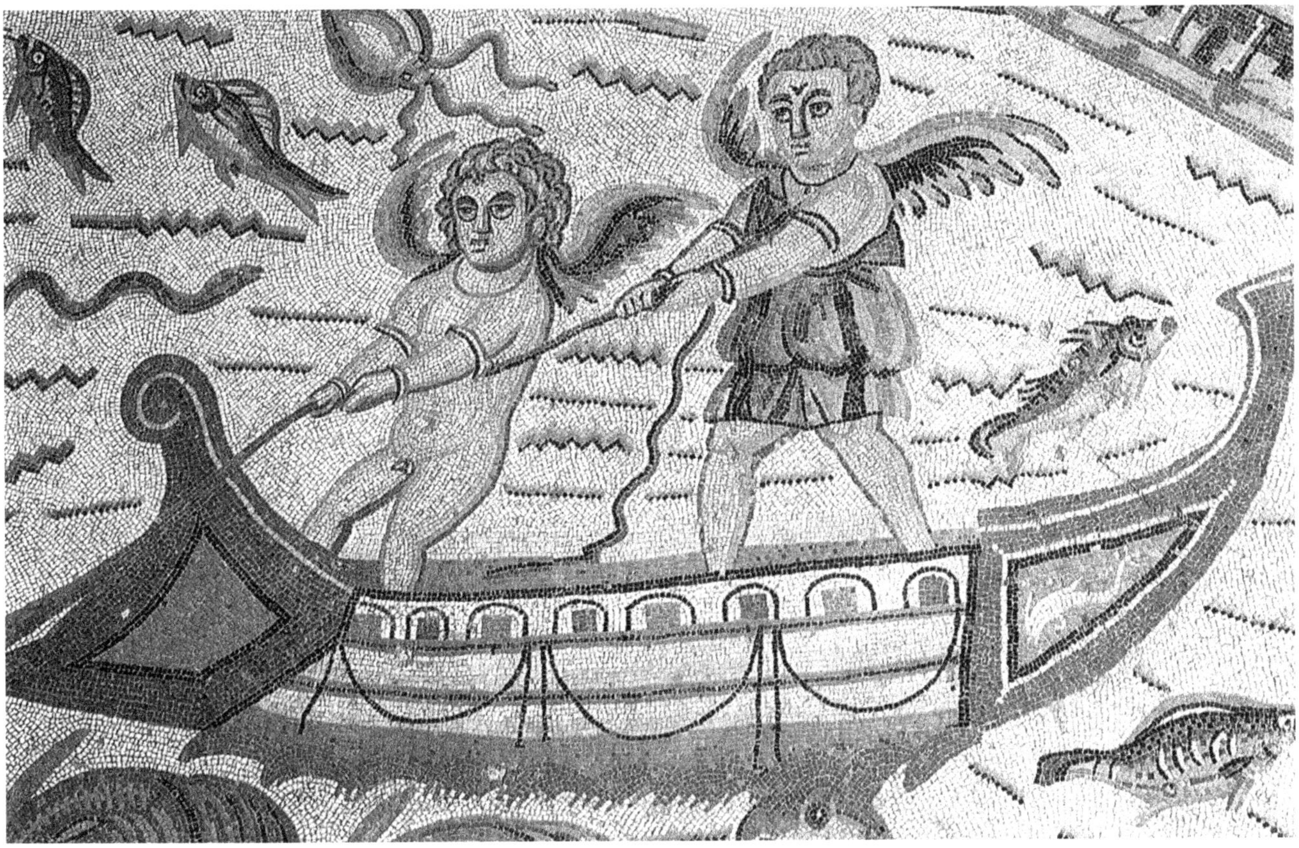

Fig. 3.8.17: Boat 4 - atrium

strips. This pattern is similar to Boat 3 (fig. 3.8.16), which may represent stylized strops used to mount and secure the oars to the gunwale. There are no oars in the boat and therefore these loops are more like decorations. The shallow deck beneath the gunwale is deduced from the hidden feet of both putti. They stand on the deck, facing forward and pulling the other line of the seine net full with fish (fig. 3.8.17).

Boat 5:[30] This boat is identical to Boat 2 (fig. 3.8.15), but in reverse, with its transom prow pointed to the left (fig. 3.8.18). Within the semi-circular frame that follows the shape of the transom board is depicted a white hippocampus. Due to the photo montage the boat is slightly distorted. The top strake is decorated with short undulating waves in a perspective view. The rounded upraised stern is surmounted by a vertical sternpost adorned by a fishtail tip. The deck is at the gunwale level with a rectangular opening in the quarter (fig. 3.8.18). The standing putto on the prow, depicted in almost full face, looking over the port stem is in the action of throwing a cast net laid on his right arm; the securing line of the net is looped in his left hand (fig. 3.8.18). The feet of the second putto seated on the quarter go through the square opening and rest on the bottom of the boat. He holds the loom of the port oar while the starboard oar hangs free on the quarter.

[30] This boat was studied from the black and white picture produced in Carandini *et al.*, Pl. XXXVII.

Boar 6:[31] The bow points to the left. The hull of this boat is somewhat longer and slimmer than the boats described above (fig. 3.8.19). The concave stem is finished with a projecting cutwater whose tip is submerged beneath the rippled waves. The forward projecting stempost is adorned with an inward-turned volute. The top plank is decorated with a zigzag pattern of *acanthus* leaves similar to that of Boats 2 and 4 in Room 29 (figs. 3.8.9, 3.8.11). Within the trapezoidal frame on the port stem is inscribed a white dolphin with its head projecting forward. The rounded stern is surmounted by a high vertical sternpost adorned by a fishtail tip. The deck is at the gunwale level. The putto standing on the fore deck facing the stern holds a fish by its tail. The second putto *en face* empties the fish on to the deck from a rounded basket trap (fig. 3.8.19).

The Great Hunt Hall

This section of the villa's complex comprises a 60m long corridor with an apse at either end (fig. 3.8.1/*36b*). The mosaic floor depicts complex scenes of capturing wild beasts in North Africa, Egypt and India, and followed by their embarkation in merchant ships to be transported to Rome for the games held at the Colosseum and Circus Maximus. The scenes depict a great range of wild animals, such as a bull, an elephant, a hippopotamus, a rhinoceros, leopards, tigers, horses, ostriches, etc. that are driven into

[31] Ibid., Pl. XXXVIII. The atrium is closed to the public due to the poor condition of the mosaics. Since 2007 the mosaic undergoes restorations.

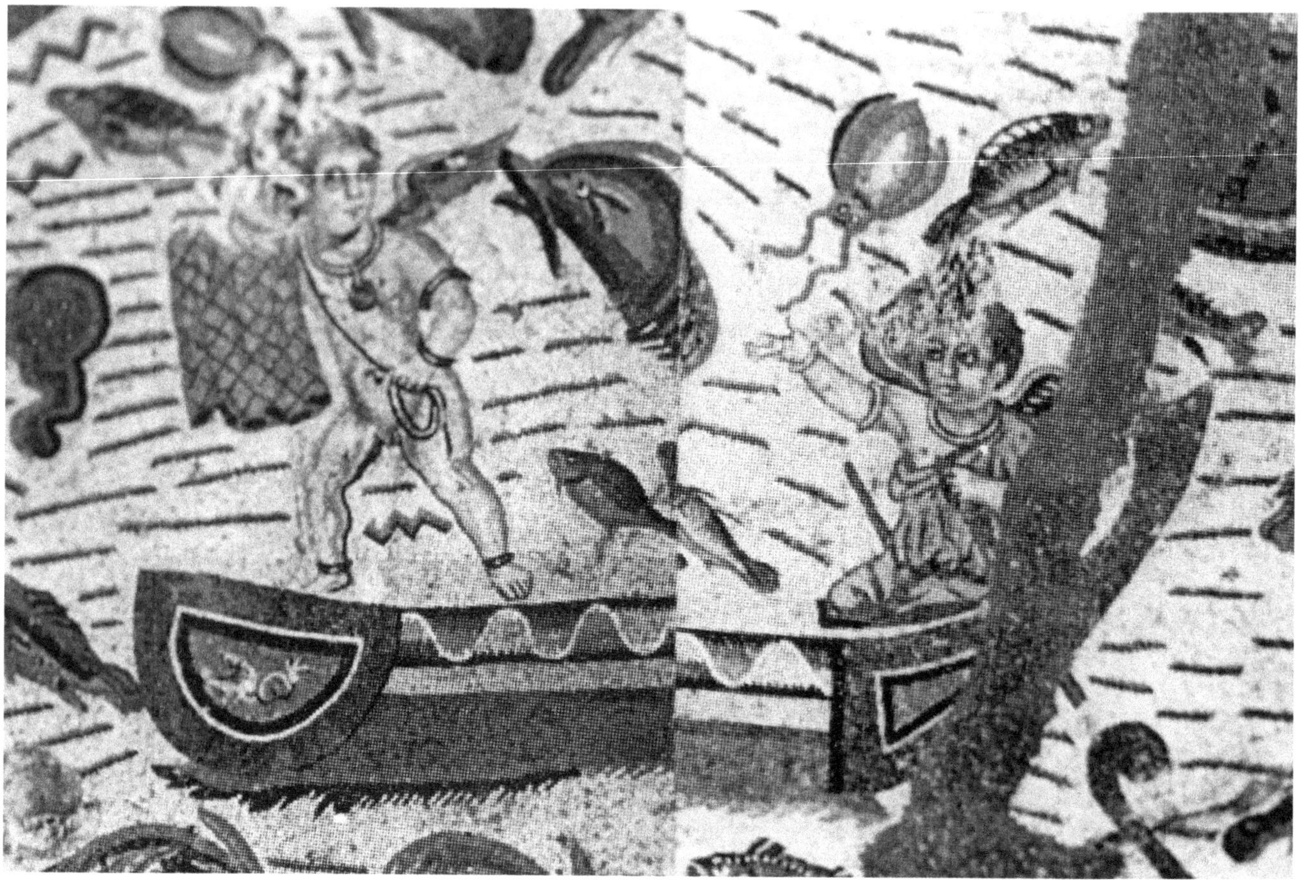

Fig. 3.8.18: Boat 5 - atrium

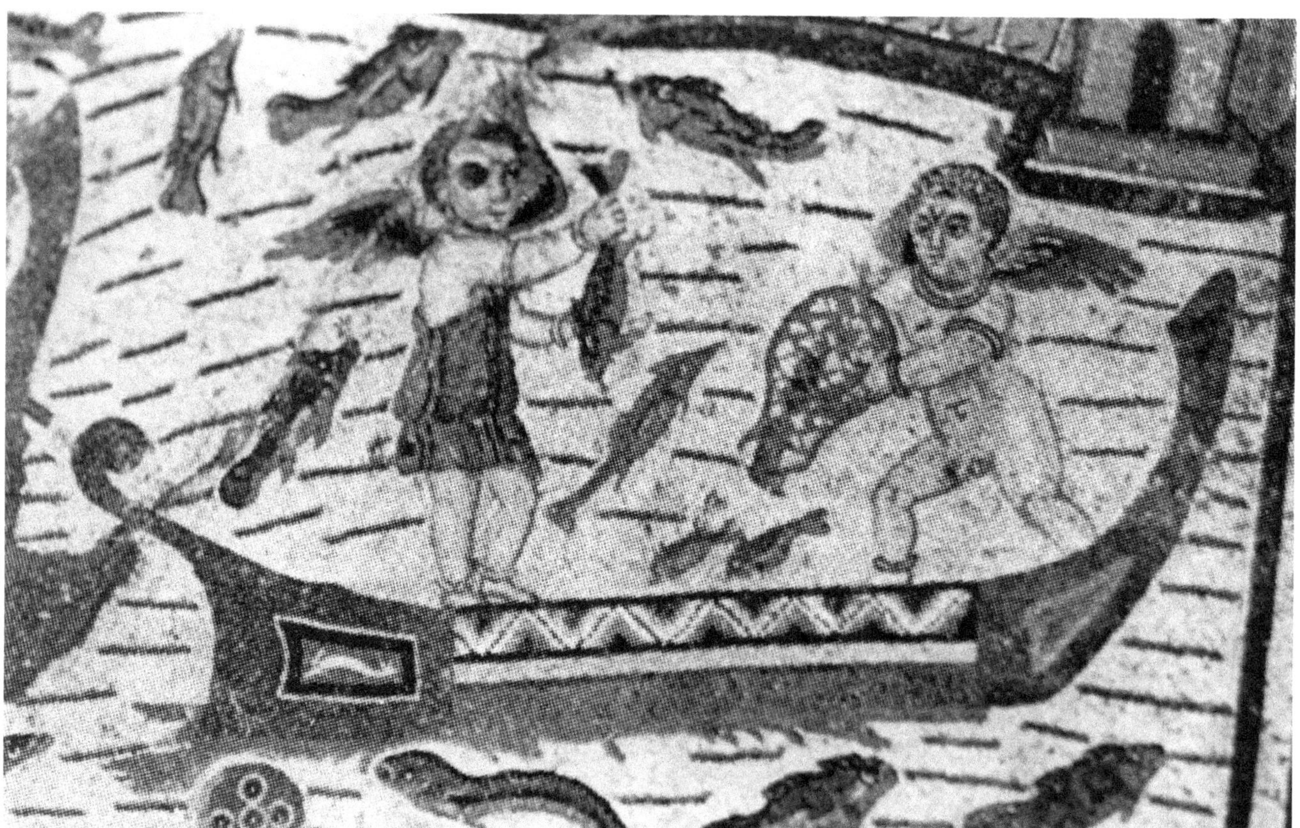

Fig. 3.8.19: Boat 6 - atrium

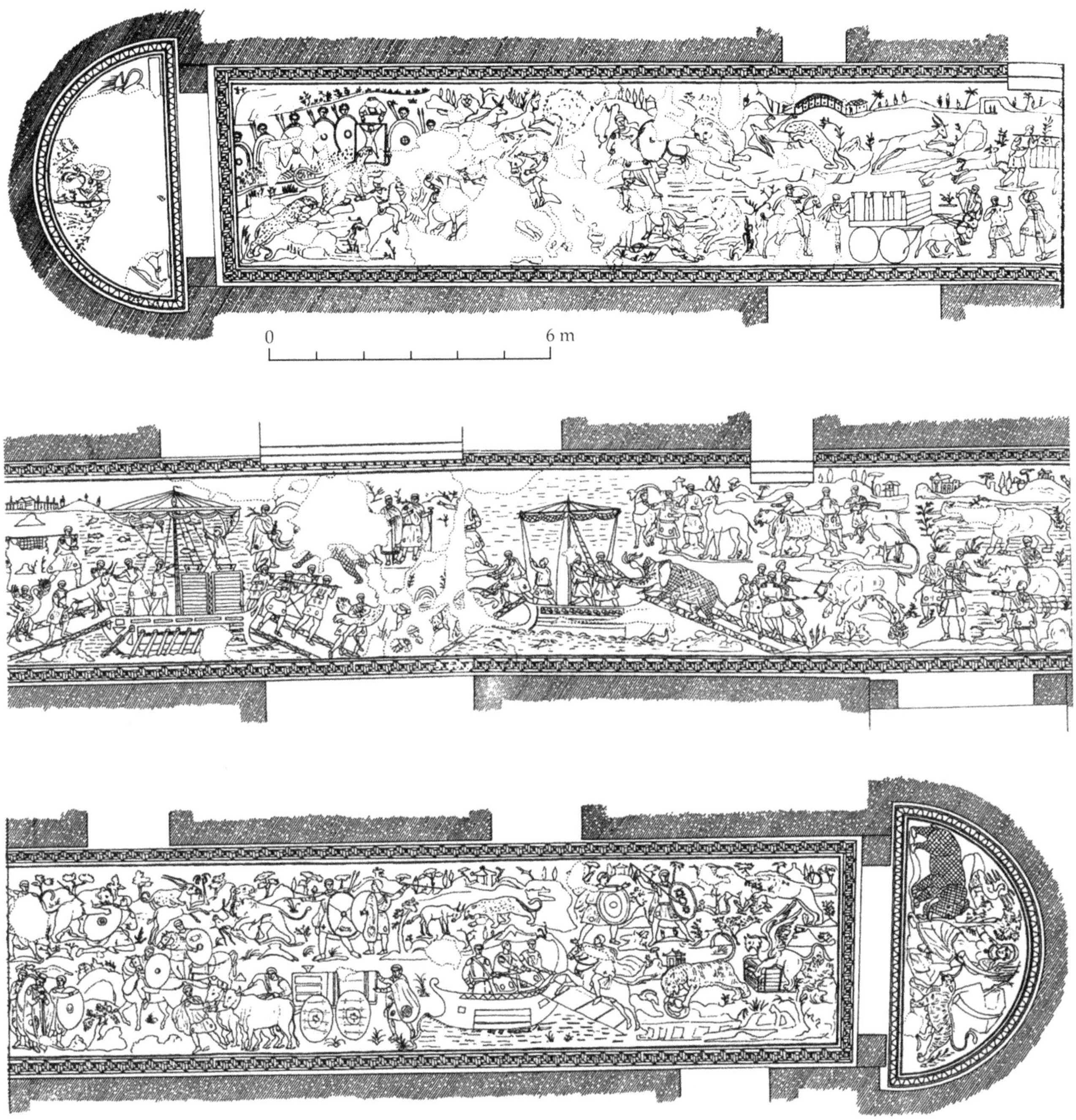

Fig. 3.8.20: Plan of the Great Hunt mosaic

captivity by different methods.[32] This mosaic known as the Great Hunt Hall is a long narrative strip divided into three distinct parts (fig. 3.8.20):

1. In the left part are represented the five provinces from the Dioceses of Africa: Mauritania, Numidia, Tripolitania, Proconsulare and Byzacene.[33]
2. The central part shows two sailing ships being loaded with the captured animals on one side and simultaneously unloading them on the other side at the destination port. This scene is a very sophisticated concept, summarizing the sea voyage from its start in Africa and ending in the home port of Ostia (fig. 3.8.20, the middle strip).
3. The right side depicts the hunting of exotic animals in India and Egypt, their driving to the merchantmen anchored in the harbors at Alexandria and Carthage, to be transported to Ostia or Rome (fig. 3.8.20, the lower strip). The third ship, minus its sailing or rowing rig seems to transport soldiers and horses to supervise the capturing and transportation of the animals.

[32] For detailed descriptions of the hunting scenes, see: Wilson, p. 24.
[33] Ciurca and Bologna, p. 49.

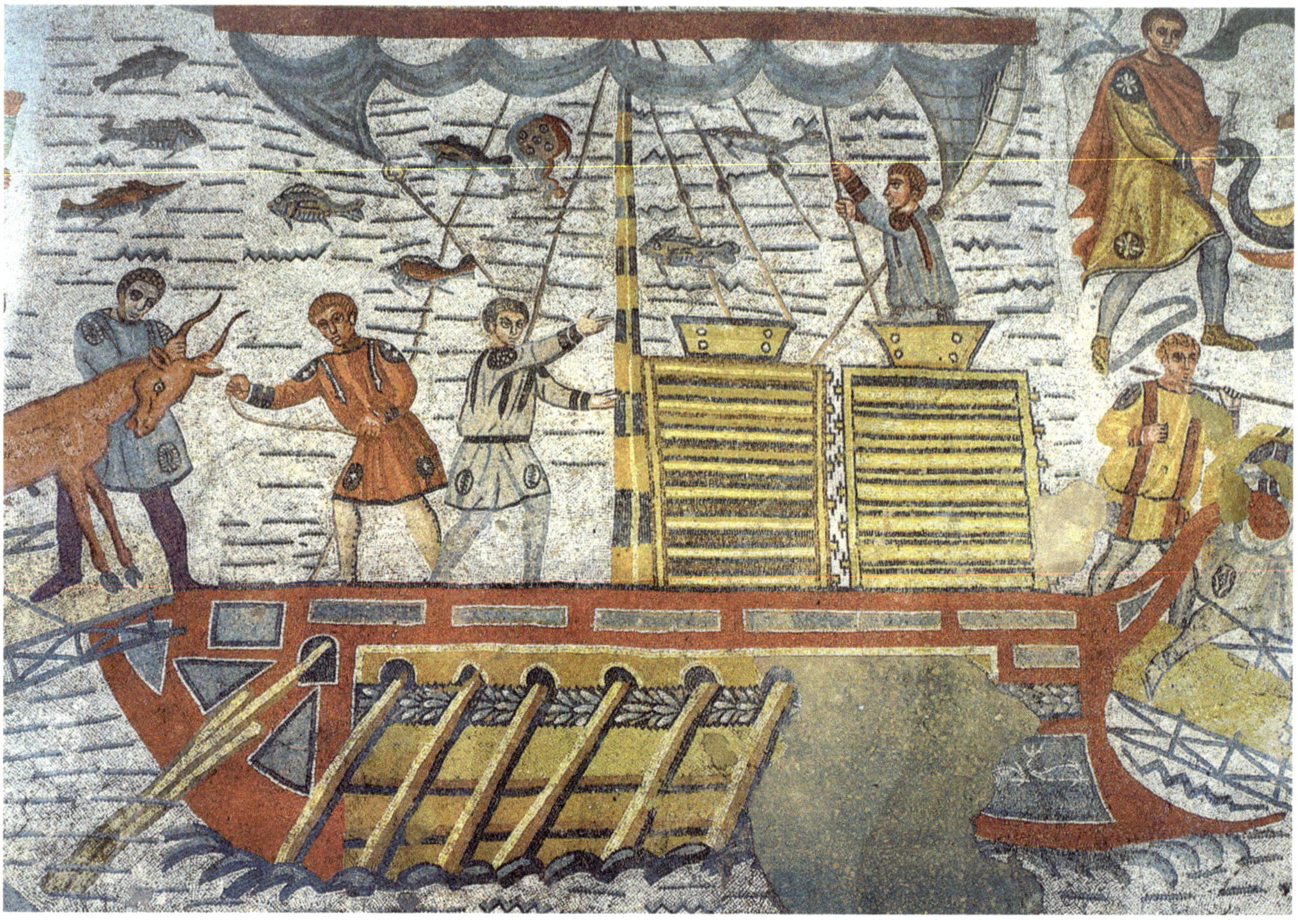

Fig. 3.8.21: Sailing Ship 1

Sailing Ship 1 (left-hand): The prow points to the right. This large merchant ship was adjusted to transport exotic animals caught in the Roman North African provinces (fig. 3.8.21). The ship is probably anchored in the port of Carthage and the animals were brought to be embarked there. The hull is very high and broad, probably comprising at least two decks. The background paint of the hull is red where the strakes and other decorations are applied (fig. 3.8.21). The concave stem is finished with a projecting pointed cutwater above the ripple waves, thus emphasizing the sea environment. The forward angled stempost is adorned with an outward-turned volute. In the surviving rectangular frame depicted on the starboard stem are rendered two white dolphins following each other with their heads pointed forward. The upraised broad stern finishes at the gunwale level. Five thin rectangular bands made with blue tesserae are depicted on the top plank. They may symbolically indicate the projecting ends of the through beams that supported the top deck. On the upper yellow strake are depicted six circular oarports that the shafts of the row-oars projecting through. The fore part of the hull is damaged and the oars and oarports are missing. Most probably three more oarports and oars were represented on this section. Beneath the oarports, along the starboard hull is rendered a band with three rows braded laurels leaves (fig. 3.8.21). On the fore deck are depicted two high wooden rectangular containers, made with horizontal planks joined by "dovetail" tenons. On top of each container is a reversed trapezoidal rim, indicated by the short planks that their angled edges are locked with wooden treenails or bronze nails. They probably represent the cages used for the captured animals to be transported. Similar containers may have been in the hold of the ship, or the animals may have been caged in compartments divided by wooden walls.

The rigging comprises one mast, yard, the furled sail with several lines, six (or nine) oars on the starboard side and one large rudder. The tall, tapered pole projecting above the deck is the composite mast[34] stepped amidships. It is depicted with alternate yellow and black bands (fig. 3.8.21). The white strip alongside the face of the mast represents the halyard worked by the man standing closely to the left side of the mast. He appears to rise the yard and the furled sail, thus to prepare the vessel for its departure. The mast is secured in place by one backstay (the extreme left line stretches from the masthead to the back of the standing figure on the quarter to his right foot). The second line stretching from the *parrel* or sheave-block to the left shoulder of the standing man working the halyard probably represents the starboard shroud. On the right side of the mast are shown three lines with a block or deadeye attached to their upper part (fig. 3.8.22). The right-hand line (extreme)

[34] See Glossary.

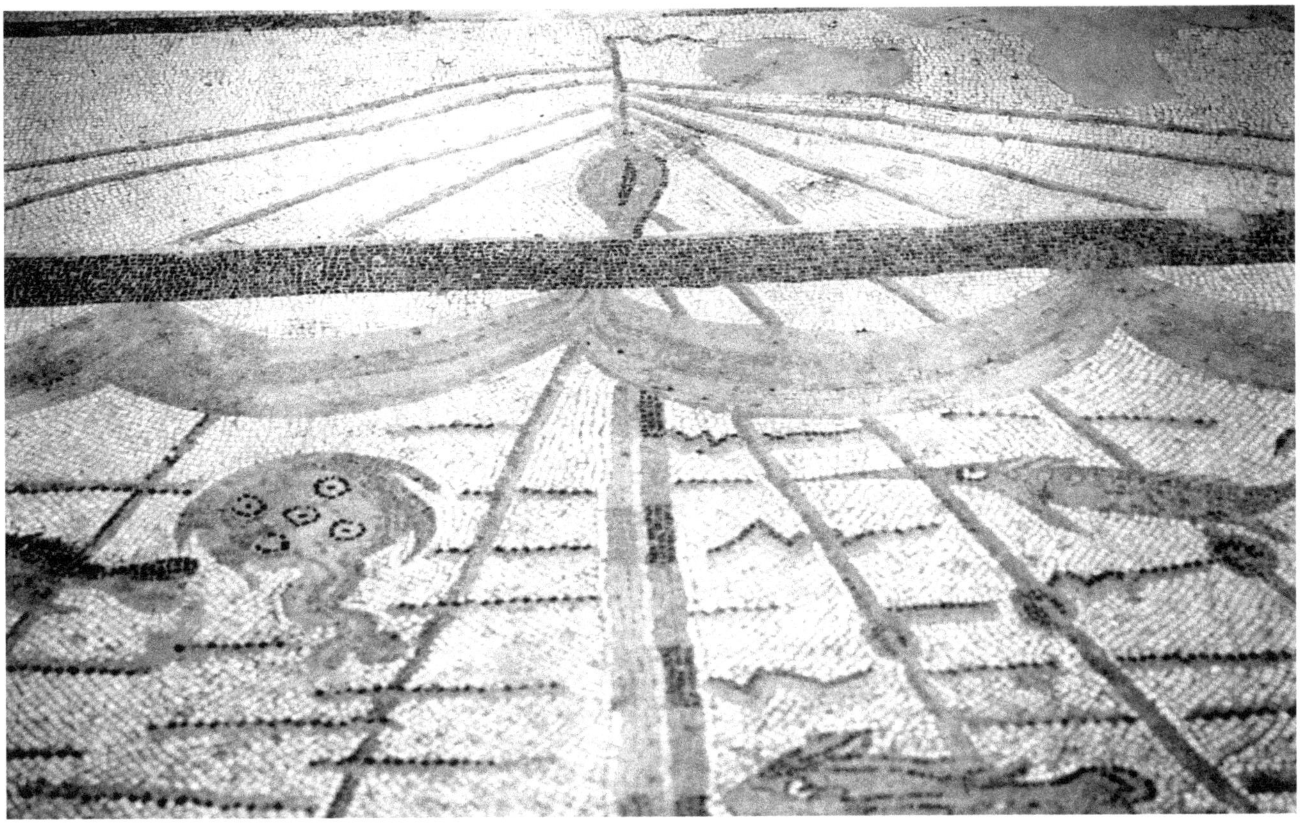

Fig. 3.8.22: The parrel/block, masthead and lifts

stretching from the masthead towards the right leg of the man standing on the fore-container is probably the forestay. The other two lines stretching from the *parrel*/block to the edge of the second container may represent the double-lines port shroud that probably their lower ends were tight around the bitts on the port stem gunwale. Apparently the mosaicist did not stretch the lines to the quarter and the stem because he would have crossed the figures and therefore he desired to keep a clear view of men and their work (fig. 3.8.21). The *parrel*/block and the masthead project above the yard. The length of the yard almost equals the length of the ship. It is depicted with purple tesserae. The yard is secured by four lifts stretching from the masthead to the port yardarm and five lifts to the starboard yardarm fig. 3.8.22). The furled sail beneath the yard is tight by robands.[35] The line hanging from the port clew to the head of the figure close to the mast and working the halyard may represents the port sheet. The line stretching from the starboard clew behind the back of the man standing on the container is the starboard sheet (fig. 3.8.22). Apparently each sheet was passed through a deadeye or a wooden fairlead (figs. 3.8.21, 4.24).[36] The deadeyes attached to the standing rigging provided better and easier maneuverability of the tackle lines. The man standing on the fore container works one of the brails. Both edges of the fore bunt and their leeches are turned backwards. On either face are depicted three reinforcing-bands (fig. 3.8.23).

Besides its sailing gear, the ship is equipped with six (or nine) row-oars represented by their thick rounded shafts projecting through the circular starboard oarports with a backward slanting (fig. 3.8.21). The blades are submerged beneath the waves. The same arrangement of oars may was true on the port side. The shaft of the long starboard rudder projects through a square oarport with an arched top is cut into the top quarter plank. The shaft bisects longitudinally the tapered blade. Both shoulders of the wings are rounded. The upper wing seems to be higher than the lower one, probably emphasizing the slanting of the rudder, or the mosaicist wanted to give a perspective view to the rudder. A similar arrangement was true on the port side.

On the stern and the stem is set a gangplank by which animals are loaded and unloaded (fig. 3.8.24). The man standing on the quarter and holding the end of a line tight around the horns of an antelope pulls the animal from the gangplank to the ship. The man standing at the end of the gangplank helps with the loading by pushing the animal by its horns. Following in line are two ostriches walking up the gangplank and supervised by their keepers. Four men carrying a wild boar on a stretcher on their shoulders descend the gangplank placed on the port stem (fig. 3.8.24). The men working the rigging of the ship were probably the crew. Their clothing and those of the hunters point to Roman soldiers who were appointed for capturing and transporting the African animals to Rome for *venations*.

[35] Ibid.
[36] Ibid.

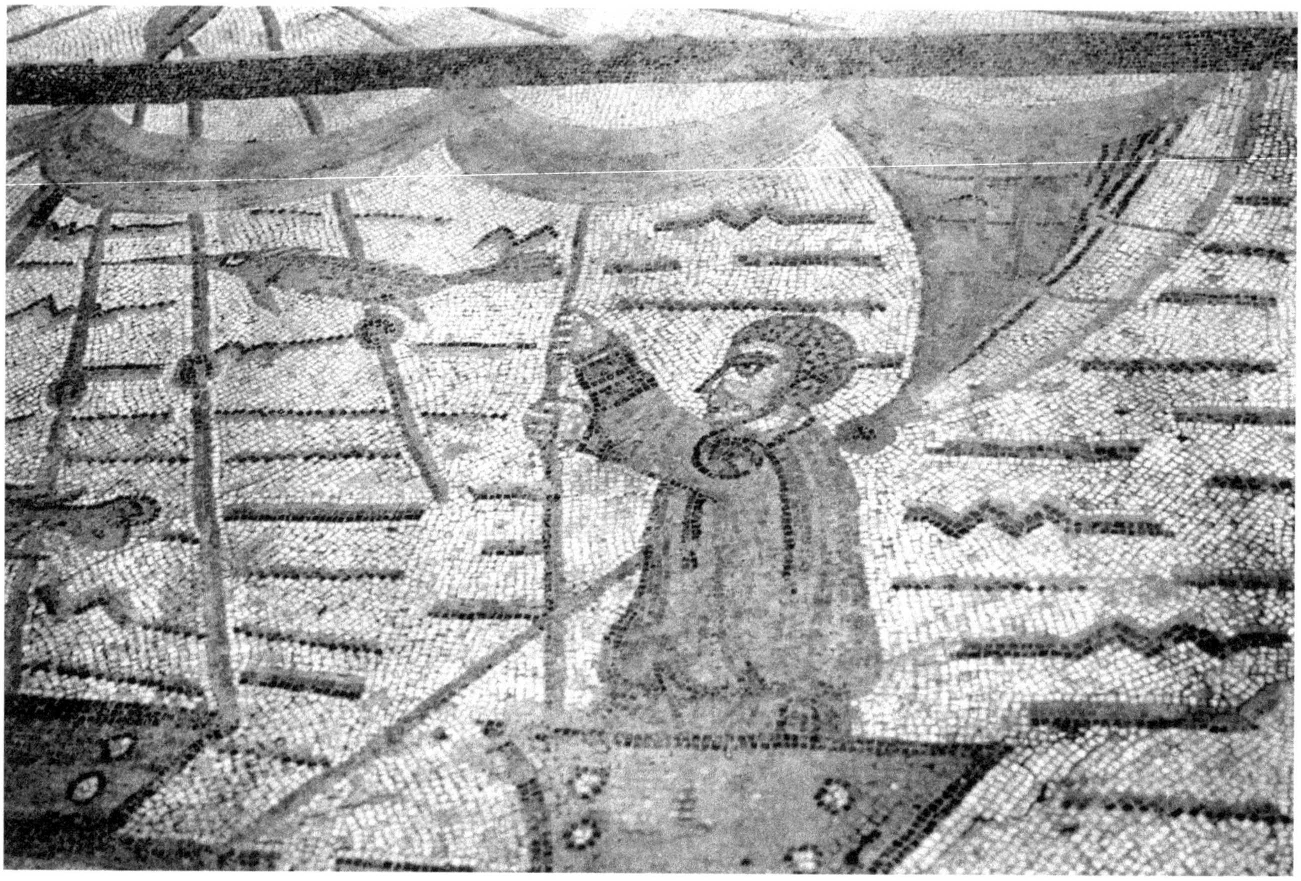

Fig. 3.8.23: Close-up of the sail's starboard corner

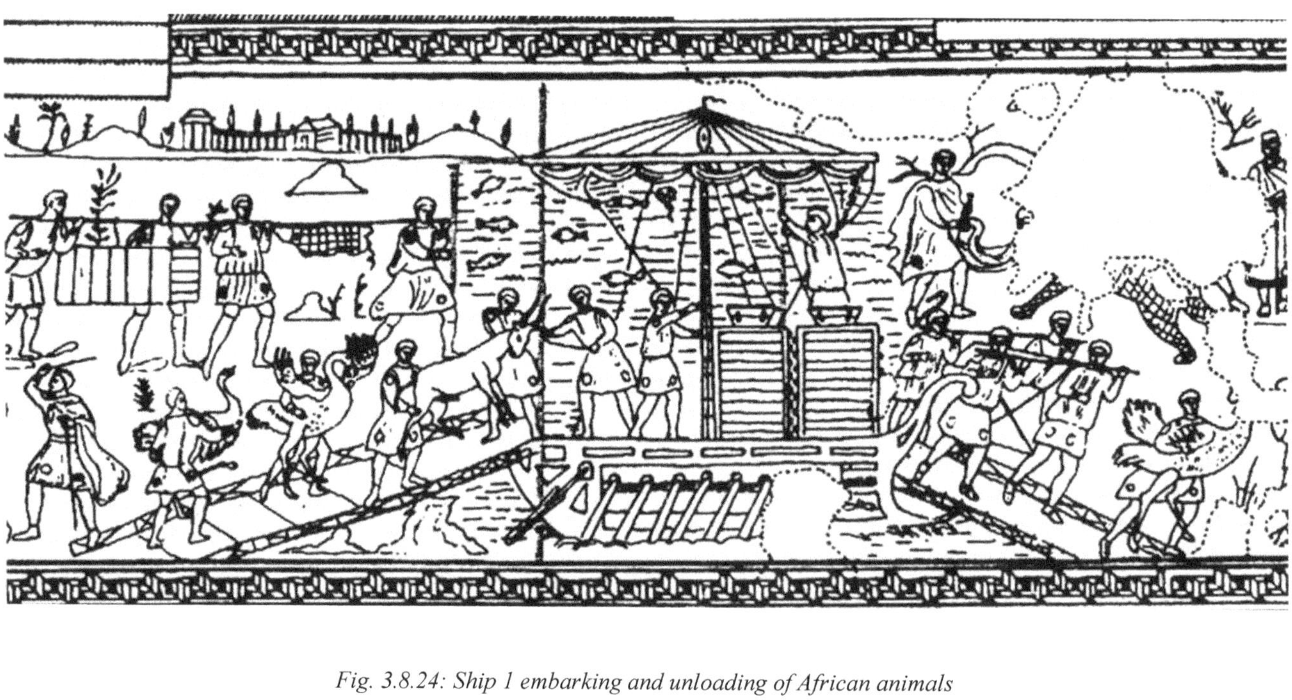

Fig. 3.8.24: Ship 1 embarking and unloading of African animals

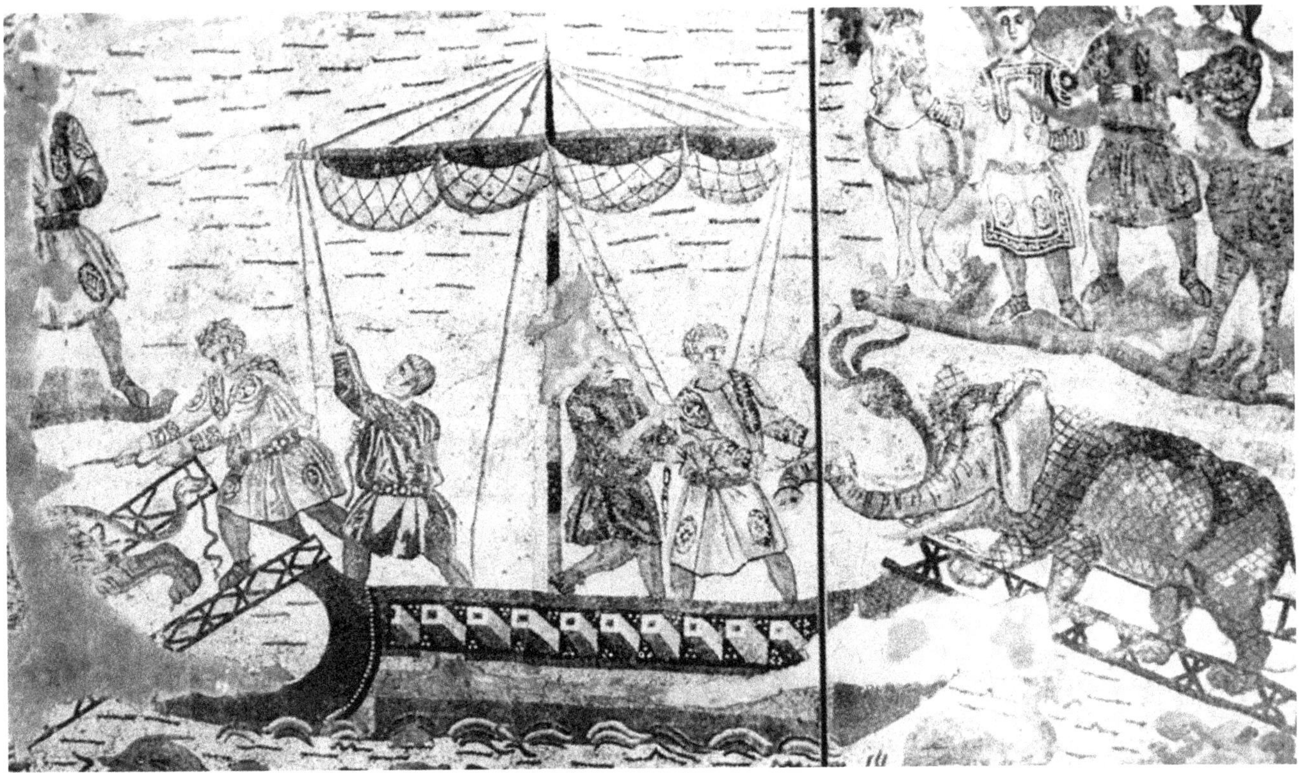

Fig. 3.8.25: Sailing Ship 2

Sailing Ship 2 (middle):[37] The hull is similar to that of Ship 1 but slightly smaller (fig. 3.8.25). The prow points to the left. The concave stem is finished with a projecting pointed cutwater that its tip, and part of the stern and port quarter are damaged. The top plank is decorated with a pattern that symbolically represents the ends of ten projecting through beams in perspective (seen from beneath), which probably suggests the supporting frame of the top deck. The gangplank set above the port stem hides the upper part of the forward-curved stempost. The upraised broad stern is surmounted by an inward-curved sternpost adorned by an *aphlaston* with a globular base and open three branches (fig. 3.8.25). A second gangplank is laid on the port quarter.

The rigging comprises one mast, yard, the furled sail and several lines. The tapered pole depicted with alternate black and yellow bands indicates a composite mast; it is similar to that of Ship 1 (fig. 3.8.21). The forestay stretches from behind the yard to the left leg of the standing man on the bow (fig. 3.8.25). The backstays seem to be represented by a rope ladder or ratlines, stretching from beneath the yard to the back of the figure standing on the quarter. He is engaged in pulling the elephant into the ship by the ends of the rope passed around his neck. The yard is made with light, brownish-gray tesserae and the lower part with purple stones. Four lifts stretching from either side of the masthead to both yardarms secure it to the mast. The masthead projects above the yard. Remain of another masthead with a small flag attached to its tip is seen above the masthead of the Sailing Ship 2 (fig. 3.8.25). Apparently this ship replaced an earlier one that may have been damaged. The sail is furled beneath the yard and it is tight with three robands. On the visible face of the bunt is depicted a checker pattern that may symbolically represent several brails and reinforcing-bands (fig. 3.8.25). The man standing on the bow works one of the double-line starboard brace. A similar double-line brace is attached at the tip of the port yardarm. The lines hang with a slight forward rake behind the back of the man on the quarter pulling the elephant from the gangplank into the ship. The figure standing near the right side of the mast aids to load of the elephant by holding the ends of the rope passed around the neck of the animal. The man walking down on the upper end of the fore gangplank unloads a leopard by holding the rope passed around the neck of the beast. This ship seems to be anchored at Alexandria where elephants and leopards were loaded and then transported to Ostia where they were unloaded to be transferred to the Colosseum or Circus Maximus at Rome. The ship lacks its steering gear. Probably the mosaicist did it intentionally when he placed the gangplank of the stern, or when this ship replaced the previous one then the steering oars were omitted.

Ship 3 (right-hand): This is the smallest of all the ships in the hunting scene (fig. 3.8.26). It is very similar in shape to Ship 2 but much smaller. The bow points to the left. Almost the entire hull is made with reddish-purple tesserae. The lower hull is decorated with one yellow and two yellowish-green strips. They may represent the lower strakes or the lead or copper sheathing to protect the hull against *teredo navalis* (ship worms) or other physical damages. The

[37] This ship is studied from the black and white picture produced in Carandini *et al.*, Pl. XXIX. Due to the position of the mosaic and the narrow passage, I could not take a proper photograph of this vessel.

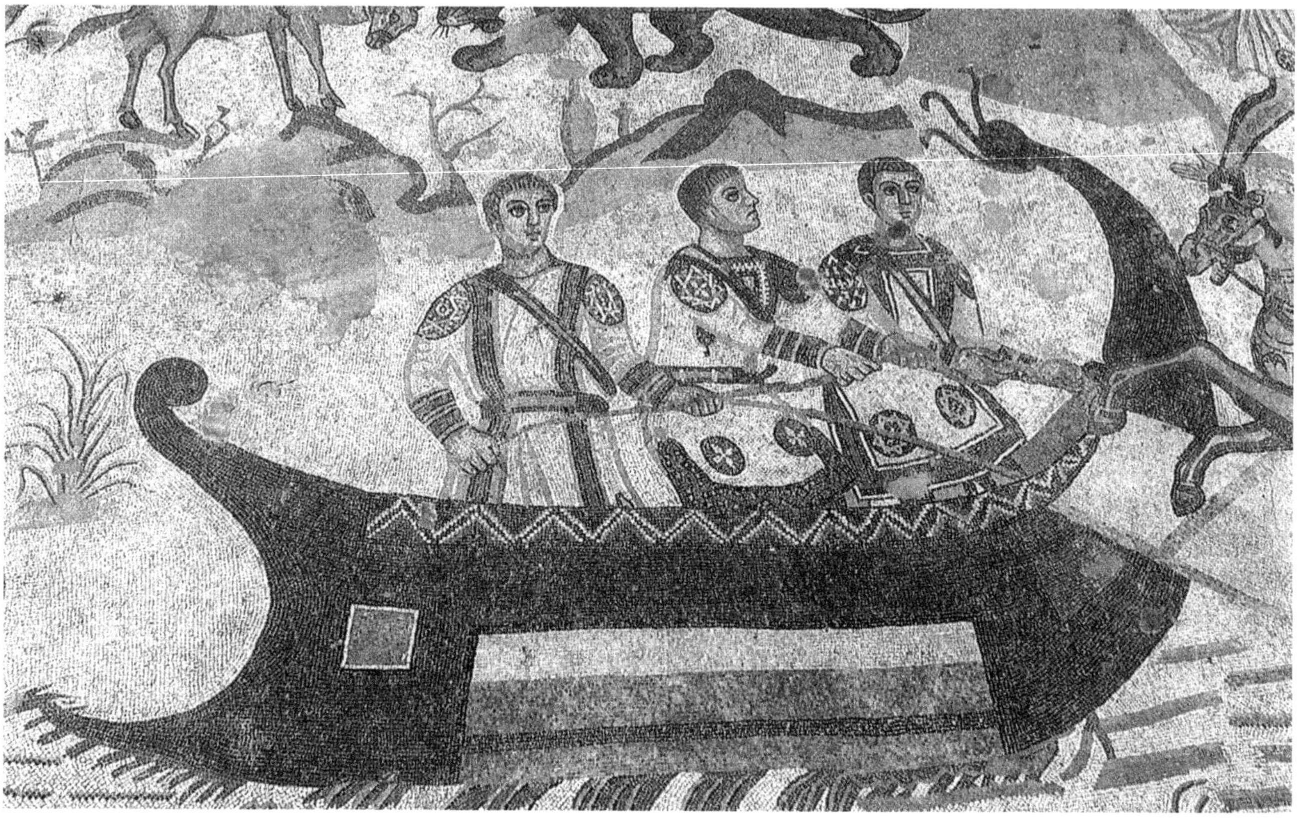

Fig. 3.8.26: Ship 3

bulwark strake is decorated with a zigzag pattern that also covers the upper port quarter. The concave stem ends in a projecting pointed cutwater partly submerged in the ripples. The forward projecting stempost is adorned with an inward-turned small volute. The upraised rounded stern is surmounted by an inward-curved sternpost, adorned with an *aphlaston* with a globular base and open three-branches similar to Ship 2 (fig. 3.8.25). Three figures stand on the floor of the vessel. On the port quarter is set a gangplank and the horse with its rider are pulled into the ship by the ropes held by all three men standing on the deck. They are soldiers deduced from their tunic adornments.

Discussion

The maritime mosaics in the Armerian villa described above are divided into two themes:

1. Fishing scenes in open waters are depicted in the mosaic floor of the *frigidarium* (fig. 3.8.2). Similar scenes carried out in enclosed waters within harbors faced by waterfront colonnades are depicted in Room 29 (fig. 3.8.7) and the semi-circular *atrium* (fig. 3.8.13).
2. Exotic animals captured in North Africa, Egypt and India for *venations*, are embarked in the harbors of Carthage and Alexandria and then were transported to Ostia and Rome for games held in the Colosseum and Circus Maximus (fig. 3.8.20).

Almost all the boats in the fishing scenes have similar hulls, depicted with a concave stem and a projecting pointed cutwater.[38] These boats are rigged with one pair of oars used for rowing and steering (figs. 3.8.3, 3.8.4, 3.8.8, 3.8.15, 3.8.18). Both rowing oars in Boat 4 in the *frigidarium* are distinct from the steering one (fig. 3.8.6). They have thick shafts while the blades probably are submerged in the water; the shaft of the steering oar is thinner and has a distinct tapper lower end that suggests the blade. Two putti occupy each boat in the *frigidarium* and *atrium*, while three putti, all engaged in different fishing activities, inhabit the boats in Room 29. The fishing methods are varied:

1. A simple fishing rod (figs. 3.8.5, 3.8.9, 3.8.10, 3.8.14).
2. A trident harpoon (figs. 3.8.4, 3.8.5, 3.8.9, 3.8.15).
3. A cast net held on one arm (fig. 3.8.18).
4. Basket traps (figs. 3.8.10, 3.8.19).
5. A seine net between two boats (figs. 3.8.12, Boats 1 and 4; 3.8.13, Boats 3 and 4).

Two boats in the *atrium* are depicted with a transom stem (figs. 3.8.15, 3.8.18), which is a rare representation in any arts and especially in mosaics. A boat with a transom stern and an inward-turned stempost, dated to the early Roman period is depicted in a fresco with a Nilotic scene from Pompeii (fig. 3.5.22; 1st century CE) The first depiction in mosaic of a similar boat with a transom stern or stem appears in the "*Catalog of Ships*" from Althiburus, Tunisia, dated to the beginning of the 3rd century CE (fig. 3.8.27/*20*). Boats 2 (fig. 3.8.15) and 5 (fig. 3.8.18) depicted in the

[38] The cutwater is a structural element providing better hydrodynamics and hydrostatics to the vessel; see Glossary.

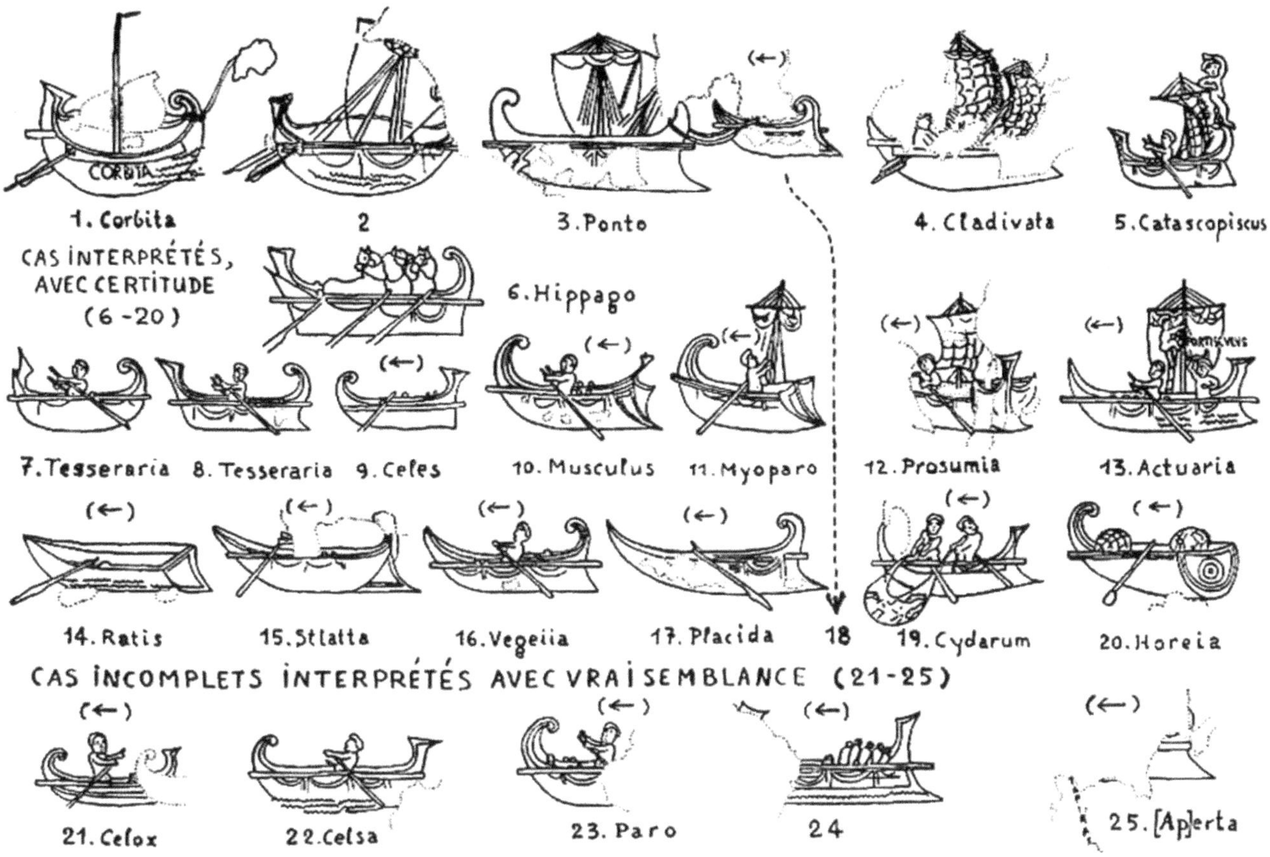

Fig. 3.8.27: Catalog of Ships *in the Althiburus mosaic*

atrium seem to be the only examples with a transom prow depicted in the Piazza Armerina mosaics. The assumption that the Armerina boats have a transom stem is based on the fact that all the other boats in the same mosaic have a distinct rounded upraised stern surmounted by a vertical sternpost with a fishtail-tip. The boats with transom ends also have a similar stern and sternpost. The boat in the Althiburus mosaic (fig. 3.8.27/*20*) is quite similar to the boat in the Nilotic scene from Pompeii (fig. 3.5.22). The stempost of both boats seem to be finished with an inward-turned stempost. In the Pompeii boat, the oars are clearly placed on the quarters, indicating that the distinct transom board forms the stern. The boat in the Althiburus mosaic most probably has also a transom stern, as we may deduce from the position of the oars mounted amidships with a rearward inclination, thus suggesting that they are pulled (fig. 3.8.27/*20*). All the rowing boats in the Althiburus mosaic are rowed by one figure seated with his back to the stem and working the oars with a rearward slanting. Both Armerina boats in the *atrium* have their oars mounted on the stern and each standing putto on the transom prow is engaged in fishing with a trident (fig. 3.8.15) and a cast net (fig. 3.8.18).

The fishing boats depicted in the *frigidarium*, Room 29 and *atrium* belong to a single class. They are small vessels 3 – 6m long, probably with a beam of 1 – 1.5m, and mostly used for fishing and carrying light cargo. The crew comprised two to three putti. In either case, with two or three men in a boat, one is always in charge of the oars and the others with fishing. The boats probably represent the *skiff* class mainly rowboats used on rivers, lakes and even sailing close to the shore. We may suggest some specific names associated with the Armerina boats:

horia, a one-man fishing boat that could accommodate two men;
horeia, a small rowing boat with a transom stern or stem (fig. 3.8.27/*20*);
kydaron (in Latin cydarum), appears in the Althiburus mosaic as a two-man fishing boat with a projecting pointed cutwater (fig. 3.8.27/*19*);
mydion-musculus, literally "little mouse";
ratis/ratiaria, also depicted in the Althiburus mosaic is a rowboat with a concave prow ending in a projecting pointed cutwater and rowed from amidships (fig. 3.8.27/*14*).

The subject of putti fishing and waterfront colonnade porticos and known as *villa marina*, appear in majority of the North African mosaics repertoire.[39] The closest parallel to a fishing scene with a *villa marina* on the top, and putti in boats similar to those in the Piazza Armerina is found in a mosaic from Carthage, now in the Bardo Museum, Tunisia.[40] This mosaic is quite damaged, but there are

39 Wilson, p. 59.
40 Ibid., p. 55, fig. 34.

enough preserved sections to show the similarities to the Armerina mosaics and the boats. The Carthage mosaic is probably dated to the end of the 3rd – beginning of the 4th century CE or it may be contemporary to those in Piazza Armerina. Apparently North African mosaicists with great skills and practical experience were brought to Piazza Armerina to carry out the work. They may have brought with them the patterns from their homeland workshops and used them as guidelines for the complex mosaic works at the Armerina villa.

The ships within the Great Hunt Hall represent two large sailing merchantmen (figs. 3.8.21, 3.8.25; Ships 1, 2) and one smaller ship (fig. 3.8.26; Ship 3) that were either propelled by at least one sail and/or oars. The hulls of all ships are identical with a concave stem followed by a projecting pointed cutwater and a rounded upraised stern. Ships 2 (fig. 3.8.25) and 3 (fig. 3.8.26) have an inward-curved sternpost adorned with an *aphlaston* with a globular base and open three-branches atop. This type of sternpost adornment is a characteristic of Greek war-galleys dated to the 6th – 4th centuries BCE.[41] This element became a common feature on Hellenistic war and merchant vessels as well.[42] The *aphlaston* decoration also appeared on Roman ships dated to the 2nd – 1st BCE,[43] as well as on the warships found on the Column of Trajan (2nd century CE).[44]

The red coloring of the Sailing Ship 1 hull and the decoration band with three braded strips of laurels leaves (fig. 3.8.21) point to an imperial vessel employed for transportation of African animals to Rome for the games held in the Colosseum and Circus Maximus. The purple coloring of Ship 3 hull (fig. 3.8.26) indicates an auxiliary vessel in the imperial fleet. The fancy tunics of the soldiers with applied rounded emblems on the shoulders and the lower edge of the tunics probably belonged to the imperial units appointed to the capture and the transportation of exotic animals to Rome. The first clear evidence of the military involvement in capturing and transportation of African animals comes from the early imperial period.[45] Some contemporary sources suggest that hunting indeed formed an important part in the duty of the soldiers.[46]

The sailing rigging of Ships 1 (fig. 3.8.21) and 2 (fig. 3.8.25) consists of a tapered mast, depicted with alternate yellow and black bands, an indication of a composite mast, a furled sail beneath the yard and the lines associated with the standing(fore and back stays, shrouds), as well as the running rigging (sails, brails, braces and sheets). Only Ship 1 has additional gear comprising rowing oars (six or nine oars per side) and one rudder mounted on either quarter. The other ships lack this gear. It appears that when a ship entered or left a harbor it used rowing oars because the open sail would have been cumbersome. Ship 3 (fig. 3.8.26) is lacking the sailing or steering rigging, may suggest that when a smaller ship was loading cargo, its mast and sail were removed. The soldiers within the ship that are pulling the ropes to bring in the horse and its rider may indicate a military ship, carrying troops to Africa for catching the animals on imperial order, as well as patrolling the large merchantmen *en route*.

All the Armerina ships represent *navis oneraria* (merchant vessels) that are large and broad vessels with a beam/length ratio of 4:1 or 3:1. The ships may also be considered as military vessels used to transport soldiers and horses not only for combat but also for hunting exotic animals in Africa and India to be brought to Rome for the games and pleasures of the emperor and the nobility. The ships may be associated with a more specific type of *cercuri* (Latin) or *kerkouroi* (Greek), equally suited for commerce and warfare. They were in use in the Mediterranean from the beginning of the 5th to the middle of the 1st centuries BCE.[47] Such vessels were rigged with one sail and also with six oars. In *P.Cairo Zen.* 59053 (257 BCE) are mentioned the dimensions of a *kerkouros*: 45 cubits (20.6m) long, 7 cubits (3.2m) wide, and had 10 oars per side;[48] such a ship was of a moderate size (80 – 200 tons). There were larger types that were 50m long and 7.7m abeam and held 450 - 500 tons burden, whereas of the total length only 70% would be available for the cargo.[49] The Armerina ships may also relate to *pontos* that were merchant ships used in the 1st century BCE.[50] Such an example appears in the Althiburus mosaic (fig. 3.8.27/*3*). There is no reference to indicate when such vessels ceased being used.

It has been mentioned above that the mosaics at Piazza Armerina show strong links with the mosaics from North Africa, especially Carthage. Ship 2 (fig. 3.8.25) shows an extraordinary similarity to the Dermech Hunt mosaic at Carthage (fig. 3.8.28).[51] The scene in the Dermech mosaic is more realistic and better proportioned. The figures are shown in full view, while the ship is less realistic, being depicted in a quite distorted view as seen slightly from the beneath the starboard quarter. This ship depicts all the elements found within all three ships in the Armerina mosaic. The decoration of the upper plank in the Dermech Ship is similar to that of Armerina Ship 2 (fig. 3.8.25). Each symbolic end of the through beams (depicted in perspective from bellow) in the Dermech ship projects through a square frame (fig. 3.8.28). The ends of the through beams in the Aremrina Ship 2 are depicted alongside in the same plane (fig. 3.8.25). The bow of the Dermech Ship has a slight forward angle and is finished with a projecting cutwater, decorated with a lion head at its tip. Probably, this decoration represented a bronze ram, deduced from

41 Basch, 1987, figs. 473, 579, 580, 581, 586-92, 627.
42 Ibid., figs. 737-39; 782-83; 790; 803; 804/*10, 19, 22b, 27, 41*; 874 A, B.
43 Ibid., figs. 902,904, 909-11, 923, 928, 944, 962-63, 969 B, 976.
44 Ibid., figs. 978, 984, 998, 1001.
45 Epplett, 2001, p. 211.
46 Ibid.
47 Torr, 1964, p. 110.
48 Casson, 1971, p. 164, n. 41.
49 Ibid., p. 166, n. 43.
50 Torr, p. 121.
51 This mosaic measures 8.5 x 7m; originally it belonged to a rectangular room with an apsidal end. It was found in 1965 and it is now displayed at the Carthage Museum; Ben Khader and Soren, 1987, pp. 213, 215.

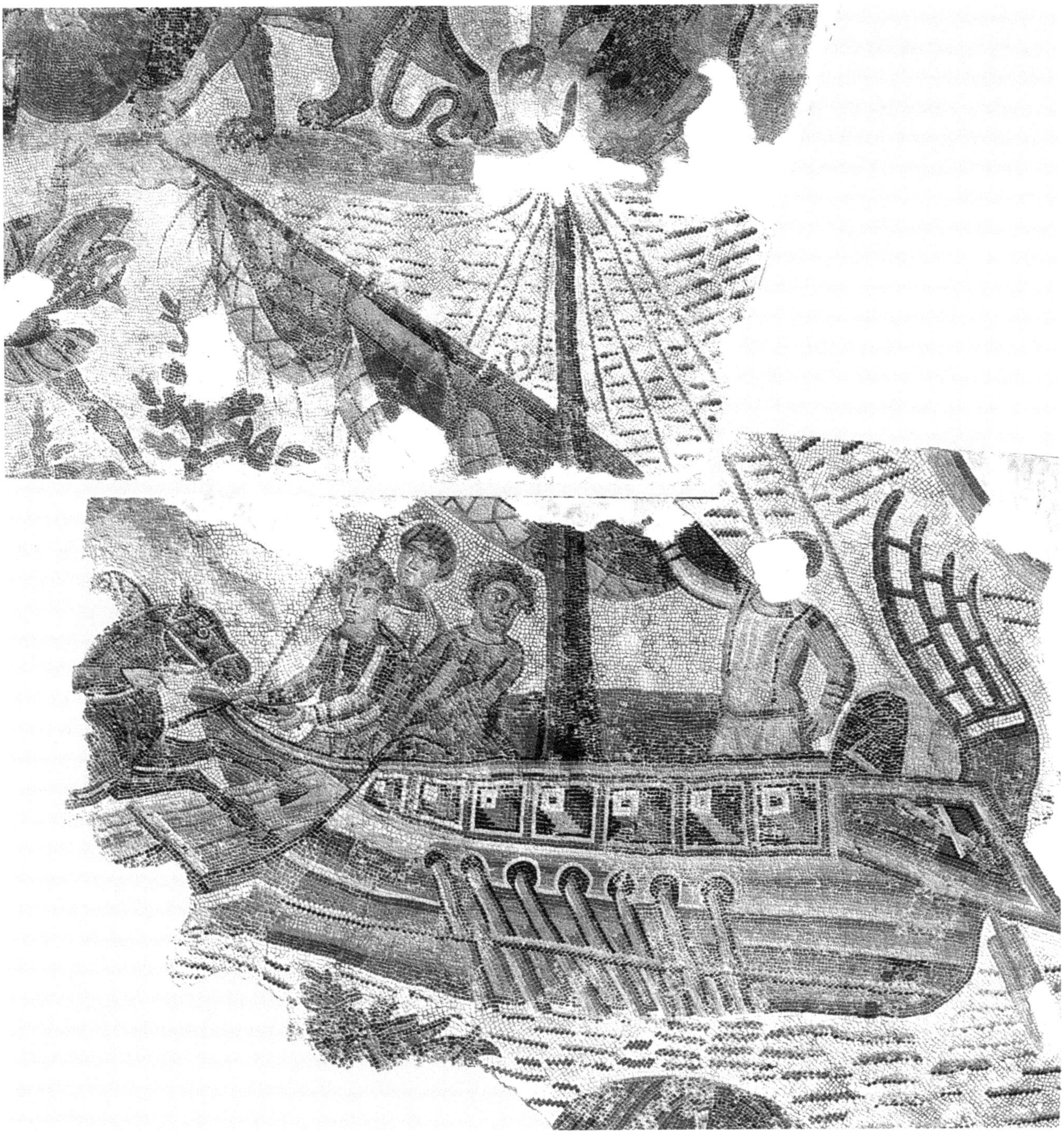

Fig. 3.8.28: Sailing ship in the Dermech mosaic from Carthage

the light and a darker hue of brown tesserae. The stempost is slightly higher than the fore gunwale. The stern seems to have a square shape and the almost vertical sternpost is adorned with a large four-branched *aphlaston*. The aft-wing of the top strake is depicted as a trapezoidal frame though revealing the upper part of the rudder's shaft. A similar depiction but more clearly shown appears in the monumental statue of Odysseus' Ship from Sperlonga, Italy, dated to the 1st century BCE or CE.[52] Although, the aft end of the top strake in the Dermech Ship is distorted it still provides the picture of how the rudder was housed and worked. A tiller was inserted perpendicularly into the head of the rudder's loom (fig. 3.8.28). The position of the port rudder seems to indicate an aft mount where the shaft rested on the aft side of the through beams supporting the aft-wing. The elongated blade of the Dermech Ship has straight cut shoulders and its lower end is rounded, while in the Armerina Ship 1 each shoulder of the wings is rounded (fig. 3.8.21).The standing figure on the quarter (with the damaged face) is probably the helmsman working the port rudder by the tiller and also some of the tackle lines.

The sailing rigging of the Dermech Ship (fig. 3.8.28) is similar to that of Armerina Ship 1 (fig. 3.8.21). It comprises

[52] Casson, fig. 170; White, 1986, p. 149, fig. 152.

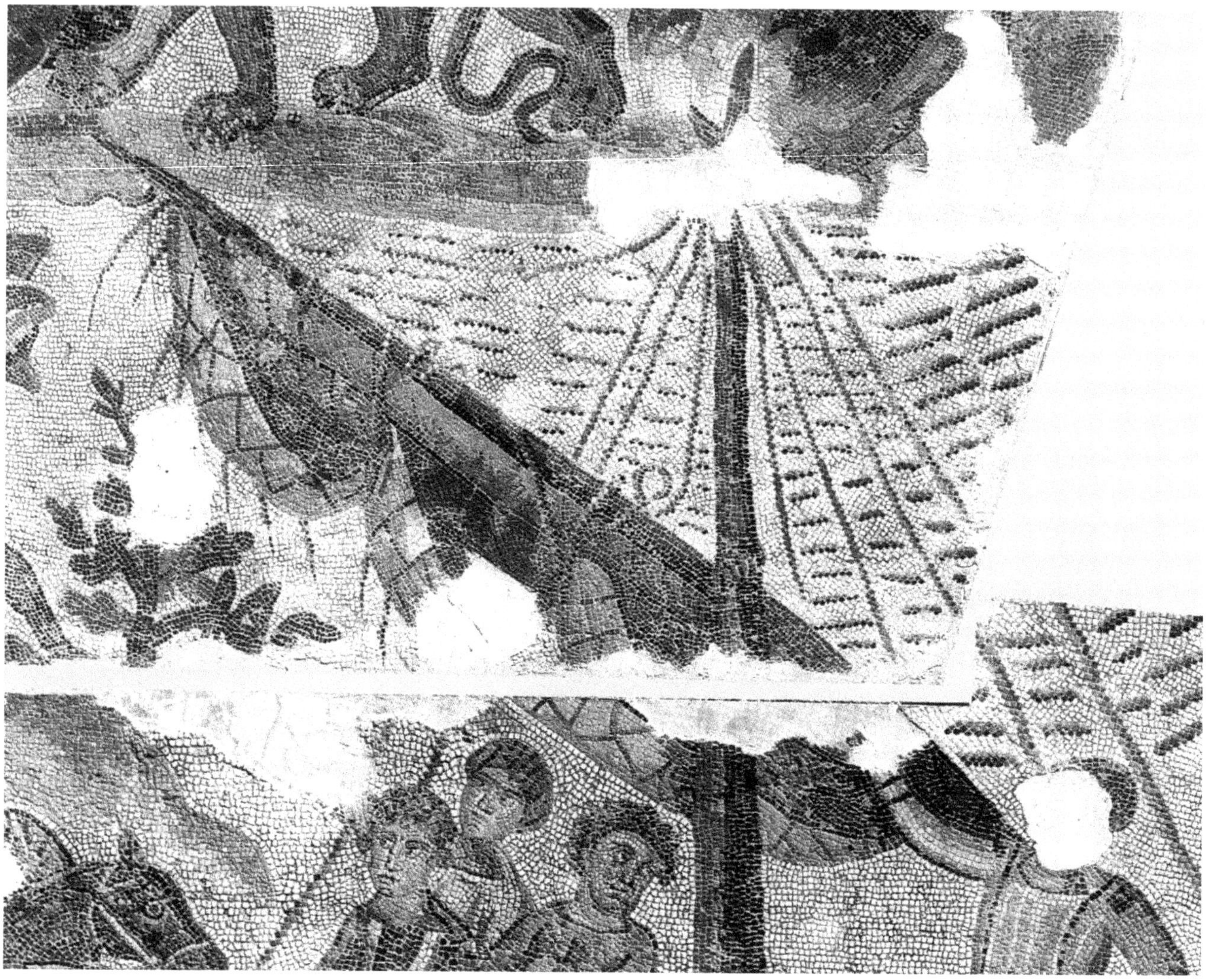

Fig. 3.8.29: Close-up of the sail rigging

a tapered composite mast, a furled sail beneath the yard, seven rowing oars on the port side and one rudder. The yard is in the process of being raised and is shown at an angle to the mast, at about its mid-height (fig. 3.8.29). A rounded *parrel* is found just beneath the masthead. Three lifts on the portside of the masthead and two on the starboard side secure the yard. The right-hand line (extreme) stretching from the masthead to the inner side of the *aphlaston* is the backstay. The line closer to the backstay and stretching to the head of the helmsman probably is the starboard shroud (misplaced), or it is one of the lifts. The free hanging line on the left side of the masthead with is lower end attached to yards, probably is the halyard. On the exposed face of the furled sail is depicted a similar checker pattern as that on the sails of Armerina Ships 1 and 2 (figs. 3.8.21, 3.8.25). It may suggest symbolic brails intersecting the reinforcing-bands. The small hanging strips are the reef-points used to tie the sail to the yard when it was completely lowered (fig. 3.8.29), in a similar way as is done today in modern yachts.

The thick rounded shafts of the rowing oars in both Armerina Ship 1 (fig. 3.8.21) and the Dermech Ship (fig. 3.8.28) project through circular oarports. In both mosaics three men on the bow help the horse and its rider to board the ship in a similar way (figs. 3.8.26, 3.8.28). This vessel also is an imperial ship, similar to Armerina Ship 3 (fig. 3.8.26) escorting *en route* the merchantmen transporting captured exotic animals to Rome for the games held in the Colosseum and Circus Maximus.

Both Dermech and the Great Hunt mosaic at Piazza Armerina are very similar in their depiction and style. They strongly suggest that artists of the same workshop produced both mosaics. The Dermech mosaic is somewhat earlier than the Piazza Armerina. Two coins of the Emperor Maximian (297 – 302 CE) found in the mortar of the Dermech mosaic suggests that it was laid in the first years of the 4th century CE. The Great Hunt mosaic at Piazza Armerina was laid in 310 – 315 CE, as deduced by two coins found above the mosaic and the pottery remains.[53]

[53] One of the coins is of Maxentius (306 – 312 CE) and the second of Flavius II (312 – 363 CE); Wilson, p. 36.

Chapter 4
Ship Archaeology

"The sailor should scan his course from the shore, if so be that he has the power and means; for once he is on the high seas he must run with whatever comes".[1]

4.1 Introduction

This chapter contains a summary of several shipwrecks that were found off the coasts of Italy, Sicily, France and Turkey. They are concomitant to the mosaics with ship depictions studied in this monograph. While ship depictions in any arts provide us with an overall view (not at scale) of the characteristic features of vessels and their propulsion and steering gear, shipwrecks reveal details of construction techniques, the cargoes carried on board, an indication of the sea route they followed, as well as an estimated time when the sinking occurred. Neither shipbuilding manuals nor sailing charts of the ancient mariners have survived.[2] The combined study of ship depictions especially in mosaics and shipwrecks data facilitate better understanding and bring a more complete picture for marine and nautical archaeologists, or anyone interested in this subject. Excavated shipwrecks become living testimony to the majority of ships that are represented in mosaics and other arts, as well as those recorded in ancient written sources, namely merchantmen or war galleys.

4.2 General Definitions

Mediterranean cultural diffusion is the result of rafts and boats development that permitted the transportation of people and goods from place to pace. The water vessels may be classified into two main groups. The first group of boats is a generic name referring to small crafts with no decking and usually propelled by oars or one sail, or in modern times by an inboard engine. Some exceptions to this definition are the fishing boats, sometimes decked or half-decked and propelled by oars, and nowadays by an inboard diesel engine.[3] The second group consists of ships generically defined as seagoing vessels. The stages of development of a ship from its first tentative seagoing example to the various types in use today relate more to the development of their propulsion, rather than to particular changes in the shipwright's art.[4] The Egyptians were the first ancient society to use sails on the Nile and seagoing vessels. The Greeks brought the art of rowing to a level of perfections that was not surpassed by any ancient civilization. Most of our knowledge of ancient vessels comes from ships represented in different arts and ancient writings. Until the 19th and the beginning of the 20th century, the occasional discovery of shipwrecks and their cargoes threw some light on the existence of ancient ships. The development of aqualung equipment and underwater archaeology during the past 30 – 40 years and the most recent development of the ROV (Remote Operated Vehicle) have provided important information about ancient ships and shipping, cargoes carried on board and also an estimate of the period when the sinking occurred. Sails, masts, yards, oars and ropes are the most vulnerable and easily destroyed gears of a ship when it wrecks. Very few remains of such gear have survived. The best preserved sections of a shipwreck are the lower parts of the hull which were protected by the cargo and later by silt and debris.

4.3 Shipbuilding Materials and Techniques

The wooden ships of classical periods, whether warships or merchantmen, were built with the same "shell-first" technique, where the planks were assembled by mortise-and-tenon joints and locked by treenails, wooden pegs or bronze nails. Frames and half-frames were inserted afterwards within the hold to strengthen the hull. Apparently the ancient shipwright built his ship with the skills he learned throughout generations that were passed on from father to son. Homer, describing how Odysseus built his boat/raft on the Island of Ogigia (where he was detained by the beautiful nymph Calypso), may be referred as the earliest written source of building a vessel using the shell-first technique of assembled planks by mortise-and-tenon joints. This technique was known to Odysseus from his homeland Ithaca. Then he continued his work by placing the frames within the hold of the vessel. Odysseus built his craft with the tools that Calypso gave him: an ax, an adze and a drill, and the trees from the forest appointed by her:

> *"He felled twenty trees; adzed them into planks; bored them all and fitted them to each other; hammered it with dowels and joints, making his craft as wide a good shipwright will layout the lines for the bottom of a heavy merchantman; then setting up the decks by fastening them to close-set frames; and*

---

1. Athenaeus; *Deipnosophistae*, xv.695.8.
2. White, 1986, p. 141.
3. Kemp, 1988, p. 98.
4. Ibid., p. 781.

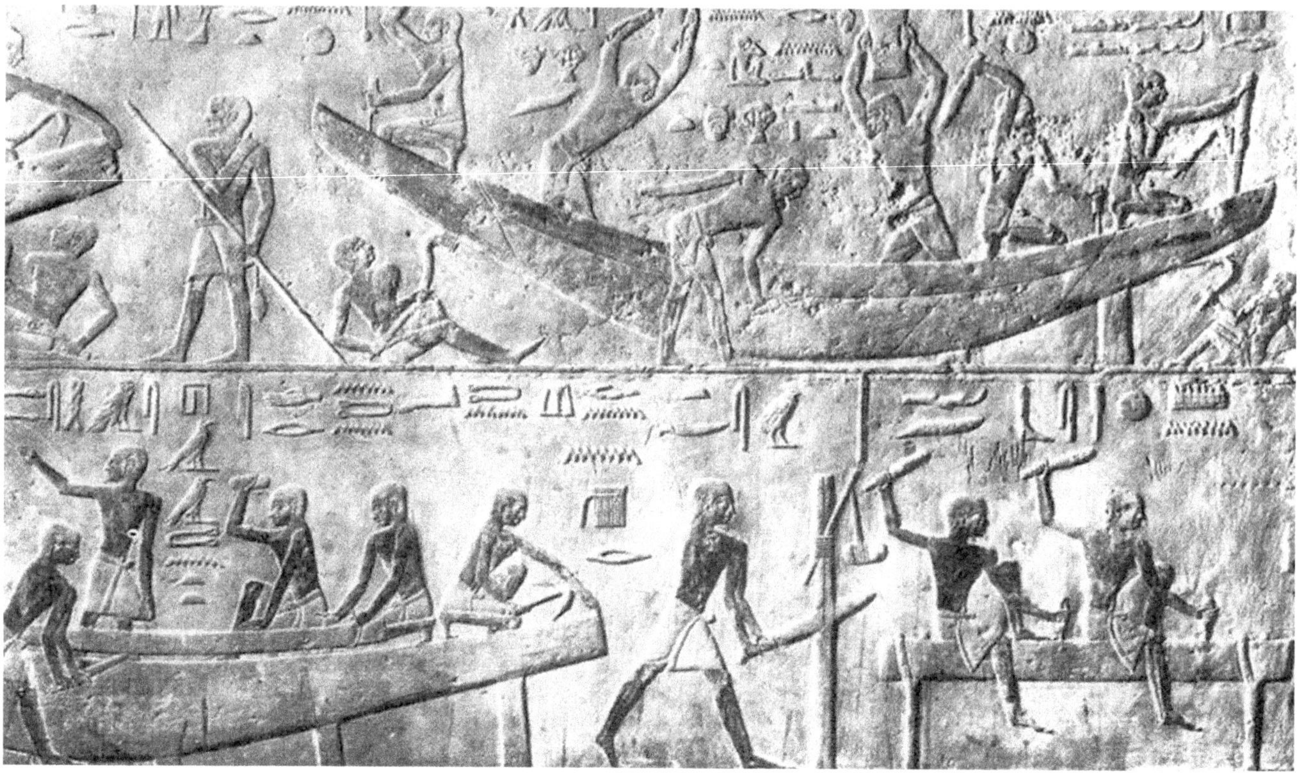

Fig. 4.1: Wooden planks shipbuilding from the Tomb of Ti at Saqqara

finished up with long pieces after completed the hull, he then turned to the fitting: mast and yard, rudder, latticed bulwark".[5]

Excavated shipwrecks revealed species of timber and their building technique. The earliest Egyptian vessels were built of papyrus (*Cyperus papyrus*) bundles tight with cords and ropes. Pliny (*NH* 7.56; 13.21) mentions that rafts of papyrus, reeds and rushes were made in the Nile Delta, whereas the inner bark of the papyrus stems also were woven into sailcloth, as well as turned up into ropes.[6] Ropes were also made of willow.[7] Egypt was not blessed with forests to provide the necessary timbers for shipbuilding and therefore she had to import cedar logs (*Pinus cedrus*) from Lebanon, Cyprus or Turkey. The local woods for shipbuilding in Egypt were: acacia (*Acacia nilotica*), sycamore-fig (*Ficus sycamorus*), date-palm (*Phoenix-dactylifera*), dom-palm (*Hyphaena thebaica*), persea (*Mimusops shimpery*) and tamarisk (*tamarix* sp.), but they did not produce planks longer than 6m and acacia was a hard wood to work.[8] To build ships, the Egyptians had to import cedar wood from Lebanon that was an easy workable material and could provide planks of 20m or longer.[9] The import of timber is well evidenced by the story of Wenamon who was sent to Byblos (*c.*1000 BCE) to buy cedar logs for the building of the sacred barge of Amon.[10] One of the earliest illustrations of wooden plank boat building is found in the Tomb of Ti at Saqqara, dated to the 5th Dynasty (fig. 4.1). Herodotus gives a nice account of the construction of a planked cargo ship of acacia wood, when he visited Egypt in the 5th century BCE:

"The boats in which they carry cargo are made of acacia, which is in form most like to the lotus of Cyrene, and its sap is gum. Of this tree they cut logs of two cubits length and lay them like courses of bricks, and build the boat by making these two long cubit logs fast to long and close-set stakes; and having so built they set crossbeams athwart and on the logs. They use no ribs. They caulk the seams within with byblus. There is no rudder passing through the hole in the boat's keel. The mast is of acacia wood and the sail of byblus. These boats cannot move upstream unless a brisk breeze continues; they are towed from the bank".[11]

The Mediterranean vessels were built with varied timbers provided by the rich forests in Lebanon, Cyprus, Turkey, the Alps, etc. The planks were mainly cut from Allepo pine (*Pinus halepensis*) or cedar (*Pinus cedrus*). Oak (*Quercus cerrus*) was used for planks, fastenings (tenons, treenails) or keel, false-keel. Other species of timber used in ancient shipbuilding were: olive (*Olea* sp.), or pistacio pine (*Pistacia Palaestina*).[12]

The earliest evidence of pegged mortise-and-tenon

5 Casson, 1964, p. 61.
6 McGrail, 2001, p. 49.
7 Ibid., p. 185.
8 Ibid., p. 6.
9 Ibid., p. 16
10 Raban, 1987, p. 120.
11 Herodotus, II.96.
12 Ibid., p. 136.

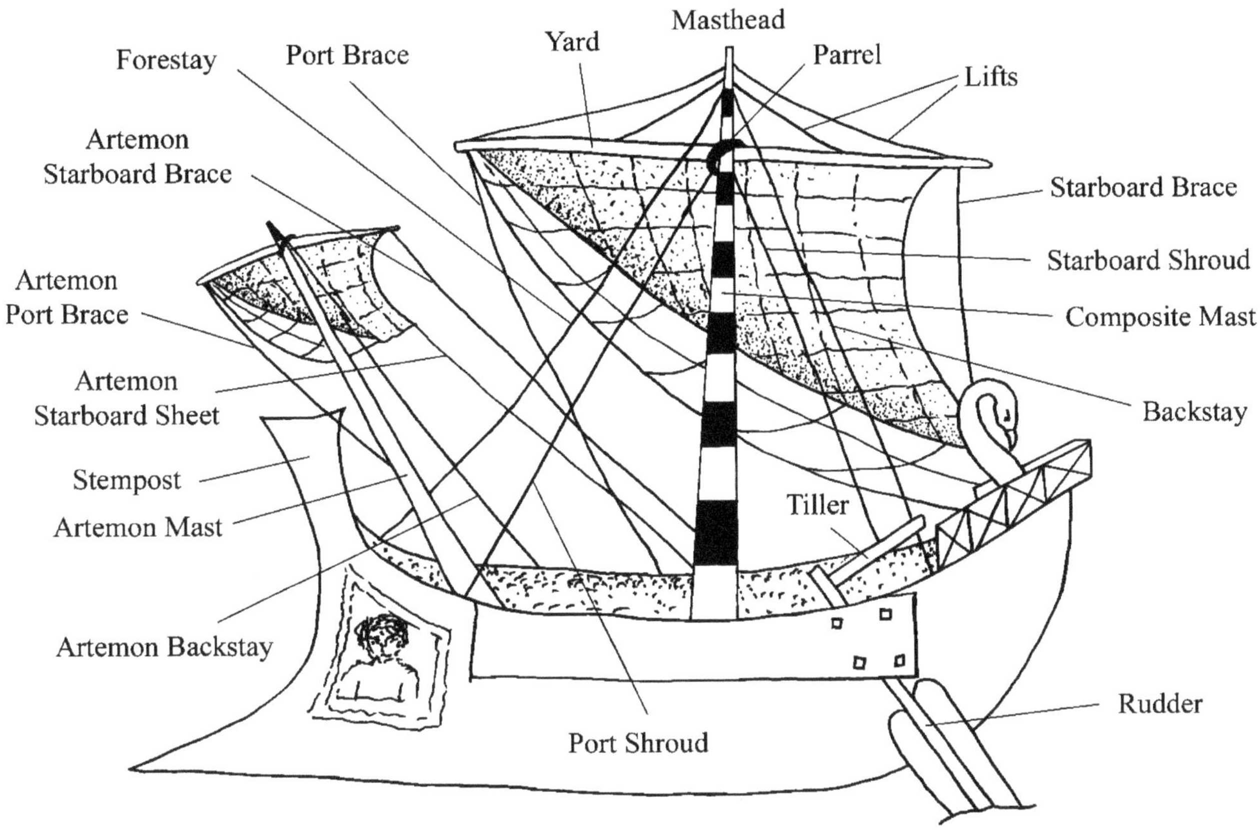

Fig. 4.2: General view of the ship's elements

joints hull come from the Bronze Age merchant ship at Uluburun, Turkey (14th century BCE).[13] The hulls of the Greek seagoing ships were strengthened by *hypozoma* (undergird) that provided them the resistance, while they were under heavy stress of the rough seas. This device was a heavy cable running along outside of the hull, from stem to stern, which could be tightened when needed by means of windlasses.[14] This device is rarely shown in any ship depiction.[15]

The Greco – Roman vessels comprise of two basic forms:

1. The "*long ship*" *(navis longa)*[16] was essentially a warship or a pirate vessel, designed for rowing at high speed in naval battles or incursions. It also could be fitted with a sail when cruising along the coasts or for long voyages. Its slender hull had a width/length ratio of 10:1.[17] The bow was fitted with a projecting ram and the vessel was steered by a pair of steering oars, one oar mounted on either quarter.
2. The "*rounded ship*" (*navis oneraria*)[18] was a merchant vessel with a beamy hull and propelled by a large square sail with a series of brails that controlled the surface of the sail. The lower part of the sail was controlled by two strong ropes known as sheet, attached to each clew (corner). The yard was raised or lowered by the halyard and the braces attached to the tips of the yardarms turned it around. Several lifts stretching from either side of the masthead supported the yard. The mast was secured in place by the standing rigging, the fore-and-back stays, and laterally by the shrouds (fig. 4.2). Occasionally such vessels could use oars for rowing on rivers or entering/leaving harbors, or in varied harbor works. The width/length ratio was 5.5:1 or 6.5:1.[19] In order to increase the cargo capacity the bow and stern were broad and rounded. In many ship depictions the vessels are fitted with a projecting cutwater, which is a constructional element that improved the hydrostatics and hydrodynamics of the vessel. The steering gear comprised a pair of steering oars or rudders, with an elongated rectangular blade. The shaft was mounted behind the extended aft-wing of the top plank or wale, or projecting through oarports cut into the quarter upper plank. The oars were steered almost from a vertical position with the shaft serving as the turning axis.[20] Several depictions of ships show a tiller inserted perpendicularly into the head of the loom to work the rudder or the steering oar (fig. 4.2).

13 Bass, 1985, p. 25; Delgado, 1997, p. 430. The shipwreck was excavated between 1984 and 1994; the estimated overall length is 15 – 16m. Within the cargo there were 10 tons of copper and bronze ingots, and one ton of tin.
14 Landels, 1978, p. 138.
15 Morrison and Coates, 1996, figs. 7 a – c, 10 b, 12 a, 13 a, 23 c – e.
16 Casson, 1971, n. 7, p. 159; ibid., 1995.
17 Ibid., 1971, p. 158.
18 Ibid., p. 159 and n. 7.
19 Ibid., p. 158.
20 Ibid., p. 225.

Archaeological evidence shows that at times the hulls of merchant were sheathed with lead to protect the wood from *teredo navalis* (ship worms) and also from physical damage caused by shallows or rocks. The discovery of the 3rd century BCE, Marsala Punic ship revealed that also the hulls of warships were sheathed with lead or copper. The forefoot of the Marsala "Sister Ship" was bound with a fabric impregnated by a resinous substance. Some small fragments of copper were found on the cutwater, thus indicating that at least the forefoot was sheathed with copper or bronze.[21]

4.4 Size of the Merchantmen

Ancient literature and the archaeological finds of shipwrecks led to a better classification of ancient merchantmen mostly based on the cargo remains. These vessels can be grouped into three classes:

1. Small ships with a cargo capacity from 5 to 80 tons.
2. Medium sized with a cargo capacity from 80 to 130 tons.
3. Large freighters with a cargo capacity from 130 to 500 tons.

Super-freighters with a cargo capacity of 1,000 – 3,650 tons were rare and are only known from written sources. Two known examples are the *Isis* Ship described by Lucian, who mentions that it carried grain from Alexandria to Rome. *En route* the ship anchored in the port of Piraeus, where it aroused the curiosity and admiration of the locals.[22] The second giant vessel was *Syracusia* or *Alexandris*, described by Athenaeus.[23] The ancient texts and the archaeological finds show that ancient ships carried mixed cargo. One example of a merchantman carrying a mixed cargo is augmented by the shipwreck found at La Cavaliére (*c.*100 BCE), France, with a capacity of 20 tons. It appears that its journey began in North Africa and made the route along the western coast of Italy, picking up wine in Apuleia, wine and oil in Campania and pigs in Northern Italy.[24] Vessels of 100 to 150 tons capacity were in common use, while those of 350– 500 tons of cargo were rare. The Emperor Claudius granted certain privileges to shipwrights to build grain-carrying ships of 70 tons capacity (*c.*10,000 *modii*), thus indicating that such small vessels were preferable to the authorities.[25] Apparently the measurements of the harbor basins and expediency in the organization of the harbors dictated the size of the ships. In the 4th century BCE, the Greek port of Thasos issued a decree specifying that only ships of 3,000 talents (80 tons) could anchor within the boundaries of the outer harbor and no ships under 5,000 talents (130 tons) could anchor within the inner harbor.[26] The issuing of such decree limiting the inner harbor to large vessels shows that there were many smaller ships which caused congestion and it was necessary to exclude them from the protected inner harbor.[27]

At the end of this chapter are presented three tables that include the wrecks mentioned in this research, with reference to their size, period, cargo hold and the location where they were found, as well as referred in ancient literature. The tables also includes some other wrecks that were not described in this monograph, but are presented as compares for their size, ratio of width/length and cargo capacity. This data should be viewed as a basis for comprehension of the sizes and types of ships depicted in the mosaics presented above.

4.5 Archaeological and Literature Evidence of Ancient Ships

Most of the wrecks discovered in the Mediterranean are remains of merchantmen dating from the Bronze Age to the modern times. A summary survey of varied vessels will relate to wrecks from the Hellenistic to the late Byzantine era that were found off the costs of Italy, Sicily, France and Turkey. These wrecks relate to the period of the mosaics with ship depictions described above, as well as to literature documentation concurrent to the periods of the mosaics.

Egypt

Iconography of ships from the Pharaonic periods in Egypt is varied, but it will not be discussed in this monograph. Later representations showing preserved Egyptian shipbuilding traditions are illustrated in the Palestrina Nile mosaic, late 2nd century BCE. Athenaeus, Strabo, Herodotus and Pliny referred to the construction and sailing of varied types of vessels in their writings. Archaeological evidence of Hellenistic, Roman and Byzantine shipwrecks is rare in Egypt.

Representations of papyrus boats in Hellenistic and Roman period arts are attested by the historical writings of Pliny, showing that such rafts were still used in the Delta and in the Nile, in the 1st century BCE. He also wrote that the inner bark of the papyrus stem was woven into sails and also turned into ropes,[28] the essential gear for sailing vessels. Besides papyrus boats there was the coracle "made of wicker and hide*s sown round it*".[29]

Syracusia, an impressive merchantman from the 3rd century BCE, is known from Athenaeus's reference. This ship, built by Hieron II, tyrant of Syracuse, under the supervision of Archimedes was of such enormous proportions that there was no harbor in the Mediterranean that could provide proper mooring facilities for her. Athenaeus gives us a wonderful account of its building and its size.[30] The ship was 61.5m long, abeam 15.4m and had a load capacity of 1,700 tons of grain, while its total load capacity was 3,650

21 Frost, 1981, p. 71.
22 Houston, 1987, p. 444; Lucian, The Ship or The Wishes 1–4.
23 Athenaeus, V. 206–209; Turfa and Steinmayer, 1999, pp. 105 – 125.
24 Houston, 1988, p. 557 and Table 4.
25 Casson, p. 172, n. 23.
26 Houston, p. 559.
27 Ibid., p. 559, n. 32.
28 Pliny, NH, 7.56.13-21.
29 Ibid., 7.56.206.
30 Athenaeus, V.206 – 209.

tons (cargo, crew and passengers).[31] The vessel was also a very luxurious ship and the floors of all three decks were paved with beautiful mosaics depicting mythological scenes from the Iliad. It was estimated that the mosaic stones weighed 1.13 tons.[32] The hull was sheathed with lead that also increased its draft because of the additional weight.[33] *Syracusia* was a super-freighter whose sailing life was very short with only a single major voyage to Alexandria, when Hieron II offered it as a present to Ptolemy II during a severe famine in Egypt. The Roman ship of Caligula transporting the Egyptian obelisk to Rome was considered as a symbol of the emperor's absolutism. After fulfilling its purpose, the vessel was used as the foundation of the lighthouse at the Claudian harbor. The wooden hull had completely disintegrated but its imprints that survived into the concrete permitted to reconstruct the original dimensions of the vessel, 91m long and 21m abeam.[34] It appears that *Syracusia* and the obelisk ship of Caligula were impractical. Their draft was so great that most of the Mediterranean harbors could not provide adequate mooring facilities and therefore these vessels could not be considered as a pattern for building large freighters; their importance was in being a 'status symbol' of the absolute rulers.[35]

Thalamegos: Besides the great grain freighters that arrived to Egypt and the local continuous use of traditional papyrus boats, we learn from the historical records of Athenaeus that Egypt of Ptolemy Philopator was very prosperous, which also was expressed by the king's ownership of different vessels.[36] In addition to them he built a luxurious river boat, the so called *thalamegos* (cabin-carrier). It had a length of three hundred feet (99m) and at beam was fifty-five feet (14.85m). The height, including the pavilion was sixty feet (19.8m). The vessel did not resemble a war-galley or a rounded bottomed merchantman, but it had been altered in draught to suit its use on the river, with a flat and broad bottom. It had a double bow and double stern,[37] which resemble a megalithic catamaran. The mast height was one hundred and five feet (34.65m). The main sail was woven of fine linen and a purple topsail was set above the yard,[38] similar to traditional seagoing vessels.

The Nemi "Floating Palaces"

The Nemi Ships are the closest archaeological evidence for the *thalamegos* (cabin-carrier) described by Athenaeus. They were built as pleasure barges sailing on the Lake Nemi, Italy. These two outsized vessels, more like pleasure barges or floating palaces, built by the Emperor Caligula during his reign (37 – 41 CE), were probably used as ceremonial ships for the cult of the Egyptian goddess Isis. They were not intended to be seagoing ships but had been built to sail/float on Lake Nemi. It seems that the vessels had quite

[31] Turfa and Steinmayer, Table I, p. 106.
[32] Ibid., p. 117.
[33] Ibid.; for the lead weight see Table IV, p. 118.
[34] Testaguzza, 1964; 1970.
[35] Duncan-Johns, 1977, p. 332.
[36] Athenaeus, V.204.
[37] Ibid.
[38] Ibid., V.206.

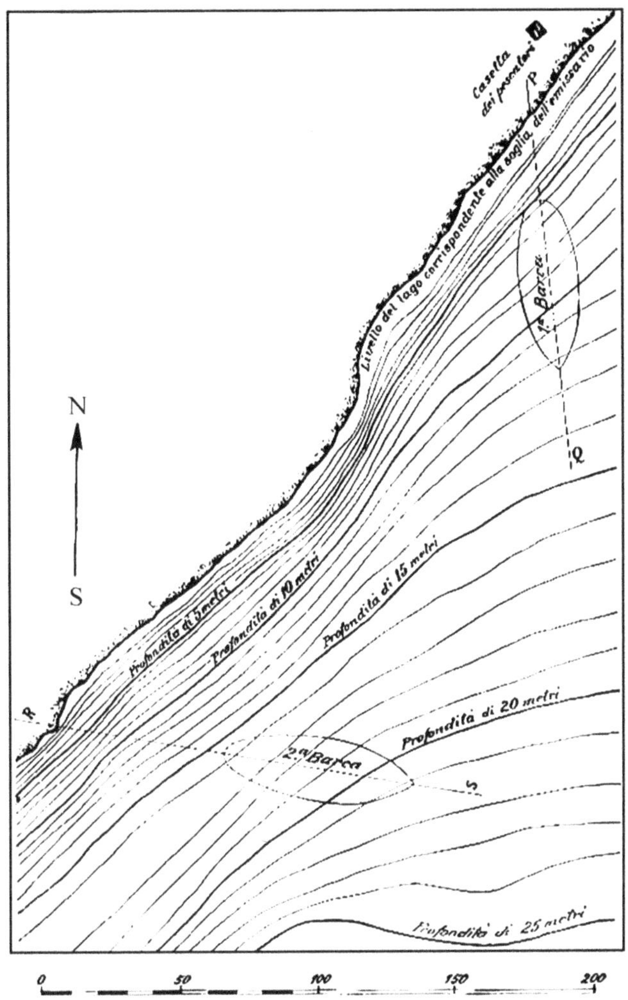

Fig. 4.3: Nemi Ships site plan

a short life, being sunken in the Lake about two decades after their building. The first ship, with its bow pointed to the south lay in shallow waters (5 to 12m) and the second ship with the bow turned to the east, laid in deeper waters (15 to 22m) (fig 4.3). The search for the vessels started in 1446 and continued through 1535, 1827 and 1895. During these periods, only parts of the ships' timbers and various artifacts were recovered, and additionally two anchors: one of iron 4m long, and the second was 5m long made of wood with lead stock and collar. The surviving parts (bronze fittings) are now displayed in Museo Nazionale at Palazzo Massimo, Rome. The final process of excavation and recovery of the ships was made between 1928 and 1932, when the Lake Nemi was completely drained by Guido Ucelli, a hydrological engineer. The first ship measured 71.3m in length and 20m abeam, the upper hull, the deck and bulwark were missing (fig. 4.4). The second ship was slightly larger, 73m in length and 24m abeam (fig. 4.5).[39] Five keels were set to sustain the strength of the very broad flat-bottoms of the vessels (fig. 4.6), which were sheathed with lead, similar to seagoing ships (fig. 4.7). The planks were joined by mortise-and-tenon fastenings. Copper nails

[39] Delgado, p. 233.

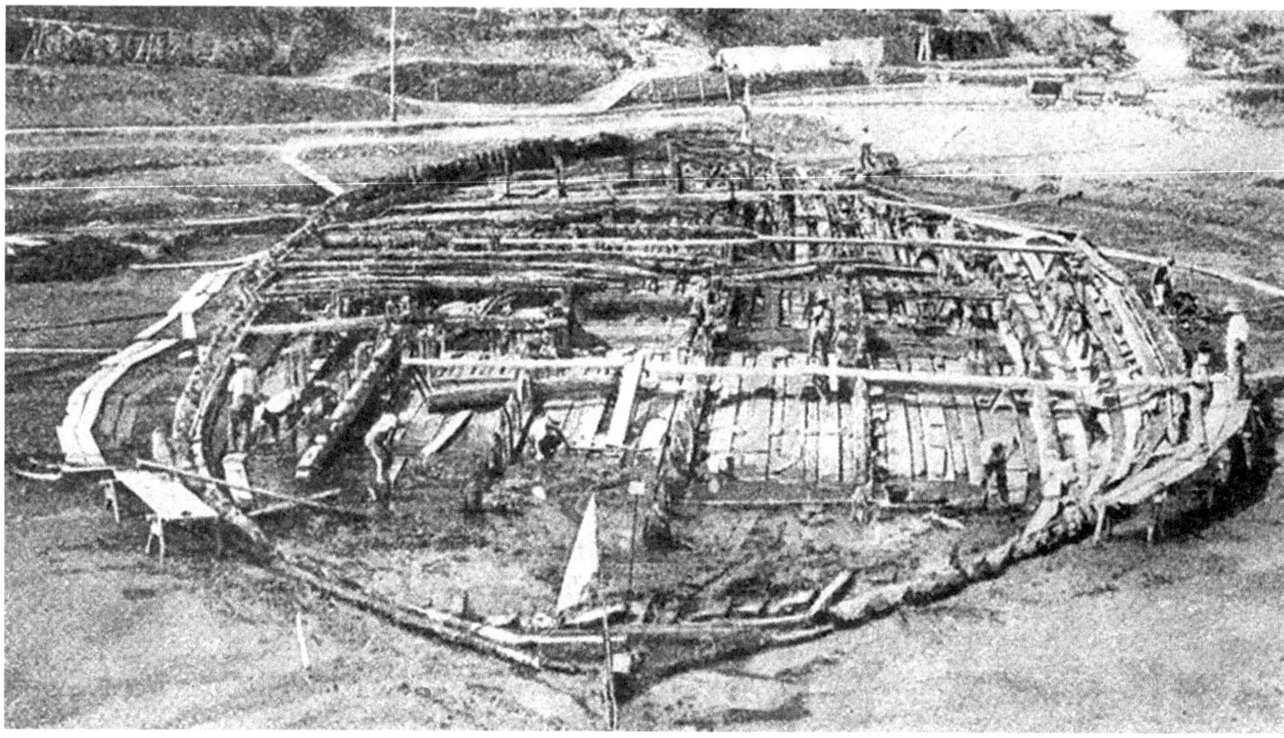

Fig. 4.4: Nemi Ship 1

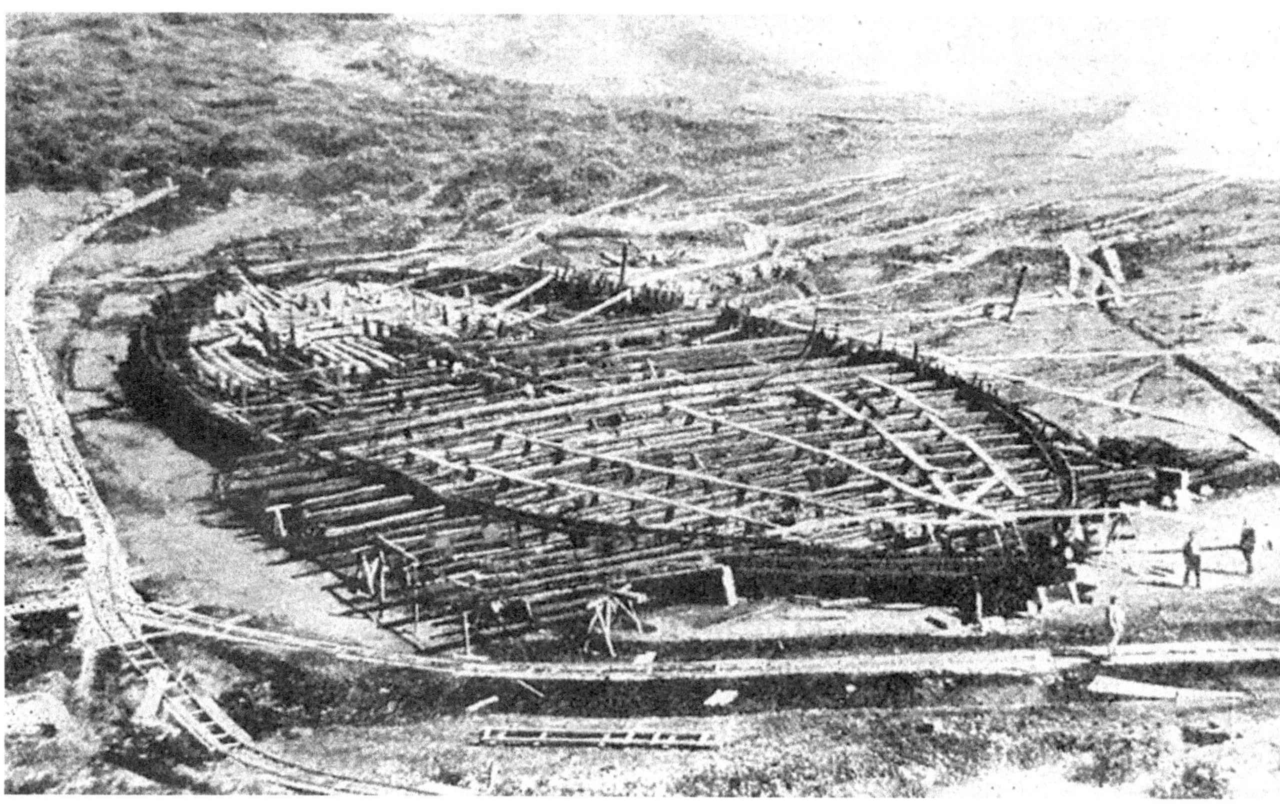

Fig. 4.5: Nemi Ship 2

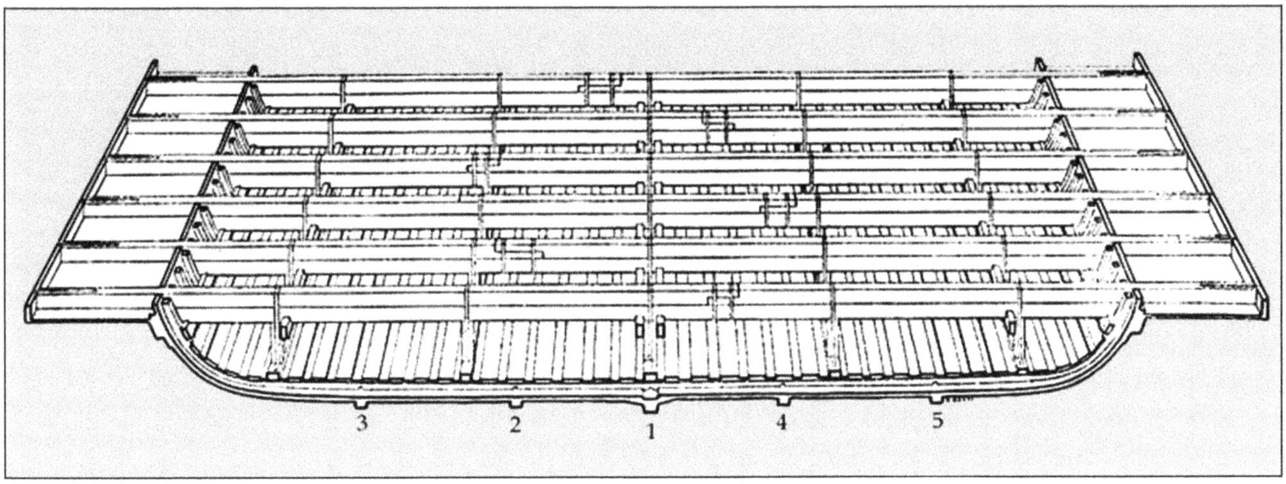

Fig. 4.6: Cross section of the hull showing the five keels

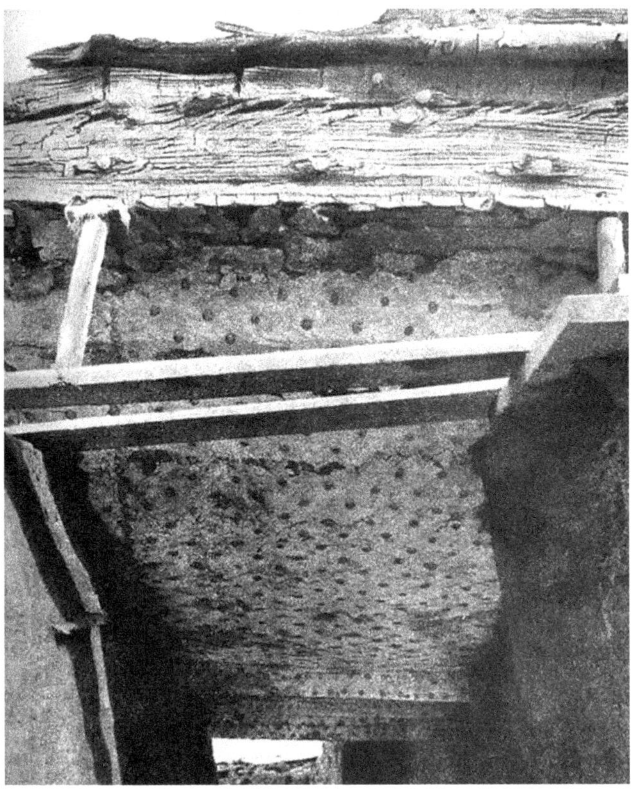

Fig. 4.7: Close-up of the lead sheathing on Nemi Ship 1

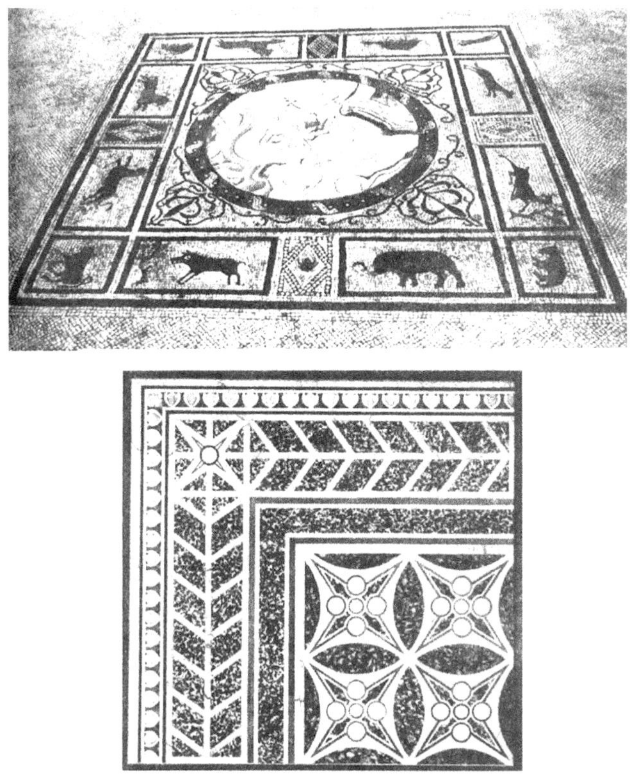

Fig. 4.8: Mosaic decorations from Nemi Ship 2

and bronze spikes were used to fasten the frames to the hull. On the deck of the Ship 1 were found remains of structures with gilded copper tiled roofs.[40] The walls and floors were decorated with different hues of marble *opus sectile* and polychrome mosaics (fig. 4.8). The style of the mosaics and various coins found within the ships indicate that they were still in use during the reign of Nero (54 – 68 CE). It is considered that the Nemi Ships were sunken not long after the senate declared Nero a public enemy and followed by his later suicide.[41]

The stempost of Ship 1, quite a long section of which has survived, indicates that it was concave and ended in a forward projecting cutwater (fig. 4.9). The starboard rudder of the same ship also survived. The shaft and a small fragment of the tiller were preserved. The blade is missing. Remains of the mortise-and-tenon fastenings on both sides of the shaft indicate that the blade was made of two wings each being attached alongside the lower shaft (fig. 4.10). The rudder was mounted between two projecting crossbeam (fig. 4.10a). The same arrangement was true for the port rudder, and similarly in Ship 2.

---

40 Ibid.

41 Carlson, 2002, p. 31.

Fig. 4.9: Stempost of Nemi Ship 1

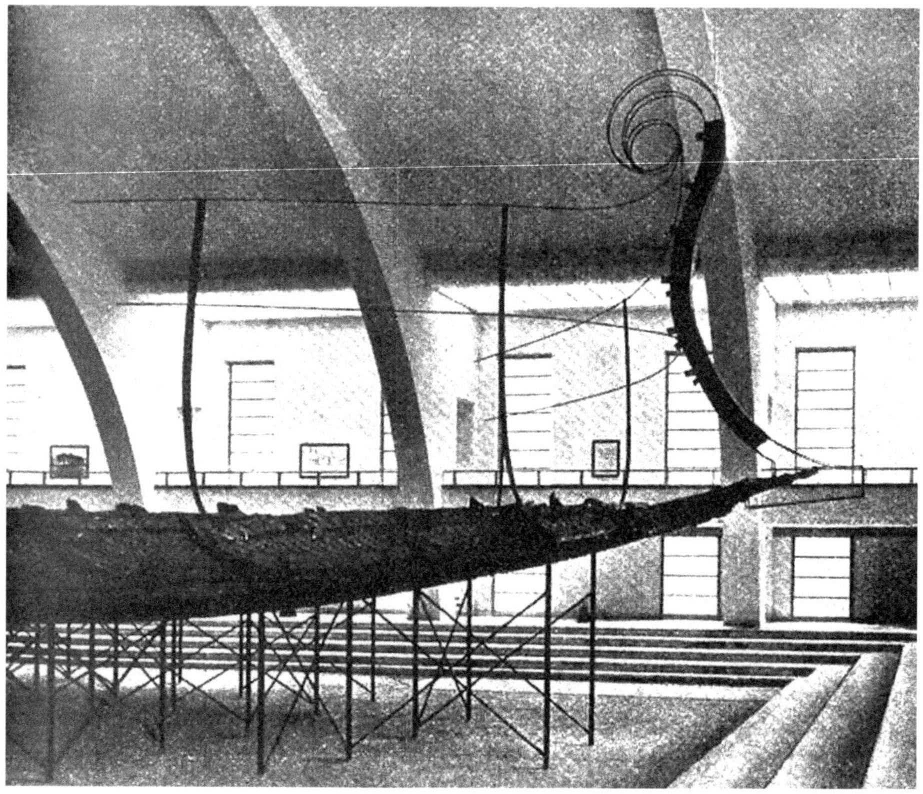

Fig. 4.10: The starboard rudder and its housing; Nemi Ship 1

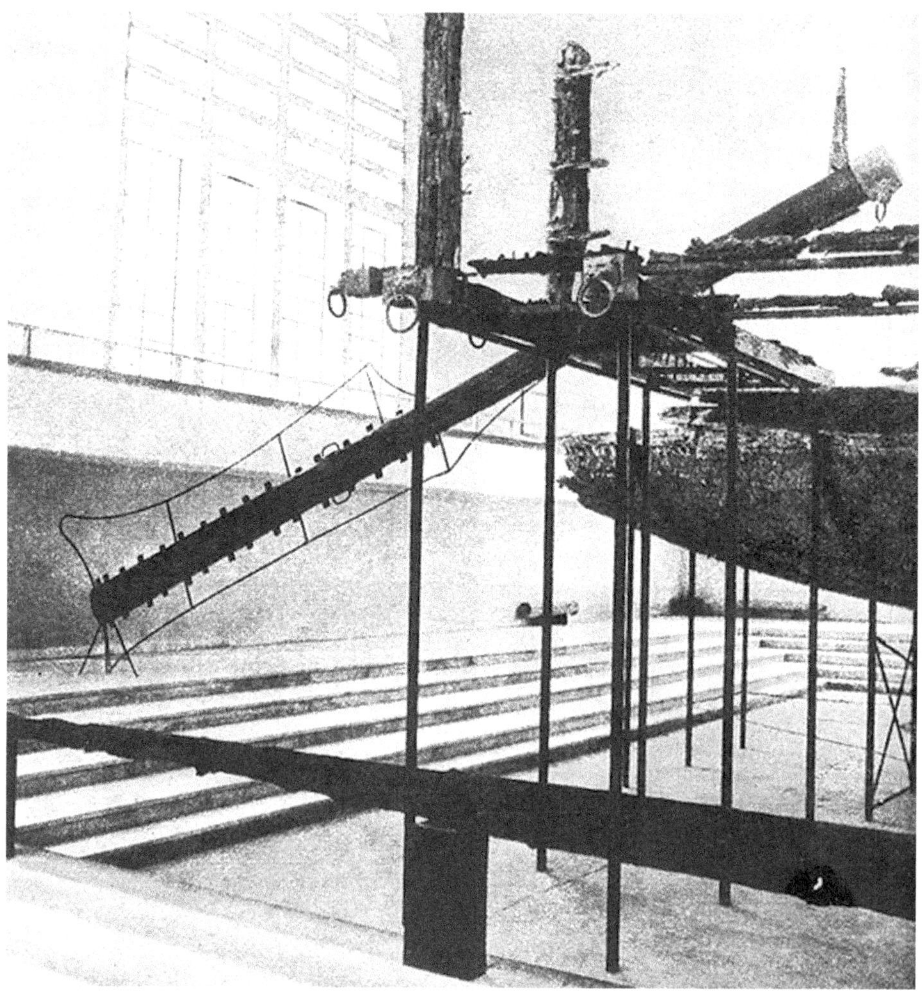

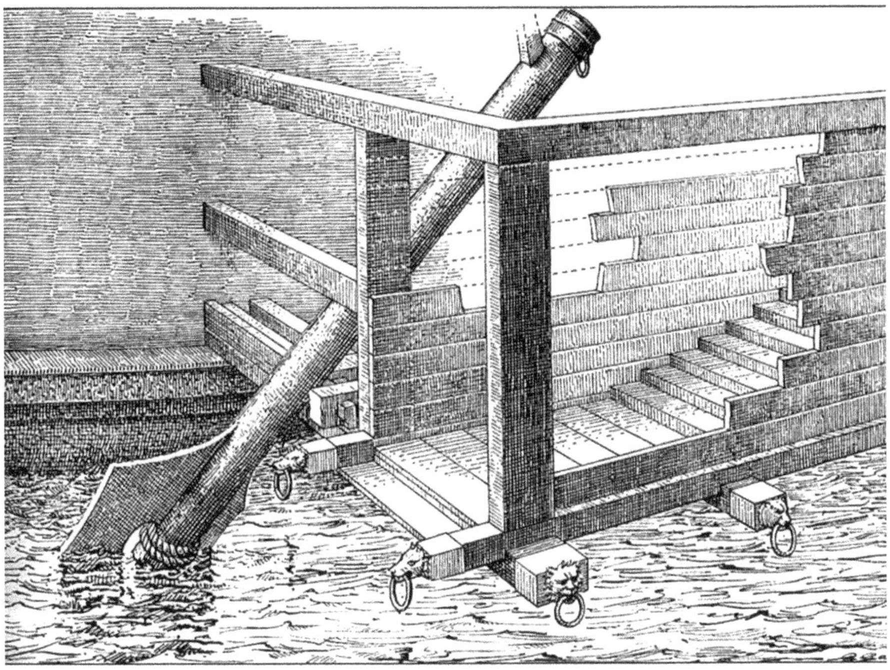

Fig. 4.10a: Reconstruction of the rudder's mounting

The main interest in the Nemi Ships (although they did not have commercial functions) is because they show that the Romans were capable of building huge ships as is known from historical records (1st century CE). The vessels were destroyed by fire on May 31, 1944, probably deliberately, in protest against Mussolini who was in alliance with Germany during the Second World War. Since 1999 the building of a full-scale replica is underway.[42] The Nemi Ships were not seagoing vessels but they may be considered an evidence to illustrate the *thalamegos* (cabin-carrier) described by Athenaeus. The luxurious vessel engaged in hippopotami hunting depicted in the Palestrina mosaic may be considered a smaller version of a *thalamegos* (fig. 3.6.17).

Italian Shipwrecks

The seagoing ships depicted in the mosaics described in this monograph may be compared with some shipwrecks found in the Mediterranean and closely dated to the periods of the mosaics. Not all the ships found around the coasts of Italy will be described below, but only some examples which are relevant to the mosaics with ship depictions discussed above.

The Isis Wreck, Sicily:[43] The late-Roman wreck is located about 120 km North of Tunis and the ancient Carthage, *c.*160 km west of Trapani, on the NW tip of Sicily, and about 80 km north of Skerki Bank.[44] The wreck was the first one to be found and surveyed with the latest technology of ROV (Remote Operated Vehicle) in 1989.[45] It lies at a depth of 800m and was discovered during the JASON Project by Dr. Robert Ballard of the Woods Hole Oceanographic Institute.[46] Due to the great depth of the site the shipwreck could not be excavated, but several objects and coins were removed and considered for the research of the wreck, pointed to the dating of the last quarter of the 4th century CE.[47] The *terminus post quem* of "*Isis*" Ship is provided by a copper coin of Constantinus II, dated 355 – 361 CE, which was embedded in a pot of pine tar.[48] Apparently the tar was used on deck to maintain the ship. The vessel was built of white oak and Mediterranean cedar, in the ancient shell-first technique with planks joint by mortise-and-tenon fastenings. It was estimated that the ship's length was between 12 -15m, and 4m abeam, with a load capacity of *c.*30 tons.[49] The outer hull was not sheathed with lead but lead patching was observed, probably of partial sheathing or maybe repairs in the damaged areas of the hull.

A study of the artifacts and the location of the wreck suggest that her homeport or the last port of call was probably Carthage and that *Isis* Ship was *en route* to Rome. The coin found embedded in the tar, though of Alexandrian mint, could have been in circulation at Carthage.[50] The cargo of amphorae and other objects in the wreck attest that in the immediate pass before leaving Carthage the ships probably sailed in the eastern Mediterranean. Its final destination probably was Ostia and Rome. The wreck provides evidence of a trade route over open sea in the western Mediterranean which resulted in her wreckage about 80 km N of the Skerki

42 Rosario D'Agatta, former public relations director of an Italian oil company, established the "Association Diana Lacus", to reconstruct one of Caligula's Nemi Ships. It is estimated that the total coast of the project will be $10 million, and it will probably take about two years of work. The central keel, stem and stern are made of oak, and their cost is $50.000; Carlson, p. 26.

43 The shipwreck was name "Isis", for the benefit of the television program for children; McCann and Freed, 1994, p. 3.

44 McCann and Freed, p. 3.

45 Ibid.

46 Ibid., fig. 4, p. 5, fig. 6, p. 6.

47 Delgado, p. 208.

48 McCann and Freed, fig. 18a, p. 17, fig. 18b, p. 18.

49 McCann and Freed, p. 49; Delgado, p. 208.

50 McCann and Freed, p. 55.

Fig. 4.11: Comacchio Wreck

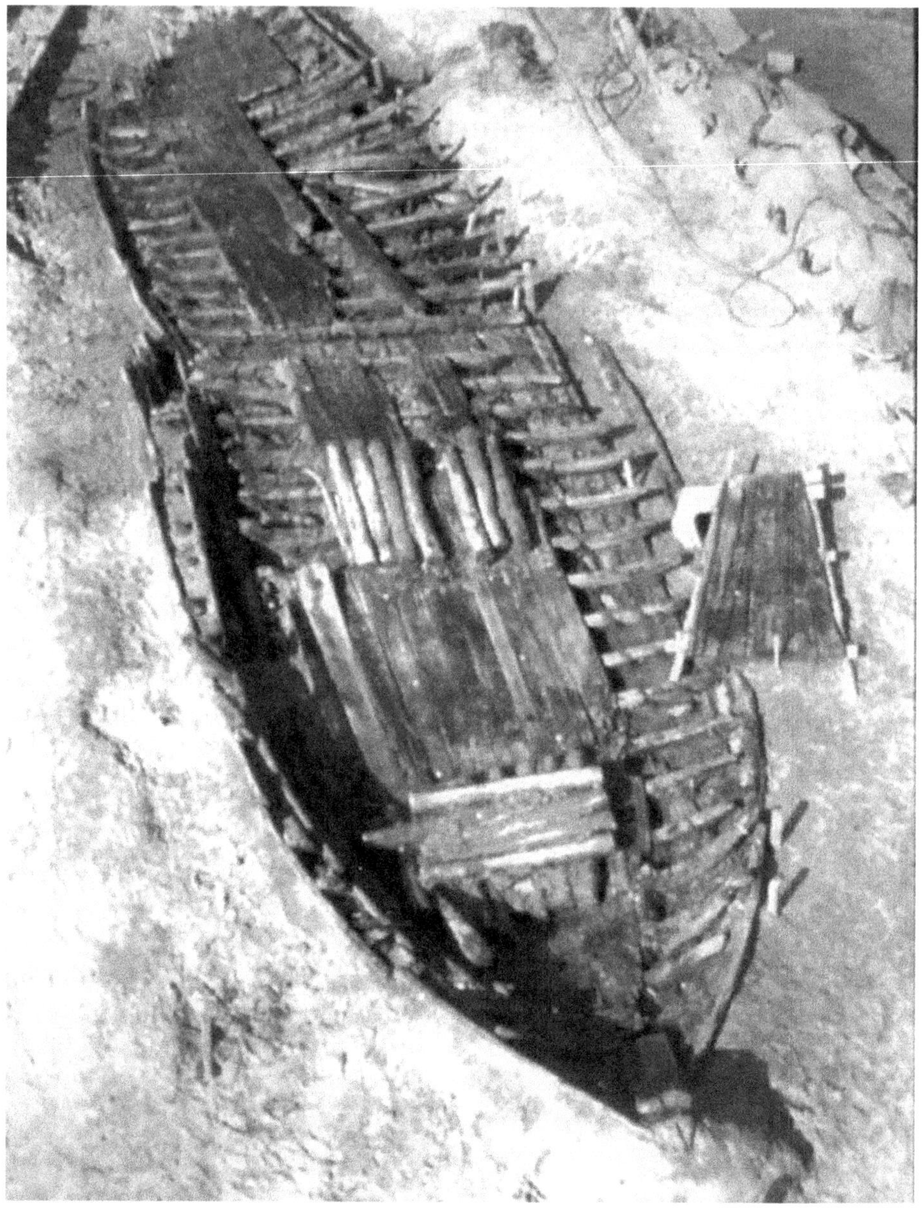

Bank, a long reef that lies only 1.8m bellow the sea surface. It has caused many shipwrecks, both ancient and modern.[51]

Comacchio Wreck: Several timber fragments of the wreck were revealed in the fall of 1980 by the Comacchio archaeologists, during dredging operations in the Valle Ponti canal, on the outskirts of Comacchio, north of Ravenna. The wreck was covered by a layer of 3.5m of silt. The upper portion of the wreck and its cargo were recovered during an extensive excavation season in the summer of 1981 (fig. 4.11).[52] The internal planking and frames were removed between the end of the autumn of 1986 and the winter of 1987, while the hull planks and futtocks of the port side were left in place. The recovery of the entire ship took place during the winter of 1988 - 1989. The length of the preserved hull was about 20m. Part of the port side was still connected to the stern gripe up to the wale, while towards the bow (identified by the presence of an anchor) the planks were detached and overlapping.[53] The bow and the adjacent section, as well as the starboard side, were missing. The research of the hull revealed that the vessel had a flat bottom. Its original dimensions were 21m in length and 5.62m in width. When fully loaded, it must have weighed 130 tons. The ingots, the dendrochronological analysis of the ship's timbers and samples of the boxwood timber cargo, as well as the ceramic shipment, dated the ship to the last decade of the 1st century BCE.

The primary cargo consisted of 102 lead ingots (3 tons weight), stamped with the name of Agrippa (Emperor Augustus's son–in-law),[54] boxwood logs, and amphorae for food transport. *Sigillata* pottery originating in North Italy was also carried on board, as well as six tiny lead foil votive temples. Among various materials belonging to the ship supplies was a bronze balance, probably used for weighing merchants, or fish for trade. Numerous items of

51 Ibid., p. 3.
52 Berti, 1998, http://www2.rgzm.de/Navis/Home/NoFrames.htm
53 Ibid.
54 Delgado, p. 105.

clothing, containers, and leather shoes belonged to the crew and passengers. Auxiliary items comprised, a series of tools for ship's maintenance (mallets, an axe, a plane), and for its steering (sheave blocks, a boat bailer, an iron anchor).[55] The Comacchio ship is a rare evidence of a merchantman transporting wooden logs for trade.

The vessel was wrecked by wind and currents during a sea storm, that pushed it near the waterline, at the river's mouth. The superstructures were destroyed by waves. The vessel sank into the sand, which was aggravated by the considerable weight of the cargo.

The Fiumicino Ships

Several ships of different sizes came to light during the construction of the Leonardo da Vinci International Airport of Fiumicino in 1958; the ships are now displayed at the "*Museum of Roman Ships*" within the compound of the airport. The black appearance of the hulls when revealed is a result of carbonization or reduction process activated by microorganisms living within the sedimentation layers.[56] Parts of the submerged sections of the hulls, which were not yet covered by sand and lime, befell to the wood-boring organisms such as *teredo navalis* (ship worms). The first ship, Fiumicino 2 (*oneraria maggiore* II), was revealed in 1958 and immediately, extensive excavations were carried out under the direction of Dr. Santa Maria Scrinari, which lasted until 1965. The wrecks were found adjacent to the right mole of the Claudian harbor, in a marginal area of the harbor basin.[57] It is assumed that in ancient times the site may have been a "cemetery" where boats and old ships or in poor condition to be in service were abandoned there.[58] The majority of the lower hull and the bottom of the vessels survived. Some of the ships still had their upper hulls, but during the excavations the timbers left exposed to the open air, suffered significant degradation and also pillaging for firewood. The ships were preserved by a mixture of resin at the *Instituto Centrale del Restauro* (Central Institute of Conservation) in Rome, and afterwards their hulls were consolidated. Steel frames support the ships that are displayed in the museum.

The ships were built with stone pine (*Pinus pinea*), cypress (*Cupressus sempervirens*) for the planking. Oak (*Quercus sp.*) was also used for planking, but especially it was for the keel and keelson, and the frames. Walnut (*Juglans regia*) was identified in some frames in Fiumicino 4 ship. The holm (*Quercus ilex*), the ash (*Fraxinus excelsior*) and olive tree (*Olea europaea*), were used for the tenons and pegs. The black pine (*Pinus cf. nigra*) was also identified in many of the pegs. The willow (*Salix sp.*) was used for the pre-inserted treenails connecting frames to the planking.[59]

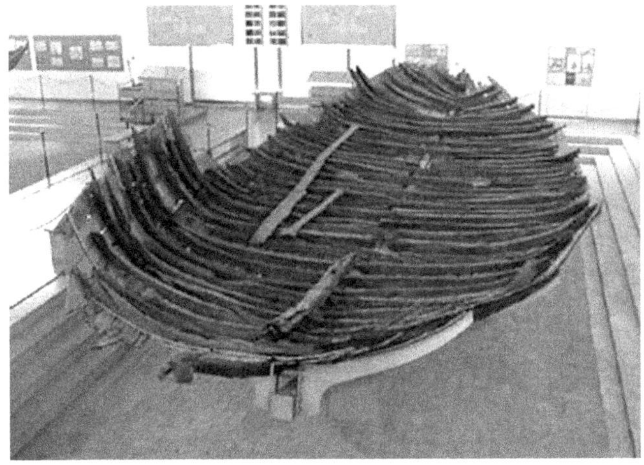

Fig. 4.12: Fiumicino 1/oneraria maggiore

Fiumicino 1 (inv. 37797) (*oneraria maggiore*) was excavated in 1959.[60] The hull was salvaged in 1961 and after conservation with a mixture of resins it was exhibited in 1979, in the "*Museum of the Roman Ships*" (fig. 4.12). The preserved length is 13.8m (originally it may have been 21m) and its width is 4.57m (originally 7m). The height of the starboard side was preserved at 1.47m (originally, 3.5m).[61] The surviving keel was made of two pieces, the sterngripe and the keel. The hull is single planked and carvel built, in the shell-first fashion with tenon-and-mortise joints. Numerous repairs of the hull have been fixed to the frame only by iron nails, while in the adjacent planks are with the mortises fastenings. The ship was found with 42 frames, rectangular or trapezoidal in section (6/10 cm wide and 7/12 cm high), while the floor timbers aftward are higher (from a low of 13 to a high of 18 cm).[62] In general, the frames are arranged in the common pattern of floor timbers alternating with half frames though their distance is irregular (on average 19 cm). On the starboard side, the preserved futtocks are not fastened to the floor timbers. Rectangular limber holes (5 cm wide and 3 cm high) were cut into the bottom of the frames. They were cut one in a central position along the keel and two laterally along the third strake. Where the half frames are very narrow there is a half limber hole on each element.[63] The mast was stepped forward of the center of gravity in the fore part of the vessel. The maststep has an internal slide, thus indicating that it could be lowered on the deck when the vessel was anchored. Two displaced fragments of floor planks were found atop the frames. The longer one is 5.29m while the second measures 1.71m; both are 14 cm wide and 3.5 cm thick. The very rounded stern, the elongated stem, and the mast which was stepped forward of the centre of gravity in the Fiumicino 1 Ship indicate a typical river barges known as *caudicaria navis*.[64] It is assumed that the ship was towed

[55] Berti, 1998, http://www2.rgzm.de/Navis/Home/NoFrames.htm
[56] Boetto, 1998, http://www1.rgzm.de/navis/Musea/Ostia/Fiumicino_English.htm
[57] Ibid.
[58] Ibid.
[59] Scrinari, 1979, pp. 21, 26, 27, 33, 37, 43; Boetto, http://www2.rgzm.de/Navis/Home/NoFrames.htm.

[60] Ibid., p. 37.
[61] Scrinari, Tav. VI, p. 57; Boetto, 2000, p. 99.
[62] Boetto, http://www2.rgzm.de/Navis/Home/NoFrames.htm.
[63] Ibid.
[64] Boetto, 2000, p. 102.

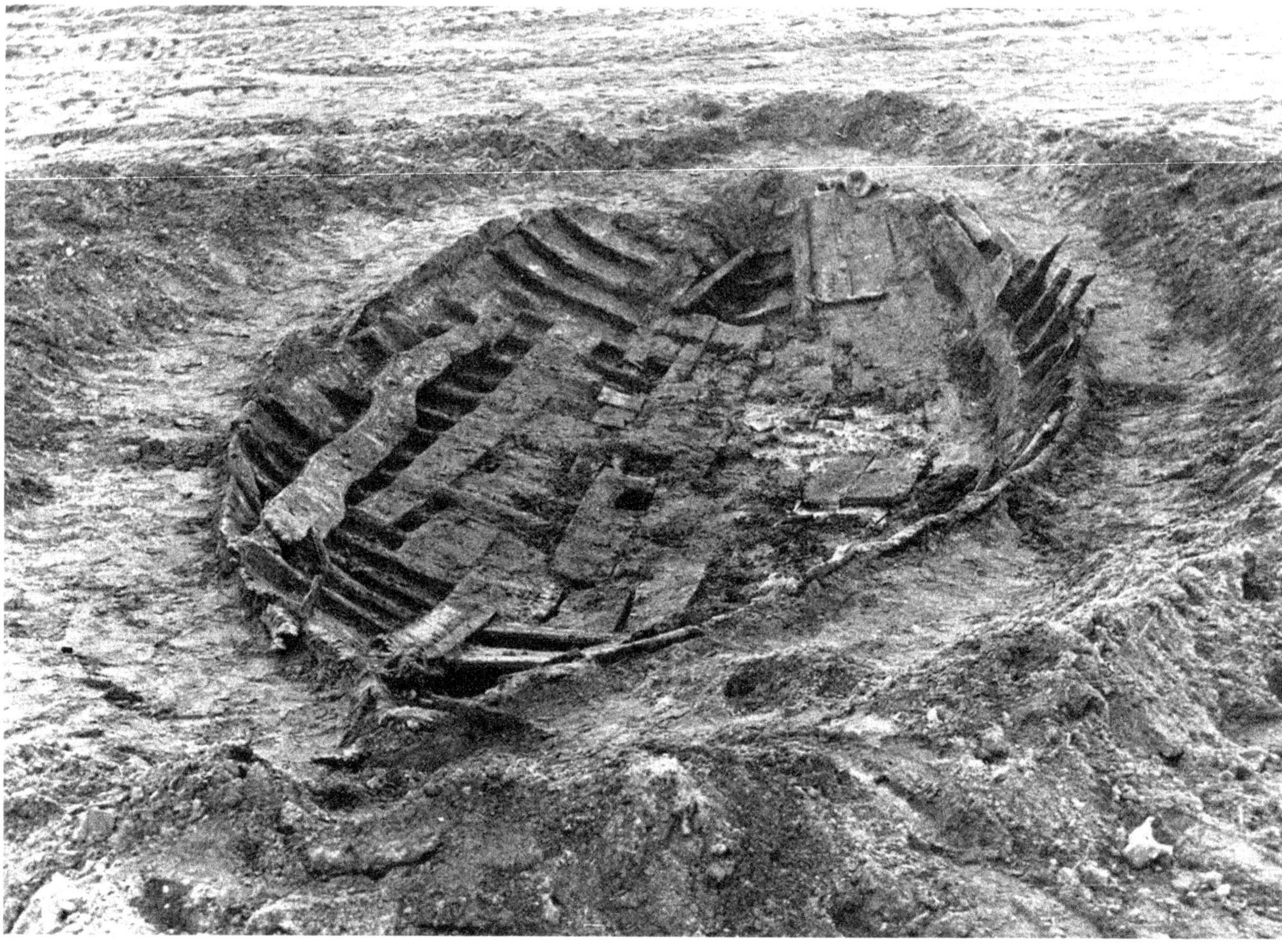

Fig. 4.13: Fiumicino 4/oneraria minore

on the Tiber up to Rome.[65] The displacement of the ship when fully loaded was 80 – 100 tons; the cargo capacity on board was 50 – 60 tons. The vessel was dated to the 3rd – 4th century CE.[66]

Fiumicino 2 (inv. 37798)[67] was the first and the largest ship found in 1958 that was recovered in 1959. After the conservation process, made with a mixture of resins and some restorations with modern pieces of wood, the wreck was exhibited in 1979 in the "*Museum of the Roman Ships*". Fiumicino 2 is a flat bottomed ship which is very similar to the Fiumicino 1 wreck. Its preserved length is 14.5m (originally 24m), the width is 5m (originally 8m), and the port aft side survived to a height of 0.85m (originally 4m).[68] The planking is heavily damaged and the transversal woodwork has been subject to many repairs. The ship was construction is shell-first technique deduced from the general homogeneity of the planking.[69] The ship had 35 frames that were connected to the planking with iron nails (squared shaft of 1/1.3 cm; head diameter 4/5 cm) driven through pre-inserted treenails (1.2/1.5 cm).[70] Three floor timbers were connected to the keel with iron bolts similar to the big nails driven in from the bottom of the keel (diameter 2.2/2.4 cm). The displacement of the ship when fully loaded was 80 – 120 tons, whereas the cargo capacity was probably 50 – 80 tons. The general structure of the hull, with the curved stern and the elongated stem suggests that Fiumicino 2 could be a *caudicaria navis* type. The ship was dated to the 3rd – 4th century CE.[71]

Fiumicino 3 (inv. 37799)[72] when discovered had a preserved length of 7m (originally 12m), and a width of 2.66m (originally 4m). Within the wreck 19 frames were preserved. The hull is much damaged, whereas the stem and stern posts are missing. The construction of this vessel is similar to all Fiumicino ships, shell-first with mortise-and-tenon fastenings. This vessel is a small cargo ship (*oneraria minoris*) with a cargo capacity of 22 – 28 tons, probably used as an auxiliary to transport cargo from the large merchantmen anchored in open sea, up the Tiber to Ostia and Rome, or for fishing. The wreck is dated to the 3rd century CE.[73]

Fiumicino 4 (inv. 37794)[74] was found in 1965 at the site of

65 Ibid., p. 101-102.
66 Scrinari, p. 57, Tav. VI.
67 Ibid., p. 43.
68 Ibid., p. 43; Boetto, http://www2.rgzm.de/Navis/Home/NoFrames.htm.
69 Boetto, http://www2.rgzm.de/Navis/Home/NoFrames.htm.
70 Ibid.
71 Scrinari, p. 56, Tav.VI.
72 Scrinari, p. 26.
73 Ibid, p. 57, Tav. VI.
74 Ibid., p. 27.

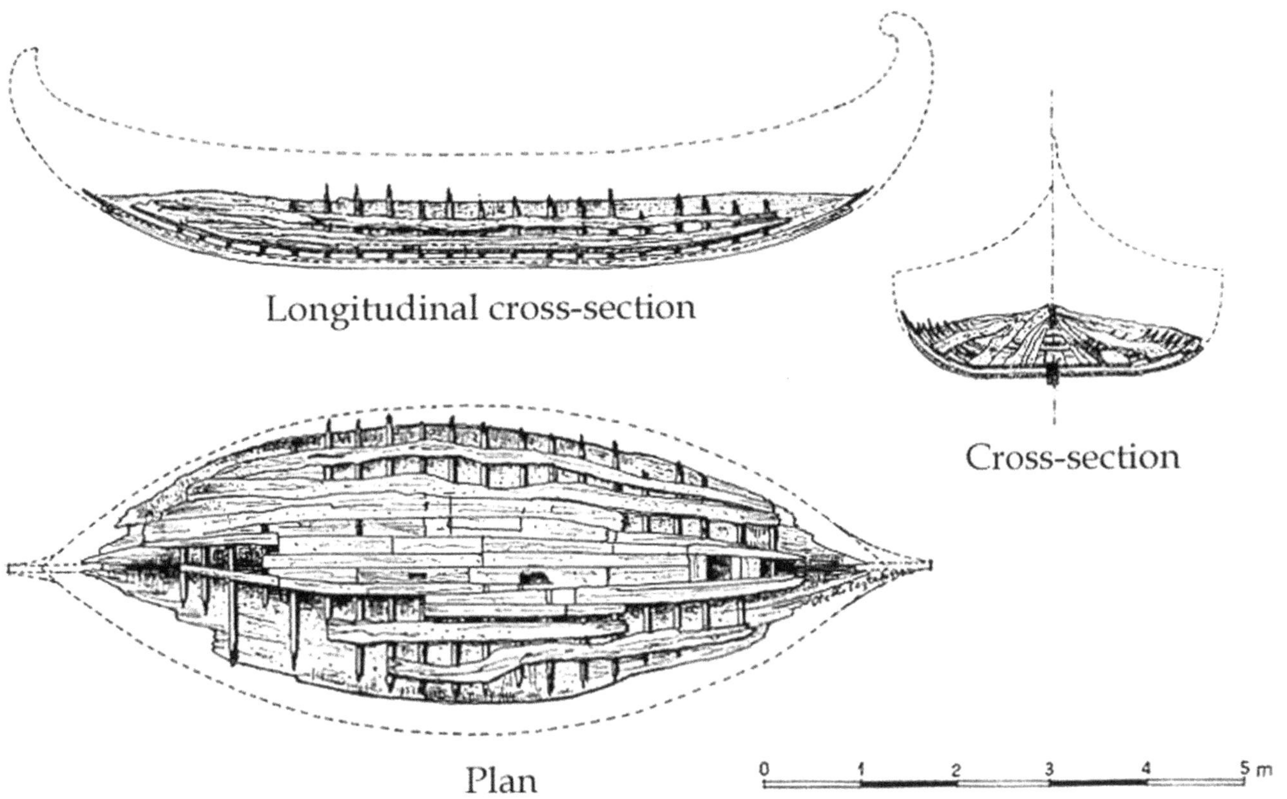

Fig. 4.13a: Fiumicino 4 plan

the Portus ancient harbor basin. The wreck was salvaged in 1968 in very good condition (fig. 4.13). After a conservation process made with a mixture of resins, it was exhibited in 1979 in the "*Museum of the Roman Ships*". Fiumicino 4 has a slander shape with a concave-convex section (fig. 4.13a). It has survived at a length of 7.96m (originally 13.5m), and a maximum width of 2.79m (originally 4.5m). The starboard side is preserved at a maximum height of 0.77m (estimated height amidships 2.25m), while the port side was broken along the knee. The stern is missing, probably damaged during the pillage following the discovery of the ship. The stern-and-stem gripes have rectangular sections (6/6.5 cm wide and 13/14 cm high) with two rabbets to fit the garboards. The hull was single planking and carvel built. There were a total of 39 planks, of which 23 were on the starboard side and 17 on the port side. The ship was found with 22 frames that their average distance was 25 cm (their cross section is rectangular, squared or rhomboidal due to the deformation of the wood).[75] The frames were fastened to the planking by wooden treenails (diameter 1.3/1.5 cm). In the starboard side were found 5 futtocks nailed only to the floor timbers. The mast step (1.17m long, 1.46m wide and 0.12m high) was set parallel to the keel and had a rectangular recess with a slide for the foot of the mast. Aftwards there is a recess to insert a stanchion to support the mast vertically.[76] A pump to remove bilge water was located in a socket within the crutches flanking the mast step. No other Fiumicino vessel was fitted with a bilge-pump. Fiumicino 4 was a small cargo ship (*oneraria minoris*) with a cargo capacity of 28 – 31 tons.[77] The vessel was used in coastal trade or as a fishing ship. The propulsion system comprised a single squared sail supported by a mast set slightly forward in the ship. Fiumicino 4 was dated to the 3rd – 4th century CE.[78]

Fiumicino 5 (inv. 37792)[79] is a unique fishing boat, revealed in 1959 that has survived almost complete (fig. 4.14). The hull was salvaged in 1961 and after conservation it was exhibited in the "*Museum of the Roman Ships*" in 1979. Its preserved length is 5.2m (originally 5.5m), abeam 1.5m (originally 1.8m) and the surviving height of the remaining hull is 0.55m (originally 1m). The boat is single-planked and carvel built. The hull comprised 6 strakes on the port side and 7 on the starboard side. Inside the hull 18 frames were connected to the planking by treenails (diameter 0.9/1 cm).[80]

The troncopyramidal fish-well with a squared base (1 x 1m), a box made of very thick planks (5 cm) assembled with mortise-and-tenon joints is placed amidships (figs. 4.4, 4.14a). The corners are stiffened with iron nails. The bottom of each transversal side is shaped with the curvature of the hull and was fastened to the planking by copper nails driven in from the outside. The tank measures 1.3m in length, 1.05m in width and 0.53m in depth, and on the

75 Boetto, http://www2.rgzm.de/Navis/Home/NoFrames.htm.
76 Ibid.

77 Scrinari, p. 57, Tav.VI.
78 Ibid.
79 Ibid, p. 21.
80 Boetto, http://www2.rgzm.de/Navis/Home/NoFrames.htm.

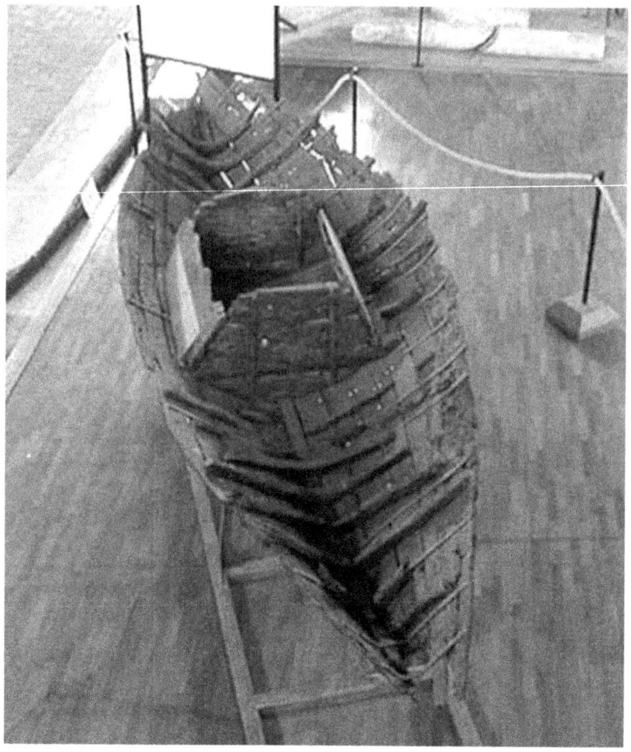

Fig. 4.14: Fiumicino 5 – fishing boat

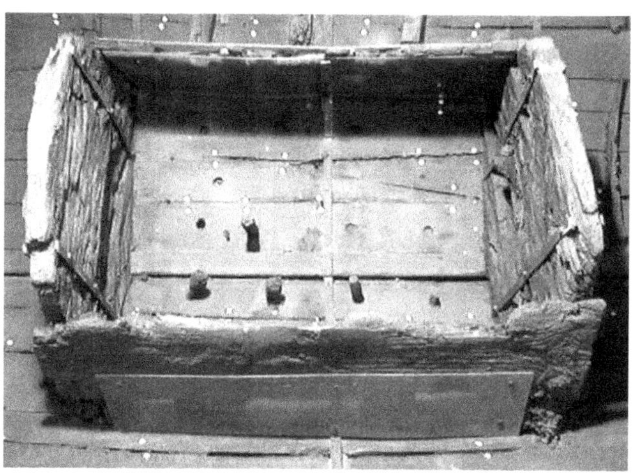

Fig. 4.14a: Close-up of the fish tank

bottom of the well are 19 holes (diameter 2.5/3 cm). When the tank was filled with sea water, the holes were closed with wooden conical wooden pegs, 7 cm long and 2 cm in diameter.[81] This tank was designed to transport caught fresh fish to the market. The boat was propelled by oars; it was dated to the 2nd century CE.[82]

Fiumicino 1, **2** and **3**, having similar constructional characteristics, of flat bottom and broad hull are features which indicate vessels used for river transport.

[81] Scrinari, p. 21.
[82] Ibid. p. 57, Tav. VI.

They probably were auxiliary vessels known as *naves caudicariae*, and carrying cargoes unloaded from the large merchantmen anchored in open sea, at close proximity to the Tiber mouth, and then were sailed up to Ostia and Rome, as well as used for harbor works. They must have been towed by animals along the right bank of the Tiber by a propulsion system known as "tracking", still in use along the river in the nineteenth century.[83] The hull of Fiumicino 4 indicates a small to medium size vessel sailing along the coast. Fiumicino 5, a unique find of the Roman period, is a special fishing boat adjusted with a fish-well at its center to store the freshly catches and keeping it alive in sea water, to be transported to the market.

The Pisa Ships

In December 1998, during the work to lay the foundations of the new railway station in Pisa, at San Rossore, the remains of the ancient Etruscan and Roman harbors were uncovered. The excavations carried out within the Roman period harbor basin, revealed 39 ancient shipwrecks buried under nine centuries of silt; 16 vessels have been partially or fully excavated so far, and their date range from the end of the 1st century BCE to the late 5th or the beginning of the 6th century CE (fig. 4.15).[84] The vessels are very well preserved, due to the abundance of groundwater at the surface and the absence of oxygen in the sandy layers.[85] The excavated wrecks comprise three cargo ships, one oared vessel, three boats probably used on the Arno River, and the remains of a ship that had overturned. The cargoes of most of the ships contained sealed clay amphorae filled with olives, nuts and wine. One amphora contained the ancient Roman fish condiment known as garum. Objects of the sailors' personal possessions were also found in these ships. Many mariners' skeletons were discovered in the area of the shipwrecks; one crushed under the weight of a capsized ship.[86]

Researchers say that starting around the 6th century BCE the docks of the port of Pisa were accessed by a canal that made a loop connecting it to the open sea. Pisa also was an important Roman naval base. Ancient sources mention fleets sent from Pisa to Gaul and Spain during the Second Punic War (218 - 201 BCE). The gradual silting of the harbor was probably due to the accumulations of the silt brought by the Auser River that evidently started from the middle of the 1st century BCE.[87] New harbor installations were built in the silted southern part of the basin during the reign of Tiberius. Actually every hundred years or so over the course of nearly a thousand years *tsunami*-like waves violently flooded the waterway capsizing and burying ships, their cargoes, along with their passengers and crew, even tiny birds and animals, and uprooted trees .[88] The sinking of the Pisa vessels probably occurred for several reasons

[83] Boetto, http://www2.rgzm.de/Navis/Home/NoFrames.htm.
[84] Bruni, 2000, p. 41.
[85] Ibid., 22.
[86] Ibid., p. 37.
[87] Ibid., p. 39.
[88] Ibid.

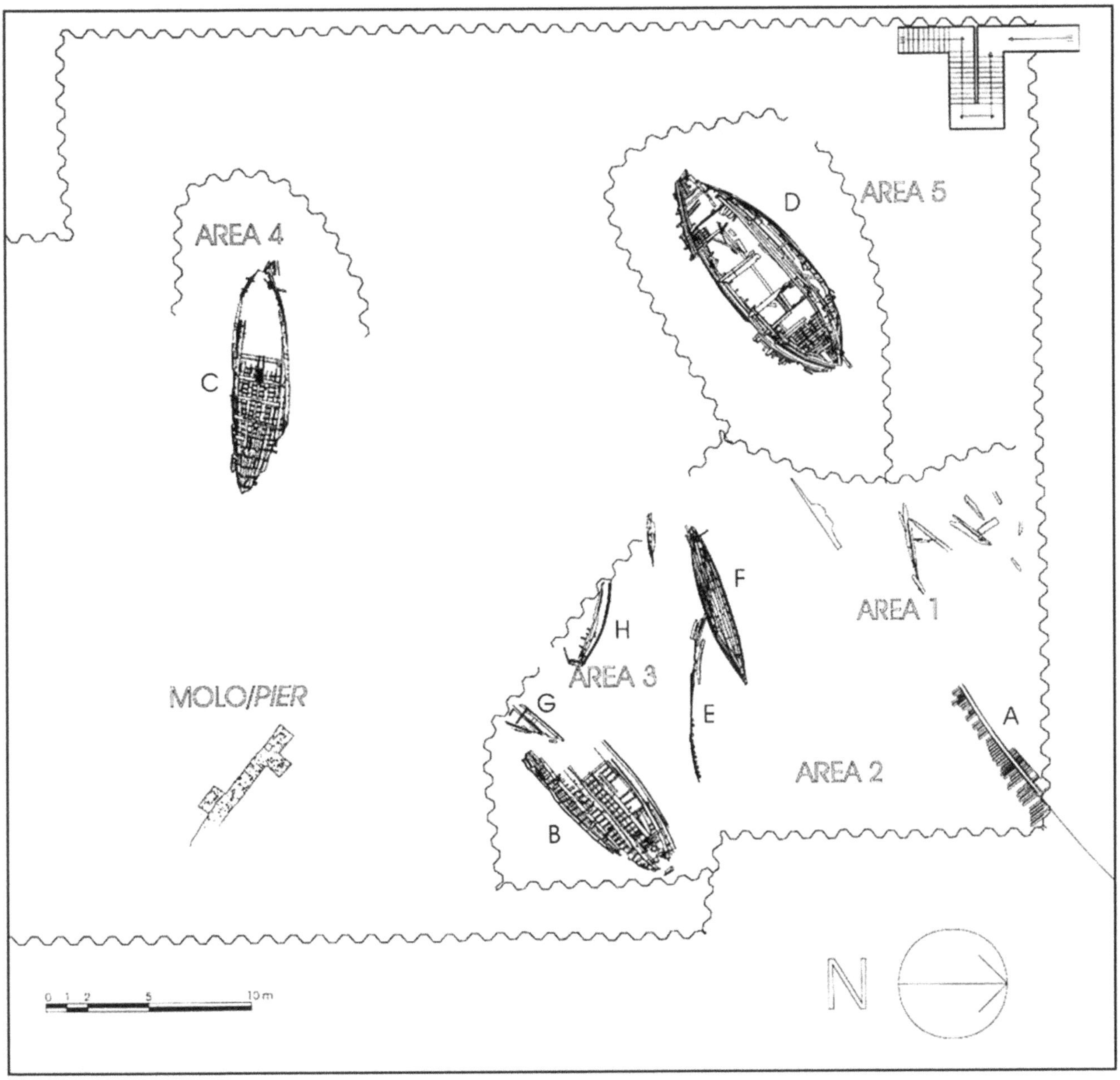

Fig. 4.15: Plan of the Pisa Wrecks

connected with navigational error, poor atmospheric conditions, or sudden flooding of the Arno River.[89]

The Hellenistic Ship probably sank in the early decades of the 2nd century BCE, due to the destroyed pier running northeast – southwest.[90] The sinking of the ship was a particular event as deduced from human bones discovered with the remains of the cargo, and most probably they suggest that some sailors of the ship's crew lost their life at that time.[91] The cargo comprised Greco-Italic and Punic amphorae which contained wine. The contents of the Punic jars contained pieces of butchered pork. Some of the animal bones found together with the amphorae comprised the food of the crew, while the remains of three horses and an adult lioness are regarded with the cargo of the ship. The presence of the lioness[92] indicates that the vessel sailed from North Africa to Pisa, following a route with stops along the coast of Sicily and Campania.[93] The massive timbers of the wreck (one measured over two meters in length) indicate that originally the ship was a very large *navis oneraria* (merchantman).

Wreck A was the first to be excavated. It was found in the northeast corner of the site (fig. 4.15). The ship is a large *navis oneraria*, deduced from its preserved dimensions, 15m in length starting from the stern; apparently the original

[89] Ibid.
[90] Ibid., p. 37.
[91] Ibid.

[92] Lions were used in the Roman world from the beginning of the 2nd century BCE, in gladiator shows of important events, such as triumphs or other celebrations; Bruni, p. 38.
[93] Bruni, p. 38.

length of the vessel was 25 – 30m.[94] Some of the cargo was still on the bottom of the vessel laid on the planking. Few item directly related with the wreck, and the stratigraphy analysis of the area indicate that the vessel sank around the end of the 2nd or the beginning of the 3rd century CE.[95] The varied cargo and personal belongings on board suggest that the ship was sailing from the Adriatic Sea to the North African coasts. Presumably it returned from the Iberian Peninsula, and passed through Gallia before reaching Pisa, where it sunk.[96]

Wreck B appears to be the oldest, as dated to the 1st century BCE – CE. It was a large cargo ship of 10m in length (originally *c*.20m long), and its preserved width is 4.3m. The wreck was found as lying on its right side (fig. 4.15). The hull built with oak and fir-wood was double planked, whereas many frames were still in *situ*, as well as some stanchions.[97] The cargo on board and around the wreck comprised Dressel 6 and Lamboglia 2 Adriatic amphorae.[98] Some jars contained fruits (cherries, peaches, plums, olives), nuts (walnuts, chestnuts), and sand originating from Campania and Vesuvius. Marble statue fragments and small blocks of Vesuvian lava rock formed the ballast. Some human bones have been recovered in the area of Wreck B. One skeleton of an adult male (aged 35 – 40) was found with a small dog placed next to man's left forearm.[99] He probably drowned with his pet while still being on board of the vessel. Almost all the found skeletons were men, but one femur seems to belong to a woman.[100]

Wreck D was the best preserved vessel found at the excavation (fig. 4.15). The ship was mainly built with oak planks. When revealed it was in an upside-down position whereas its bottom and lower part of the hull were missing: e.g. the keel probably was cut off in antiquity. The vessel is about 14m long and its maximum width is 6m. The construction of this vessel is different from the vessels described above. Hulls with double planking nailed on the frames from the outside are evidenced by ships dated from the 6th – 7th century CE onward.[101] In the vicinity of the stern and stem of Wreck D were found pairs of projecting structure. These were set on large brackets attached above a parallel series of three beams, probably connected with the mast-stays.[102] The structures were probably were the hatchways providing access into the hold and the deck.[103] Apparently the mast-heel was still set in the socket and anchored with cables.[104] The external bulwarks (the extension of the sides of the ship above the deck) were reinforced with double gunwales.

[94] Ibid., p. 45.
[95] Ibid.
[96] http://www.cantierenavipisa.it/Scavo_NaveA.html.
[97] Camilli, 2002, p. 22.
[98] Ibid.
[99] Bruni, p. 344.
[100] Ibid., p. 345.
[101] Ibid., p. 51.
[102] Sedge, 2002, p. 75.
[103] Bruni, p. 49.
[104] Ibid.

Besides the wrecks indicating large *navis oneraria* (A, B, D), were revealed five smaller ships which probably were used on the Arno or Auser Rivers and the lagoon:

Wreck E: The remains of this small ship did not allow the excavators to identify its original size (fig. 4.15). The cargo found nearby comprised Hispanic, south-Gaul, and Corsica amphorae. The wreck was dated to the 2nd century CE, based on the type of amphorae found nearby.[105]

Wreck F was preserved at a length of 8.2m and just over one meter in width. When unearthed it laid on its side and partly on Wreck E (fig. 4.15).[106] The boat was built with oak planks and with few alder insertions.[107] The straight sides and raised ends of this vessel indicate that this craft was used in river and the lagoon. Therefore it may be associated with the *linter* class. The wreck was dated to the early 2nd century CE, as indicated by an *as* minted during the reign of Hadrian (117 – 138 CE).[108]

Wreck G was partly laying on Wreck B (fig. 4.15). It is a small flat bottom boat that its length survived at *c*.8m. The materials recovered from the overlaying layers indicate that the boat dates to the 1st – 2nd century CE.[109]Apparently, this boat is one of the earliest examples used in Arno and Serchio Rivers, as well as in the lake in northwest Tuscany. Actually the construction of the boat did not change and is still preserved to the present days.[110]

Wreck H (fig. 4.15): The remains of the flat bottom and straight lower flanks indicate that it belongs to the fluvial boats type that was punt. This type is still used in the present days for hunting in the marsh of Fucecchio.[111]

Wreck I had a preserved length of 12m, and a width of about 2.5m. It has a mixed building technique of mortise-and-tenon joins, but also iron nails were extensively used, in a similar way as in Wreck D described above. This vessel indicates a fluvial boat known as *linter* type. It probably was used as a transporter of cargo unloaded from the large vessels anchored in open sea and then sailed up the Arno River to Pisa and the surroundings. Similar boats are still used today in the Padule of Fucecchio.[112] The boat was dated between the 4th and the 5th century CE upon the African amphorae cargo on board and in the vicinity of the wreck.

Wreck C has been revealed in the southernmost part of the harbor basin and it is the only rowing vessel found within the complex of the Pisa Ships (fig. 4.15). Evidently, the ship sank while moored to a large post alongside the ship. The mooring line was still secured by a knot to a large iron

[105] http://www.cantierenavipisa.it/Scavo_NaveE.html
[106] Bruni, p. 45.
[107] http://www.cantierenavipisa.it/Scavo_NaveF.html
[108] Bruni, p. 46.
[109] Ibid.
[110] Ibid.
[111] http://www.cantierenavipisa.it/Scavo_NaveH.html
[112] http://www.cantierenavipisa.it/Scavo_NaveI.html

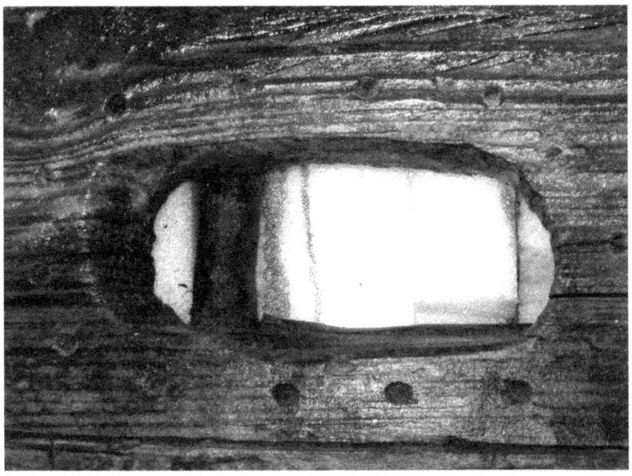

Fig. 4.16: Oarport – Pisa Wreck C

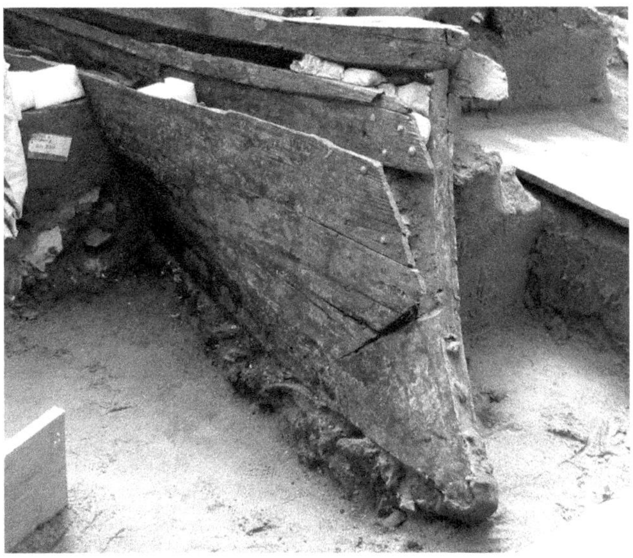

Fig. 4.17: Close-up of Pisa Wreck C pointed cutwater

ring attached to the upper side of the post.[113] The ship sank to the bottom in line with its natural inclination (the prow was much lower than the stern), thus it revealed the great shock of the waves impact.[114] The ship is in a very good state of preservation. Large patches of red or red-ochre paint applied in antiquity over a coat of white lead hue still survived on the side of the prow.[115] The vessel has a preserved length 11.7m, and 2.8m abeam. Much of its inner hull and the prow preserved all their structural elements: keel, keelson, timbers, mast partners and step, top bulwark and prow, as well as six rowing benches set at regular intervals.[116] In correspondence of the benches are set the elliptical oarports (the vertical tholepin set at 2/3rd of their width), which were cut through the upper plank, beneath the gunwale (fig. 4.16). Around the outer-rim of the oarports a series of small nails were found, which nailed the leather sleeves (*askoma*). It appears that externally the bulwarks were reinforced with two square gunwales running parallel above the quickwork.[117] The concave bow was fitted with a pointed cutwater (fig. 4.17) very similar to those we see in different art depictions. Probably it was coated with lead or bronze plates to protect the wood from the *teredo navalis* (ship worms) and other physical damages. Several items found inside the hull, near the prow, suggest that the vessel dates to the last reign of Augustus or the early Claudian period (1st century CE).[118]

A Greek graffito was incised on the first rowing bench of the Wreck C. The term, *ALKDO* (O D A) seems to be a Greek transcription of the Latin term *alcedo*, meaning gull.[119] We may assume that this graffito either suggests the name of the vessel, the name of the rower's seating place, or the sailor documented the bird he saw flying above or around the ship.

The Naples Shipwrecks

In the late 1990's – early 2000's during the construction of a new subway line in Naples, Italy, the archaeologists had the occasion to survey the revealed waterfront of the ancient town. A section of the ancient *Neapolis* harbor[120] was revealed in Piazza Municipio, near the modern Statione Maritima and the Aragonese Castel Nuovo.[121] Under 17m of sediment deposits, three Roman ships in quite well preserved condition were revealed (fig. 4.18).[122] The stratigraphy investigation indicates that two ships "Napoli A" and "Napoli C" were abandoned around at the end of the 1st century CE, near the mole facing east-west orientation.[123]

Napoli A was a seagoing merchantman (*navis oneraria*) that its length preserved at 11.77m, a width of 3.32m and a depth of 0.88m.[124] All the frames and some floor-timbers were still in *situ*, within the bottom of the ship.

Napoli B apparently sank at the end of the 2nd or the beginning of the 3rd century CE. Its preserved hull is 7.3m long and the width is 2m.[125] Remains of lime and calcareous stones comprised the cargo of the ship.

Napoli C has a preserved length of 13m, a width of 3.7m and a depth of 0.8m. Forty-five futtocks and fifty frames were found in *situ*, four strakes of the floor planks on each side and some internal boards lying parallel to the keel, and also ten strakes per side (fig. 4.19). Two internal posts associated with the transom bow are the most significant

113 Bruni, p. 47.
114 Ibid.
115 Ibid.
116 Ibid.

117 Ibid.
118 Ibid., p. 48.
119 http://www.cantierenavipisa.it/Scavo_NaveC.html
120 The harbor of Neapolis was established around the late 4th century BCE and remained active use until the 5th century CE, when the harbor basin silted up completely; Boetto, 2009, p. 290.
121 Boetto, p. 290.
122 Ibid.
123 Ibid.
124 Ibid.
125 Ibid., p. 291.

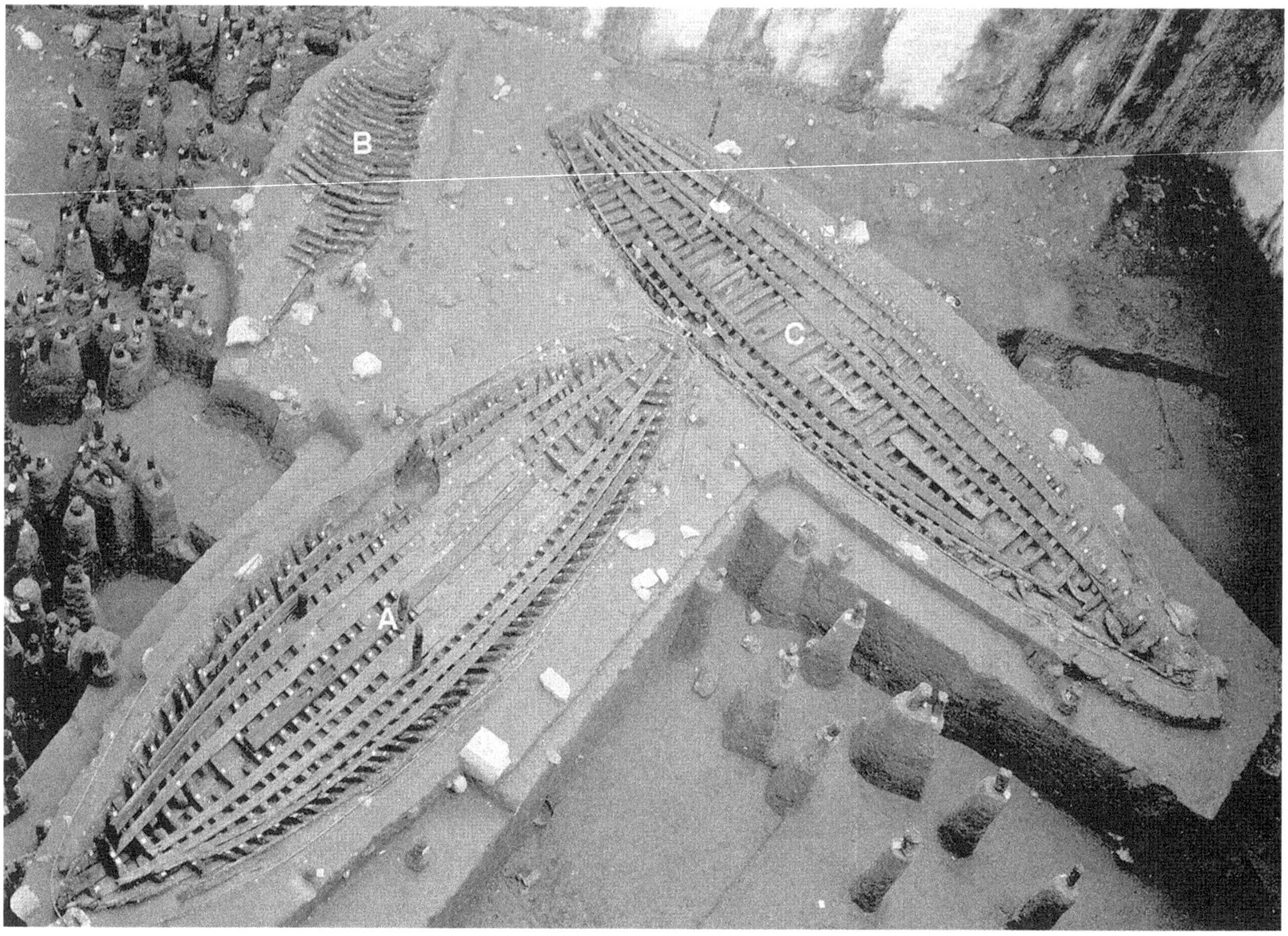

Fig. 4.18: The Napoli Ships

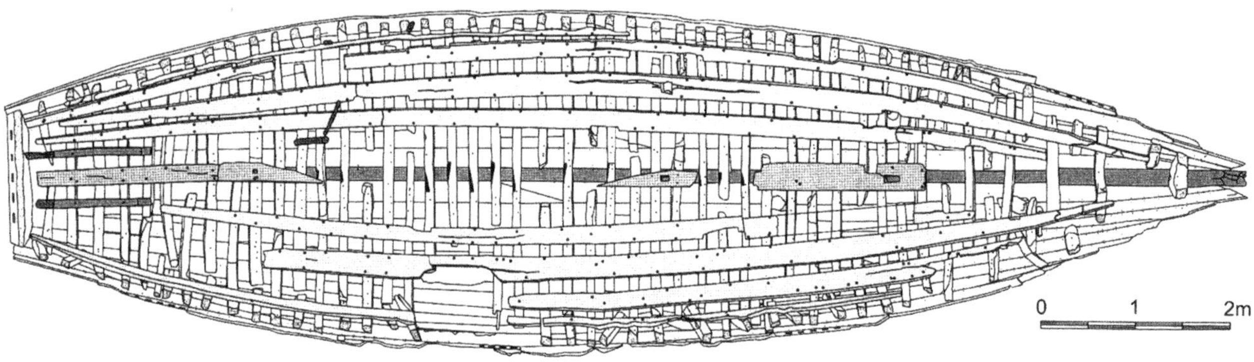

Fig. 4.19: Plan of Napoli C Ship

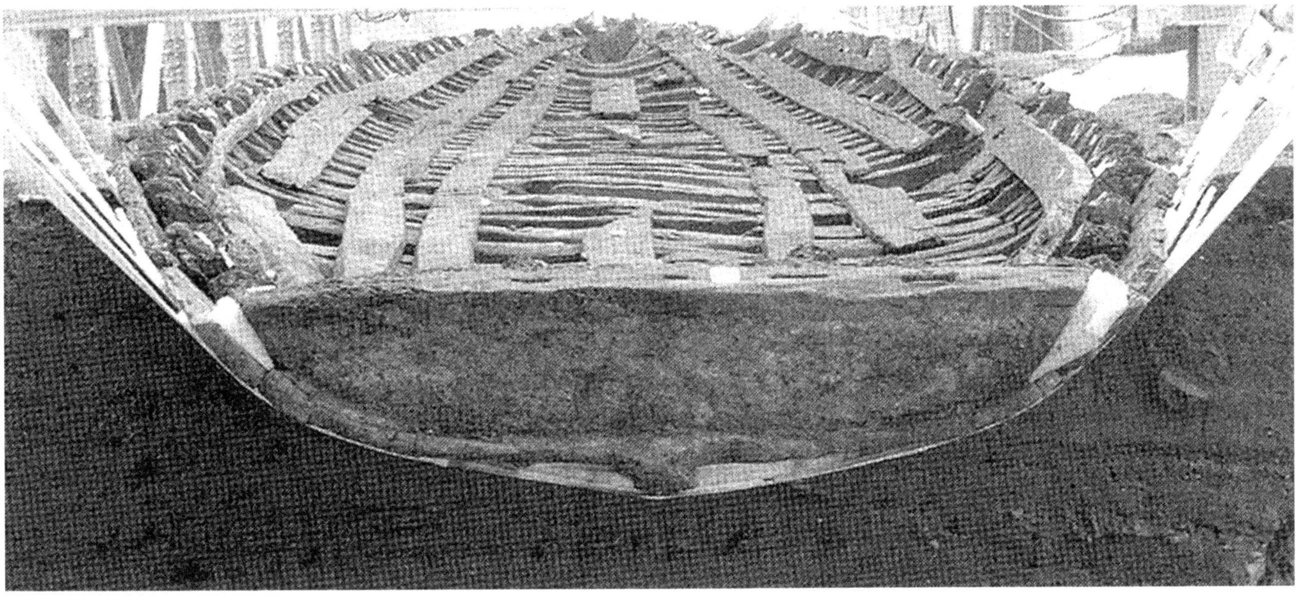

Fig. 4.20: Transom end of Napoli C Ship

finds (fig. 4.20).[126] Napoli C is one of the rare ancient ships to be found with a transom prow. A keelson of 6m long and 13 to 30 cm wide may have been provided with a mast-step (fig. 4.19).[127]

France

The southern coast of France houses a cluster of classical wrecks that when revealed, threw more light on ancient shipbuilding, shipping and trade in the western Mediterranean. Five wrecks will be described bellow with a date ranging from the 1st century BCE to the second half of the 4th century CE.

Dramont A wreck was discovered in 1956 by Claude Santamaria, at a depth from 25m to 40m under the sea. Several excavations campaigns were carried out from 1957 to 1961, while systematic excavations were undertaken from 1971 to 1979.[128] The amphorae cargo comprised Dressel I B type with different potter's marks. The hull was preserved to a length of 23.5m and a width of 4 – 5m. The vessel was dated to the end of the 1st century BCE. Apparently the port side and the keel preserved in good condition. The reconstructed dimensions of the original vessel are 25m long and 7m beam; the estimated cargo capacity was 150 tons. Besides the keel part of the stem-and-stern posts were also preserved in *situ*. The hull was double planked with mortise-and-tenon joints. The framing comprises alternated floor-timber and haft-frames, except for the middle part where a succession of eight floor-timbers strengthened the structure under the mast-step timber.[129]

Dramont F wreck belongs to a small vessel that was found WNW of the Ile d'Or, an islet about 100m off Cape Dramont. It lies at a depth of 58m bellow the sea, on gray, muddy sand bottom.[130] The roughly built vessel is 10–12m long and 4–5m wide. The keel is slender and the planks were assembled in the shell-first technique with mortise-and-tenon joints. The cargo comprised 120 amphorae of the Dressel 27 type, which permitted the archaeologist to date the wreck the middle to the 4th century CE.[131] All the amphorae were loaded with pine resin and they were stopped by hay and twigs, both materials were also well preserved by the resin.[132] The special interest in this wreck is because it augments the local trade in pine resin collected from the forests of Maure and Esterel, which may indicate that the ship sailed to the east.[133]

Madrague de Giens: This Roman wreck was found off the small fishing port of La Madraque, located on the NW coast of the Giens peninsula, south of Toulon in France. The shipwreck was discovered at a depth of 20m by divers of the French Navy, in 1962. The almost entirely preserved length is 40m and its beam was 12m when found. This wreck is one of the largest ancient vessels to be excavated yet.[134] Systematic excavations were carried out from 1972 to 1982, by a team from the Centre National de la Recherche Scientifique (CNRS) and the University of Provence under the directions of Andre Tchernia and Patrice Pomey. The cargo of the ship comprised 6,000 – 6,500 amphorae of Dressel I B type stacked in staggered rows of three layers. Many jars were marked with the potter's stamp of which the most frequent was that of Publius Veveius Papus, originating at Terracina to the south of Rome.[135] The cargo,

126 Ibid.
127 Ibid.
128 Delgado, 1997, p. 131.
129 Ibid.
130 Joncheray, 1977, p. 3.
131 Ibid., p. 6.
132 Ibid., p. 7.
133 Ibid.
134 Delgado, p. 252.
135 Ibid.

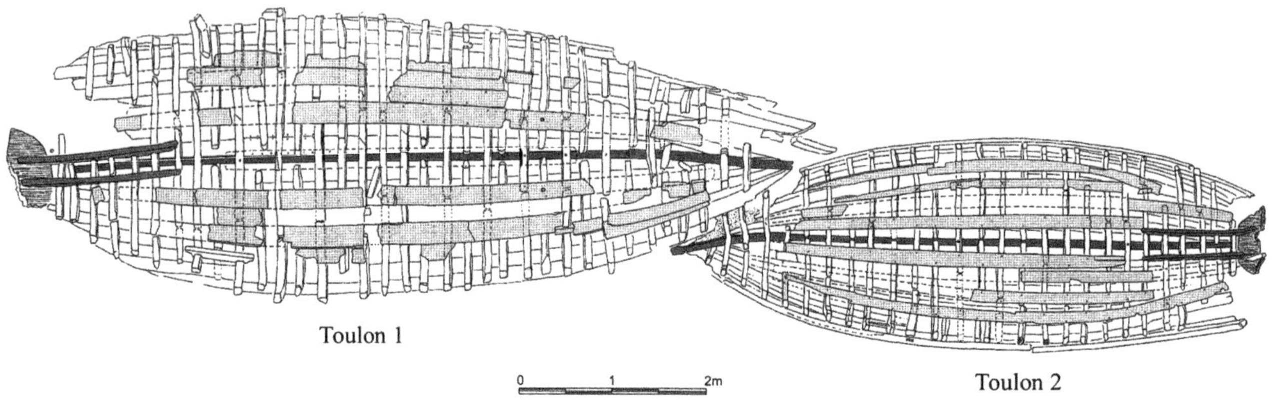

Fig. 4.21: Plan of Toulon Ships 1 and 2

different artifacts and especially the coins found on board dated the ship between 75 and 60 BCE.

The preserved hull is double planked, built in the traditional shell-first technique with mortise-and-tenon joints, pegged from inside. The preserved dimensions of the wreck permitted the archaeologist to reconstruct its original size: 40m long, 9m wide and the depth of the hold 4.5m. The cargo capacity was 400 tons; when fully loaded the vessel had a displacement of 520 tons.[136] The concave stem was fitted with a projecting cutwater and the outer hull was sheathed with lead. Originally the ship was rigged with two (or even three) masts and square sails, probably similar to Ship 1, depicted in Station 23, at Ostia (fig. 3.7.21). Madrague de Giens may be considered a very large merchantman used in antiquity and it may be compared to a "*myriophoros*" type, carrying 10,000 amphorae.

Toulon Shipwrecks: The construction of a modern commercial center in Toulon, France, revealed the remains of the Roman harbor in the ancient town of *Telus Martius*.[137] Two boats, dated to the end of the 1st century CE were revealed in 1987 (fig. 4.21), during the construction of a modern commercial center. They were reused in the building of a quay in the harbor.[138] Both vessel present a special interest because each one of them preserved evidence of one transom end. **Toulon 1** has a preserved length of *c*.8.5m, a width of *c*.3.5m and a depth of 0.35 to 0.45m. **Toulon 2** has a preserved length of 6.3m, the width is 2.2m and the depth is 0.32 to 0.45m.[139] The hulls of both boats survived to the gunwale height and each one has a flat bottom. **Toulon 1** has 35 frame and two internal posts related to the transom end; **Toulon 2** has 27 frames and similar posts for the transom end.[140] Several floor planks still survived in *situ*.

Turkey

The coasts of Turkey are time capsule of ancient wrecks that have a time span which begins with the Bronze Age, Uluburun (14th century BCE) and Cape Gelidonia (12th century BCE), up to the Ottoman period. Below will be described three wrecks dating from the 5th century BCE to the late Roman and late Byzantine periods. The earliest Greek vessel was found at Tektaş Burnu and the other two ships were located in the same area of Yassiada (Flat Island), an island in the Aegean situated between the Dodecanese Island of Pserimos and the Turkish coast.[141]

Tektaş Burnu Shipwreck was found along the rugged coast SW of Siğacik (ancient Toes) and east of Chios, at a depth of 40m.[142] The remains of this 5th century BCE wreck were first discovered by the Institute of Nautical Archaeology (INA) in 1996. The excavations were carried out in 1999 and 2001. The cargo comprised about 200 wine amphorae known as pseudo-Samian type.[143] The contents of 9 – 10 jars comprised pine tar. The ceramic cargo originates from Mende, east Greek, Chios, and Samos. The typology of jars found within the remains of the ship's hold were dated to 450 – 425 BCE. The main concentration of the amphorae cargo indicates that the ship was about 10 - 12m long and probably 3 – 4m wide. The orientation of the vessel upon the seabed comes from two rare marble disks found in the loose sand at the top of the wreck. These disks, each 13.8 and 13.5 cm in diameter are the *ophthalmoi*, the ship's eyes.[144] The disks have a flat roughly-finished inner surface and a convex external face (fig. 4.22). A rounded hole of *c*.13 mm in diameter was pierced in the center of each disk and a lead spike of ten sq.mm cross-section was driven through.[145] The exterior convex surface of the disks was incised, each with two concentric circles, and

[136] Ibid.
[137] Boetto, p. 291.
[138] Ibid.
[139] Ibid.
[140] Ibid.

[141] Bass, 1972, p. 137; Delgado, p. 460; Carlson, 2003, pp. 581 – 600; Nowak, 2001, pp. 86 – 94.
[142] Carlson, 2003, p. 581.
[143] Ibid.
[144] Ibid., p. 595.
[145] Nowak, 2001, p. 86.

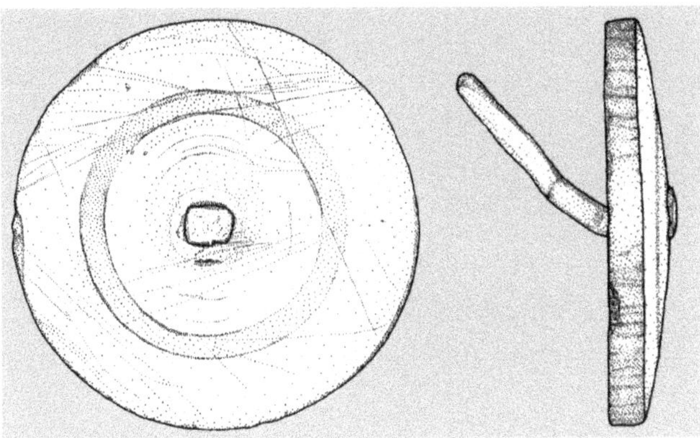

Fig. 4.22: T1 - Tektaş Burnu - marble ophthalmos

remains of reddish paint between these lines indicate the decoration.[146] The first disk T1 still preserved the original lead nail, 8.6 cm long and bent at 45 degree to the flat inner surface.[147] Seven *ophthalmoi* were recovered at Piraeus, in the 19th century, at the shipshed that housed the Athenian triremes, and three more were found at the Athenian Agora, in a well nearby the Tholos.[148] The Piraeus and Athenian eyes are more natural, in an almond shape, likewise as they are depicted on oared ships illustrated in Greek vases. The Tektaş Burnu shipwreck, is the first example to reveal circular *ophthalmoi* adornments of marble painted disks on the sides of the prow.

The 4th Century Yassiada Wreck was found at a depth of 47m with a cargo of about 1,100 amphorae. It was established that the ship sank in the second half of the 4th century.[149] Excavations of the cargo and part of the preserved hull began in 1967 and were completed during the summer of 1969. The port hull was well preserved up to the wale line. The remains of the wreck permitted to reconstruct the ship's original size: the overall length was 20m with a beam of c.8m, and the keel was 7m long. The hull seems to be quite symmetrical with its maximum width at or very near amidships. The strakes were joined with mortise-and-tenon fasteners. Four pairs of wales were found on the sides of the hull between the waterline and the deck level. The lower two wales run horizontally, while both upper ones curved upwards at the stern and the stem. Apparently, a pair of half-frames has been erected amidships only after five assembled planks.[150] Perhaps they indicate that some frames may have been inserted amidships to assist the shipwright in shaping the hull.[151] At the stern there was a spacious galley, going down to the hold, where a stone hearth was found. Ceramic vessels for cooking and tableware, copper coins, a casting net and some lamps were also found in the galley, probably the belongings of the crew. It is assumed that the hull shape of this vessel may be associated it with the Roman *corbita* type.[152] The cargo comprised three kinds of jars: bag-shaped, cylindrical and smaller ovoid jars.[153] The content analysis of the jars revealed that they were filled with oil and wine.

The 7th Century Yassiada Wreck (late Byzantine) was completely excavated between1961 and 1964. The length of the preserved hull was 20.7m and its beam was 5.02m; it had a sharp bow, a full stern, deep wine glass shape in section and a load capacity of about 60 tons.[154] The planks were not fastened to the frames by wooden treenails and clench nails as in the earlier traditions, but by light nails barely penetrating from the outside halfway into the frames.[155] The planks were joined with very small mortise-and-tenons (compared to earlier first-shell building), widely spaced, and not pegged; apparently they were found mostly below the waterline.[156]

The main mast seemed to be braced against the back of a through-beam, which protruded on either side of the hull just forward of amidships.[157] Evidence for two closely spaced through-beams near the ship's stern, with their projecting ends probably formed the box-like mounting of the steering oars.[158] There is no evidence to suggest what kind of sailing rig the ship had. On board were found 11 iron anchors, 7 were compacted on the deck, forward amidships and the other 4 were stowed close by on the gunwale ready for use, two on either side. It was concluded that each of the 6 anchors weight 73.5 kg and each of the other three weight 129 kg.[159]

Similar to the 4th century wreck, this ship also had a galley at the stern. The cooking was done on a tiled hearth supported by a set of iron bars embedded in clay.[160] Wares for cooking and storing food were also found in the cooking area. The water supply was provided by jars that stood within a small storage area between the galley and the sternpost. Among the artifacts found on board were a gravel for the ship's boat and some tools (two axes, a pickaxe, two bill-hooks and a shovel) used by the crew while gathering firewood and water on land.[161] Needles and spare fishing weights for preparing fishing-nets were stored on board. The owner or captain of the vessel was named Georgios,[162] deduced from the inscription on a large steelyard, which was part of the weighing equipment, found together with the balance-pans.[163] Another smaller steelyard, a box of silver inlayed weights and also 20 oil lamps were kept in the galley's cupboard. There were 16 gold and more than 50 copper

146 Carlson, p. 595.
147 Nowak, 2006, p. 30.
148 Carlson, 2009, pp. 348 – 349, 350 - 352.
149 Bass, 1972, p. 137; Delgado, p. 470.
150 Delgado, p. 470.
151 Ibid.
152 Ibid.
153 Bass, p. 147.
154 Delgado, p. 469.
155 Ibid.
156 Ibid.
157 Bass, p. 140.
158 Ibid.
159 Ibid.
160 Ibid., p. 141.
161 Ibid., p. 143.
162 Georgios was a priest and the ship may have belonged to the church, and was designated to transport both cargo and churchmen; Delgado, p. 470.
163 Delgado, p. 470.

coins that were also kept in the cupboard. The Rhodian Sea-Law stipulates that any passenger who came aboard and held gold or silver coins had to deposit them with the captain. If he did not do so, and in case of subsequent theft, the passenger was not refunded.[164] The coins found on board indicated that the ship sank either in 625 CE or shortly thereafter.[165] Within the hold there were 700 bag-shaped jars packed in three files and about 100 smaller cylindrical amphorae placed horizontally between the necks of the top layer.

Israel

In 2008, a lifeguard diving at Yavne-Yam next to the Palmachim beach, south of Tel-Aviv, found a rare marble disk with flat inner surface and the outer convex face (he handed it to the Israel Antiquities Authority). The disk 20 cm in diameter had a perforation for a spike that was driven through. Remains of two concentric circles are painted around the center of the disk.[166] This disk dated to 500 – 400 BCE is one *ophthalmo*, the ship's eye, which was set one on either side of the stem. The Yavne-Yam disk is the second to be found in Israel. The first marble disk 23.5 cm in diameter was found off the Carmel coast, near Haifa, several years ago, within a Hellenistic assemblage dated to the 2nd century BCE.[167]

4.6 Rigging of the Vessels

No wreck has been yet found with its complete sailing or steering gear. Wreck D found in the Black Sea at a depth of 320m is a rare find that survived almost intact. The new technologies using the sidescan sonar and the ROV (Remote Operated Vehicle) permitted this discovery at such a depth.[168] The wreck was found with its entire mast still stepped amidships in its vertical position about eleven meters above the silt covering the deck. The wooden sample taken from a timber as a rudder support, identified as fur and analyzed by C14 indicates that the ship was dated to 410 – 520 CE.[169] The vertical mast and all the wooden components of the ship are completely preserved without the damages of *teredo navilis* (ship worms). The lower part of the mast was supported by a short upright timber (bracing-timber or mast-partner) possibly fixed in the bottom of the hull and projecting to the deck level (fig. 3.4.14).[170]

Ships depicted in mosaics and in any arts are mostly represented with a fully open square sail, a vertical mast and steering oars. Although they have open sails these vessels are not sailing, rather they are at anchor in open waters deduced from lack of anchor or mooring lines, their static position, and lack of human figures with them. Apparently, such representations were a common convention, especially

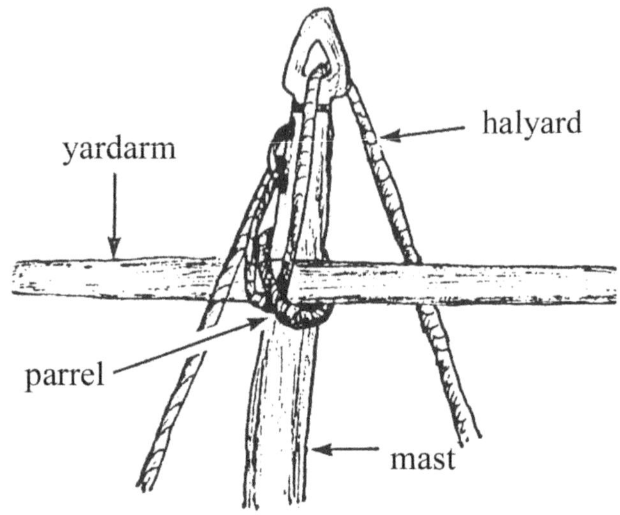

Fig. 4.23: Parrel holding the yard to the masthead

to illustrate merchantmen. The sailing ships depicted in Roman - Byzantine mosaics are simplified, being represented with the general shape of their hulls, the sailing and steering gears.

The basic rigging of an ancient sailing ship was the mast pole stepped (usually) amidships or slightly forward, in a step carved in the keelson, and then secured in place by the fore-and-back stays; further lateral support was provided by the port and starboard shrouds (fig. 4.2).[171] In iconographic representations the mast in general is rendered with alternating white and black horizontal bands, which indicates a composite mast of wooldings at fixed intervals.[172] A long yard, almost equal to the length of the ship (made of one, two or three spars fished together) was attached to the mast by a *parrel* (fig. 4.23) and secured by two or more lifts running from either side of the masthead to the yardarms. A square sail was set on the yard by brail-fairleads or robands and a series of brails running through brail-rings worked the bunt. These rings of metal or rope-loops were sewn on the fore face of the bunt (fig. 4.24). The reinforcing-bands or seams were sewn horizontally to reinforce the bunt and also provided the base for the brail-rings/loops. The brails and the reinforcing-bands formed a checker pattern on the bunt. Most of the ships depicted in different arts are shown with a fully open sail and the checker pattern on the face of the bunt. The edges of the sail are reinforced by boltropes and the corners where they wear were reinforced with patches of leather (fig. 4.24).[173] The foot of the square sail was free and could be reefed by the brails at any angle suited to the wind conditions for optimal sailing. The reefing could be done from the middle of the foot as depicted in the Palestrina mosaic (figs. 3.6.4, 3.6.5) or from either side of the sail. The sheets attached to the clews in the lower corner of the sail

[164] Bass, p. 143; Ashburner, 2001, C14, p. 62; C13, p. 94.
[165] Bass, p. 143.
[166] http://www.israelnationalnews.com/News?News.aspx./126823; 2008
[167] Galili and Rosen, 2008, p. 360.
[168] Ward and Ballard, 2004, p. 2.
[169] Ibid., p. 6.
[170] Ibid., p. 11.

[171] Since ratlines cannot be used when shrouds are set up in ancient ships, there was a rope ladder abaft the mast for getting aloft and working the yard and the sail; Casson, p. 210.
[172] Casson, p. 231.
[173] Ibid., p. 234.

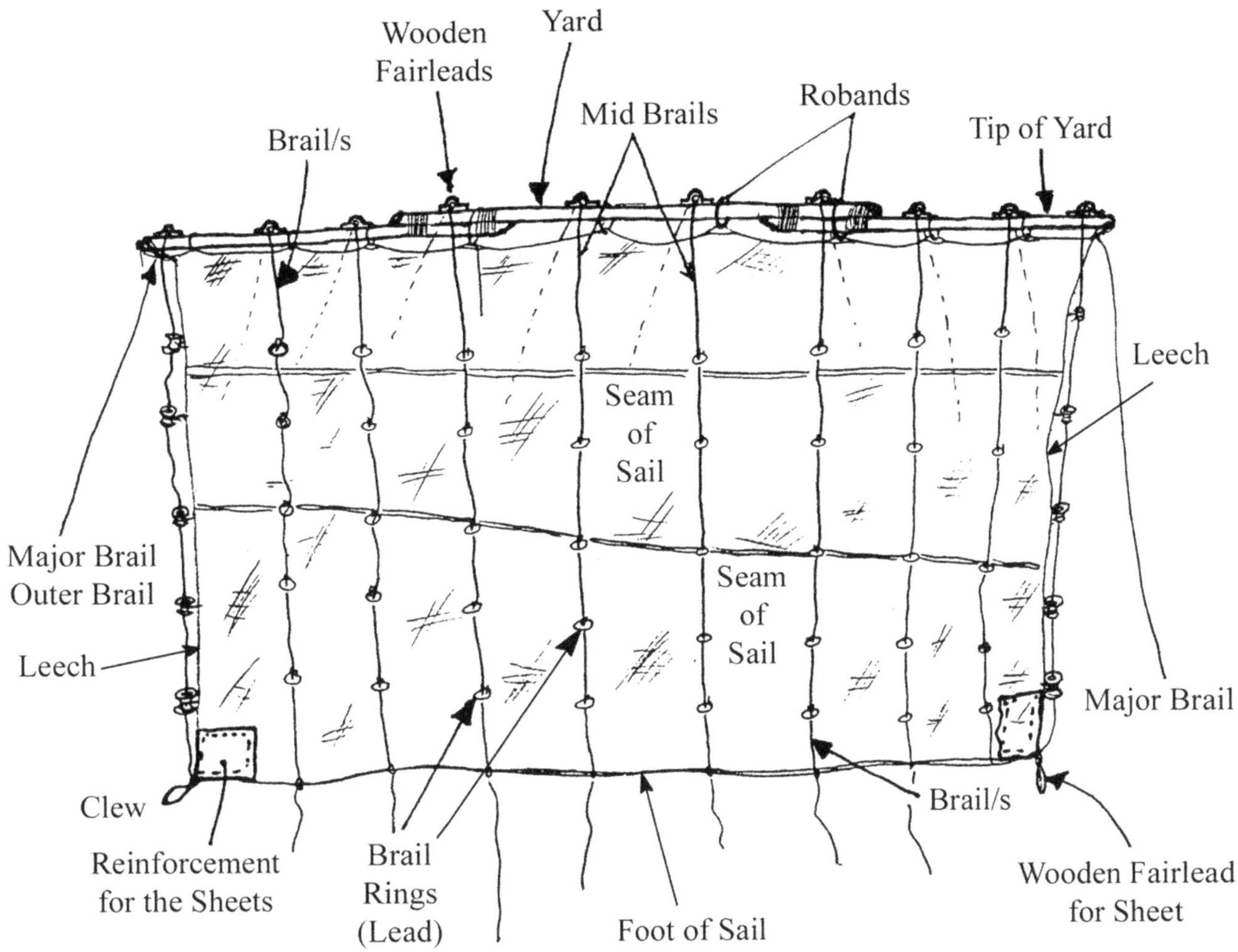

Fig. 4.24: Detail of a square sail and its means of working

worked the lower sail to adjust it to the wind conditions. The yard and the sail could be lowered or raised by the halyard. They were also worked by one or double-lines braces attached to either tip of the yardarms, similar to the depiction of Armerina Sailing Ship 2 (fig. 3.8.25). When ancient ships sailed over open waters they also raised a topsail over the yard and the main sail. It had a triangular shape and was made from one or two triangles with their vertical edge adjacent to the masthead. The topsail appears in any Roman ship from the last decade of the 2nd century CE and are commonly depicted on ships from the 3rd and 4th centuries CE. In most of the ships there is a second sail and mast stepped forward on the stem; it is the *artemon* or the fore rigging, slightly smaller than the main rigging. The giant merchantmen were rigged with three masts and sails, where the aft rig was known as the *mizzen*. The only pictographic representation of such a vessel especially in mosaics is found at Ostia, in the office of the shippers from Syllectum, in North Africa (fig. 3.7.21). Apart from this depiction, we learn about a three masts ship from historical writings, especially Athenaeus, when he described the super-freighter *Syracusia*.[174]

[174] Athenaeus, V.208.

Almost all the ancient vessels were decorated with a sign or identification device consisting of a flag or light,[175] as illustrated by varied ship depictions. Pennants (bands) were attached to the sternpost of Hellenistic ships of all types (fig. 3.6.17). In Roman merchantmen, pennants or small flags were attached to the masthead and also to the tips of the yard (figs. 3.7.11, 3.7.14, 3.7.15, 3.7.21, 3.7.29, 3.7.30, 3.7.32, 3.7.33, 3.7.35, 3.7.36, 3.7.38, 3.7.39). They probably served both for identification but mostly for checking the wind direction. When the mast was lowered or missing, signal flags or pennants could be flown from the *stylis*.[176]

Sailing vessels and warships had at least one ship's boat towed astern. Large sailing ships could have more than one. Such small vessels were usually rigged with rowing oars and a pair of steering oars, but they also could be rigged with sailing gear comprising one mast and a square sail, as augmented by the mosaics from Kelenderis (fig. 3.4.7) and Kenchreai (figs. 3.5.9, 3.5.14, 3.5.4a).

[175] Casson, p. 246.

[176] Ibid., p. 247.

4.7 Steering Gear

Ancient ships were steered by one or two steering oars mounted on the quarters. This oar was larger than a rowing oar and hung at a slant, in such a way that allowed it to be pivoted.[177] The Egyptians first introduced the quarter rudders in the New Kingdom period. The obelisk carrier and the Punt ships of Queen Hatshepsut (1500 BCE) are illustrated with giant rudders. The massive oars, with long thick shafts and very large tapered blades were secured to the quarter by metal thongs and a rope girdled between the upper part of the loom and the projecting ends of the quarter through-beams.[178] From the 4th Dynasty (Old Kingdom) onward, a tiller bar was inserted into the loom of the steering oar making the work of the helmsman much easier. The tiller bar is even used in modern yachts and small sailing boats. The shaft of the steering oar formed the turning axis allowing the proper distribution of the amount of blade forward and abaft.[179] Pulling or pushing the tiller made the loom pivot within its fastening and this put the blade at an angle to the hull and thus directed the ship.[180] The Romans adapted the quarter rudder and adjusted it to different types of vessels. The iconographic evidence of seagoing and river vessels show that the aft part of the top plank was extended into a wing-like projection on either quarter, which was supported by the short through-bars and therefore it housed the steering oars or rudder. The quarter mount holds the rudder in place at two points along its length and permits it to rotate around the axis of its shaft.[181] In small boats, the rudder is a sculling oar that steers the vessel in calm conditions or when sailing in shallow waters along the coast. The quarter rudder in larger ships could weight several tons. The ability to move it along its axis brought to adjustments when the rudder lost its buoyancy through the absorption of water.[182] Representations of Roman ships show that quarter rudders were mounted at an angle of 30 and 45 degrees from the vertical.[183] The quarter rudder mounting of ancient ships had to fulfill three requirements:

1. It had to hold the rudder firmly at two points or along a section of the shaft;
2. It had to permit the rudder to be moved up and down in the mount, in a direction parallel to the shaft;
3. It had to allow the rudder to be removed for maintains.[184]

The mounting of quarter rudders may be deduced from many iconographic representations. They are not very detailed but provide some elements, which facilitated some reconstructions.

Brace Mount: In this mounting system, the shaft is inserted between the through-beams; further the rudder was pulled up, the tighter it sat in the mount because the rudder's buoyancy would decrease, thus causing it to pivot downward around the lower beam and push the upper shaft into the upper through-beam (fig. 4.25).[185] This kind of mounting would require less lashing or none at all. A similar type of mounting having the ability to raise and lower the rudder rapidly was used in merchant ships of varied sizes. In many aft-wing representations, the end or ends of the through-beams are visible. This aft plank hides the shaft of the oar, whereas its position may suggest the mounting system, as also deduced from the angled shaft.

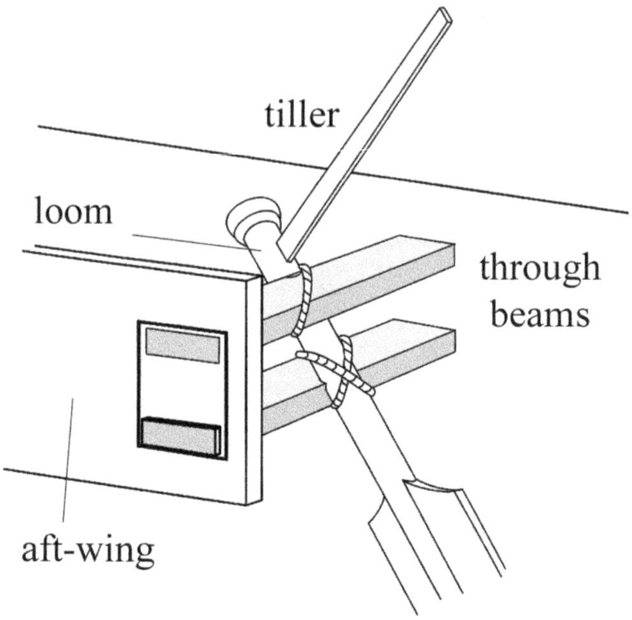

Fig. 4.25: Reconstruction of a rudder brace-mount system

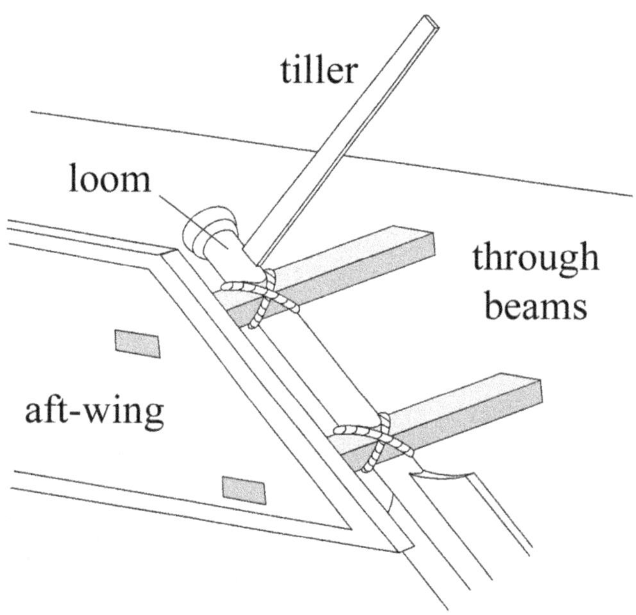

Fig. 4.26: Reconstruction of a rudder fore-mount system

[177] Casson, p. 224.
[178] Ibid., figs. 14, 18.
[179] Ibid., p. 225.
[180] Ibid.
[181] Mott, 1997, p. 10.
[182] Ibid., p. 11.
[183] Ibid., p. 17.
[184] Ibid., p. 18.
[185] Ibid., p. 20.

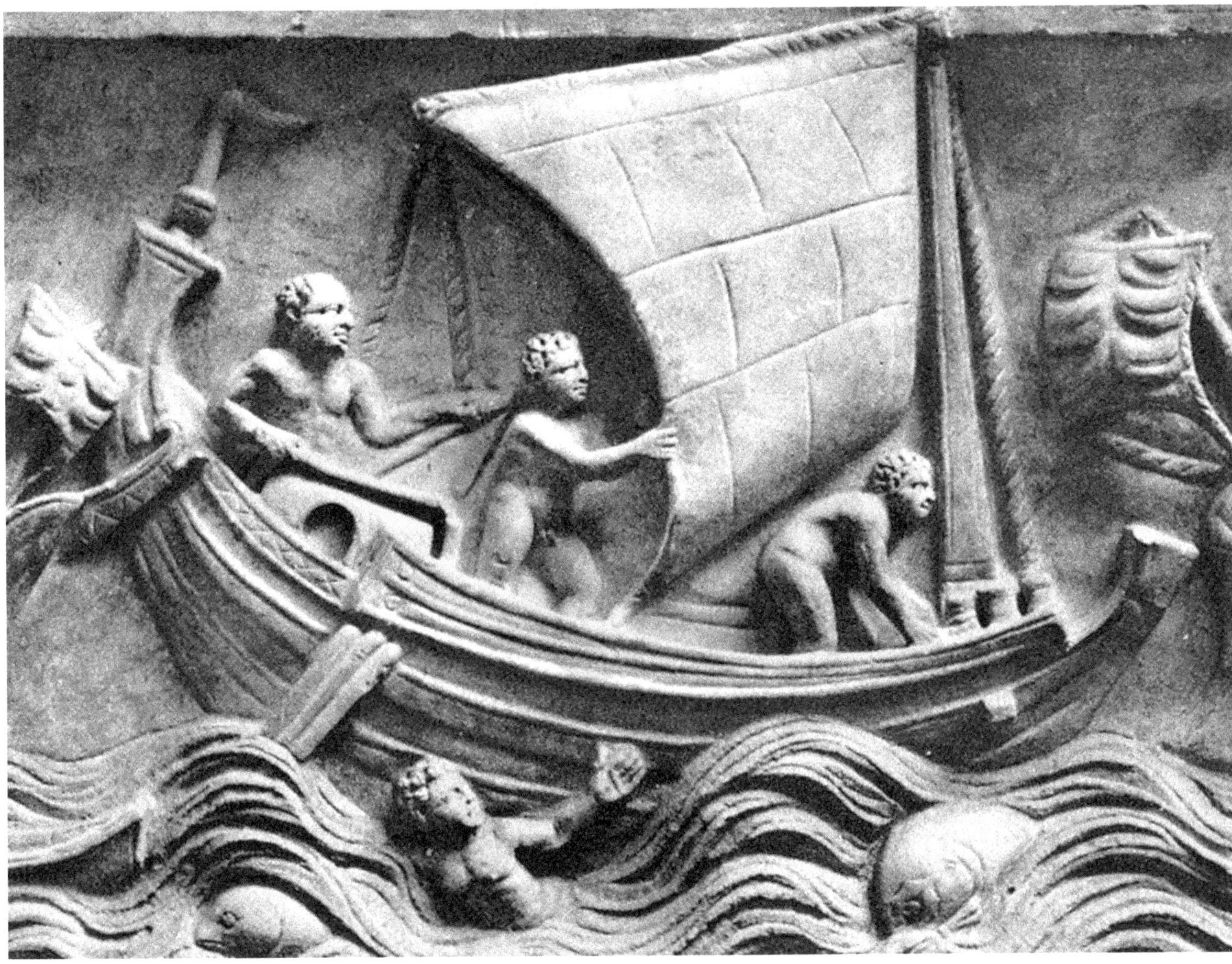

Fig. 4.27: Fore-mount rudder; Copenhagen sarcophagus middle ship

Aft-Mount System consisted of two through-beams that supported the fore side of the shaft, namely, it was laid in notches curved on the aft part of the beams and then held in place by lashing (fig. 3.7.46). A similar mounting method can be seen today on the *pisini* boats of Indonesia.[186] A relevant depiction associated with this system is suggested by the left-hand vessel in the Copenhagen sarcophagus (fig. 3.7.47). The position of the shaft and the ends of the through-bars supporting the aft-wing indicate the mounting system. An additional example is nicely represented on the monument of Odysseus' Ship at the Sperlonga Museum, Italy.[187] The aft-mount permitted much easier work when pulling the rudder upwards or removing it completely from the water when the oar was damaged. It appears that this method survived throughout the Middle Ages.[188]

Forward-Mount: This system is very similar to the previous method except that the shaft is mounted against the foreside of the through-beams and then lashed by a rope (fig. 4.26). Such a system may be deduced from the middle vessel in the Copenhagen sarcophagus, dated to the 3rd century CE (fig. 4.27). It appears that this system was mostly used in small vessels while the aft-mount was more suited to large ships. The advantage of the fore-mount method was that the lower bracket bore all the stress from the rudder's drag and the waves, and not the lashing, as was the case for the brace-mount system.[189]

Box-Mount: This system consisted of two closely spaced through-beams, placed at the same level near the waterline. The beams were spaced so that the rudder shaft could barely pass between them at an angle.[190] The upper part of the shaft was either lashed to a bitt or was passed through a slot in the side of the hull. The shaft was prevented from moving laterally by wooden crosspieces on either side. Together with the through-beams, these crosspieces formed a box with the shaft passing through and holding the rudder firmly in place, while the other systems permitted the rudder to rotate upwards to a horizontal position if it struck an object.[191] The box-mounting system had some disadvantages, especially when the rudder hit submerged objects and there was nothing to prevent the oar from breaking. The rudder mounted by this system could not be removed and therefore when the ship anchored, the water

[186] Ibid., p. 23.
[187] Casson, 1971, fig. 144.
[188] Mott, p. 27.

[189] Ibid., p. 31.
[190] Ibid., p. 32.
[191] Ibid., p. 33.

had to have enough depth so as not to damage the oar. The box-mount was used until the 15th century.[192] There are other ways of mounting a rudder but we shall not discuss them here because they are not relevant to the ships depicted in the mosaics presented in this monograph.

4.8 Discussion

This chapter will discuss varied types of vessels mentioned in ancient literature and augmented by found shipwrecks dated from the 5th century BCE to the 7th century CE. Generally there are two main types of vessels:

1. The long ships (*navis longa*) that are swift vessels, built for incursions or piracy. They mostly were rowed but also could be rigged with a mast and a square sail when the wind conditions were favorable. The ratio of W/L of such vessels is 10:1 or 8:1. The bow was fitted with a projecting forefoot sheathed with copper or lead, but usually they carried a bow-ram.
2. The rounded ships (*navis oneraria*) are merchantmen with a broad hull and propelled by a square sail set on a mast stepped amidships or slightly forward. They also were rigged with *artemon* mast and sail, known as the fore-rigging. The mast was stepped forward and slightly angled over the stem. Very large merchantmen were fitted with three masts and square sails: the main stepped amidships, the *artemon* and the *mizzen* (back rig). The ratio of W/L of such vessels was 6:1 or 3:1. Merchantmen were classified upon their load capacity, varying from 5 – 10 tones to 500 tons, whereas the giant merchantmen carried from 1,000 to 3,650 tons cargo.

The vessels in discussion are built in the classical tradition of "shell-first" with mortise-and-tenon joints locked by wooden treenails and bronze nails. In the 7th century ships are found combined shipbuilding techniques, the "shell-first" is only partially used bellow the waterline, while the hull above the water level was completed with planks nailed from the outside to the frames inserted within the hold, which apparently were carved accordingly to the shaped of the hull. The archaeological evidence of the shipwrecks described above were used as comparative material for the ships depicted in mosaics and studied in this monograph.

In Table I are listed the shipwrecks described above along with some other wrecks that are brought for comparing data in relation to their cargo capacity, size, W/L ratio and their period to be concurrent to the mosaics described in this monograph. The preserved wrecks are seagoing vessels or coasters. The length of the seagoing merchantmen listed in Table I vary from 7.3m (Napoli B) to 40 – 45m (Albegna, Caesarea, Madrague de Giens). The cargo capacity also is an indicator for the size of merchantmen: light (20 – 60t), medium (60 – 120t) and large (120 – 450/500t). The load capacity of the ships was also evaluated by the amphorae cargo. The very large merchantman carrying 6,000 - 6,500 amphorae was known as *myriophoros* or *myriagogos*.[193] The ratio W/L 2.3:1 (Pisa D), 2.5:1 (Yassiada I) or 2.8:1 (Pisa C, La Cavaliére), indicates that the hull of these vessels was broad, probably square-like shape or had very rounded ends (stem-and-stern). These ships may be associated with a more specific type, *corbita* or *kerkouros*. Smaller *kerkouros* carried a cargo from 225 to 250 tons, and the largest as much as 450 tons.[194] Such ships probably were 50m long and 7.7m wide, and the W/L ratio was 6.5:1.[195] The merchantmen in the Mediterranean were mainly engaged to transport grain from North Africa, Egypt and Sicily to Rome. The grain was carried in bulk, sacks or stored in amphorae. The cargo had to be stored above the bilge-water in the hold as well as being protected from the water spays and waves on the deck.[196] Other merchants brought to Rome were fine wines from France and Spain, olives, olive oil and *garum* (fish-spice) from North Africa and Greece. One of the most important merchants needed for shipbuilding, constructions and heating was the wooden logs brought to Rome from Lebanon, Cyprus, Turkey or the Adriatic areas. The large merchantmen could be adjusted to transport exotic captured animals in North Africa, Egypt or India, for entertainments in Circus Maximus and Colosseum at Rome.

When ships entered or left the Claudian and later the Trajan harbor the cargo was inventoried and registered by a *tabulator* who wrote his records on wax-tablets that had to be stored at the custom-office. These tablets were thick pieces of wooden boards with recesses filled with bee-wax. The writing was done with a metal *stylus* on the wax surface (fig. 3.7.40; Station 51, Ostia). The tablets were bound together and then a seal impression of the witness was placed over the binding in groove.[197] The Rhodian Sea Law (10) refers to the condition of loading and recording of cargo on ships. The ship-owners received the cargo on board "*per apertum scriptum*". Weigher appointed by the parties and paid by the ship's owner weighed the cargo, in a similar way as indicated by weighing the ingots depicted in a mosaic from Sousse, Tunisia (fig. 4.28). The ship's *scribanus* had to write the weight in the *cartularium* (ship log).[198]

In Table II are listed the wrecks which indicated river and/or harbor vessels engaged in transporting cargo unloaded from large merchantmen anchored in open waters and then transported up river to Ostia, Rome, or Pisa, or used in varied harbor works at Ostia, Naples or Toulon. Strabo (5.3.5) and Dionysius of Halicarnassus (III.44.3), inform us that oared ships and merchantmen up to three thousand bushels (three thousand amphorae, *c.*108 t) entered at the mouth of Tiber where their cargoes were unloaded or loaded into river vessels which were rowed or towed up to Rome. The rigging of such merchantmen comprised one mast and a square sail that could be retracted when the ship anchored

[192] Ibid., p. 34.

[193] With a capacity of 10,000; Casson, p. 396.
[194] Casson, p. 164 and n. 40.
[195] Ibid., p. 166 and n. 43.
[196] Rickman, 1980, p. 265.
[197] Adkins and Adkins, 1994, p. 209.
[198] Ashburner, 10, p. cliii.

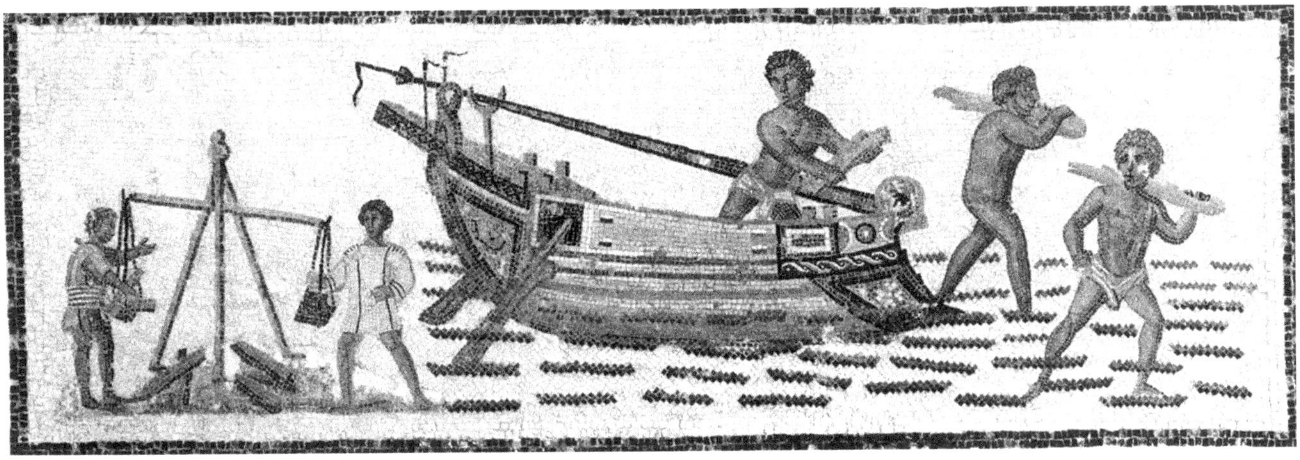

Fig. 4.28: Weighing ingots cargo

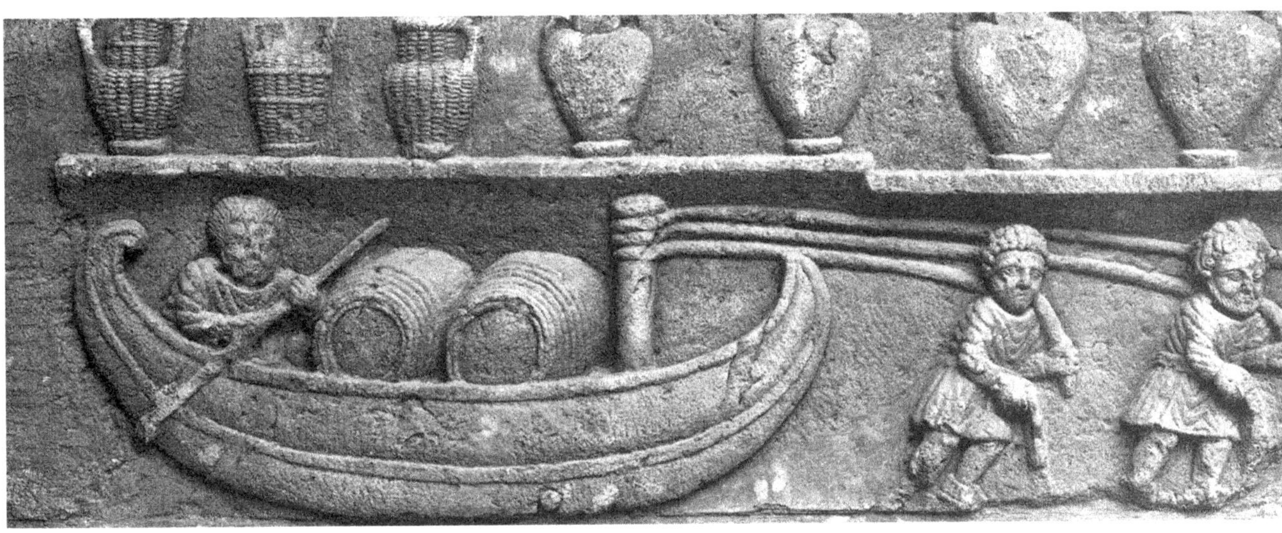

Fig. 4.29: Small river craft hauled by a towline fastened around the top of tow-mast

or the wind conditions failed. They also could be rigged with an *artemon* mast and sail. This rig was preferred when entering or leaving harbor, giving the vessel a better maneuverability, whereas the main rig would be a burden in operating the ship until it reached the open waters. Several pairs of oars were also found with the rigging, and when used they were lashed with throbs against the tholepins set above the gunwale. They aided in entering/leaving harbors or when the sailing conditions were not favorable. The towline in such vessels was fastened to a capstan set on the poop[199] and/or around the mast stepped forward amidships (figs. 3.7.24, 4.29). These vessels are known by varied name: *skaphae*, *lembos*, *lenunculi*, *horeia* or *ratiaria*.[200] Another type of vessel that was towed on the Tiber was *caudicarria*.

The generic Latin name of river/harbor boat is *linter*, which could be propelled by oars, or sail, or both.[201] We may suggest that Fiumicino 3 and 4, Pisa G and I may be associated with *linter*, engaged in carrying the cargoes unloaded from large merchantmen anchored in open sea and then towed up on the Tiber or Arno.

Pisa F with W/L ratio 8:1 indicates a swift vessel, either used for rapid incursions or as a messenger boat sailing on the river or along the coast. Such vessels are known in ancient literature as *vectoria*, whose prime service was to meet newcomer vessels into the harbor and check their documents.[202]

Napoli C (fig. 4.19) and Toulon 1 and 2 (fig. 4.21) are very rare vessels to be found with one transom end, stem or stern. Such vessels appear in some ship depictions dated from the 1st century BCE to the 4th century CE. In the Althiburus mosaic (3rd century CE) is depicted a boat with a transom stern, propelled by a pair of oars mounted amidships, whereas the inscription bellow indicates the type of boat as *horeia* (figs. 3.8.27/*20*, 4.30). Therefore we may suggest that Napoli C and Toulon 1 and 2 augment the *horeia* type,

199 Casson, p. 332.
200 Ibid., p. 330.
201 Ibid., p. 333.

202 Ibid., 1965, p. 35.

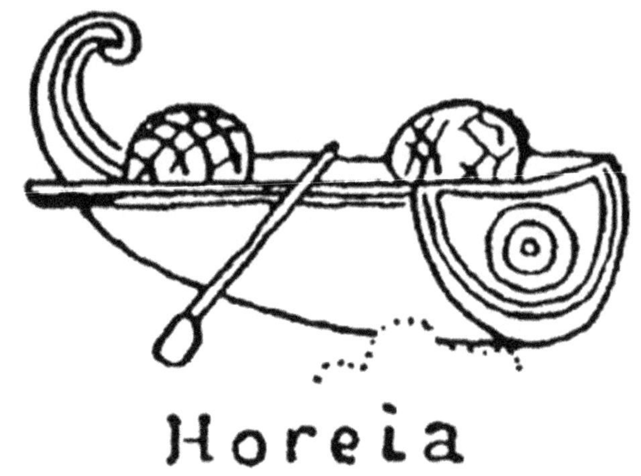

Fig. 4.30: Horeia with transom stern

used for varied harbor works and transporting cargoes on rivers as well as along the Mediterranean shores.

In Table III are presented the super-freighters known from ancient literature and also from excavations. These vessels were built to sail or float on the river or lake, or had a single oversea voyage in the Mediterranean (Caligula's obelisk ship, *Syracusia*). The Nemi Ships provide good examples for understanding of the oversize ships described by Athenaeus. We may assume that when Caligula built his ships, he may have used Athenaeu's reference as a guide. There are some similarities between the Nemi Ships and Athenaeus' *Syracusia*. All three ships were pleasure vessels built to float rather than sail in open waters. They were adorned with luxurious cabins, temples, and the floors and walls were decorated with polychrome mosaics and colorful marble surfaces. The size of the ships and their W/L ratio shown in Table III indicates that they follow a ratio of 6.7:1 to 3:1 similar to seagoing merchant vessels. The Nemi Ships, the Ptolemaic barge and Caligula's obelisk ship may be considered as monuments to the power of the ruler who had ordered their building and only used them for their own pleasure, rather than as a model for super-freighters building.

No wreck was found yet with a complete rudder mounting, but only suggested by some remains of spars or notches at the stern. The rudder mounting in the ships depicted in the mosaics studied in this monograph is assumed to comprise all four systems described above. These arrangements are difficult to discern; therefore they may be deduced from the position of the shaft behind the aft-wing plank. The rudder blades of the merchant ships are similarly shaped to those of the rowing oars, but rather larger. The rudders of the Lod Ships seem to be worked from a rudder house with an arched roof and fenced by a lattice screen (figs. 3.2.3, 3.2.5). Such a steering system is the only example to be found in mosaics or any arts dated to late Roman – early Byzantine period. The shaft of Lod Ship 1 is mounted forward of the central stanchion of the fence which also forms the tholepin (fig. 3.2.3). The shaft of Lod Ship 2 is placed between two stanchions that support the fence and probably make the tholepins (fig. 3.2.4). This arrangement resembles the box-mount system. The rudder of Nemi Ship 1 passed between the projecting bars supporting the outrigger aft-plank (figs. 4.10, 4.10a). This mounting apparently indicates a brace-mount system (fig. 4.25). Athenaeus did not describe the steering arrangement of the Ptolemaic *thalamegos* or *Syracusia*. The port rudder mounting behind the aft-wing extension of the top plank in the Palestrina *thalamgos* (fig. 3.6.17) may suggest an aft-mount system (fig. 3.7.46).

The small fishing boats depicted in the mosaics may be compared in size with the Fiumicino 5 fishing boat (fig. 4.14). Although we do not see a fish tank in boat depictions, it does not mean that such boats were not fitted with a similar device. Boats represented with a deck platform at the gunwale level and rectangular openings in the fore or aft part, probably indicate the existence of a fish tank. The Armerina boats (figs. 3.8.10, 3.8.15) are mostly shown with such decks and the openings may suggest a fish tank in the hold.

Ships in any culture are believed to contain spirits that protect them and the mariners from the dangers of the waters. Therefore shrines, deity models, or *ophthalmoi/ oculi*, the ship's eyes, set on either side of the prow were used as *apotropae* for the vessel and its crew. Both marble disks found with the Tektaş Burnu shipwreck (5th century BCE) are conclusive evidence of adornments on the prow of merchantmen. Such "eyes" are especially known from ship iconography. Two marble disks found off shore the Israeli coast, at Yavne-Yam (5th century BCE) and Megadim (2nd century BCE) are also associated with the *ophthalmoi* of ships. The Zea and Athenian almond-shaped eyes and dated from 475 - 465 to 425 - 400 BCE are assumed to be votive objects and/or ship's *ophthalmoi*. In his research of the *ophthalmoi* Nowak concluded that the almond-shape eyes adorned the upper stempost of the late-6th and early-5th century BCE warships. The circular eyes adorned the prow of merchantmen dated to the 6th – 5th century BCE.[203] The lead spike/nail bent at 45 degree preserved within T1 Tektaş Burnu disk evidences the securing of the eyes to the side of the prow. The use of lead as fastenings through the center of the disk was preferred upon iron or bronze because these materials could scar or break the marble eye. The lead corrosion expends less than those of iron or bronze.[204] Rendering of eyes on the prow of ancient ships were not used as decorations but rather functioned as epiphany to represent a supernatural consciousness that helped the ship to avoid hazards. Philostratus when describing the Tyrrhenian pirate ships noted:

> *"Strike terror into those they meet and may look to them like some sort of monster; it (ship) is painted with bright colors, and it seems to see with grim eyes set into its prow".*[205]

[203] Nowak, 2006, p. 112.
[204] Ibid., p. 43.
[205] Philostratus, 19.23-24.

The Tektaş Burnu and the Israeli marble disks augment the *ophthalmoi* mounting on the stem of Greek - Hellenistic ships.

Table I: Seagoing and/or Coastal Merchantmen

Country	***Name of the ship***	***Overall L**** (meter)	***Overall B***** (meter)	***Ratio W/L***	***Cargo***	***Period***
Italy	Albegna	40	8 – 10	4:1	1,1000 jars 450 t	1st BCE
"	Comacchio	21	5.62	4.1:1	130 t	1st BCE
"	Fiumicino 1	21	7	3:1	50–60 t	3rd-4th CE
"	Fiumicino 2	24	8	3:1	50–80 t	3rd-4th CE
"	Napoli A	11.77	3.32	3.5:1		end-1st CE
"	Napoli B	7.3	2	3.6:1		2nd-3rd CE
"	Napoli C	13	3.7	3.5:1		end-1st CE
"	Pisa A	25-30				2nd-3rd CE
"	Pisa B	c.20	4.3	4.65:1		1st BCE-CE
"	Pisa D	14	6	2.3:1		6th-7th CE
"	Pisa C	11.7	2.8	4.2:1		1st CE
Sicily	Isis	15	4	3.75:1	30 – 40 t	4th CE
Israel	Caesarea Maritima	40 – 45	10	4.5:1	300-400 t?	18-15 BCE
Turkey	Tektaş Burnu	10-12	3 - 4	3:1 (4:1)		450-425 BCE
"	Yassiada I	20	8	2.5:1	1,100 jars 30 t	4th CE
"	Yassiada II	20.7	5.02	4.1:1	60 t	625 CE
France	Dramot A	25	7	3.6:1	55 – 60 t	1st BCE
"	Dramot F	10 - 12	4 - 5	3:1	1,200 jars	mid-4th CE
"	La Cavaliére	13	4.5	2.8:1	20 t	c.100 BCE
"	Grand Conglue	23	6.8	3.4:1	3,000 jars 108 t	2nd BCE
"	Titan	20	5 – 6	4:1	55 t	1st BCE
"	Madrague de Giens	40	9	4.4:1	6,500 jars 400 t	75-60 BCE

Table II: River/Harbor Vessels

Country	***Name of the ship***	***Overall L**** (meter)	***Overall B***** (meter)	***Ratio W/L***	***Cargo***	***Period***
Italy	Fiumicino 3[205]	12	4	3:1	22–28 t	3rd CE
"	Fiumicino 4[206]	13.5	4.5	3:1	28–31 t	3rd-4th CE
"	Fiumicino 5[207]	5.5	1.8	3.06:1		2nd CE
"	Pisa F	8.2	1	8.2:1		early-2nd CE
"	Pisa G	8	1.5	5.3:1		1st–2nd CE
"	Pisa I	10 - 12	3 - 4	3:1 (4:1)	200 jars	4th-5th CE
France	Toulon 1	8.5	3.5	2.4:1		1st CE
	Toulon 2	6.3	2.2	2.8:1		1st CE

Table III: Giant Vessels

Country	***Name of the ship***	***Overall L**** (meter)	***Overall B***** (meter)	***Ratio W/L***	***Cargo***	***Period***
Egypt	*Thalamegos*[208]	99	14.85	6.7:1		3rd BCE
"	*Syracusia*[209]	61.5	15.5	4:1	1,700/3,650 t	240 BCE
Italy	Nemi 1	70	20	3.5:1		37-41 CE
"	Nemi 2	73	24	3:1		37-41 CE
"	Caligula's Obelisk[210]	104	20.3	5:1	1,300 t	1st CE
Sicily	*Isis*[211]	55.5	13.9	3.9:1	1,200-1,300 t	2nd CE

A cargo of 50 amphorae would roughly be 2.5 tons[212]
* **L** = length; ** **B** = beam

205 Scrinari, p. 57, Tav. VI.
206 Ibid.
207 Ibid.
208 The estimated metric units are considered upon 1 foot=0.33m.
209 Athenaeus V.206-209; Turfa and Steinmayer, 1999, Table I, p. 106.
210 Turfa and Steinmayer, Table I, p. 106.
211 Lucian: The *Ship or The Wishes 1 – 4; Landels, 1981, p. 160; Casson, 1971, p. 186*
212 Houston, 1988, p. 557, note d.

Chapter 5
Ship Interpretation in Mosaics

As previously stated ships depicted in the mosaics and researched in this monograph were studied with the purpose of determining what can be learned from them and if they can be considered a reliable source of information for studying ancient ships and also complementing the data of known Roman – Byzantine shipwrecks. Although the ships are not produced to scale in the mosaics or any other arts, they show the general shape of their hull, propulsion and steering gear, and suggest their sailing environment: Nile Delta, river/harbor and open sea. The study of ship depictions in mosaics has not been previously carried out from the point of view of marine archaeology and therefore was paid less attention. Since mosaic surfaces are a very durable art form they may be considered as valuable preserved pictographic information of the ancient ships whose dating may be concurrent to the period of mosaics production with such themes. The information we gather from ship iconography in mosaics also help us to follow ancient shipbuilding traditions that survived from the Hellenistic to the late Byzantine periods. These mosaics furthermore inform us about different fishing techniques and harbor activities. Some of the vessels are manned by figures engaged in different activities, but most of them lack human figures. Although ships depicted in the mosaics are shown with fully open sails generally they are static and not sailing, rather being anchored, nonetheless that such device is not depicted. Apparently when ships are depicted with fully open sails it is a symbolic convention referred to merchantmen. The small fishing boats inhabited by putti are either rowing the boat or are engaged in various types of fishing.

Almost all the vessels described above are sailing ships with one or two sails, and suggest their sailing environment; *in harbors*: Kelenderis, Turkey; Kenchreai, Greece; Piazza Armerina, Sicily; Ostia, Italy; *sailing in open water*: Lod, Israel; Ostia, Italy; in *the Nile Delta*: Palestrina, Italy. Also this category comprises special types: the *thalamegos* (cabin-carrier) propelled by oars (fig. 3.6.17), the *papyriform* sailing ship (figs. 3.5.19, 3.5.20, 3.6.5) and the warships (figs. 3.1.1, 3.1.4, 3.6.10).

5.1 Shipbuilding Materials

Shipbuilding materials are varied and can be traced through the mosaics presented in this monograph. These materials can not be identified by their exact species, but rather kind of material in general. The coloring of the hulls may suggest the building material of such vessels. Boats are built of papyrus/reeds, wooden log and planks. The wood is not represented by any species in ship iconography rather their species and building technique can be identified from found shipwrecks concurrent to the period of mosaics with ship depictions. The purpose of this research is to identify the generic kind of shipbuilding and type of ship. Papyrus and reeds are the earliest and quite simple building materials to be found in abundance on the banks and in the marshes of the Nile and Tigris-and-Euphrates. Such boats have been built since pre-historic times and they are still found nowadays in Africa (Lake Chad), Ethiopia (Lakes Zwai and Tana), in Sardinia and as far as the Eastern Islands, in the Pacific Ocean, Mexico and Lake Titicaca (between Peru and Bolivia).[1] Papyrus boats are depicted only in the Palestrina mosaic discussed above (figs. 3.6.13, 3.6.14, 3.6.15, 3.6.16). The longitudinal black or dark-brown strips on the hull of such boats indicate the papyrus bundles, while the vertical black strips are the cords or ropes to tie the bundles. The light ochre or yellowish-brown hues of the hull indicate the bundles of dried papyrus used for their building. The boats are depicted with black and dark brown hues on the deck and the inner hull, thus indicating the bitumen/pitch coating to make them watertight. Hide-boats were (and still are) built of a light frame of woven young branches of willow (or other flexible tree branches or bushes) and dried hides of sheep or cattle that are stretched over the frame. The *coracle* in the Palestrina mosaic is the only example of such a boat depicted in the mosaics studies above (fig. 3.6.9). The boat depicted in the Yakto mosaic, probably indicates a dugout, made of a half split log (fig. 3.3.6). The elongated, slim and uniformly colored port hull, shallow hold, the flat bottom and the outward angled ends, suggest such a vessel.

The common building material for river or seagoing vessels is timber. The hulls of small and large vessels were built with wooden planks/strakes. Classical ships were built in "shell-first" technique (carvel fashion), where the planks were set alongside one another and joined by mortise-and-tenons that were locked by wooden treenails and bronze

1 Heyerdahl, 1971, figs. 1 – 5.

nails. To strengthen the hull, frames and half-frames were inserted within the hold and fastened to the hull by long spikes and nails made of bronze or iron. The hulls of most ships illustrated in the mosaics studied above are indicated by long strips of different hues of brown (figs. 3.2.3, 3.2.5, 3.5.7, 3.5.9, 3.5.10, 3.5.11, 3.5.12, 3.5.13, 3.5.15, 3.5.17, 3.5.18, 3.5.19, 3.5.20, 3.5.21, 3.6.4, 3.6.5, 3.6.10, 3.6.17, 3.8.3, 3.8.4, 3.8.5, 3.8.6, 3.8.8, 3.8.9, 3.8.10, 3.8.11, 3.8.14, 3.8.15, 3.8.16, 3.8.17, 3.8.18, 3.8.19, 3.8.21, 3.8.25, 3.8.26, 3.8.28). In the black-and-white mosaics, the seams or the edges of the planks are indicated by white strips of tesserae (figs. 3.7.7, 3.7.9, 3.7.11, 3.7.15, 3.7.16, 3.7.18, 3.7.21, 3.7.22, 3.7.24, 3.7.25, 3.7.26, 3.7.28, 3.7.30, 3.7.32, 3.7.33, 3.7.35, 3.7.36, 3.7.38, 3.7.39, 3.7.40, 3.7.43, 3.7.44, 3.7.45).

The sailing rig of the vessels presented in this monograph comprises at least one mast and yard that are made of wood. The mast in most of the ships is a tapered pole, in some examples it is depicted with intercalated black and white or yellow wide bands, thus emphasizing a composite mast. Such masts are mainly prominent in the large merchant ships (figs. 3.2.5, 3.7.14, 3.7.15, 3.7.27, 3.8.21, 3.8.25).

The rowing oars, punt poles and steering gear are also made of wood, as we may deduce from their rendering with light and dark brown tesserae. The rowing oars are generally represented by a long and thin shaft without blades that probably were submerged beneath the waterline (figs. 3.6.8, 3.6.9, 3.6.14, 3.6.15, 3.6.17, 3.8.3, 3.8.4, 3.8.6, 3.8.8, 3.8.15, 3.8.18, 3.8.21). In small vessels the oars have dual function, rowing and/or steering. The oar mounted on the port side of the Yakto boat (dugout) is represented by a long and thin shaft with a rectangular blade, while the second oar is indicated only by its loom (fig. 3.3.7). Both oars seem to be used for rowing or steering the boat. In general steering oars are much larger than the rowing oars and are sometimes identical. They are shown with a long shaft ending in an elongated rectangular blade, bisected longitudinally by the lower end of the shaft (figs. 3.2.3, 3.2.5, 3.3.6, 3.4.7, 3.5.7, 3.5.19, 3.7.11, 3.7.14, 3.7.15, 3.7.16, 3.7.27, 3.7.28, 3.7.29, 3.7.30, 3.7.32, 3.7.33, 3.7.35, 3.7.36, 3.7.38, 3.7.40, 3.7.43, 3.7.44, 3.8.21). Some blades are depicted with rounded shoulders (figs. 3.4.7, 3.7.32, 3.7.33, 3.7.35, 3.7.36, 3.7.40, 3.8.21), or rounded lower end (figs. 3.4.7, 3.7.19, 3.7.21, 3.7.27, 3.7.32, 3.7.35, 3.7.36, 3.7.40, 3.7.44). Apparently steering oars are identified by the shape of their blades with their upper and lower ends angled towards the shaft (figs. 3.2.3, 3.2.5, 3.3.7), or the external edges have a concave shape (figs. 3.2.3, 3.2.5, 3.7.27, 3.8.21). The steering oars of the vessels depicted in the Kenchreai panels have elongated and trapezoidal blades that some are bisected longitudinally by the lower end of the shaft (figs. 3.5.8, 3.5.10, 3.5.12, 3.5.13, 3.5.16, 3.5.19, 3.5.20). The steering oars in some of the vessels studied in this monograph show only a long pole, thus suggesting that the blade is submerged beneath the water line (figs. 3.6.8, 3.6.17, 3.7.5, 3.8.4, 3.8.15, 3.8.16). The large rudders were worked by a tiller inserted into the head of the loom, which provided a better maneuverability of the oar. The rudder of Nemi Ship 1 still bearded a fragmentary tiller (fig. 4.9). Several mosaics with ship depictions studied above show such a device (figs. 3.6.4, 3.6.17, 3.7.19, 3.7.22, 3.7.33). Lucian gives a very nice description of how effective the tillers were:

> *"And all this (the ship), a little old man (helmsman) has kept from harm by turning the huge rudders with a tiller".*[2]

Two of the papyrus boats in the Palestrina mosaic are punted from the bow by a long wooden pole (figs. 3.6.14, 3.6.15) and one boat is paddled by the seated man on the deck (fig. 3.6.15).

5.2 Classes of Vessels Depicted in Mosaics

The vessels depicted in the mosaics discussed above can generally be classified according into three main categories: small boats engaged in different activities, merchant ships and warships. These vessels also suggest their sailing environment: Nile Delta, river/harbor and open waters (the sea). The small boats are engaged in different fishing techniques and personal transportation that take place in the Nile Delta, and in harbors fronted by colonnade porticoes. These scenes appear in the Palestrina Nile mosaic, in the *frigidarium*, Room 29, and the semi-circular *atrium* in the Piazza Armerina mosaics. Larger vessels are anchored in open waters (Lod, Ostia), anchored or sailing in/out harbors (Kelenderis, Kenchreai, Ostia), associated with hippopotamus hunting in the Nile Delta (Palestrina), or embarking and shipping of African exotic animals (Piazza Armerina).

Small boats

The papyrus boats have slim, elongated hulls with both broad rounded ends that have a slight inward turn (figs. 3.6.13, 3.6.14, 3.6.15, 3.6.15, 3.6.16). They are punt or paddled. The *coracle* (hide-boat) has an elliptical shape and is rowed from amidships by a seated figure (fig. 3.6.9). A dugout, probably carved out from a half-log appears in the Yakto Thetis mosaic (fig. 3.3.6). The hulls of most wooden boats resemble copies of the large merchant vessels. There are two distinct shapes:

1. *Rounded ends* and partially decked at the prow and stern, appear in the Kelenderis mosaic as towed astern the large merchantman. They are the "ship's boats"[3] and were propelled by oars as well as by sails (figs. 3.4.10, 3.4.11). The stem-and-stern posts finishes at the gunwale level. There are other boats with a slightly raised stem and stern or with a bow missing the stempost (figs. 3.5.10, 3.5.12, 3.5.14-small sailing boat, 3.5.15, 3.5.20). Boat 3 in the *frigidarium* at Piazza Armerina is depicted with both rounded ends surmounted by vertical stem-and-stern posts (fig. 3.8.5).
2. *Concave prow finished with a pointed cutwater and a*

[2] Lucian, The *Ship* 5 - 6.
[3] Casson, 1995, p. 248 and n. 93, p. 249.

rounded stern: In this type of boat the stempost has a slight outward angle and is surmounted by an inward/outward turned rounded tip or a small volute (figs. 3.8.3, 3.8.4, 3.8.6, 3.8.8, 3.8.9, 3.8.10, 3.8.11, 3.8.14, 3.8.16, 3.8.17, 3.8.19). The raised sternpost of these boats is finished with a rounded tip (figs. 3.8.3, 3.8.6). The stern of other boats is surmounted by an upraised sternpost adorned with a stylized fishtail tip (figs. 3.8.4, 3.8.8, 3.8.9, 3.8.10, 3.8.14, 3.8.16, 3.8.17, 3.8.19).

In the *atrium* at Piazza Armerina two boats are depicted with a rounded stern surmounted by a vertical sternpost adorned with a fishtail tip, while the prow ends with a semi-circle transom board (figs. 3.8.15, 3.8.18). Such boats are rarely depicted in any arts.

Types and names of small boats are mentioned in many ancient written sources. The only mosaic representing different vessels followed by a written name is found in the Althiburus *Catalogue of Ships* mosaic (3rd century CE) (fig. 3.8.27); not all the names in the mosaic are evidenced in the written sources. The boats discussed in this monograph are small crafts worked by one to three putti who either work the oars or are engaged varied fishing activities:

1. A simple fishing rod (figs. 3.6.13, 3.8.5, 3.8.9, 3.8.10, 3.8.14).
2. A trident harpoon (figs. 3.8.4, 3.8.5, 3.8.9, 3.8.15).
3. A cast net held on one arm (fig. 3.8.18).
4. Basket traps (figs. 3.8.10, 3.8.19).
5. Fish net (3.3.6)
6. A seine net between two boats (figs. 3.8.12, Boats 1 and 4; 3.8.13, Boats 3 and 4).

Some of the names found in the written sources and in the Althiburus mosaic may be associated with the boats described in this monograph:

Horia:[4] a one-man fishing boat; it also could accommodate two men.

Horeia appears in the Althiburus mosaic as a rowboat with a transom stern or stem (fig. 3.8.27/*20*).

Kydaron (*cydarum* in Latin) is depicted in the Althiburus mosaic as a two-man fishing boat with a concave prow and a pointed cutwater (fig. 3.8.27/*19*).

Lembos (*lembus* in Latin)[5] appears for the first time in the 4th century BCE, as a small boat towed behind a merchantman, necessarily since large ships anchored off shore. It also is mentioned that *lembos* was invented or perfected by the Illarians in the 3rd century BCE, and used as a pirate vessel, designed for speed and armed with a ram.[6] Such vessels were built for speed as well as for carrying men and equipments across open waters and rivers.[7] These boats were used as naval auxiliary, whereas the larger type was rowed by twenty-five men aside, and the smallest by eight aside.[8] The term also applied to "ship's boats", harbor/river craft and fishing boats.[9] Some *lemboi* could carry fifty men or twenty men and two horses.[10]

Pakton[11] originally was a small vessel built of flexible branches that resemble woven-work. It probably was like a *coracle* used on the Lower Euphrates, and was coated with pitch over to make it watertight. This type of boat was used on the Nile to carry passengers to the Island of Philae near the 1st cataract. Later such boat was made of willow wood, it was *c*.6m long, and had a single steering oar and tow sculls. It was used to transport cargo and people, and still plied the Nile up to the 8th century CE.[12]

Ratis/ratiaria:[13] first appears in the Althiburus mosaic (fig. 3.8.27/*14*); it is a rowboat with projecting cutwater.

Skaphae, uperetika:[14] this vessel was found in a fleet, being towed astern by a larger vessel. The name refers to a "ship's boat". It also was used as a harbor barge known as "*levamenta*" a lighter, or as sizable river vessels.[15]

Stlatta[16] appears in the Althiburus mosaic as a rowboat with a pointed cutwater (fig. 3.8.27/*15*).

Large vessels

We may assume that these vessels depicted in mosaics were built using the shell-first technique with mortise-and-tenon joints. Although these depictions do not show the internal construction of the ships, they reveal the outer shape of the hull as well as their means of propulsion and steering. Some examples do reveal the nailing of the outer planks (fig. 3.7.24). Many found shipwrecks in the Mediterranean throw light on the construction technique of ancient ships, which also allow identifying the species of wood used in their construction. The cargo carried on board permitted to reconstruct the original size of the vessels but not their initial propulsion and steering gear. The ship iconography in mosaics helps us to view the general shape of the hull (sometimes distorted, and never at scale), which may complement the overall view of the hull shape of found shipwrecks and suggest for their sailing and steering gears.

In Greek, sailing ships are known as *strongyla ploia* (rounded ship),[17] *holkades*[18] or simply *ploia* that differ from *navis longa* (warship). The Latin term for an ancient sailing cargo ship was *navis oneraria* (ship of burden).[19] All these vessels were rigged with one or two square sails and occasionally with three, thus indicating a giant merchantman (fig. 3.7.21). When sailing in open water, a single or two-part triangular topsail was raised above the yard. The artemon sail was the efficient gear to maneuver

4 Casson, 1971, p. 330, n.11.
5 Torr, p. 115.
6 Morrison and Coates, 1996, p. 263.
7 Casson, p. 162 and notes 35-36.
8 Morrison and Coates, p. 264.
9 Casson, p. 162.
10 Torr, 1964, p. 116.
11 Strabo, 17.1.50.
12 Casson, p. 342 and n. 70.
13 Ibid., p. 330 and n. 7.
14 Torr, 1964, p. 115.
15 Casson, p. 336 and n. 44.
16 Ibid., p. 333, n.26.
17 Ibid, p. 169.
18 Ibid., notes. 12 - 13.
19 Ibid., p. 169.

the vessel when entering or living a harbor as also aided by some rowing oars; the main sail would be a burden.

The vessels were used for different purposes, but primarily for carrying grain, wood, transport of captured exotic animals from Africa and India, as well as transport of passengers. Their load capacity varied from small (20 – 80t) to large freighters (350 – 500t) that were used from the 5th century BCE to the 2nd – 3rd centuries CE.[20] Super-freighters are mentioned in ancient literature as carrying from 1,200t grain (*Isis*)[21] to 3,650t of mixed cargo *(Syracusia)*.[22] Such large vessels were rigged with three masts, as indicated by Athenaeus[23] and also by Ship 1, Station 23 at Ostia (fig. 3.7.21). No discovery has yet been made of such a great ship; Madrague de Giens shipwreck probably carried three masts, deduced from its large size, but not from physical remains.

The ships in the mosaics are mostly shown in profile, thus revealing the shape of the hull and their sailing and steering gears. The survey of these ships permits a general classification of their hull shape that groups them into three main categories:

1. Hull with rounded stern and stem, nearly crescent shape. The prow ends with a simple stempost and the stern is adorned with a goose/duck-head (fig. 3.7.22). Vessels in this category may also have a horizontal, forward projecting stempost and a latticed hanging poop, as illustrated by both Lod Ships (figs. 3.2.3, 3.2.5).
2. The hull is similar in shape to the first category, but the rounded stem and stern are surmounted by simple stemposts, and the sternposts sometimes have a fishtail/fin tip (figs. 3.4.7, 3.5.7, 3.5.10, 3.5.12, 3.5.15, 3.7.15, 3.7.25). Some vessels in this category have raised block-shape stem-and-stern posts (figs. 3.7.4, 3.7.5, 3.7.7, 3.7.11, 3.7.16, 3.7.22, 3.7.29, 3.7.30, 3.7.33, 3.7.35, 3.7.39, 3.7.40, 3.7.43, 3.7.44, 3.7.45).
3. Rounded hull with a concave stem and a projecting pointed cutwater. The stempost finishes with a rounded tip, an inward/outward turned volute or a short forward projecting post. The stern is adorned with a lattice-screen poop with projecting goose/duck head, a simple inward turn or even an *aphlaston* sternpost decoration (figs. 3.7.9, 3.7.19, 3.7.21, 3.7.27, 3.7.32, 3.7.36, 3.7.38, 3.8.21, 3.8.25, 3.8.26).

The merchantmen or transport vessels in the Palestrina mosaic are wooden river craft. Their hull shape indicates a development of papyrus boats and some kind of barges. The wooden Sailing Ship (fig. 3.6.5) shows a strong link to the Egyptian Dynastic *papyriform* vessels. The hull is painted in the conventional scheme of such boats: green and dark brown, indicating the papyrus bundles, while the black lines represent the cords used to tie the bundles. Apparently when the Egyptians built the earliest wooden vessels, they copied the hull shape from the papyrus boats and this resulted in a new type, the *papyriform* boats. The oldest excavated planked vessel in Egypt comes from a pit near the pyramid of Cheops (Khufu), the 4th Dynasty, *c.*2600 BCE (fig. 5.1). This vessel shows that the Egyptians were able to build planked boats before 2660 BCE when copper tools became available during the early Dynastic periods.[24] Apparently the Cheops Ship is the first wooden *papyriform* vessel whereas the hull was not painted in the later color scheme rather it shows the natural wooden planks (fig. 5.1). Both its raised ends are shaped as papyrus umbels. In later depictions of *papyriform* boats, the coloring scheme of the hull was green amidships and gold or yellow for the ends. The yellow stem-and-stern posts were originally imitations of canvas or leather sheaths sown around the ends to give them a more elegant form.[25] *Papyriform* vessels shown in many wall paintings and reliefs from the Old Kingdom (2686 – 2180 BCE) till the end of the New Kingdom (1570 – 1070 BCE) were used in different activities: sacred rites, fishing, hunting, transport of passengers, etc. The *papyriform* ship depicted in the Nile mosaic (fig. 3.6.5) may be considered as a conclusive evidence of a preserved shipbuilding tradition that still sailed in the Nile and the Delta in the later Hellenistic period (late 2nd century BCE). The rounded stem is surmounted by the typical Hellenistic inward-turned stempost finished with a rounded tip or suggested open volute. The short vertical sternpost is finished with a rounded tip does not follow the characteristic Dynastic umbel ends. Two vessels depicted in the Kenchreai opus sectile panels probably represent *papyriform* sailing boats (figs. 3.5.19, 3.5.20). The bow of both boats with a slim hull and a flat bottom is damaged, while the broad and rounded stern is surmounted by an inward-turned sternpost. The shape of both sterns resembles the stern of the papyrus boats depicted in the Palestrina mosaic (figs. 3.6.13, 3.6.15, 3.6.16). This Egyptian influence also may be concluded from the Nilotic theme depicted in other Kenchreai opus sectile glass panels that were not discussed in this monograph because they are not relevant to this research.

The Palestrina Sailing Ship 1 (fig. 3.6.4) may be considered as a typical river vessel. The narrow elongated hull has an angled raised prow ended by a transom board. The rounded broad stern is surmounted by an inward-turned fishtail sternpost. Hulls with at least one transom end are seen in Egypt since the 5th Dynasty. They became more common during the 6th Dynasty.[26] The ship depicted in the Palestrina mosaic indicates that such vessels still sailed in the Nile and the Delta during the later part of the Hellenistic period. The fishtail sternpost indicates a Greek influence.

Warships

Most of the vessels depicted in the mosaics studied above are merchantmen and small boats. Warships are rarely depicted in mosaics, rather they are found on coins, frescoes,

20 Ibid., n. 25, p. 172.
21 Lucian, *The Ship* 1– 4.
22 Athenaeus, V.206–209.
23 Ibid., V.208.
24 McGrail, 2001, p. 23.
25 Landstrom, 1971, p. 28.
26 Ibid., p. 119

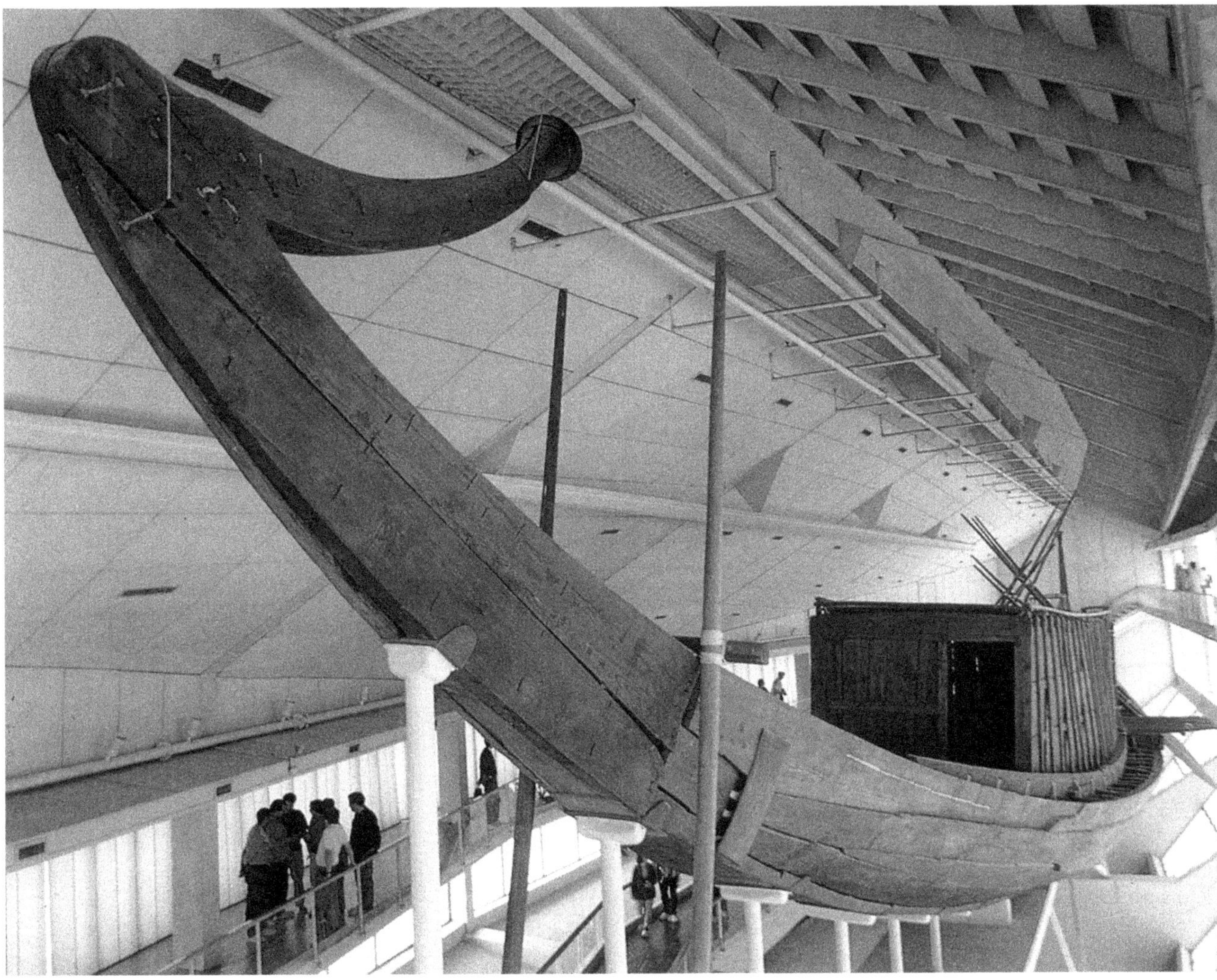

Fig. 5.1: The Sacred Ship of Cheops

reliefs, graffiti, or monuments. There are three mosaics in this work that depict a part ship and one whole vessel. The ships in both Berenike mosaics represent a headdress shaped as the prow of a warship (figs. 3.1.1, 3.1.4). They are not rigged with oars but show some characteristic features of a Hellenistic warship. The concave stem is finished with the projecting three-bladed ram (resembling the Atlit Ram)[27] and an inward-turned rounded tip adorning the highly raised curved stempost. Beneath the stempost is found the *proembolion* (top smaller ram) shaped as a two-bladed ram (figs. 3.1.3, 3.1.6). A projecting outrigger oarbox is shown on either fore side of the hull. The *caduceus* a distinctive element of a typical Hellenistic warship is depicted on the upper stem of both Berenike ships (fig. 3.1.3, 3.1.6).

An entire warship is depicted in the Palestrina Nile mosaic (fig. 3.6.10), showing its fore and the entire port side that are represented with all the typical elements of a Hellenistic warship:

1. The concave stem is finished with a projecting ram (probably three-bladed), submerged beneath the water line.
2. A rounded tip or an inner-turned volute adorns the upraised curving stempost.
3. *Ophthalmos* depicted within a square frame on the port stem, also is an indicator of a Hellenistic warship. A similar adornment was true on the starboard stem (not visible). Oars arranged in two rows in *en echelon* protrude from oarports cut into the side of the projecting outrigger oarbox along the hull's port sides.
4. The rounded stern is surmounted by an inner-turned sternpost adorned with open branches *aphlaston* crossed by an angled *stylis* (fig. 3.6.12).
5. The crew comprises: *prorates* standing on the bow and blows in a trumpet laid on his right arm, *keleutes* who beats the rhythm for the rowers and the *hoplites* or *epibatai* (warriors) standing on the deck equipped with oval/elliptical shields and long lances (fig. 3.6.10).

The ship is not engaged in sea a combat rather it has a festive appearance while entering the harbor or an anchorage in the Nile Delta to participate in the flood and Isis festival.

27 Morrison and Coates, fig. 22, p. 222.

5.3 Special Elements and Devices

The study of ship iconography in the mosaics discussed above is done with the purpose of learning about details of ancient ships that will help us to understand their types and how ancient ships were really built and rigged in antiquity. Some ships are shown with unique elements that could not have been produced in mosaics without some nautical knowledge of the mosaicist or the patron, or from direct observation of ships anchored in harbors or anchorages. A rare element depicted in mosaics is the *parrel* (figs. 3.2.3, 3.4.7, 3.4.10, 3.8.22, 3.8.29), which is a rope collar or hook that secures the yard to the mast (fig. 4.23). Only a person who understood or knew the function of such a detail could produce it when rendering a ship in a mosaic. Not even in wall paintings, reliefs, models, etc., is possible to distinguish this element. The bowsprit was a spar projecting forward over the stem of large sailing vessels. It was an additional gear used to secure the fore-lines rigging of the square sails. This element is depicted only in two mosaics from Ostia (figs. 3.7.33, 3.7.39)

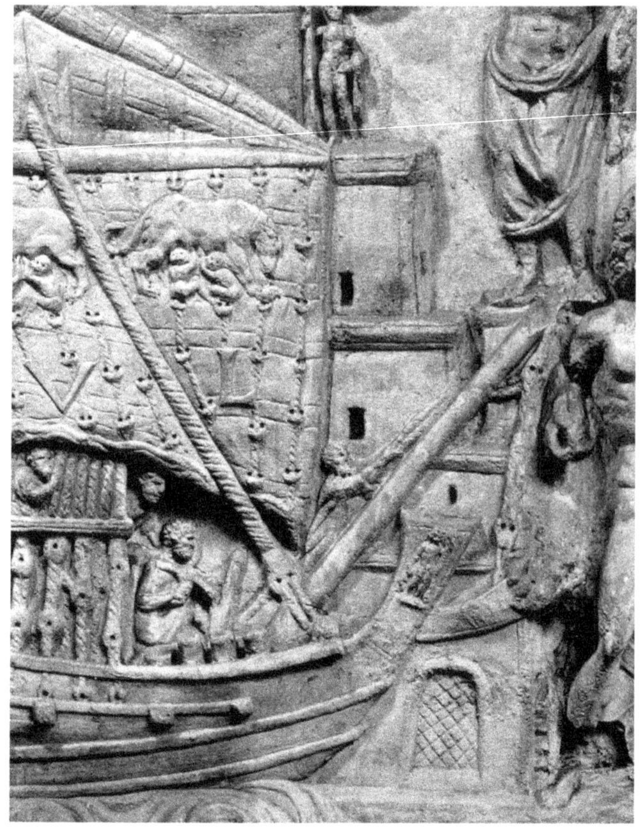

Fig. 5.2: The artemon mast used as a crane in the Torlonia relief

Block sheaves or deadeyes attached to the standing and running rigging of sailing ships are also rare elements in ship depictions, especially in mosaics. They were important devices that permitted an easier and faster way to tighten or loosen the shrouds, the fore-and-back stays when securing and holding the mast in place, or working the sail. Blocks were used to secure the lower ends of the fore-and-back stays and the shrouds to the gunwale, as we can see in the Lod Ship 1 (fig. 3.2.4), Kelenderis Sailing Ship (fig. 3.4.9), and Armerina Ship 1 (figs. 3.8.21, 3.8.22, 3.8.23). Blocks also could be attached to the masthead for working the halyard in a similar way as is depicted in the Armerina Ship 1 (fig. 3.8.22).

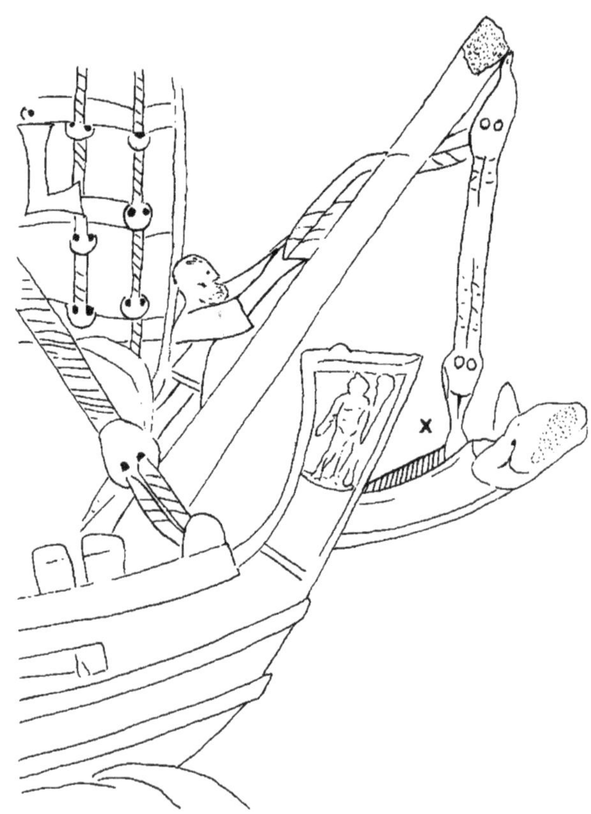

Fig. 5.2a: Detail of the artemon mast/crane

Loading or unloading ships when they are moored closely to the quay was done by at least one crane placed near the warehouse and/or in the ship. Appears that the artemon mast was well suited to serve as a crane, as indicated by the ship from Narbon, in Station 32 at Ostia (fig. 3.7.26). This mosaic is a rare example showing such an important task carried out in the vicinity of a warehouse in the harbor. Such a representation was possible to be produce only from a direct observation of similar works that were performed at Ostia or in the home harbor of the shipper. A comparable example is provided by the Torlonia ship (3rd century CE), where the artemon mast clearly shows its converted function as a crane (figs. 5.2, 5.2a).

When ships were towed the towlines was passed around the mast and/or a capstan set on the poop. The use of a capstan is nicely illustrated by Ship 1, Station 25 (fig. 3.7.24) and also suggested in Ship 2, Station 46 (fig. 3.7.33). The presentation of the capstan on the quarter deck probably resulted for two reasons:

A. The artist did not set the capstan on the bow because if he did so, he would not be able to show the stevedore crossing the gangplank set on the stem while carrying a jar from one ship to another (fig. 3.7.24).

B. Probably capstans were also set on the stern, where the helmsman, when not steering, could control the towline. Apparently the *caudicaria* was fitted with a capstan on the poop. This device enabled the relative heavy craft to winch itself upriver to a point where the going was particularly hard on the towing team.[28] This device also could be used to pull a fish net, in a similar way as seen in modern fishing trawlers. Capstans in ships were necessary devices when a ship was pulled out of the water on a slipway for repairs or maintenance.

5.4 Types of Vessels Mentioned in Ancient Written Sources

The vessels described in this monograph may be divided into two main groups: small boats and large vessels; the majority of them are merchant ships, sailing in open water (the sea), on rivers, or are engaged in harbor activities. Written sources mention varied names for types of vessels. The *Catalogue of Ships* in the Althiburus mosaic (fig. 3.8.27) provides some illustrative information of vessels followed by a written name that also are mentioned in ancient literature. Both sources will help to classify the ships described in this monograph:

Akatos (*actuaria* in Latin):[29] The Greek *akatos* vaguely means boat, while in Latin it means merchant galley. This vessel was propelled by sail, as well as by oars. It is depicted in the Althiburus mosaic (fig. 3.7.27/*13*). The term *akatos* could be applied to small vessels propelled by oars, sculled in pairs, or vessels large enough to suit 50 oars and rowers.[30]

Cauducaria:[31] type of coastal and river vessel. It was rigged with a mast that also could be used to fasten the towline around its tip or on the lower part. A capstan was also fitted on the poop.

Corbita:[32] indicates a merchant vessel of a great size. The Romans used it in the 2nd – 1st centuries BCE. Such a vessel appears in the Althiburus mosaic (3rd century CE), with a very broad hull and both rounded ends, and rigged with a tall tapered mast (fig. 3.8.27/*1*). The name is written beneath the top plank.

Hippos: was a Phoenician merchant vessel with horse figurehead. Some *hippoi* ships were of considerable size. Apparently one such figurehead (*c.*112 BCE) was brought to Egypt from a wreck that was found on the east coast of Africa and was attributed to a ship from Cadiz.[33]

Horeia[34] appears in the Althiburus mosaic as a fishing or merchant boat with a transom stem (fig. 3.8.27/*20*).

Keles (*celox* in Latin) was a built for speed.[35] The vessel was single banked with few oars, rather small and carried a modest cargo. It is depicted in the Althiburus mosaic as a small rowboat, rowed from amidships by a single seated man (fig. 3.8.27/*21*).

Kerkouros (*cercurus* in Latin)[36] was equally suited for warfare and for commerce. They were in use in the Mediterranean from the 5th to the middle of the 1st centuries BCE. The name is perhaps an adaptation by the Phoenicians of the Hebrew word *kirkarah* (carriage). This type was a standard vessel for carrying grain on the Nile.[37]

Kubaia (*cybae* in Latin)[38] was a merchant vessel much broader than the *kerkouros* with a square bow. Its name "cubic" or "boxlike" also indicates the shape of the hull. It was used on both open water and rivers to carry grain and other cargoes. Oars could also be used to propel such ships.[39]

Kydaron (*cydarum*, in Latin) appears in the Althiburus mosaic as a boat with concave prow and a projecting pointed cutwater, engaged in fishing with a net (fig. 3.8.27/*19*). In *P.Oxy*. 1197 (211 CE) is mentioned *kydaron* with a load capacity of 3¾ tons.[40]

Lenunculus is a type of a small boat used in harbor works. It was a special vessel engaged in two types of services:

1. Tugboat for warping big sailing ships about the entrance of a harbor, similar to the tugboats used today to guide large vessels to the quays in a harbor;

2. Unloading cargo and ferrying it either to harbor warehouse or upriver.[41]

Linter is a name for the general class of small boats, propelled by oars or sail, or both, which were frequently found on rivers or in shallow waters in harbors.[42]

Phasolos (*phaselis* in Latin): this vessel was suited for carrying passengers and cargo. They were in use in the Mediterranean in the 1st centuries BCE - CE. It relied on sails and had no oars.[43] The larger version served as a man-of-war.[44] There are no historical references about the period when such vessels cheesed to sail in the Mediterranean.

Thalamegos: a houseboat or pleasure vessel of extraordinary size and splendor was built by Ptolemy Philopator for his voyages on the Nile. The term *thalamegos* was sometimes replaced by *lusoria* or *cubiculata*.[45] Later it became a working craft.[46] By the 4th century CE, *thalamegos* was a small, light galley, which the Roman navy had adopted as the standard unit for the forces it maintained on rivers such as the Rhine and the Danube.[47]

Verctoria: was a vessel for carrying passengers and not for carrying cargoes. This name is known from the 2nd century CE.[48]

---

28 Casson, p. 333.
29 Ibid., p. 160, notes 12 - 13.
30 Torr, p. 106.
31 Casson, pp. 332, 390,
32 Torr, p. 112.
33 Ibid., p. 113.
34 Casson, p. 331 and n. 12.
35 Ibid., p. 160, n. 17.
36 Torr, p. 110.
37 Meijer, 1996, p. 322.
38 Casson, p. 166, notes 48-52.
39 Ibid., p. 166, n. 48.
40 Ibid., p. 331, n. 11.
41 Ibid., p. 336.
42 Ibid., p. 333.
43 Torr, p. 120.
44 Casson, p. 168, n. 58 and n. 59.
45 Torr, p. 124.
46 Casson, p. 333, n. 27.
47 Ibid., p. 334, n. 28.
48 Torr, p. 124; Caesar (*de Bello Gallica*, V.8) used this type of vessel for carrying troops and stores.

5.5 Discussion

The vessels depicted in the mosaics described above and found in different architectonic settings can be classified according to their building material, size and type which also suggest their sailing environment. The building materials used for the simplest vessels were papyrus (figs. 3.6.13, 3.6.16), flexible branches and hides for the *coracle* (fig. 3.6.9) and a wooden log for the dugout (fig. 3.3.6). These boats were propelled by oars or punted by a pole (figs. 3.6.14, 3.6.15). The small-planked boats with a concave stem and a projecting pointed cutwater resemble copies of large merchantmen. The rounded sterns are surmounted by a vertical or inward-turned sternpost, or a fishtail-tip. Such boats are inhabited by one to three putti and propelled by oars (figs. 3.8.3. 3.8.4, 3.8.6, 3.8.8, 3.8.10, 3.8.11, 3.8.14, 3.8.16, 3.8.17, 3.8.19). Two rare boats with a transom bow depicted in the *atrium* at Piazza Armerina (figs. 3.8.15, 3.8.18), are very similar in building to the boats mentioned above. Two vessels (a rowboat and a sailing boat) depicted in the Kelenderis mosaic have both rounded ends and are partly decked on the bow and the stern, with two to four thwarts indicating that they were used to carry passengers, being the "ship's boat" (figs. fig. 3.4.7, 3.4.10, 3.4.11). They were propelled either by rowing oars or by the wind power when a mast was stepped amidships with a square sail.[49] Both boats are towed astern of the large vessel anchored in the harbor. Tow boats rigged with a square sail also appear in the Kenchreai maritime panels, which were probably used to carry passengers and light cargo (figs. 3.5.9, 3.5.14, 3.5.14a). The ship's boat in both Kelenderis and Kenchreai mosaics indicate that they were able to step a mast and sail and be propelled by the wind power, which was faster and more efficient.

Some of the names mentioned in ancient literature and listed above can be related to the boats depicted in the mosaics studied in this monograph, consequently based on their appearance and activity. The Yakto log-boat (fig. 3.3.7) may be associated with a *horia* (two-man fishing boat). The boats towed astern of the large ship in the Kelenderis mosaic (figs. 3.4.7, 3.4.10, 3.4.11) and in the Kenchreai panels (figs. 3.5.9, 3.5.14, 3.5.14a) may be associated with *uperetika*, *lembos*, or the general name *skaphae*, being "ship's boats" that were propelled either by rowing oars or a square sail. The boats with the concave stem and pointed cutwater depicted in the mosaics in Piazza Armerina may be associated with *kydaron* (*cydarum* in Latin), *ratis/ratiaria* or *stlatta* (figs. 3.8.3, 3.8.4, 3.8.6, 3.8.8, 3.8.9, 3.8.10, 3.8.11, 3.8.14, 3.8.16, 3.8.17, 3.8.19). Boat 3 in the *frigidarium* (fig. 3.8.5) may be associated with a *horia*. The Armerina boats with a transom bow (figs. 3.8.15, 3.8.18) illustrated in the semi-circle *atrium* may be associated with a *horeia*, comparable to the boat depicted in the Althiburus mosaic (fig. 3.8.27/*20*). The *coracle* in the Palestrina mosaic (fig. 3.6.9) may be associated with a *pakton*, if we consider the literary description of such vessel. Most probably this craft is a hide-boat, built with a light frame and hides stretched over the frame. Bitumen or pitch was coated inside the boat to make it watertight. The same coating was also used on the deck and inner hull of the Palestrina papyrus boats (figs. 3.6.13, 3.6.14, 3.6.15, 3.6.16).

Large sailing ships are found in the Nile Delta, harbors and open waters. The vessels depicted in the Palestrina Nile mosaic may be considered as evidences of preserved shipbuilding traditions known since the Pharaonic Egypt that continued to be built and sailing on the Nile and the Egyptian Mediterranean coast as late as the Hellenistic period (late 2nd century BCE). The luxurious cabin-carrier apparently is a smaller version of a *thalamegos*, which also resembles a Phoenician *hippos* ship deduced from the stempost horse figurehead (fig. 3.6.17). The fishtail sternpost indicates a Greek influence. The *thalamegos* was not built as a typical seagoing ship, rather as a pleasure boat sailing on the Nile and the Delta or in enclosed waters (lake) as evidenced by both Nemi Ships (figs. 4.4, 4.5). The Sailing Ship 1 (fig. 3.6.4) and the *papyriform* sailing ship (fig. 3.6.5) seem to represent typical Nilotic river vessels. Both are types developed from similar ships known since the Dynastic Egyptian periods and show that not much changes took place in their hull shape since. The highly raised and inward-turned stempost with a rounded tip of the Palestrina *papyriform* ship shows a Hellenistic adoption. The raised angled stem with a transom end and a flat bottom of Ship 1 (fig. 3.6.4) represents a typical river boat that served more as a barge to transport cargoes and passengers, sailing with the currents of the Nile and not suited to sail in the open sea. The rounded stern adorned with a fishtail sternpost shows a Greek influence. Philostratus, when he described the Tyrrhenian pirate ships mentioned that the sternpost has a fishtail adornment "*... and the stern curves up in a thin crescent like end of a fish's tail*"[50], which augments the sternpost of the Hellenistic ships depicted in the Palestrina mosaic (figs. 3.6.4, 3.6.17).

The seagoing vessels studies in this work have a broad hull with a concave stem finished by a projecting cutwater, or both ends are rounded. The Lod ships with broad hull and both rounded ends (fig. 3.2.3, 3.2.5) may be associated with the *kerkouros* or *corbita* type, especially used to carry grain from the provinces to Ostia and Rome. The large Kelenderis ship (fig. 3.4.7) and two Kenchreai vessels (figs. 3.5.9, 3.5.14, 3.5.14a), represent *phasoloi* ships especially suitable for carrying passengers. When Josephus describes one of his journeys from Caesarea Maritima to Rome, he wrote that the ship in which he sailed (*phasolos*) wrecked in the Adriatic Sea.[51] He mentions that in the ship were six hundred passengers and crew who had to swim for their lives all the night after the wreckage. Josephus and other seventy nine people were lucky to be rescued by a ship from Cyrene.[52] Synesius (404 CE) describes in his *Letters* a short voyage from Alexandria to Cyrene in a *phasolos* that nearly ended with the wreckage of the ship and its passengers near

[49] Casson, p. 248 and n. 93.

[50] Philostratus, 19.23-24.

[51] Josephus, *Life*, 3.

[52] Ibid.

the shores of Libya. We learn that the vessel had a crew of thirteen and carried on board fifty passengers:

> *"The crew of twelve with the skipper made it thirteen. More than half of them were a collection of peasants, who even as nearly as last year had never handled an oar... We had taken on board, more than fifty passengers, about a third of them women..."*[53]

Some vessels depicted in the mosaics at Piazzale delle Corporazioni and Kenchreai can be grouped in two categories: river/harbor ships and coasters, and seagoing vessels. The first category comprises vessels with both rounded ends (figs. 3.4.7, 3.5.10, 3.5.12, 3.5.12, 3.5.15, 3.5.15a,3.5.18, 3.7.4, 3.7.5, 3.7.7, 3.7.11, 3.7.14, 3.7.15, 3.7.24, 3.7.29, 3.7.30, 3.7.33, 3.7.35, 3.7.39, 3.7.40, 3.7.45). The second group comprises vessels with concave stems and pointed cutwaters, which are harbor and seagoing vessels (figs. 3.7.9, 3.7.19, 3.7.21, 3.7.25, 3.7.27, 3.7.32, 3.7.36, 3.7.38, 3.8.21, 3.8.25, 3.8.26). All the vessels are rigged with one or two sails and sometimes carried oars as auxiliary equipment (figs. 3.7.24, 3.7.29, 3.8.21, 3.8.28). Ships engaged in towing were rigged with a tapered mast with cleats on either side or a capstan set on the stern (figs. 3.7.24, 3.7.33). Such vessels were used to transfer the cargoes from the large merchantmen anchored in the open sea at the mouth of the Tiber or Arno, to Ostia, Rome or Pisa. The names that can be associated with the tow or tug vessels are *keles, linter, lembos,* or *caudicaria.* The operators of tug-boats as evidenced by many inscriptions are named *lenuncularii tabularii auxiliarii* "boatmen for the service of checking and control".[54] The barges were operated by *lenuncularii pleromarii auxiliarii* "boatmen for cargo services".[55] Vessels employed in harbor work were built for speed but also could carry a modest cargo (figs. 3.7.7, 3.7.11, 3.7.14, 3.7.15, 3.7.30, 3.7.35). The wrecks with their transom stems found in Naples and Toulon (figs. 4.18, 4.20, 4.21), revealed that they were harbor vessels and represent *horeia* type, as evidenced by the boat depicted in the Althiburus mosaic (fig. 3.8.27/*20*). Some of the harbor ships may also be associated with *vectoriae*, mostly known from the 2nd century CE.[56] These mosaics, although dated from the 4th to the 6th centuries CE, indicate that such vessels were still in use in later periods. Wrecks C and F from Pisa (fig. 4.15) may be associated with the *vectoriae* that fetched the incoming merchant ships into the harbor and therefore checked the papers, or they were used for rapid incursions along the Mediterranean coasts.

The seagoing vessels have varied load capacity varying from 20 - 150 tons to150 – 500 tons. The sailing rig of these vessels comprises one to two masts and square sails. The giant merchantmen were rigged with three masts and sails (fig. 3.7.21). These vessels may be associated with *kerkouros* that mostly carried grain to Ostia and Rome (figs. 3.7.9, 3.7.16, 3.7.19, 3.7.21, 3.7.22, 3.7.25, 3.7.25, 3.7.30, 3.7.36, 3.7.40). Ships carrying large cargoes of amphorae were known as *myriogogos* or *myriophoros*, with a "capacity of 10,000" (*c*.450 t) [figs. 3.7.21, 3.7.25, 3.7.40].[57] Smaller vessels may be associated with *kubaia*, deduced from their short, broad and squarish like ends (figs. 3.7.29, 3.7.43, 3.7.44, 3.7.45).

The large vessels depicted in the Great Hunt mosaic at Piazza Aremerina are *navis oneraia* (merchantmen) that were adapted to carry mixed cargoes. The larger vessels had at least two decks where exotic animals from Africa and India were embarked to be transported to Rome for games held in the Colosseum and the Circus Maximus. Ships 1 and 2 (figs. 3.8.21, 3.8.25) may represent a very large *kerkouros* (500t), propelled by a very large square sail, while the oars probably being used as auxiliary equipment aiding when the ship entered or left a harbor or when sailing conditions were not favorable for using the sail. Ship 3 (fig. 3.8.26) may represent a *phasolos* or a larger *lembos* that carried passengers as well as horses, being an auxiliary in the navy.

Ship 3, Harbor IV, depicted from the prow indicates that when a vessel left a harbor it did not use its sail (fig. 3.5.21). The ship either was towed or it was rowed by several pairs of oars until reaching the open water where it was possible to raise the sail and to power the vessel on its journey.

Warships are rarely depicted in mosaics. They appear in three mosaics studied in this monograph: two mosaics of Berenike from Thmuis, Egypt (figs. 3.1.1, 3.1.4) and one entire ship is represented in the Palestrina Nile mosaic (fig. 3.6.10). Both Berenike figures adorned with ship headdress probably were intended to emphasize the Ptolemaic domination of the sea, while the Nile ship has a festive appearance and is ready to take immediate participation in the flood and Isis festivities. We learn about the Ptolemaic sea domination from an oration of Dio of Prusa to the Alexandrian audience:

> *"Not only have you monopoly of the shipping of the entire Mediterranean because of the beauty of your harbor, the magnitude of your fleet, and the abundance and marketing of the products of every land, but also to outer waters that lie beyond are in your grasp, both the Red Sea and the Indian Ocean... For Alexandria is situated, as it were, at the crossroads of the whole world, of even its most remote nations, as it if were a market serving a single city, bringing together all men into one place, displaying them to one another and, as far as possible, making them of the same race".*[58]

---

[53] White, 1986, p. 156; Casson, 1995, p. 268 and n. 1.
[54] Casson, p. 337.
[55] Ibid., p. 336.
[56] Torr, p. 124.
[57] Casson, 1971, p. 396.
[58] Dio, *Or. 32.36*, in: Bowman, 1996, p. 218.

Chapter 6
Conclusions

The purpose of the this study is to evaluate mosaics with ship representations, and bearing in mind the shortcomings of the *milieu*, examine whether they can be used for the understanding of ancient ships and their gearing. In order to further appraise the data gathered in the study comparisons to shipwrecks found in the Mediterranean concurrent more or less with the mosaics were undertaken. The art of mosaics, while durable, poses some limitations, which the artists had to overcome. In most cases, they did so in one way or another. They utilized different colors and different materials in the mosaics as well as broader perspectives in the designs. In the Hellenistic period, artists tried to stay as true to paintings, using available space to achieve a third dimension in any pattern. In the Roman-Byzantine period, however, vessels were represented in a two dimensional view, with less details and on the whole the mosaicists achieved to symbolize ships rather than battle with the rendering of the particular details of specific types of ships. Obviously, in the black and white technique, while a faster and probably cheaper method of mosaic production the artists had to use more sophisticated solutions to represent the ships and their rigging in anything resembling the true forms.

The mosaics depicting ships and boats chosen for this research span a period of about 800 years, from the late 3rd century BCE (both Berenike panels) to the 6th century CE (Kelenderis). While the mosaics might have been produced as decorations they were also utilized emblematically to the viewer. The decoration theme of ships could convey a message about the function of the room or the significance of the activities that took place there. They also were designed to complement visually the rest of the decoration or the architectural setting. The fact that the mosaics are durable and water-resistant medium made them as suitable for baths and atrium. The black-and-white mosaics, relatively cheaper in the technical production than the colorful ones, are found in varied structures some of which served for commercial functions (Piazzale delle Corporazioni), or in the decoration of porticoes and baths, as well as private quarters.

The vessels depicted in the mosaics studied above were not figments of the imagination. Some were copied from pattern books or other sources, whereas others indicate strong links to sketches made from vessels engaged in hunting (hippopotami) or fishing in/and around harbors. Depictions of ships in mosaics provide us with valuable information about the ships' construction material, types and propulsion gear, but do not show the building techniques as provided by known ancient shipwrecks. Such depictions also suggest the sailing environment. Ship portrayals in mosaics either appear as individual items or more frequently as part of the surface decoration.

Most of the mosaics described above are not as well known as many others, yet they reveal to us the preserved ship building traditions from the early Pharaonic Egypt to the much later 6th century CE. They present us with the Hellenistic perfection and elegance (as in the Berenike and Palestrina examples), as well as the standard merchantmen and the small boats which did not change their hull construction or the means of propulsion for many a century. The mosaic themes with ships rendering seem to be influenced by their architectural settings: palace (both Berenike are royal portraits), villas (Lod, Yakto, Kelenderis, Piazza Armerina), religious purpose: the Isis temple (Kenchreai), and the Square of Shippers and Traders (Piazzale delle Corporazioni) that had a distinct commercial and public function. The ships also may be associated with *apotropaic* attributes related to the safe return of the vessel and as a celebration for such a homecoming.

The vessels represented in this monograph revealed that they may be classified by their material construction, size and sailing environment. The hull shape and their material construction may be deduced from varied hues of brown tesserae while in black-and-white technique a thin white strip indicates the seam between the planks. The viewer is not shown their internal construction of the frames, half frames, keelson, floor timbers, etc. The hull shape and the means of propulsion permit the identification of a specific type of vessel. They are grouped in two main categories: small boats engaged in different activities and large merchantmen. Warships are rarely depicted in mosaics; they mainly appear on coins, frescoes, monuments, models, etc. The small boats are found in harbors fronted by colonnades (Kelenderis, Turkey; Kenchreai, Greece; Piazza Armerina, Sicily), in the Nile Delta (Palestrina, Italy), rivers or close to the seashore (Palestrina, Italy; Yakto, Turkey). The large merchantmen comprise river vessels (Palestrina), harbor crafts (Kenchreai; Piazzale delle Corporazioni) and seagoing ships (Lod, Kelenderis, Kenchreai, Piazzale delle Corporazioni, Piazza Armerina).

Representations of waterfront villas and boats engaged in varied fishing activities were common themes in North Africa mosaics.[1] Such themes are known from the writing of Pliny and from wall paintings in Roman villas (37 BCE):

> *"... walls painted with fish ponds, canals, rivers, coasts together with sketches of people going for a stroll or sailing in different boats and also people fishing".*[2]

This description seems to fit some mosaics in this study, showing boats manned by putti and engaged in different activities. The boats are built of papyrus (figs. 3.6.13, 3.6.16; Palestrina), a hide boat or *coracle* (fig. 3.6.9; Palestrina), a log-boat or dugout (fig. 3.3.6; Yakto), and wooden planked (figs. 3.4.10, 3.4.11; Kelenderis ship's boats; 3.8.3, 3.8.6, 3.8.8 – 3.8.11, 3.8.14 – 3.8.19; Piazza Armerina). They are propelled by rowing oars that also can be used as steering oars (figs. 3.3.6, 3.6.9, 3.8.3, 3.8.4, 3.8.6, 3.8.8, 3.8.15, 3.8.18), paddled (fig. 3.6.16), or punted by a long pole from the stem (figs. 3.6.14, 3.6.15). The boats also are used for transportation, carrying light cargo or are engaged in different techniques of fishing:

1. Fishing rod (figs. 3.6.13, 3.8.5, 3.8.9, 3.8.10, 3.8.14).
2. A trident harpoon (figs. 3.8.4, 3.8.5, 3.8.9, 3.8.15).
3. Cast net held on the arm (fig. 3.8.18).
4. Basket traps (figs. 3.8.10, 3.8.19).
5. Fishing net (figs. 3.3.6, 3.8.6).
6. A seine net stretched between two boats (figs. 3.8.12, Boats 1 - 4; 3.8.13, Boats 3 - 4).

Among the boats described above appear two rare examples with a transom bow that are typical river crafts (figs. 3.8.15, 3.8.18). Boats with a transom bow are an exceptional depiction in any arts and seem to augment a preserved tradition known from the Old Kingdom, 5th – 6th Dynasties in Egypt. The boats in the semi-circular atrium at Piazza Armerina indicate that such crafts still sailed on rivers or shallow waters as late as the 4th century CE. A similar boat appears in the *Catalogue of Ships* mosaic at Althiburus (fig. 3.8.27/*20*). Both mosaics show the Egyptian influence that passed through sea trade connections around the Mediterranean from North and East Africa to Sicily. Napoli C and Toulon Ships 1 and 2 (figs. 4.19, 4.21) with a transom stem evidence actual vessels used in harbor works and coaster along the southern shores of the Mediterranean. The simple papyrus boats (figs. 3.6. 13 – 3.6.16) and the *coracle*/hide boat (figs. 3.8.9) show that they did not change their hull construction or the method of propulsion from the pre-Dynastic period, and were still used in the later Hellenistic era in the Nile Delta. The *coracle* is a useful craft that is found even today on the Tigris-and-Euphrates, in northern Europe and in some Eskimo communities, as well as in Lake Titicaca, Bolivia - Chile. The log or dugout boat shows that such a simple craft was still plying on the Orontes River in the 4th century CE (fig. 3.3.6). The boats described above may be associated with names mentioned in ancient literature and some illustrative examples in the Althiburus mosaic. The assumed names of boat types are *horia, horeia, kydaron, uperetika* or the generic name *scapha*.

Some of the maritime scenes described above seem to follow descriptions found in ancient literature. Vitruvius informs us that maritime scenes were mostly used in covered promenades:

> *"... in covered promenades, because of the length of the walls, they used for ornament the varieties of landscape gardening, finding subjects in the characteristics of particular places; for they paint harbours, headlands, shores, rivers, springs, straits..."*[3]

St. Augustine in his work "*De Civitate Dei*" describes mosaic pictures on the esplanade in the harbor of Carthage that were probably were made in the picturesque style of the Alexandrian landscape[4] and almost certainly produced in the same manner described by Vitruvius. We may assume that the harbor and fishing scenes depicted in the Yakto (figs. 3.3.4, 3.3.6), Kelenderis (fig. 3.4.6), Kenchreai (figs. 3.5.5a, 3.5.5b, 3.5.6a, 3.5.6b) and Piazza Armerina (figs. 3.8.2, 3.8.7, 3.8.13) mosaics can be considered as following Pliny's, Vitruvius' and St. Augustine's descriptions. The Nilotic scene during the flood period and the hippopotami hunt in the Palestrina mosaic (fig. 3.6.3a) seem to follow the descriptions of Strabo (17.1.3-4), Diodorus (I.35.9-10; I.36.7-10) and Herodotus (II.92).

Ships and seascapes represent the everyday and symbolic universe of different social groups. For the elite the ships are lucrative and commercial resources acquiring raw materials, a way to wage wars, or an instrument in political strategy.[5] The relationship of the elite to the sea is indirect, as they do not spent extended times aboard ships. For the second group (shipbuilders, sailors, soldiers, etc.) the navigation was a way of life and always a potential for dangers, whereas their relation to the sea was direct. Therefore both groups needed to represent ships as a basic element of their *imaginaire*.[6] The elite employed artisans to build, paint, and carved vessels used for transportation and war or funerary ceremonies. The workers group, those whose way of life was fishing, building ships and sailing them did not have the resources to employ artisans. Their artistic skills made them to produce more sophisticated representations. The images

1 Blanchard-Lemeé, et al., 1996, p. 121.
2 Pliny, NH, XXXV.17.116.
3 Vitruvius, VII.V.2
4 Anthony, 1968, p. 55.
5 López-Bertran et al., 2008, p. 346.
6 Ibid.

produced by the seafarers and those who commissioned artisans reflect the direct and indirect relation to the sea.[7]

The ships depicted in the mosaics studied above, although not to scale, provide valuable information of the hull shape, sail rigging and the steering gear, which helps us understand varied types and ancient shipbuilding traditions that survived from the early Pharaonic to the Byzantine periods, as well as augment nautical cultural diffusion in the Mediterranean. The large vessels represent two categories: seagoing ships and river/harbor types. The river vessels comprise three types, which show a continuity of ancient shipbuilding traditions in Egypt. The Palestrina *papyriform* sailing ship and two vessels in the Kenchreai glass opus sectile panels do not have the typical umbel ends like the Pharaonic vessels but an upraise and inner turned stem or no stempost, a characteristic feature of the Hellenistic vessels (figs. 3.5.19, 3.5.20, 3.6.5). This vessel may be associated with a *phasolos* or *kerkouros*, carrying cargo and/or passengers on the Nile and the Delta, and the Mediterranean coasts. The Palestrina Sailing Ship 1 with a flat bottom and a raised prow finished with a transom board, illustrates a typical river vessel that could "take the ground" on the banks of the Nile (fig. 3.6.4). Its appearance in the Palestrina mosaic indicates that such vessels originating in the early Dynastic periods in Egypt were still in common use during the Hellenistic era. This ship may represent a *phasolos* as well as a *kerkouros*. The largest ship engaged in hippopotami hunting in the Nile Delta is a *thalamegos* or a cabin-carrier (fig. 3.6.17). Athenaeus wrote that Ptolemy Philopator built such a vessel as a pleasure boat, under the supervision of Archimedes. Later this type was reduced in size and converted into a luxurious yacht for the Romanized Egyptian nobles who used such vessels for pleasure hunting and fishing in the Nile Delta, or carrying officials from the port at Schedia on the Canopic Canal to the Upper Egypt.

The seagoing vessels are grouped into two types as indicated by their hull shape. The typical hull was elongated and finished with a projecting cutwater that was a constructional element to provide better hydrodynamics and hydrostatics to the ship. The second hull type has both rounded ends. Both types are broad with a W/L ratio 3:1 or 4:1. Usually, they are propelled by a single large square sail (figs. 3.4.7, 3.7.14, 3.7.15, 3.7.33, 3.7.43, 3.7.44, 3.8.21, 3.8.25, 3.8.28), and also topped by one or two triangular topsails (figs. 3.2.5, 3.5.7, 3.5.10, 3.5.11, 3.5.15a, 3.5.19, 3.7.39). These sailing vessels also were rigged with a secondary mast and sail, the *artemeon*, stepped forward and angled above the prow; it provided better maneuverability of the vessel, and especially was the essential rig when entering or leaving a harbor (figs. 3.5.12, 3.7.4, 3.7.5, 3.7.7, 3.7.9, 3.7.11, 3.7.16, 3.7.19, 3.7.22, 3.7.27, 3.7.29, 3.7.30, 3.7.32, 3.7.35, 3.7.36, 3.7.38, 3.7.40). The large freighters were rigged with three masts and sails (fig. 3.7.21), as also attested by Athenaeus (V.206 – 209) when he described the super-grain carrier *Syracusia*. One shipwreck that was probably rigged with three masts is Madrague de Giens (75 – 60 BCE).

The sailing rig and the steering gear of found shipwrecks did not survive. The assumption of such gears is based mostly on the size of the vessel and the cargo loads that the ships carried prior their sinking. The Roman seagoing ships also are distinguished by the latticed screen poop and the projecting head of a goose, swan or duck on the stern (figs. 3.2.3, 3.2.5, 3.7.19, 3.7.21?, 3.7.27, 3.7.32). The *ophthamos/oculus* decoration on the stem of Greek – Hellenistic warships (fig. 3.6.10) was later used on Roman merchant vessels. The dolphin and *oculus* decorations on both sides of the stem also were used as an *apotropae* for the ships protection from hazards at sea and safe return to the home port (figs. 3.7.21, 3.7.25, 3.7.32, 3.7.36).

The study of ships depicted in mosaics allows us to note specific details that could not have been produced in mosaics without some nautical knowledge of the patron or the mosaicist or from sources depicting such details. The standing rig (fore-and-aft stays, shrouds) were fastened to the gunwale through a series of block sheaves or deadeyes; although, they are depicted schematically in the Lod Ship 1 (fig. 3.2.3), the Kelenderis Sailing Ship 1 (figs. 3.4.7, 3.4.9) and Armerina Ship 1 (figs. 3.8.21, 3.8.22) we can understand their function and placement within the sail rigging. This method was still in use in the late 19th – early 20th century sailing ships. Apparently the halyard was also worked through a block attached to the masthead, in a similar way that we can see in Sailing Ship 1 in the Great Hunt mosaic (fig. 3.8.22). Another detail shown in the mosaics is the *parrel*, a collar of rope that kept the yard attached to the mast (figs. 3.2.3, 3.4.7, 3.4.10). Bowsprits used to secure the fore lines rigging of the sail are shown in three mosaics from Ostia (figs. 3.7.22?, 3.7.33, 3.7.39). Pennants or small flags usually attached to the masthead and also to the tips of the yard were used as decorations but primarily indicated the wind direction (figs. 3.2.3, 3.2.5, 3.4.7, 3.4.10, 3.7.11, 3.7.14, 3.7.15, 3.7.21, 3.7.25, 3.7.29, 3.7.30, 3.7.32, 3.7.33, 3.7.35,3.7.36, 3.7.38, 3.7.39, 3.8.22, 3.8.25).

The steering gear comprised a pair of very large oars, one set on either quarter. When the oar was mounted behind the extended aft-wing of the top strake, it formed a quarter rudders with the shaft providing the rotating axis of the oar. Some ships show that the rudder was worked by a tiller inserted perpendicularly into the head of the loom, thus giving a better maneuverability to the oars (figs. 3.6.4, 3.6.17, 3.7.19, 3.7.22). The blades shape of the steering oars in the ships studied above may be considered an indicator of a rudder; there were two types:

1. A long shaft ending in an elongated rectangular blade, bisected longitudinally by the lower shaft. The shoulder of the wings is straight, rounded or angled towards the shaft. The lower end of the blade is straight, rounded or can also angle towards the shaft. The longitudinal sides of the blade have a concave cut (figs. 3.2.3, 3.2.5, 3.4.7, 3.7.19, 3.7.21, 3.7.27-the left-hand oar, 3.7.29, 3.7.32, 3.7.41, 3.8.21).
2. A trapezoidal blade inserted into the lower shaft. This

[7] Ibid.

type appears only in the Kenchreai ships (figs. 3.5.8, 3.5.10, 3.5.12, 3.5.15a, 3.5.19).

Rudder blades bisected by the lower shaft may indicate that it comprised two wings that were attached to the shaft by mortise-and-tenons, as attested by the giant rudder of Nemi Ship 1 (fig. 4.10) or they were inserted into a longitudinal groove cut on either side of the lower shaft and then fastened by wooden treenail and bronze nails (fig. 3.3.5).

The vessels presented above are typically seagoing *navis oneraria* (merchantmen) from small cargo capacity (20 – 80t) to the largest (150 – 500t) or even super-freighters of 1,500 – 3,650t. The smaller vessels in the mosaics discussed above may be associated with the *Fiumicino Ships* 1and 2, *Isis*, *Yassiada* I and II, *Dramot A* and *F*, *Titan* and *La Cavalière*.[8] The medium size vessels were probably similar to *Comacchio*, *Grand Conglue*, while the largest merchantmen were like *Albegna*, *Caesarea* and *Madrague de Giens*.[9] We may assume that the vessels depicted in mosaics were mostly used to carry grain from the provinces in the Mediterranean (Egypt, North Africa, Sicily) to Rome, as referred in ancient literature and especially emphasized by the vessels depicted in the offices of the shippers and traders in Piazzale delle Corporazioni at Ostia. These vessels also could be adjusted to transport exotic animal African and Indian animals to Rome for the games held in the Colosseum and Circus Maximus (figs. 3.8.21, 3.8.24, 3.8.25).

The second group of ships studied in the mosaics described above comprises river/harbor vessels and coasters. They mainly have broad hulls with both rounded ends (figs. 3.4.7, 3.5.10, 3.5.12, 3.5.12, 3.5.15, 3.5.15a,3.5.18, 3.7.4, 3.7.5, 3.7.7, 3.7.11, 3.7.14, 3.7.15, 3.7.24, 3.7.29, 3.7.30, 3.7.33, 3.7.35, 3.7.39, 3.7.40, 3.7.45). All the vessels are rigged with one or two sails and sometimes carried oars as auxiliary equipment (fig. 3.7.24). Ships engaged in towing were rigged with a tapered mast with cleats on either side and/or a capstan set on the stern (figs. 3.7.24, 3.7.33). Such vessels were used to transfer the cargoes from the large merchantmen anchored in the open sea at the mouth of the Tiber or Arno, to Ostia, Rome or Pisa. The names that can be associated with the tow or tug vessels are *keles, linter, lembos,* or *caudicaria*. The operators of tug-boats as evidenced by many inscriptions are named *lenuncularii tabularii auxiliarii* "boatmen for the service of checking and control".[10] The barges were operated by *lenuncularii pleromarii auxiliarii* "boatmen for cargo services".[11] Vessels employed in harbor work were built for speed but also could carry a modest cargo (figs. 3.7.7, 3.7.11, 3.7.14, 3.7.15, 3.7.30, 3.7.35). Some of the river/harbor vessels presented above may be associated with *Fiumucino* 3, 4, or *Pisa* F, G, I. The wrecks with their transom stems found in Naples and Toulon (figs. 4.19, 4.20, 4.21), revealed that they were harbor and coastal vessels and represent *horeia* type, as evidenced by the boat depicted in the Althiburus mosaic (fig. 3.8.27/*20*). Some of the harbor ships may also be associated with *vectoriae*, mostly known from the 2nd century CE.[12] Pisa Wrecks C and F (fig. 4.15) may be associated with the *vectoriae* that fetched the incoming merchant ships into the harbor and therefore checked the papers, or they were used for rapid incursions along the Mediterranean coasts. Vessels facing each another in the mosaics at Piazzale delle Corporazioni may symbolize such activities carried out at the proximity of the Tiber's mouth (figs. 3.7.3, 3.7.6, 3.7.13, 3.7.20, 3.7.23, 3.7.28, 3.7.31, 3.7.34, 3.7.37, 3.7.42).

Warships are rarely depicted in mosaics (they are mostly found in reliefs, wall paintings, graffiti, and especially coins). Part of a ship as a headdress appears in both Berenike mosaics (figs. 3.1.3, 3.1.6) and one entire ship is depicted in the Palestrina mosaic (fig. 3.6.10). The ships are adorned with the characteristic elements of typical Hellenistic warships: *caduceus* (fig. 3.1.3, 3.1.6), three-bladed ram and an upper ram or *proembolion* (figs. 3.1.3, 3.1.6), *aphlaston* crossed by a *stylis* (fig. 3.6.10), projecting outrigger oarbox (figs. 3.1.3, 3.1.6, 3.6.10) and a crew comprising *prorates*, *keleutes*, as well as auxiliaries such as *epibatai* or *hoplites* (fig. 3.6.10). Two rows of oars arranged in *en echelon* projecting through outrigger oarbox are also typical arrangements in Hellenistic warships (fig. 3.6.24). Such warships powered by two rows of oars on either side represent a *dikrotos* (Greek) or a *bireme* (Latin). The Berenike and the Palestrina warships are not engaged in sea combat, rather they are associated with a headdress probably an indication of the Ptolemaic sea power (*thalassocracy*), and a festive ceremony during the Nile's flooding period and the Isis festival.

Ship iconography in mosaics also provides important information about harbor activities, such as loading or unloading ships by stevedores (fig. 3.7.23; Station 23, Ostia), the use of a crane, whereas the artemon mast could be used such as (fig. 3.7.26; Station 32, Ostia; 5.2, 5.2a), and recording of the cargo in a merchant ship before it was unloaded or shipped away (fig. 3.7.40; Station 51, Ostia). Kenchreai Ship 3 in the Harbor Scene IV indicates that when a ship left a harbor (fig. 3.5.6b), it did not use its sailing gear but was either towed or it used rowing oars until reaching the open sea. Sailing Ship 1 in the Great Hunt mosaic from Piazza Armerina suggests that the vessel was rowed out of the harbor, while the crew worked the sail for sailing overseas (fig. 3.8.21). The embarking and disembarking of African animals in Ships 1 and 2 in the Great Hunt mosaics at Piazza Armerina, represents a unique concept of symbolism, the summary of a sea voyage from its start in Africa (Carthage and Alexandria) and the final destination in the port of Ostia (figs. 3.8.20-middle strip, 3.8.21, 3.8.24, 3.8 .25). The Kelenderis and Kenchreai mosaic show that ship's boats were used for communication between large ships anchored in the outer harbor and the shore (figs. 3.4.7, 3.5.9, 3.5.14, 3.5.14a). Piazzale delle

8 Table I, p. 277 above.
9 Ibid.
10 Casson, p. 337.
11 Ibid., p. 336.
12 Torr, p. 124.

Corporazioni is a unique site, gathering at Ostia shippers and traders from the Roman provinces with the purpose of enabling the emperor to exercise centralized control, especially over the import of grain and the Mediterranean trading.

The shipwrecks found in the Mediterranean and presented in this work (Tables I, II) are concurrent time wise with the mosaics with ship depictions, extending from the 5th century BCE to the 7th century CE. The wrecks provide information on the construction techniques of ancient ships and their cargo capacity (20 to 500t), but due to their state of preservation we don't possess complete information of the original hull shape and their rigging. Ships iconography in mosaics has much to contribute for an overall view of similar vessels and their means of propulsion. The W/L ratio of the reconstructed ships in Tables I and II, indicate that the vessels were quite broad in their beam and carried mixed cargoes. Ships depicted in mosaics represent symbols of different types that may be considered as a relevant source of visual information for the study of preserved ancient ships and shipbuilding traditions. Despite the fact that these depictions are mostly shown in a two-dimensional view and sometimes distorted as a result of the limited competence of the mosaicist to render ships in perspective, they augmented found shipwrecks concurrent to the period of the mosaic surfaces with such depictions. One of the most relevant confirmations for this link between ship iconography and an excavated vessel is attested by one of the Pisa wrecks. The projecting cutwater depicted in all kinds of ship representations was considered to be a decorative element. The almost intact stem of Pisa Wreck C (fig. 4.17) proves that such an element was part of the ship's construction and was designed to improve its hydrostatics and hydrodynamics.[13] Comparative studies of mosaics with ship depictions and archaeological finds will provide us with more valuable information on ancient shipbuilding and their propulsion gears.

The Palestrina Nile mosaic, may be considered as one of the best pictographic evidence for preserved shipbuilding traditions that survived through centuries in Egypt from early Pharaonic till the later Hellenistic periods and that had influences around the Mediterranean. The earliest papyrus/reed boats were already in use in prehistoric Egypt, Mesopotamia and even Greece. They were propelled by punting either from the bow or stern, or by paddling. The papyrus boats in the Nile mosaic indicate that they still survived during the Hellenistic period (figs. 3.6.13 – 3.6.16). Flat-bottomed vessels with a transom bow are seen in Egypt from the 5th Dynasty (Old Kingdom). River craft with transom stems were still used in the Nile Delta in the later Hellenistic period as attested by the Palestrina mosaic (fig. 3.6.4), and/or harbor/coasters as evidenced by Napoli C (figs. 4.19, 4.20) and both Toulon Ships (fig. 4.21). The Armerina fishing Boats 2 and 5 (figs. 3.8.15, 3.8.18), each with a transom stem, were probably made after copies or models that originated in Egypt. The Kenchreai Ship 4 in the Harbor Scene II (figs. 3.5.15, 3.5.15a) with a transom stern may have been made after models of Egyptian origin or they indicate that similar vessels with slight changes also formed seagoing ships and plied the seas in the 4th – 6th centuries CE.

The horse figurehead stempost of the *thalamegos* ship, in the Palestrina mosaic (fig. 3.6.17) demonstrates a strong link to the Phoenician *hippos* ships. Horse head decorations of the stem and/or stern were characteristics of Phoenician merchant vessels from the 9th to the 7th centuries BCE. The fishtail/fin sternpost decoration of Ship 1 (fig. 3.6.4) and the *thalamogos* (fig. 3.6.17) indicate a Greek influence. The earliest Greek fishtail sternpost decoration appears in the mid-8th – beginning of the 7th century BCE. Similar decorations on Egyptian vessels show a tradition that was transmitted through sea trade connections in the Mediterranean. The sternpost or stern goose/swan head decoration appeared on Greek ships from 540 BCE. It became a trademark of seagoing merchant ships from the late 6th century BCE and it was a common pattern on the Roman merchantmen from the 1st to the 4th centuries CE. Caligula, when he built both "floating palaces" known as the Nemi Ships probably may have used the *thalamegos* described by Athenaeus (V.204 – 206) as a guideline and also adopted elements of seagoing vessels.

Ships in any culture are believed to contain spirits that protect them and the mariners from the dangers of the waters. Therefore *ophthalmoi/oculi*, the ship's eyes, painted or attached on either side of the prow was used as *apotropae* for the vessel and its crew. Both marble disks found with the Tektaş Burnu shipwreck (5th century BCE) are conclusive evidence for adornments on the prow of merchantmen. Such "eyes" are especially known from ship iconography. Two marble disks found off shore the Israeli coast, at Yavne-Yam (5th century BCE) and Megadim (2nd century BCE) are also associated with the *ophthalmoi* of ships. The Zea and Athenian almond-shaped eyes and dated from 475 - 465 to 425 - 400 BCE are assumed to be votive objects and/or ship's *ophthalmoi*. In his research of the *ophthalmoi* Nowak concluded that the almond-shape eyes adorned the upper stempost of the warships dated to the late-6th and early-5th century BCE. The circular eyes adorned the prow of merchantmen dated to the 6th – 5th century BCE.[14]

Ideas may be transmitted from place to place not only by passengers and crews, but also by the vessels themselves (Lucian, *The Ship* 1 – 4, Athenaeus, V.206 – 209). The Mediterranean vessels reaching places through trade or even sea combats, were seen, admired and copied by locals who later adapted these techniques to their shipbuilding. The similarity of ship types and decorations strongly indicate that such influences could be seen on ships sailing in the eastern and western Mediterranean. Most similarities of hull shapes are probably due to structural, hydrodynamic and

[13] Few of the found shipwrecks (the Punic ship of Marsala, Kyrenia, Madrague de Giens, etc.) had a forward extension of the keel, thus indicating a projecting cutwater. Up today Pisa Wreck C is the only example with a complete cutwater.

[14] Nowak, 2006, p. 112.

hydrostatic requirements, rather than a specific transfer of technology.[15] Overhanging bows were useful for riverboats but became impractical on seagoing vessels. Flat-bottomed ships can take the ground in a foreshore, while a rounded hull with a projecting keel (cutwater) is impractical in a vessel that has to be worked windward.[16] The ship iconography in the mosaics researched in this monograph show links and influences in their hull construction and propulsion that survived through centuries by seaway connections and nautical cultural diffusion in the Mediterranean. They are conclusive and important sources of pictographic information for ancient shipwrecks found during occasional excavations and surveys.

[15] Anthony, p. 55; McGrail, 2001, p. 435.
[16] McGrail, p. 435.

Glossary

abaft: means behind, beginning from the stem and continuing towards the stern
abaft the beam: implies any direction between a supported transverse line amidships and the stern, whether in or out of the ship
abeam: on either side of the middle area of the vessel
aft: at or towards the stern, or after part of a ship, as a word either of position or motion
aftdeck: the deck at the after end; close to the stern
aisle: a longitudinal division, as in church, separated from the main area by an arcade
akroterion: sternpost
alexandrinum: a kind of mosaic especially for pavements of rooms. The distinctive feature of this mosaic is that the lines or figures composing the design are in two colors only, the prevailing ones being red and black on a white background
amidships: in the central part of the vessel, either longitudinally or transversally
amphora: a large oval, two-handled vase, used for storage, etc.
aphlaston: sternpost, ornament atop the sternpost
apotropaic: intended to ward off evil; something that averts evil
apse: the termination of a church; it is generally of semicircular form, and surrounded by a semi cupola, but there are instances of rectangular apses
artemon: foresail, bow-sail
askoma: oarport cover (leather)
athwartship: from one side of the ship to the other; transversely
atrium: ***1.*** the main or central room of an ancient Roman house, open to the sky at the center; ***2.*** a courtyard, flanked or surrounded by porticoes, in front of an early Christian or medieval church
attire: clothes or apparel, esp. rich or splendid garment
backstay: part of the standing rigging; supporting lines running from the mast aft
beam: term used for the breadth of a ship; also a strong horizontal timber, running from side to side, that supports the deck
bilge: ***1.*** ridges between the sides and the bottom of a boat; ***2.*** part of the floor in a ship on either side of the keel which is closer to being in a horizontal rather than in a perpendicular direction and begins to take its rounded shape upwards; ***3.*** (in a hull with a double bottom) an enclosed area between frames at each side of the floors, where seepage collects
billow: swell out, to rise, puff up or roll in; to cause to rise, surge, swell, or the like
bireme: two-banked war galley
bitts: two strong pieces of straight timber, fixed upright in the fore-part of a ship and bolted securely to the beams, on which can be fastened the cables of fore-rigs, anchor-lines, etc.
blade or wash of an oar: the broad, flat, lower part of an oar, which is plunged into water when rowing or steering
block: a wooden or metal case in which one or more sheaves are fitted. Blocks consist of four parts: the shell or outside wooden part, the sheave on which the rope runs, the pin on which the sheave turns and the strop, which is the rope or wire spliced round the shell and by which the block can be attached whenever it is required
bolt-rope: rope along the edge of a sail to give it strength
bow/prow: the front part of a vessel
bowsprit: a large spar projecting over the stem, used for securing the foremast. The bowsprit itself is held in place by shrouds secured at each side of the bow
brace: a line attached to the end of the yard whose use is either to square or traverses the yard horizontally
brace the yard: is to bring it to either side by means of braces
brails: lines for controlling the area of the sail exposed to the wind; ropes used to furl a sail rapidly
bulwark: balustrades along the upper deck
bunt: main body of a sail; belly of a sail
caduceus or caduceum: a wand or laurel or olive, given by Apollo to Mercury in exchange for the lyre invented by the latter. Mercury, seeing two snakes struggling together, separated them with his wand, whereupon the snakes immediately twisted themselves around it. It was always an attribute of Mercury, who thence received the name of *Caducifer*, or caduceus-bearer. The caduceus was an emblem of peace
caprail: a timber attached to the top of a vessel's frame
capstan: a winch with upright spindle (in ancient ships) set either on the forecastle deck or quarter deck and used for heavy lifting work, particularly when working anchors

and cables
caulk: to insert material between two members and thus make the junction watertight
chiaroscuro: treatment or disposition of the light and shade in a picture
chlamys: a short light mantle, which was worn by Greek youth until they achieved manhood. The chlamys is seen in representations of men hunting beasts or fishing, as a shield wrapped around the left arm, the right grasping the spear
cingulum: a girdle or fastening round the waist
clew or clue: in a fore-and-aft sail the lower aftermost corner, in a square sail the lower corners
composite mast: mast girdled with wooldings at fixed intervals
coracle: small boat, occasionally circular but more often rectangular with rounded corners, constructed of wickerwork and made watertight originally with animal hide and more recently with pitch or other watertight material, used for river and costal transport. It is light enough to be carried easily on a man's back
crossbeam: a timber extending across the vessel
cuirass: cingulum, lorica, pectorale, thorax
cutwater: forward-most part of the stem that parts the water
deck: a horizontal platform placed across the interior of a hull
dikrotos: two-banked galley
draft: the depth to which a hull is immersed
emblema: the central motif in a mosaic, often surrounded by a frame, finer and more detailed than the background
exedra: an assembly room or hall for discussion or conversation, forming part of a gymnasium, palaestra, or private house. In many cases exedrae were in the open air. When it was covered in, one of the sides often terminated in a circular apse
fairleads: rings, eyes or loops attached to the outer or lee face of a sail to guide the brails for working the sail surface
fibula: a clap, buckle or brooch, made of gold, silver bronze, ivory
foot: the bottom side of a sail, whether square or triangular
forestay: part of the standing rigging; a line running from the mast fore-wards
forward: towards the front
frame: a transverse timber, or line or assembly of timbers, that described the body shape of a vessel and to which the planking and decks were fastened
frigidarium: a cool room in a Roman bathhouse
gangplank: plank temporarily extended from ship to shore for embarking and disembarking
graffito: a rough drawing scratched on a wall or other suitable surface
grommet: ***1.*** a ring made from a single strand of rope, used to fasten the upper edge of a sail to its stay in different places and by which the sail is hoisted or lowered; ***2.*** strand(s) of rope laid in the form of rings, used to secure an oar to the gunwale
guilloche: an ornament in the form of a braid pattern
gunwale: the uppermost course of planking on a ship's side, which covers the heads of the timbers between the main and fore drifts
gunwale of a boat: a piece of timber going round the upper sheer-strake as a binder for its top work
halias: a coastal craft that used sail and oars; it also was used for fishing with nets
halyard: line to hoist and lower yard and sail
head: the top edge of a four-sided sail
head-and-lacing: the attachment of the sail to the yard by means of loops (rope or leather) or rings of metal
heel: the lower end of the mast, boom or bowsprit in a sailing vessel
helm: the steering device of a vessel
helmsman: a person who steers the ship or boat; a steersman
hemiolia: one-and-half, with reference to the oar arrangement in galley
hold: ***1.*** the whole interior cavity of a ship, or the part between the floor and the lower deck throughout her length; ***2.*** a large compartment below the deck in a ship mainly used for the stowage of cargo but in earlier days used also for stowing provisions for a voyage, and often ship's gear
horia: a small boat; a one-man fishing boat
horiola: a small rowboat
hull: body of a ship
hypozoma: undergird; un assembly of tightened ropes pulling two points of a hull together to reduce tensile bending stress
iconography: image description; the study or analysis of subject matter and its meaning in the visual arts
impluvium: a cistern on the floor of the atrium in a Roman house, into which rain water was conducted; the aperture in the roof of the atrium
keel: the backbone of a ship; running along the lower part of the hull from stem to stern
keelson: an internal longitudinal timber or line of timbers, mounted atop the frames along the centerline of the keel; it provides additional longitudinal strength to the bottom of the hull, similar to an internal keel
keleutes: rowing master or boatswain who controlled the oarsmen
kerkouroi: were used either in warfare or commerce
kerkouroi halegoi: salt transporter on the Nile
lee: the side opposite to that from which the wind is blowing
leech: vertical edges of a square sail
lembos: is a general class of small boats, propelled by oars or sails or both; it is also referred as a rowboat that ferried people out to ships anchored in deep waters
lenunculus: is a heavier boat manned by several oarsmen
lift: line running from the masthead to the tip of the yardarms
linter: is a name for a general class of small vessels, propelled by oars or sails or both
loom: the handle of an oar
mast: spar used to support a sail by means of associated rigging
masthead: the uppermost part of a mast above the rigging
mast-step: fitting used to locate the heel of a mast
midships: to center the helm in the line of the keel

mizzen: the name of the third, aftermost mast of a square rigging sailing ship
monokrotos: a single bank of oars on either side of the hull
mortise-and-tenon join: a union of planks or timbers by which a projecting piece (tenon) was fitted into one or more cavities (mortises) of corresponding size
oarbox: the projection on each side of a polyreme, which is required by the oar system *passim*
oarport: an opening in a vessel's side through which the looms of oars or sweeps passed
ophthalmos/ oculus: literally "eye"; a device in the form of an eye, sometimes highly stylized, painted on either side of the hull close to the stem, for reasons of religion or superstition
opus sectile: floor or wall surface of colored marbles or glass cut to form geometrical or figural patterns
opus signinum: lime mortar with aggregate of crushed brick used as floor covering
opus tesselatum: mosaic pavement made with tesserae (small square cubes of stone, glass, etc.)
opus vermiculatum: refined mosaic work composed of small tesserae cut into shapes that are often irregularly shaped and set in worm-like courses that follow the contour of the design
paddle: a kind of oar, which has a shorter, broader blade than the common rowing-oar
papyrus: a reed that grows in the Nile Delta; its stalk is used for making material like paper that is also known as *papyrus*. Ancient documents, manuscripts or scrolls that are written on this material are also called *papyrus*
parrel: collar, generally of rope that holds the yard to a mast
pennant: a flag serving as an emblem set on the masthead of a ship
pitch: tar or black paint consisting of bitumen used in caulking the ship to fill the chinks or intervals between the planks
plank: a long flat piece of timber, thicker than a board
poop: from Latin *puppies*, stern, the name given to the short, aftermost deck raised above the quarterdeck of a ship. In square-rigged ships it formed the roof of the coach or rounded house, where normally the captain had his cabin
port: the left-hand side while facing forward
proembolion: upper ram
prorates: bow officer of a war galley
quarter: either side of the ship near the stern
quarter rudder: one or a pair of rudders placed on either side near the stern. They were permanently mounted and turned about a fixed axis
ram: a heavy beak or spur projecting from the bow of a warship for penetrating the hull of the enemy ship
ram bow: any bow with a projecting forefoot or ram
reef: the operation of shortening sail in a vessel by reducing the area exposed to the wind. The operation is required when a vessel is exposed to the strength of the wind
reef-band: a strip of extra canvas added on to a sail of square-rigged sailing vessel, along the line of the reef-points to support the strain on the points where the sail is reefed
rigging: the lines fitted to mast, yard or sail
rowboat: a small boat designed for rowing
rubbing-strake: a piece of half-rounded timber or rubber running the length of a small vessel from bow to stern on either side just below the gunwale. It protects the side of the vessel when coming or lying alongside another vessel or mole, etc.
rudder: a timber or assembly of timbers that could be rotated about an axis to control the direction of a vessel underway. Until the middle of the Medieval Period, the fashion was to mount rudders on one or both stern quarters; these were known as *quarter rudders*
running rigging: the lines that control the movement of sail and spars
scapha: light-rowing boat; it also refers to ship's boat carried by merchantmen
scarf: to join two pieces of timber by sloping off the ends of each and fastening them together so that they make one piece of uniform size
seam: the narrow gap between the planks forming the sides and decks of wooden vessels, which are caulked with oakum and pitch to keep out the water
shaft: the section of an oar between the loom and the blade
sheer, sheer-line: in the profile of a ship, the upward curve towards the ends of the hull
sheet: a rope fastened to each lower corner of a square sail, used to work the sail
shrouds: standing rigging, to support the mast laterally
square sail: sail that is set athwart ship
standing rigging: rigging that supports a mast
starboard: the right-hand side while facing forward
stays: line leading from the masthead fore and backwards to support the mast
step: a square framework of timber or steel built up and fixed on the keelson of a ship to take the heel of the mast
steering oar: an oar used to steer a small vessel, either from the side or the stern. A steering oar is not a *quarter rudder* commonly used to steer ancient vessels
stelae: an upright slab bearing sculptured or painted designs or inscriptions, used as a gravestone
stem: the foremost timber of the ship, rising up from the forward tip of the keel
stempost: a vertical or upward curving timber or assembly of timbers, scarfed to the keel, into which the two sides of the bow were joined
stern: the after end of a vessel
sternpost: a vertical or upward curving timber or assembly of timbers stepped into, or scarfed to the after end of the keel or heel
stilus, stylus: instrument of bone, ivory, bronze or silver of c.10 cm long, having one end pointed and the other flattened like a spatula; the widened end probably served either to spread the wax or erase by smoothing down what had been written upon it; the sharper end served for writing upon the wax-covered tablet
strake: a single or a combination of planks that stretch from one end of the boat to the other
stylis: identification device carried at the stern
tabling: an extra strip of canvas sewn around the edges of

sails to reinforce them where the boltrope is sewn on
tackle: the gear for running rigging for handling a vessel or performing some task on a vessel
tapered: decrease in size towards one end; conical shape
teredo-teredinidae: ship worms are notorious for the damage they caused in the timbers of wooden ships, etc. They bored long cylindrical holes in the wood that they inhabited, usually in such great numbers that only a thin film of wood existed between the holes
tesserae: small cubes of stone or other material used in mosaic making
therme: a bath complex in the Roman world, which contained not only rooms of different temperatures but also an exercise area
thole: a pin projecting upwards at sheer level to provide a pivot for an oar
tholepin: pin against which the oar is worked
thwart: cross plank that served as a seat for oarsmen, support for a mast or provided lateral stiffness to the hull
tiller: a wood or metal bar, which fits into or round the head of the rudder and by which the rudder is moved as required
topsail: in square-rigged ships, the triangular sail set above the yard
transom: the athwart ship timber bolted to the sternpost of a ship to give her a flat stern
treenail: wooden pegs or through fastenings used to join two members
triclinium: the dining room of a Roman villa; the name derives from the three couches placed around the walls of a square room
trident: a three-pronged fork, the attribute of Neptune
volute: ***1.*** a spiral ornament, found especially in the capitals of the Ionic, Corinthian and Composite orders; ***2.*** having a volute or rolled-up form, as the decoration of stem or sternposts
wale: thickened strake of external hull planking acting as a longitudinal strengthening member
waterline: line on the hull that the water reaches when the vessel is floating normally
windlass: originally a small capstan-like fitting, but on a horizontal shaft, located in the fore part of a small vessel, for raising its anchor. Like the old-time capstans, windlasses were fitted with bars to be worked by manpower and had a pawl and ratchet gear which by an up and down motion of the bars provided a rotary motion to the spindle on which the windlass was mounted
yard: a large wooden or metal spar, crossing the mast horizontally or diagonally, from which a sail is set. Usually, the yard is made of two pieces scarfed together in the middle, thus providing the desired length to fit the large square sail
yardarm: either of the outer projections of the yard of a square sail

Bibliography

Adkins L and R. A. Adkins, 1994: *Handbook to Life in Ancient Rome*; FactsOnFile, An InfoBase Holdings Company, New York

Agnese G. and M. Re, 2001: *Ancient Egypt: Art and Archaeology of the land of the Pharaohs*; White Star Publishers, Vercelli

Alliata E. and de Luca S., 2000: *Lod also Lyddea, called also Diospolis* – (Lydda, Lud); Franciscan Cyberspot Navigation; http://servus.christusrex.org/www1/ofm/mad/discussion/086discuss.html

Andreae B., 2002: Il mosaico del Nilo di Palestrina; *Forma Urbis*, Anno VII. nr.3, pp. 17-32

Anthony E. W., 1968: *A History of Mosaics*; Hacker Art Books, New York

Ashburner W. (Ed.), 2001: *The Rhodian Sea-Law*; The Lawbook Exchange, New Jersey

Ashby T., 1912: Recent Discoveries at Ostia; *JRS*, Vol. II, pp. 153 – 94, Pls. X - XIV

Athenaeus (Gulick Ch. B., Eng. Tr.), 1957: *The Deipnosophists*, Vol. II; The Loeb Classical Library; Harvard University Press, Cambridge, Massachusetts

Aurigemma S., 1959: Il Restauro di Consolidamento del Mosaico Barberini concotto nel 1952; *Rendiconti*; Volumi XXX - XXXI, Anni Accademici 1957-1958, 1958-1959

Aurigemma S., 1960a: Leptis Magna : La Villa del Nilo; *L'Italia in Africa, Le Scoperte Archeologiche (a.1911 – a.1943); Tripolitania*, Vol. I – *Monumenti d'Arte Decorativa, Parte Prima* – I *Mosaici* ; Istituto Poligrafico dello Stato, Roma; pp. 45-52, Pls. 75-97

Aurigemma S., 1960b: Luttrell A. (Eng. tr.), *Italia in Africa, Archaeological Discoveries (1911-1942); Tripolitania*, Vol. I – Monuments of Decorative Arts; Part I: The Mosaics; Istituto Poligrafico dello Stato, Rome

Avissar M., 1996: Lod – The Mosaic Floor; *Hadashot Arkheologiyot*, Vol. 105, pp. 157 – 160 (Hebrew); Israel Antiquities Authority, Jerusalem

Avissar M., 2001: The Representation of Two Merchant Ships on a Late Roman Mosaic Floor in Lod (Lydda), Israel; *TROPIS* VI, Proceedings of the 6th International Symposium on Ship Construction in Antiquity, Lamia, 1996; pp. 47 - 54

Bandinelli B., R. (Green P.; Eng. tr.), 1971: *Rome: The Late Empire, Roman Art AD 200-400*; Thames and Hudson, London

Basch L., 1987: *Le musée imaginaire de la marine antique* ; Athènes

Bass G. F., (Ed.), 1972: *A History of Seafaring based on underwater archaeology*; Thames and Hudson, London

Bass G. F., 1985: The Construction of the Seagoing Vessels of the Late Bronze Age; *TROPIS* I, Piraeus, pp. 25 - 35

Becatti G., 1969: *Scavi di Ostia; Edificio con Opus Sectilae, fuori Porta Marina*; Vol. VI; Istituto Poligrafico dello Stato, Libreria dello Stato, Rome

Ben-Dov M. and Y. Rappel, 1987: *Mosaics of the Holy Land*; Adama Books, New York

Ben Khader A. B. A. and D. Soren, 1987: *Carthage: A Mosaic of Ancient Tunisia*; The American Museum of Natural History; W.W Norton & Company, New York - London

Berti F., 1998?: The Comacchio wreck; http://index.waterland.net/navis/home/Ships/

Blanchard-Lemeé M, M. Ennaifer, H. Slim, and L. Slim, 1996: *Mosaics of North Africa: Floor Mosaics from Tunisia;* George Braziller, New York

Blanckenhagen von P. H., 1957: Narration in Hellenistic and Roman Art; *AJA* 61, pp. 78-63, pls. 30-32

Bockius R. (Ed.), 2009: *Between the Seas: Transfer and Exchange in Nautical Technology*; Proceedings of the Eleventh International Symposium on Boat an Ship Archaeology, Mainz, 2006, ISBSA 11; Mainz

Boetto G., 1999a: Fiumicino 2; http://index.waterland.net/navis/home/Ships/; http://www1.rgzm.de/navis/Musea/Ostia/Fiumicino_English.htm

Boetto G., 1999b: Fiumicino 4; http://index.waterland.net/navis/home/Ships/; http://www1.rgzm.de/navis/Musea/Ostia/Fiumicino_English.htm

Boetto G., 2000: New Technological and Historical Observations on the Fiumicino 1 Wreck from Portus Claudius (Fiumicino, Rome); in: *Down the River to the Sea*; Proceedings of the Eighth International Symposium of Boat and Ship Archaeology; Gdańsk 1997; Litwin J. (Ed.); ISBSA 8, pp. 99 – 102; Maritime Museum Gdańsk

Boetto G., 2009: New Archaeological Evidence of the *Horeia*-type Vessels; in Bockius R. (Ed.); pp. 289 - 296

Bowen R. LeBaron, 1956: The Earliest Lateen Sail; *Mariner's Mirror* 42, pp. 239 - 242

Bowman A. K., 1996 (2nd ed.): *Egypt after the Pharaohs 332 BC – AD 642*; British Museum Press, London

Bruni S. (Ed.), 2000: *Le Navi Antiche di Pisa – Ad un anno dall'inizio delle ricerche* (The ancient ships of Pisa – After a year of work); Edizioni Polistampa, Firenze

Calza G., 1914: Regione I (Latium et Campania), III. Ostia; *Atti della R. Accademia dei Lincei*, Serie Quinta, Vol. XI, pp. 284 - 90

Calza G., 1933: Ostia in the Lights of Recent Discoveries; *Antiquity* VII, pp. 405 – 409, Pls. I - IV

Camilli A., 2002: Pisa Shipwrecks: The State of the Art; in: *TROPIS VIII*, 8th International Symposium on Ship Construction in Antiquity; Hydra 2002 Proceedings; pp. 21 - 27

Campbell W. A., 1934: Excavations at Antioch-on-the-Orontes; *American Journal of Archaeology*, Vol. 38, No. 2, pp. 201 - 206

Capizzi C. S. J. and F. Galati, 1998?: *Piazza Armerina, the Mosaics and Morgantina*; Artistic Guide to the Mosaics; Italcards, Bologna, Italy

Carandini S., A. Ricci and M. deVos, 1982 : *Filosofiana : La Villa di Piazza Armerina; Immagine di un aristocratico romano al tempo di Constantino* ; S. F. Flaccovio, Editore, Palermo

Cariolou G., 1997: Kyrenia II: The Return from Cyprus to Greece of the Replica of a Hellenistic Merchant Ship; in: *Res Maritimae*, Swiny *et al.* (Eds.); pp. 83 - 97

Carlson D., 2002: Caligula's Floating Palaces; *Archaeology* 55.3, pp. 26 - 31

Carlson D., 2003: The Classical Greek Shipwreck at Tektaş Burbu, Turkey; *AJA* Vol. 107, Nr. 4, pp. 581 – 600

Carlson D., 2009: Seeing the Sea: Ships' Eyes in Classical Greece; *Hesperia* 78, pp. 347 - 365

Carry E. and E. Spelmann (Eng. Tr.), 1960: *The Roman Antiquities of Dioninysius of Halicarnassus*; Vol. I; The Loeb Classical Library; William Heinemann LTD., London; Harvard University Press, Cambridge, Massachusetts

Carry E. and E. Spelmann (Eng. Tr.), 1961: *The Roman Antiquities of Dioninysius of Halicarnassus*; Vol. II; The Loeb Classical Library; William Heinemann LTD., London; Harvard University Press, Cambridge, Massachusetts

Casson L., 1951: Speed Under Sail of Ancient Ships; *TPAPA*, Vol. LXXXII, pp. 136 - 148

Casson L., 1958: Hemiolia and Trihemiolia; *JHS*, Vol. LXXVIII, pp. 14 - 18, Pls. V - VI

Casson L., 1963: Ancient Shipbuilding: New Light on an Old Source; *TPAPA*, Vol. XCIV, pp. 28 - 33

Casson L., 1964: Odysseus' Boat (*Od.* V, 244-257); *AJP*, Vol. LXXXV.337, pp. 61 - 64

Casson L., 1965: Harbor and River Boats of Ancient Rome; *JRS*, Vol. LV, pp. 31 – 39, Pls. I - V

Casson L., 1966: Studies in Ancient Sails and Rigging; in: Essays in Honor of C. Bradford Welles; *ASP* 1, pp. 43 – 58; New Haven

Casson L., 1971: *Ships and Seamanship in the Ancient World*; Princeton University Press

Casson L., 1984: *Ancient Trade and Society*; Wayne State University Press, Detroit

Casson L., 1985: Greek and Roman Shipbuilding: New Findings; *The American Neptune* XLV, pp. 10 - 19

Casson L., 1994: *Ships and Seafaring in ancient times*; British Museum Press; London

Casson L., 1995: *Ships and Seamanship in Ancient World;* The John Hopkins University Press; Baltimore and London

Cimok F., 1995: *Antioch Mosaics*; A Turizm Yayinlari Ltd. Sts., Istanbul

Ciurca S. and G. W. Bologna, 1996?: *Mosaics of the Villa "Erculia" in Piazza Armerina; Morgantina*; La Fotometalgrafica Emiliana Spa, S. Lazzaro di Savena, Bologna

Clarke L. R., 1979: *Roman Black-and-White Figural Mosaics*; New York University Press The Columbia Electronic Encyclopedia, copyright ©1994, 2000: www.infoplease.com; Columbia University Press

Comay J. and R. Brownrigg, 1980: *Who's Who in the Bible; The Old Testament and The Apocrypha, The New Testament*; Bonanza Books, New York

Cornell T. and J. Matthews, 1982: *Atlas of the Roman World*; Phaidon, Oxford

Daszewski W. A., 1985: *Corpus of Mosaics from Egypt I; Hellenistic and Early Roman Period*; Verlag Philipp von Zabern, Mainz am Rhein

Daszewski W. A.,1996: From Hellenistic Polychromy of Sculptures to Roman Mosaics; in: *Alexandria and Alexandrianism*, Papers Delivered at a Symposium Organized by The J. Paul Getty Museum and The Getty Center for the History of Art and the Humanities and Held at the Museum, April 22 – 25, 1993; The J. Paul Getty Museum; Malibu, California; pp. 141 - 54

De Graeve M., 1981: *The Ships of the Ancient Near East (c.2000-500 B.C.)*; Department Oriëntalistiek Leuven

Delgado J. P., 1997: *Encyclopedia of Underwater and Maritime Archaeology*; The British Museum Press, London

Diodorus Siculus (Oldfather C. H., Eng. Tr.), 1960: Vol. I; The Loeb Classical Library; Harvard University Press; Cambridge, Massachusetts

Diodorus Siculus (Oldfather C. H., Eng. tr.), 1960: Vol. II; The Loeb Classical Library; Harvard University Press; Cambridge, Massachusetts

Dionysius of Halicanassus (Cary E.; Eng. Tr.), 1960: *The Roman Antiquities*; Vol. I; The Loeb Classical Library; Harvard University Press; Cambridge, Massachusetts

Dionysius of Halicanassus (Cary E., Eng. Tr.), 1961: *The Roman Antiquities*; Vol. II; The Loeb Classical Library; Harvard University Press; Cambridge, Massachusetts

Dionysius of Halicanassus (Spelman E., Eng. Tr.), 1961: Vol. II, Books III and IV; The Loeb Classical Library; Harvard University Press, Cambridge, Massachusetts

Di Segni L., 1999: The Onomasticon of Eusebiseus and the Madaba Map; in: Piccirillo and Alliata (Eds.); pp. 115 - 120

Downey G. (Eng. Tr.), 1959: Libanius' Oration in Praise of Antioch (Oration XI); *PAPS* 103, pp. 652-86

Downey G., 1961: *A History of Antioch in Syria from Seleucus to the Arab Conquest*; Princeton University Press, Princeton New Jersey

Dunbabin K. M. D., 1978: *The Mosaics of North Africa: Studies in Iconography and Patronage*; Claredon Press, Oxford

Dunbabin K. M. D., 1999: *Mosaics of the Greek and Roman World*; Cambridge University Press, Cambridge

Duncan-Jones R. P., 1977: The Giant Cargo-Ship in Antiquity; *CQ* 27, pp. 331 - 32

Epplett C., 2001: The Capture of Animals by Roman Military; *Greece and Rome*, Second Series, Vol. 48, Nr. 2, pp. 210-222

Eubanks J. E., 1930: Navigation on the Tiber; *The Classical Journal*, vol. 25, no. 9, pp. 683 - 695

Evrin V., Ayaroğlu M., Őzkan K., Evrin Ç. T., Bircan K., Bircan M., Zoroglu L., 2004: Archaeological Underwater Surveys of the Cilician Coasts; *ANMED* News of Archaeology from Anatolia's Mediterranean Areas, Vol. 3

Fairbanks A. (Eng. Tr.), 1960: *Philostratus - Imagines and Callistratus – Descriptions*; The Loeb Classical Library; Harvard University Press, Cambridge, Massachusetts

Fantar H. M., *et al.*, 1994: *La Mosaïque en Tunisie*; CNRS Editions, Tunis

Ferdi Sabah, 2000: *Mosaïques des Eaux en Algérie: Un language mythologique des pierres*. Regie Sud Mediterranee, Alger

Frank Harry Thomas, 1990*: Atlas of the Bible Lands: An Illustrated Atlas of the Bible*; Hammond Incorporate, Maplewood

Fried Daniel K. (Ed.), 2003: *Hebrew-English New Covenant (New Testament)*; Hope of Israel Publications, Powder Springs

Friedman Z., 1999: *Ship Iconography on Mosaics from the Eastern Mediterranean (Israel and Jordan); 1st – 8th Centuries AD*; MA Thesis; Department of Maritime Civilizations, University of Haifa, Haifa; (unpublished)

Friedman Z., 2001: Ship Iconography on "Black-and-White" Mosaics of the 1st – 3rd Centuries CE; *TROPIS* VI, Lamia 1996 Proceedings; pp. 175 - 196

Friedman Z., 2004: The Ships depicted in the Lod Mosaic Reconsidered; *IJNA* 33.1, pp. 164 - 168

Friedman Z., 2007: The Nile mosaic of Palestrina: An illustration of shipbuilding traditions; *Archaeologia Maritima Mediterranea*, Vol. 3 – 2006, pp. 115 - 145

Friedman Z. and L. Zoroglu, 2006: Kelendeis Ship – Square or Lateen Sail? *IJNA* 35.1, pp. 108 - 116

Frost H., 1981: The Punic Ship Museum, Marsala: its interpretation and some structural observations; *Mariner's Mirror* 67.1, pp. 65 - 75

Galili E. and B. Rosen, 2008: Ancient Sounding-Weights and Navigation along the Mediterranean Coast of Israel; *IJNA* 38.2, pp. 343 - 368

Gardiner R. and J. S. Morrison, 1995: *The Age of Galley – Mediterranean Oared Vessels since pre-classical Times*; Conway's History of the Ship, Naval Institute Press, London

Gardiner R., and A. E. Christensen, 1996: *The Earliest Ships: The Evolution of Ships and Boats*; Conway Maritime Press, London

Gentili G. V., 1957: Roman Life in 1,600 years-old Color Pictures; Archaeologists Uncover Brilliance in an Ancient Villa in Sicily, Preserved for Centuries by an Avalanche; *The National Geographic Magazine* CXI.2, pp. 210 - 29

Gintzburg David, 2007: *The Holy Scripture: Hebrew and English;* Joseph House, Hitchin, Herts

Greene K., 1986: *The Archaeology of the Roman Economy*; B. T. Batsford Ltd., London

Greenhill B., 1988: *The Evolution of the Wooden Ship*; B. T. Batsford Ltd., London

Greenhill B. and J. S. Morrison, 1995: *The Archaeology of Boats and Ships, an Introduction*; Conway Maritime Press, London

Grose D. F., 1989: *The Toledo Museum of Art: Early Ancient Glass – Core-Formed, Rod-Formed, and Cast Vessels and Objects from the Late Bronze to the Early Roman Empire, 1600 B.C. to A.D. 50*; Hudson Hills Press – New York in Association with The Toledo Museum of Art

Gullini G., 1956: *I Mosaici di Palestrina*; Archeologia Classica, Roma

Gummere R. M. (Eng. Tr.), 1962: *Seneca ad Lucilium Epistulae Morales*; Vol. II; The Loeb Classical Library; William Heinemann LTD., London; Harvard University Press, Cambridge, Massachusetts

Hanfmann G. M. A., 1957: Narration in Greek Art; *AJA* 61, pp. 71-78, pls. 27-29

Haddad E. and M. Avissar, 2003: A suggested reconstruction of one of the merchant ships on the mosaic floor in Lod (Lydda) Israel; *IJNA* 32.1, pp. 73 - 77

Harden B. D. (*et al.*), 1987: *Glass of the Caesars*; Olivetti, Milan

Hermansen G., 1981: *Ostia: Aspects of Roman City Life*; The University of Alberta Press

Heyerdahl T., 1971: *The Ra Expeditions*; George Allen & Unwin Ltd., London

Herodotus (Godley A. D., Eng. Tr.), 1966: Vol. I; The Loeb Classical Library; Harvard University Press, Cambridge, Massachusetts

Hohlfelder R. L., 1976: Kenchreai on the Saronic Gulf: Aspects of Its Imperial History; *The Classical Journal*, Vol. 71, no. 3, pp. 217 – 226

Hopkins C., 1948: Antioch Mosaic Pavements; *JNES*, Vol. VII, pp. 91-87

Hornell J., 1946: *Water Transport*; David and Charles, Newton Abbot (republished 1970)

Houston W. G., 1980: The Administration of Italian Seaports During the First Three Centuries of the Roman Empire; *MAAR* XXXVI; pp. 157 – 71; American Academy in Rome

Houston W. G., 1987: Lucian's Navigium and the Dimensions of the Isis; *AJP* 108.3, pp. 444 - 50

Houston W. G., 1988: Ports in Perspective: Some Comparative Materials on Roman Merchant Ships and Ports; *AJA* 92.4, pp. 553 - 64

Humphrey J. W., J. O. Oleson and A. N. Sherwood, 1998: *Greek and Roman Technology: A Sourcebook-Annotated translations of Greek and Latin Texts and Documents*; Routledge, London and New York

Ibrahim L., R. Scranton and R. Brill, 1976: *Kenchreai*

Eastern Port of Corinth; Results of Investigations by the University of Chicago and Indiana University for the American School of Classical Studies; II. The Panels of Opus Sectilae in Glass; E. J. Brill, Leiden

Joncheray J. P., 1977: Mediterranean hull type compared: 2. Wreck F from Cape Dramont (Var), France; *IJNA* 6.1, pp. 3 - 7

Jones H. L. (Eng. Tr.), 1966: *The Geography of Strabo*; Vol. VIII; The Loeb Classical Library; Harvard University Press, Cambridge, Massachusetts Kemp P., 1988: *The Oxford Companion to Ships and the Sea*; Oxford University Press, Oxford

Landels J. G., 1981: *Engineering in the Ancient World*; University of California Press; Berkeley and Los Angeles

Landström B., 1971: *Ships of the Pharaohs, 4000 Years of Egyptian Shipbuilding*; Doubleday & Company, Inc., New York

Lassus J., 1976: Antioch on the Orontes (Antakyé) Turkey; in: Stillwell R. (ed.): *The Princeton Encyclopedia of Classical Sites*; Princeton University Press, Princeton New Jersey; pp. 61-63

Levi A. and B. Trell, 1964: An Ancient Tourist Map; *Archaeology* 17.3, pp. 227 - 36

Levi D., 1947: *Antioch Mosaic Pavements*; MCMXLVII; Vol. I, Text and Vol. II, Plates; Princeton: Princeton University Press, London: Oxford University Press, The Hague: Martinus Nijhoff

Libanius (Norman A. F., Eng. Tr.), 1992: *Autobiography and Selected Letters*; Vol. I; The Loeb Classical Library; Harvard University Press, Cambridge, Massachusetts

Liddel and Scott, 1991: *An Intermediate Greek-English Lexicon*; (7th Ed.); Clarendon Press, Oxford

Liebeschuetz J. H. W. G., 1972: *Antioch, City and Imperial Administration in the Late Roman Empire*; Clarendon Press, Oxford University Press, Oxford

Lipke P., P. Spectre and B. A. G. Fuller, 1993: *Boats: A Manual for their Documentation*; American Association for State and Local History, Neshville

López-Bertran M., A. Garcia-Ventura and M. Krueger, 2008: Could you take a picture of my boat, please? The use and significance of Mediterranean ship representations; *Oxford Journal of Archaeology* 27.4, pp. 341 - 357

Lucas A. and J. R. Harris, 1989: *Ancient Egyptian Materials and Industries*; Histories and Mysteries of Man Ltd., London - USA

Kilburn K. (Eng. Tr.), 1968: *Lucian in Eight Volumes*; Vol. VI; The Loeb Classical Library; Harvard University Press, Cambridge, Massachusetts

Marsden P., 1997: *Book of Ships and Shipwrecks*; English Heritage; B. T. Batsford, London

McCann A. M. and J. Freed, 1994: *Deep Water Archaeology: A Late-Roman Ship from Carthage and an Ancient Trade Route Near Skerki Bank off Northwest Sicily*; JRS Supplementary Series Number 13; Ann Arbor

McGrail S., 1998: *Ancient Boats in the North-West Europe. The Archaeology of Water Transport to AD 1500*; Longman, London and New York

McGrail S., 2001: *Boats of the World from the Stone Age to Medieval Times*; Oxford University Press, Oxford

Meiggs R., 1973: *Roman Ostia*; Oxford, at the Clarendon Press

Meiggs R., 1980: Sea-Borne Timber Supply to Rome; in: *MAAR* XXXVI, pp. 185 – 196; American Academy in Rome

Meijer F., 1996: Ships in Arsinoite Archive of Sitologoi; *TROPIS* IV, Athens 1991, proceedings; pp. 321 - 25

Metzger B. M., 1948: Antioch-on-the-Orontes; *Biblical Archaeologist* XI.4, pp. 69-88

Meyboom P. G. P., 1995: *The Nile Mosaic of Palestrina: Early Evidence of Egyptian Religion in Italy*; E. J. Brill, Leiden-New York-Köln

Mitford T. B., 1976: Kelenderis; in: Stillwell R. *et al.* (Eds.)

Mollett J. M., 1994: *Dictionary of Art and Archeology*; Bracken Books, London

Morrison J. S., 1980: Hemiolia, Trihemiolia: *IJNA* 9.2, pp. 121 – 26

Morrison J. S. and J. F. Coats, 1996: *Greek and Roman Oared Warships 399 – 30 BC*; Oxbow Books, Oxford

Morrison J. S., J. F. Coats, and B. N. Rankov, 2000 (2nd ed.): *The Athenian Trireme*; Cambridge University Press, Cambridge

Mott L. V., 1997: *The Development of the Rudder. A Technological Tale*; Chatham Publishing, Texans A & M University Press

Mucznik Sonia, Asher Ovadiah and Yehudit Turnheim, 2004: *Art in Eretz Israel in Late Antiquity: Collectanea*; Tel Aviv University, Tel Aviv

Neal D. S., 1976: Floor Mosaics; in: Strong D. and Brown D. (Eds.), pp. 241 - 52

Negev A. and S. Gibson, 2001 (revised and updated): *Archaeological Encyclopedia of the Holy Land*; Continuum, New York - London

Nowak T. J., 2001: A preliminary report on *ophthalmoi* from the Tektaş Burnu shipwreck; *IJNA* 30.1, pp. 86 - 94

Nowak T. J., 2006: *Archaeological Evidence for Ship Eyes: An Analysis of Their Form and Function* (MA thesis, Texas A & M University)

Parker A. J., 1990: Classical Antiquity: the maritime dimensions; *Antiquity* 64, pp. 335 - 46

Parker A. J., 1992: Cargoes, containers and stowage: the ancient Mediterranean; IJNA **21**.2, pp. 89 - 100

Pasqui A., 1914: Regione I (Latium et Campania), V. Ostia; *Atti della R. Accademia dei Lincei*, Serie Quinta, Vol. XI, pp. 98-100, 147-151

Pausanias (Johns W. H. S, Eng. Tr.), 1959: *Description of Greece*, Vol. I; The Loeb Classical Library; Harvard University Press, Cambridge, Massachusetts

Pekáry I., 1999: *Repertorium der Hellenitischen und Römanischen Schiffsdarstellungen*; BOREAS, Beiheft 18; Műnster

Pellegrino A., 2000: *Ostia Antica: Guide to the Excavations*; Edizioni Abete, Rome

Piccirillo M., 1989: *Chiese e Mosaici di Madaba*; Franciscan Printing Press, Jerusalem

Piccirillo M., 1993: *The Mosaics of Jordan*; American Center of Oriental Research, Amman

Piccirillo M., 1994: *Umm al-Rasas, Mayfa'ah I: Gli Scavi*

del Complesso di Santo Steffano; Studium Biblicum Franciscanum, Jerusalem
Piccirillo M. and E. Alliata, 1999: *The Madaba Map Centenary 1897 – 1997: Travelling through the Byzantine Umayyad Period*; Proceedings of the International Conference Held in Amman, 7 – 9 April 1997; Studium Biblicum Franciscanum, Jerusalem
Pliny (Rackham E., Eng. Tr.), 1962a: *Natural History*, Vol. I; The Loeb Classical Library; Harvard University Press, Cambridge, Massachusetts
Pliny (Rackhan E., Eng. Tr.), 1962: *Natural History*, Vol. II; The Loeb Classical Library; Harvard University Press, Cambridge, Massachusetts
Pliny (Eichholz D. E., Eng. Tr.), 1962b: *Natural History*, Vol. X; The Loeb Classical Library; Harvard University Press, Cambridge, Massachusetts
Polzer J., 1973: The Villa at Piazza Armerina and the Numismatic Evidence; *American Journal of Archaeology*, Vol. 77, No. 2, pp. 130 - 150
Raban A., 1987: The Harbor of the Sea Peoples at Dor, *BA* 50.2/June; pp. 118 - 26
Rickman G. E., 1980: The Grain Trade Under the Roman Empire; in: *MAAR* XXXVI, pp. 261 – 275; American Academy in Rome
Ringel J., 1984: *Marine Motifs on Ancient Coins*; The American Friends of the National Maritime Museum, Haifa (English and Hebrew)
Roberts O. T. P., 1995: An explanation of ancient windward sailing – some other considerations; *IJNA* 24.4, pp. 307 - 315
Robotti C., 1973: Una Sinopia Musiva Pavimentale a Stabia ; *Boll. D'Arte* 58; pp. 42 - 44
Romer F. E., 2001: *Pompeius Mela's Description of the World*; The University of Michigan Press, Ann Arbor
Rostovtzeff M., 1967: *The Social and Economic History of the Roman Empire*; Clarendon Press, Oxford
Roth M., 1988: *Ship Modeling from Stem to Stern*; TAB Books, Blue Ridge Summit
Röttgen S. and C. Przyborowski, 1994: *The Art of Mosaics: Selections from the Gilbert Collection*; Los Angeles Country Museum of Art, Los Angeles
Rowell H. T., 1948: Ostia on the Tiber; *Archaeology* 1, pp. 34 - 43
Saleh M., *et al.*, 1998: *Egyptian Treasures from the Egyptian Museum in Cairo*; White Star Publishers, Vercelli
Schulz R. and M. Seidel, 1998: *Egypt - The World of the Pharaohs*; Könemann, Köln
Scranton R. L., 1967: Glass Pictures from the Sea; *Archaeology* 20, pp. 163-173
Scranton R. L., 1976: Kenchreai, in: Stillwell R. *et al.* (Eds.)
Scranton R. L. and E. S. Ramage, 1967: Investigations at Corinthian Kenchreai; *Hesperia*, Vol. XXXVI, pp. 124-86, Pls. 33-54
Scrinari V. S. M., 1979: *La Navi del Porto di Claudio*; Tipografia Centrale, Rome
Sear F., B., 1976: Wall and Vault Mosaics; in: Strong D. and D. Brown (Eds.)
Sear F., B., 1977: *Roman Wall and Vault Mosaics*; F. H. Kerle Verlag, Heidelberg
Sedge M. H., 2002: *The Lost Ships of Pisa*; ibooks; New York
Sperber D., 1986: *Nautica Talmudica*; Bar-Ilan University Press, Ramat-Gan and E. J. Brill, Leiden
Steen G. L., 1993: *Alexandria, the Site and the History*; New York University Press, New York
Steffy R. J., 1994: *Wooden Ship Building and the Interpretation of Shipwrecks*; Texas A & M University Press, College Station
Stillwell R. (Ed.), W. L. MacDonald and M. H. McAllister (Ass. Eds.), 1976: *The Princeton Encyclopedia of Classical Sites*; Princeton University Press, Princeton
Strong D. and D. Brown (Eds.), 1976: *Roman Crafts*; New York University Press
Sutherland C. H. V., 1974 : *Roman Coins*; Barrie & Jenkins Ltd, London
Swiny S., R. L. Hohlfelder and H. W. Swiny (Eds.), 1997: *Res Maritimae: Cyprus and the Eastern Mediterranean from Prehistory to Late Antiquity*; Proceedings of the Second International Symposium "Cities on the Sea"; Nicosia, Cyprus, October 18-22, 1994
Tchernia A., P. Pomey and A. Hesnard, 1978: L'epave Romaine de la Madrague de Giens (Var), Campagnes 1972 – 1975; XXXIV supplement à *'Gallia'*; Editions du CNRS
Testaguzza O., 1964: The Port of Rome; Archaeology ***17***.3; pp. 173 - 79
Testaguzza O., 1970: *Portus: Illustratione dei Porti di Claudio e Traiano e della Citta' di Porto a Fiumicino*; Edizione Curata di R. Peliti; Jiulia Editrice
Tiradritti F. and A. De Luca, 1998: *Egyptian Treasures from the Egyptian Museum in Cairo*; White Star Publishers, Vercelli
Torr C., 1964: *Ancient Ships*; Argonaut Inc., Publishers, Chicago
Turfa J. M. and A. G. Steinmayer, 1999: The Syracusia as giant cargo vessel; *IJNA* 28.2, pp. 105 - 25
Tzalas H., 1995: On the Obsidian Trail with papyrus crafts in the Cyclades; *TROPIS III*, Athens 1989, proceedings; pp. 441 - 69
Ucelli G., 1983: *Le Navi di Nemi*; Instituto Poligraphico e Zecca dello Stato, Rome
Vitruvius (Granger F., Eng. Tr.), 1962: *On Architecture*; Vol. II; The Loeb Classical Library; William Heinemann LTD., London; Harvard University Press, Cambridge, Massachusetts
Wages S. M., 1986: A Note on the Dumbarton Oaks *Tethys* Mosaic; *DOP* 40, pp. 119-128
Ward C. and R. D. Ballard, 2004: Deep-water Archaeological Survey in the Black Sea: 2000 Season; *IJNA* 33.1, pp. 2–13
Whiston W. (Eng. Tr.) and W. S. LaSor (Foreword), 1991: *The Complete Works of Josephus*; Kregel Publications, Grand Rapids, Michigan
White K. D., 1986: *Greek and Roman Technology*; Thames and Hudson, London
Whitehouse H., 1976: *The dal Pozzo Copies of the Palestrina Mosaic*; BAR Supplementary Series 12; Oxford

Whitehouse H., 2001: *Ancient Mosaics and Wallpaintings*; Harvey Miller Publishers, London/Turnhout

Wilkinson K. and M. Hill, 1983: *Egyptian Wall Paintings – The Metropolitan Museum of Art's Collection of Facsimiles*; The Metropolitan Museum of Art, New York

Wilson R. J. A., 1983: *Piazza Armerina*; Granada Publishing, London

Zoroglu L., (in Turkish) 1994: Bir mosaik üzerinde Kelenderis Betimlemesi; 1993 Yili, *Anadoulu Medeniyetleri Müzei; Konferanslari*; Anakara

Zoroglu L., 2000: Kelenderis; http://fen.selcuk.edu.tr/arkeoloji/KelenderisIng.htm

Zoroglu L., 2007: Kelenderis, a Port between Asia Minor and Cyprus; *Proceedings of the International Symposium on Underwater Research*, pp. 68 – 73; Famagusta, YNR Cyprus

Yadin Y., 1988 (reprint): *Masada Herod's Fortress and the Zealot's Last Stand*; Steimatzly Ltd., Bnei Brak

www.ingramcontent.com/pod-product-compliance
Ingram Content Group UK Ltd.
Pitfield, Milton Keynes, MK11 3LW, UK
UKHW061847041026
15591UKWH00015B/960

9 781407 307589